HANDBOOK for
SCIENTIFIC ENGLISH WRITING

英語論文表現例集
with CD-ROM
すぐに使える5,800の例文

佐藤元志 著
CHIKASHI SATO
Idaho State University Environmental Engineering Program College of Engineering

監修
田中宏明
（京都大学大学院工学研究科教授・工学博士）
古米弘明
（東京大学大学院工学系研究科教授・工学博士）
鈴木 穣
（独立行政法人土木研究所水環境研究グループ水質チーム上席研究員）

技報堂出版

CD-ROM 版付録英語論文表現例集ソフトウェアのご利用に当たって

- 付録 CD-ROM 版英語論文表現例集ソフトウェアを使用される場合は，必ず CD-ROM 内のソフトウェア使用許諾契約書をお読みになって，その記載内容に同意の上でご利用ください．

- ソフトウェアのインストール方法は，CD-ROM 内の使用説明書（PDF 形式）を参照してください．

■動作環境

[Windows]
OS：Windows 2000，XP，Vista（日本語版）
CPU：Intel Pentium III 800MHz 以上
メモリ：256MB 以上
ハードディスク：300MB 以上
ハードウェア：CD ドライブ

[Macintosh]
OS：MacOS X 10.4.8（日本語版）以上
CPU：G4, G5，または Intel プロセッサを搭載
メモリ：256MB 以上
ハードディスク：300MB 以上
ハードウェア：CD ドライブ

■商標

Microsoft®，Windows® は米国 Microsoft Corporation の米国またはその他の国における登録商標または商標です．Apple®，Apple ロゴおよび MacOS® は，Apple Inc. の登録商標です．FileMaker およびファイルメーカーは FileMaker,Inc. の商標です．

■FileMaker ランタイム版

これ単体で動作します．レイアウト・スクリプトの変更はできません．

はじめに

　近年，国際学会での日本人研究者による発表者数や学術誌への投稿数が増えてきています。しかしながら，日本人の献身的な研究努力とその成果を考慮すれば，その数は，納得のいく数ではないはずです。私の個人的な意見ですが，日本の学会・学術誌にて発表されている論文の多くは，国際的な学術誌にも十分に受け入れられる内容だと思います。もし，それらが，英米の研究者によって投稿される英語論文と同等の英文レベルのであれば，のことですが。

　私は，1973年に渡米して以来，英作文には苦しんできました。今日でも，いまだに葛藤している状態です。同一論文で，同じ表現を繰り返し使用したりすると，研究内容がいかに立派であっても，そのような論文は元気がなくなり，ときには死んでしまうこともあります。私は，そんな論文を救おうと，異なった言い回しを捜すのに，かなりの時間を費やしてしまうのです。その一対策として，私は，学術論文のレビュー中に役立ちそうな表現に出会ったとき，その表現をインデックスカードに書き留めてきました。「塵もつもれば山となる」もので，その表現数も年々増えインデックスカードがカード箱に入りきれないほどになりました。研究論文を書いているとき，特に文の表現に行き詰まったときに，そのカード箱が役に立ってきました。私が指導している学生（修士・博士課程）の卒業論文を査読・校訂しているときも役だっています。

　2002年に，私は日本学術振興会（JSPS）からフェローシップを頂き，十ヶ月間，日本の研究所で研究をする機会が与えられました。その間に，多くの日本人研究者と接することができ，貴重な経験となりました。特に私の注意を引いたことは，多くの研究者たちが，自らの研究成果を国際的な学術誌に投稿したいという意欲をもっていたことです。そんなことがきっかけとなり，私のインデックスカード箱が，日本人研究者達が英語論文を書くときに役に立つのではないかと思い，余暇を利用して一冊の本としてまとめてきたのです。

この例文集は，用語集でもなければ，解説書でもありません。その両方のギャップを埋める役目の書と思ってよいでしょう。「用語は知っているのだが，」「文法も分かっているのだが，」なぜか英米人が書くような文章にならない。そんな悩みのある人のための参考書と考えてよいでしょう。「生きた英語」と，よく簡単に言われますが，英語を母国語としない私たちにとって，「生きた英語」は，考えて生まれてくるものでは，ありません。論文には，分野によって特有の表現があり，それを使い切るには，その分野の論文を読みこなし，書きこなさねばならず，それにはかなりの年月を要します。多くの研究者は，研究に没頭しているために，そのような時間は無いはずです。わたしは，本書が，そういった研究者の座右に置かれ，論文作成の一助となり，一つでも多くの研究論文が国際的学会で発表され，国際的学会誌に掲載されることを願うものです。

　自分としては，かなりの数の例文を収集してきたつもりではありますが，まだまだ十全とはいえません。また記述に誤った点があるかもしれませんので，広く利用者の教示をお願いする一方，今後改訂を重ねながら不足している点を徐徐に補っていきたいと考えています。

　私の分野は，環境科学・環境工学であるため，表現例文は環境に関連した例文ばかりですが，環境管理や環境法に関する例文も示されており，理工科系だけではなく，文化系の研究者にも広範に利用されることを願ってやみません。

　本書の監修については京都大学の田中宏明教授，東京大学の古米弘明教授，独立行政法人土木研究所の鈴木穣上席研究員に多大なるご負担と，ご指導を頂いたことを記して深く感謝申し上げます。また本書の出版に当たっては技報堂出版の小巻慎氏に一方ならぬお世話になったこと，ここで厚くお礼を申し上げます。

佐　藤　元　志
Idaho State University
Environmental Engineering Program
College of Engineering

監修の言葉

　わが国が世界に誇る環境・エネルギー技術，深刻な公害克服の経験や知見などを，世界の発展と繁栄に貢献するために，国際的に共有することがまさに求められています。「イノベーション 25」－開発途上国との科学技術協力の強化－（平成 19 年 6 月），21 世紀環境立国戦略（平成 19 年 6 月）にも示されているように，環境・エネルギー，水，食料，防災および感染症等をはじめとする地球規模課題に対する科学技術力の向上，開発途上国等への国際協力の期待も高まっています。また，環境リーダー育成プログラムなど意欲と能力溢れる豊富な人材を養成・活用する戦略も構築されてきています。

　このような背景のなかで，我々の学術研究や技術開発などの成果や知見は，国内だけではなく国際的な場においても情報発信が求められています。要は，わが国の学術研究，技術開発の情報を正しく伝え，評価を受けるためにはその報告書や学術論文の執筆，ホームページ作成などは英語により表現せざるを得ないわけです。その必要性は従来から大学はもちろん，公的研究機関，企業においても広く認識されていましたが，体系だった形で教育や指導が実施されている例は限られているものと思われます。その意味では，個人レベルでの英語表現に対する努力が期待されています。

　私達は，義務教育のなかで英語を習得してきておりますが，その学習成果の一つである英作文技術で，いきなり英語論文や報告書を書くにはあまりにも格差があります。英語の文章を考える場合に，多くの方は，書きたい内容をまずは日本語で考えて，それを英語に翻訳する手段を用いることでしょう。日本語の文章として適切な表現でも，その文章を単語ベースで英語に逐語訳した場合に，英語表現としてよいものになることはあまりないでしょう。また，ある日本語表現に対応しそうな英語表現が複数あることが往々にしてあります。したがって，実務的に使える，使われている表現で書くことはほとんど手探り状態ではないかと思います。英語で上手に文章表現ができるように，我々も実際に使われている英文を繰り返して真似て書くことが効率的な学習方法です。

　本書の著者は，わが国の公害問題への対応が熱く議論されていた当時，20 歳で渡米されました。そして，米国での環境工学の大学教育を受けられた草分け研究者の

一人として現在も活躍されておられます．今回の事例集は，著者が長きに渡り蓄積された貴重な知識の宝庫であり，2006年4月に図書冊子として出版されたものに，その内容をデータベース化したCD-ROMを加えたものです．

この事例集の特徴は，科学論文作成に必要不可欠なキーワード単語をアルファベット順に抽出しており，環境科学や環境工学を中心とした実際の論文で使われた文章表現例が5 800にも上って掲載されていることです．数多くの論文から選ばれた文例が豊富に掲載されているために活用できる表現パターンを容易に見つけることができるものと思います．

また，今回のCD-ROM付属版では，検索機能を有したソフトウェア化がなされたことが特筆すべき点です．英文作成はワープロ上において行うことが多いことから，印刷された図書として活用するだけでなく，パソコン上で利用できる簡易検索，複合検索，ABC検索の機能や文例のスクラップ機能が駆使すれば，効率的な執筆作業を行えるものと思われます．使いやすい検索やスクラップの機能が付加されたことで，使用者は活用できる文例や表現例を容易に見出し，論文や報告書作成などに一層利用しやすいものと思われます．

繰り返しになりますが，21世紀は，直面する人類最大の課題である地球環境問題をより明確に意識して，さまざまな課題を解決しながら環境と経済がともに向上・発展するような持続可能な社会のあり方をいかに築くかが問われている時代です．そのような時代であるがゆえに，環境にかかわる報告書や論文を作成する機会が一層増えているものと思われます．環境科学・環境工学にかかわる大学・公的研究機関・企業などの研究者や技術者，学生はもちろん，英文で科学論文などを書いておられる，あるいはこれから書きたいと考えている方々に喜んで本書を活用いただけるものと信じております．

2009年2月

田中宏明　京都大学大学院工学研究科教授・工学博士
古米弘明　東京大学大学院工学系研究科教授・工学博士
鈴木穣　　独立行政法人土木研究所水環境研究グループ
　　　　　水質チーム上席研究員

■ 本書の構成と使い方

　実際のレポートや論文で，その文がどのように使用されているかを感じ取っていただくため，あまりにも長い文を除いては，できるだけ原型を残すようにしました。なお，初心者から経験者まで広範な方々が利用できるよう，簡略化した文も提示してあります。人名，地名，河川名などの固有名詞は，できるだけ仮名にしました。

　本書は，見出し語を基に，アルファベット順に配列されていますので，辞書を引くような方法で使用するようになります。また，本書は見出し語を一段とし，三段階に構成されています。例えば，**Above** に関係する例文を捜す場合，下記のように一段目は，見出し語の **Above** が太字で表示され，日本語訳がコロン（：）の後に添えられています。

Above（See also **As** and **Below**）：上に（➡**As, Below**）

　Above は，**As** 等と一緒に使われることが多いことと **Below** と同じ様に使われるので，ここでは，**As** と **Below** も参照するように（　）内に矢印で示されています。この本は，辞書でも用語集でもありませんので，最も一般的な訳だけ（この例では，「上に」）が示されています。

　（　）を付した語は，直前の語に関連する語を示します。

＜例＞

Acceptable（Unacceptable）：受け入れられる（受け入れられない）

Accomplish（Accomplishment）：達成する（業績）

　二段目は，●印を伴って，キーフレーズが太字で示されています。**Above** の例では，下に示されているように，いくつかの英語のキーフレーズが，日本語訳を伴って示されています。

＜例＞
- **as stated above**：上述の；上述したように
- **as suggested above**：上に示唆されるように
- **as summarized above,**：上に要約したように

　セミコーロン（；）は，直前の語，または語句と入れ替わりうることを示します。

<例>

> **Accept**：受け入れる；受容する；容認する

三段目には，英語例文（または，短縮例文）が羅列され，英語例文の下には日本語訳が示されています。キーフレーズは，太字で示されています。

<例>
1. These data will be analyzed to assess the status of biological condition and to address the goals **as stated above**.
   これらのデータは，生物状態の現状を評価し，上述の目標に取り組むために分析されるであろう。
2. **As suggested above**, much more research is needed in such basic areas as sampling effort, and indicator sensitivity and variance.
   上に示唆されるように，サンプリング活動，指標の感度，分散といった基礎領域における研究はさらに必要とされる。
3. Reviewing the literature on sonochemistry of organochlorine compounds **as summarized above** did not lead to any publications on the use of ultrasound for chemical monitoring.
   上に要約したように，有機塩素化合物の 超音波化学分解に関する文献のレビューでは，化学的モニタリングとしての超音波利用に関する出版物は見あたらなかった。

キーフレーズ，または英語例文に使われている ～ は主語となる語またはフレーズを示し，… は主語以外の語またはフレーズを示します。

<例>
- ～ is the most widely accepted ..... ：
  ～ は最も広く受け入れられている … である
- ～ take account of ..... ：～ は … を考慮する

必要に応じて，解説が★印を伴って加えられています。

<例>
★感謝の表現としては thank また appreciate などもよく使われるので，それらも参照してください。
★Acknowledgement（謝辞）によく使われる例文をいくつか下に示します。

日本語索引から逆引きする場合は，まず見出し語，上記の例では「上に」を日本語索引から見つけ，それから Above が記載されているページを探すようになります。

アルファベット順
英語論文表現例

Above（See also **As** and **Below**）：上に（→ **As** および **Below**）

1. The arsenic concentration **above** the town of Redwood was 0.02 mg/L.
   Redwood の町の上流でのヒ素の濃度は 0.02 mg/L であった。
2. The concentration of lead **is well above** the upper range commonly found in soils.
   鉛の濃度は，一般に土壌に見られる（濃度）範囲の上限より更に高い。

   - **as stated above**：上述の；上述したように
   - **as suggested above**：上に示唆されるように
   - **as summarized above**：上に要約したように
   - **as noted above**：上に指摘したように
   - **as was discussed in detail above**：詳細に前述したように
   - **as demonstrated above**：上に証明されるように
   - **as shown above**：上に見られるように

3. These data will be analyzed to assess the status of biological condition and to address the goals **as stated above**.
   これらのデータは，生物状態の現状を評価し，上述の目標に取り組むために分析されるであろう。
4. **As suggested above**, much more research is needed in such basic areas as sampling effort, and indicator sensitivity and variance.
   上に示唆されるように，サンプリング活動，指標の感度，分散といった基礎領域における研究はさらに必要とされる。
5. Reviewing the literature on sonochemistry of organochlorine compounds **as summarized above** did not lead to any publications on the use of ultrasound for chemical monitoring.
   上に要約したように，有機塩素化合物の超音波化学分解に関する文献のレビューでは，化学的モニタリングとしての超音波利用に関する出版物は見あたらなかった。
6. Calibration procedures, **as noted above**, allow normalization of the effects of stream size, so index scores, such as the IBI, can be compared among streams of different sizes.
   上に指摘したように，キャリブレーションの手順は，河川の規模の影響を標準化を

可能にするので，IBI などの指数スコアが異なる規模の河川において比較されうる。
7. The choice of groups to be identified and analyzed is related to the measures that will be used, **as was discussed in detail above**.
識別・分析されるべき集団の選択は，詳細に前述したように，用いられる測定方法に関連する。

  - **above discussed**：上で論じられた
  - **outlined above**：上に概説された
  - **cited above**：前記の
  - **judging from the above**：上記から判断して
  - **refer to ..... above**：上記に参照した … では
  - **for the reasons mentioned above**：上記の理由により
  - **from the above, it will be clear that .....**：
    上記から … は明確であろう

8. The sedimentation term is calculated from the **above discussed** transport capacity.
堆積の項は上で論じた輸送容量から計算される。

9. The standard procedure **outlined above** was established so that the problems which often obscure interpretation of results are minimized.
上に概説された標準的な手順は，結果の解釈をしばしば不明瞭にする問題点が最小になるように確立された。

10. In the results **cited above**, chemical and biosurvey results agreed 58％ of the time.
前記の結果においては，化学的調査結果と生物的調査結果の一致率は，58％であった。

11. **Judging from the above**, it will be clear that ......
上記から判断して，… は明確になるであろう。

12. **Refer to** Figure **N above**, ...
上記に参照した図Nでは，…

Absence (See also **Presence**)：無い；欠如（→ **Presence**）

1. The **absence of** substantial cadmium removal may be due in part to the high solubility of cadmium ion.
本質的なカドミウムの除去が無く，その一因は，カドミウムイオンの高い溶解性のためであろう。

2. The analysis is based solely on the presence or **absence of** a group and does not take into consideration the abundance of that group.

Accept

この分析は，ある集団の有無のみに基づいており，その集団の量を考慮していない。

3. Like Metric 8, this is a negative metric and, as such, a low number (<50 individuals) or an **absence of** organisms in a sample defaults to a zero score for the metric regardless of the presence or absence of the specified tolerant taxa.
   メトリック 8 と同様にこれは，負のメトリックなので，特定の耐性分類群の有無にかかわらず，生物が少数(50 匹未満)または皆無であればデフォルトとして 0 とスコアされる。

   ● **in the absence of** ..... ：… がない場合

4. These strategies must rely on assessing various combinations of chemical-specific and WET criteria **in the absence of** instream biosurvey data.
   これらの戦略は，河川内生物調査データが無い場合，化学特定クライテリアと WET クライテリアの様々な組合せを評価することに頼らなければならない。

5. Control experiments show that the SQ dye does not degrade in TiO_2 suspensions under visible light **in the absence of** the TiO_2 particles.
   コントロール実験は，TiO_2 粒子が存在しない場合，可視光線下では SQ 染料 TiO_2 懸濁液中で分解しないことを示している。

Accept：受け入れる；受容する；容認する

  ● **It has become widely accepted** ：… は広く受け入れられてきた
  ● **〜 is the most widely accepted** ..... ：
    〜は最も広く受け入れられている … である
  ● **It is traditionally accepted that** ..... ：
    … のことは伝統的に受け入れられている

1. **It has become widely accepted** in large-scale applications.
   それは大規模な実地適用例で広く受け入れられてきた。

2. Monod's model **is the most widely accepted** mathematical model of microbial growth.
   Monod のモデルは，微生物の増殖を表わすのに最も広く受け入れられている数学的モデルである。

3. To date, qualitative methods, although used for many applications where the usefulness of the results is more important than the scientific rigor of the technique used, **have not been widely accepted**.
   今日までのところ，質的方法は，技法の科学的厳密性よりも結果の有用性が重要な多くの適用例で用いられているが，まだ広く容認されてはいない。

Acceptable (Unacceptable)：受け入れられる（受け入れられない）

- **with acceptable accuracy**：受け入れられる正確さで
- **as acceptable** .....：受け入れられている … として

1. The volatilization characteristics of the pesticide can be predicted **with acceptable accuracy**.
   殺虫剤の気化作用の特徴は，満足のいく正確さで予測することができる。
2. RME soil guideline for diesel fuel is given **as acceptable** cleanup guidelines.
   ディーゼル燃料のRME土壌ガイドラインが浄化ガイドラインとして受け入れられている。

- **unacceptable**：受け入れられない

3. The data will be evaluated to determine whether the contaminant concentrations in the soils have the potential to cause **unacceptable** concentrations in the groundwater.
   これらのデータは，土壌での汚染物の濃度が地下水を受け入れられない濃度にする可能性があるかどうか決定するために評価されるであろう。
4. There is insufficient information to determine whether contamination present at this source area may present an **unacceptable** risk.
   この汚染源地域に存在する汚染が受け入れないリスクを引き起こすかどうかを決めるための情報が不十分である。

Acceptance：容認

- ~ **is gaining acceptance**：～は容認されつつある
- **There is general acceptance of** .....：
  … が一般的に受け入れられている

1. This more holistic definition appears to **be gaining acceptance**.
   この総合的な定義は，容認されつつあるようだ。
2. **There is general acceptance of** the notion that organic compounds may play a major role in changing speciation of trace metals.
   有機化合物が微量金属の種形成の変化に主要な役割をしているであろうという概念が，一般的に受け入れられている。

Accomplish（Accomplishment）： 達成する（業績）

- ～ **is accomplished** ：～は達成される
- ～ **can be accomplished** ：～は達成されうる

1. Currently, most dredged material disposal **is accomplished** either by discharge into open water or by land disposal.
   現在，ほとんどの浚渫土砂の処分は開放水域への放出あるいは埋立によって達成されている。
2. The solution of these equations **is accomplished** using finite difference approximations in identical fashion to the Water Quality Analysis Simulation Program（WASP）.
   これらの方程式は水質解析シミュレーションプログラム（WASP）と同一方法，有限差分近似を使って解かれている。
3. These goals eventually **can be accomplished**.
   これらの目標は最終的に達成されうる。
4. This **could be accomplished** via a redirection of existing state agency resources.
   これは州機関の既存資源の方向転換によって達成されうる。

- **accomplishment** ：業績

5. The linkage of dynamic chemical runoff and instream chemical fate and transport models is **a recent accomplishment**.
   ダイナミックな化学物質の流出モデルと河川内での化学物質の衰退・輸送モデルを結びつけることは，最近の業績である。

Accordance： … に従う； … のとおり

- **in accordance with** ….. ： … のとおりに； … にしたがって

1. **In accordance with** theory, …..
   理論のとおりに，…
2. **In accordance with** principles of bacterial growth kinetics, it was assumed that ……
   バクテリアの増殖速度論の原則にしたがって，… と仮定された。
3. The results are **in accordance with** other studies on Crustacea.
   この結果は甲殻類についての他の研究のとおりになっている。
4. Assuming that the substance under consideration increases **in accordance**

with a first-order reaction, the following equation is developed.
考慮中の物質が一次反応にしたがって増加すると仮定して，次の式が導かれる。
5. The numeric water quality criteria for the designated use vary by region **in accordance with** the narrative definition for each site type.
特定利用についての数値的水質クライテリアは，各地点タイプでの記述的定義に従い，地域ごとに異なる。

According：一致

● according to ……：… によると；… に応じ

1. **According to** Suzuki(2003), …..
Suzuki(2003)によると，…
2. **According to** the model calculations, ……
モデル計算によれば，…
3. **According to** the data collected by …..
… によって集められたデータによると
4. The feed solution was prepared **according to** the medium described in Ref 8 for nitrifying bacteria with some minor revisions.
この基質溶液は，硝化菌に関して参照8に記述されている培地に従って，若干の修正を加えてつくられた。
5. Aquatic communities gradually recover with increased distance downstream from impacts, although the pattern may vary **according to** the type of impact and discharge.
水生生物群集は，影響からの流下距離が増すにつれて徐々に回復する。ただし，そのパターンは，影響および排出の種類に応じて異なる。

Account：説明する；考慮する；配慮する

● ～ account for ……：～は … を説明する

1. The model **accounts for** transport of phosphorus between lakes.
このモデルは，湖と湖の間のリンの輸送過程を考慮している。
2. Several possible mechanisms **may account for** the above observations.
いくつかの可能なメカニズムによって上の観察結果が説明されるかもしれない。
3. Differences in conditions **can account for** the apparent discrepancies.
条件の相違によって，この明白な矛盾が説明されうる。
4. The sodium, 150 mM, **accounts for** the ionic strength of both sodium and potassium ions.

Account

150 mMのナトリウムはナトリウムイオンとカリウムイオン両方のイオン強度と同じ効果がある。

5. A reaction mechanism has been proposed that **accounts for** the observed experimental data.
観察された実験データを説明する反応機構が提案されてきた。
6. When selecting these sites one **must account for** the fact that minimally disturbed conditions often vary considerably from one region to another.
これらの地点を選ぶ際には，最小攪乱状態がしばしば地域ごとに大幅に異なりうるという事実を考慮しなければならない。
7. Rapid degradation of phenol **could have been accounted for** by the release of oxidants from the plant effluent.
フェノールの速い分解は，工場排水口からのオキシダントの排出によって説明できたはずである。

● **to account for .....**：…を説明するために，…を説明すると

8. A model has been structured **to account for** the relevant factors.
モデルは関連要因を説明するよう構成されてきた。
9. The kinetic expression for growth is modified in order **to account for** environmental effects.
この成長速度の表現は，環境影響を説明するために修正されている。
10. **To account for** this in rough fashion, ...
おおざっぱな方法でこれを説明すると，…

● **by accounting for .....**：…を考慮することによって
● **without accounting for .....**：…を説明できなければ

11. **By accounting for** additional water, a depth is calculated.
追加の水を考慮することによって，深さが計算される。
12. **Without accounting for** natural geographic variability it would be difficult to establish numerical indices that were comparable from one part of State to another.
自然地理的変動性を説明できなければ，ある州の一地域と別の地域を比較可能な数値（指標）を定めることは難しいであろう。

● **～ take account of .....**：～は…を考慮する
● **taking ..... into account**：…を考慮しながら；…を考慮して
● **taking into account .....**：…を考慮しながら；…を考慮して

13. The model estimates do not **take into account** the spatial and temporal variability of the Rock Creek.

このモデル予測値は，Rock Creek の空間と時間の可変性を考慮に入れていない。
14. Aggregation effects **are taken into account** as an increase of sedimentation velocities of the particles.
    凝集効果が粒子の堆積速度の増加の原因として考慮されている。
15. The effect of dilution **must be taken into account**.
    希釈の効果が考慮されなくてはならない。
16. **Taking** all of the analysis **into accounts**, ...
    すべての分析を考慮して，…
17. Each phase **is taking into account** the interaction with the other.
    それぞれの段階が他との相互作用を考慮に入れている。
18. A mass balance is constructed **taking into account** the inflow and outflow and the various sources and sinks of constituent.
    物質収支は，流入と流出それに種々の構成要素のソースとシンクを考慮に入れて構成される。

Accuracy：正確さ

- **with sufficient accuracy** ：十分な正確さで
- **with acceptable accuracy** ：受け入れられる正確さで
- **over the accuracy of** ..... ：…の正確さに関して
- **In evaluating the accuracy of** ..... ：
   …の正確さを評価することにおいて

1. Precipitation gages and evaporation pans can be utilized **with sufficient accuracy**.
   降水量計器と蒸発パンが十分な正確さで使用できる。
2. The volatilization characteristics of the pesticide can be predicted **with acceptable accuracy**.
   殺虫剤の気化作用の特徴は容認可能な正確さで予測することができる。
3. Some concern was expressed by the author of the report **over the accuracy of** the majority of the data.
   大部分のデータの正確さに関して，若干の懸念の色が，この報告書の著者によって表された。
4. **In evaluating the accuracy of** the estimated parameter, ...
   推定されたパラメータの正確さを評価することにおいて，…

Accurate(Accurately) : 正確な(正確に)

1. **Accurate** representation of peak flow may be critical to a design study for a flood control structure.
   正確なピーク流量の表記は，洪水防止構造物の設計研究に欠かせないかもしれない。

    ● **The most accurate way to** ..... : …するための最も正確な方法

2. **The most accurate way to** decrease sample variability is to collect from only one type of habitat within a reach and to composite many samples within that habitat.
   サンプル変動性を減じるための最も正確な方法は，ある河区において1種類だけの生息場所から採集し，その生息場所における多数のサンプルを複合することである。

    ● **~ is somewhat less accurate than** ..... : ~は…ほど正確でない

3. Field measurements **are somewhat less accurate than** measurements made in a laboratory but they offer the important advantage of providing immediate results to the volunteers.
   現地測定は，実験室測定ほど正確でないが，ボランティアに直ちに結果を提示するという重要な利点を持つ。

    ● **accurately** : 正確に
    ● **as accurately as possible** : 可能な限り正確に

4. It was clear that a better tool was needed to more consistently and **accurately** characterize the aquatic communities.
   水生生物群集をより整合的かつ正確に特性把握するための優れたツールが必要なことは明らかであった。

5. Thickness of tailings deposits varies considerably and cannot be **accurately** predicted over the whole of the study area.
   選鉱滓堆積物の厚さはかなり変わり，調査全地域では正確に予測することができない。

6. A number of reference sites were sampled to ascertain **as accurately as possible** the background levels of all parameters under investigation.
   調査中のパラメータすべてのバックグラウンドレベルを可能な限り正確に確認するために，多くの参照地域でサンプルが採集された。

Achieve：…を達成する

- ~ is achieved ：~が達せられる
- to achieve ..... ：…を達成するには
- in achieving ..... ：…を達成するにあたって

1. Reasonable agreement **was achieved**.
   合理的な合意に達した。
2. Eventually a steady-state condition **will be achieved** in the reactor.
   結局，この反応槽において定常状態が達成されるであろう。
3. In the best of situation, it is possible **to achieve** an annual water balance within five percent.
   最良の状態で，年度の水収支を5パーセント内で達成することが可能である。
4. The system had approached almost compete equilibrium for mercury, but two weeks were required for complete equilibrium **to be achieved**.
   水銀では，このシステムはほとんど完全な均衡状態に近づいていたが，完全均衡に達するには2週間は必要とされた。
5. Each remedial action will be screened to determine its effectiveness **in achieving** the specified objectives.
   指定目的を達成するにあたって，それぞれの是正措置がその効果を決めるために検査されるであろう。

Acknowledge (Acknowledgement)：認める；感謝する（感謝）

1. We also **acknowledge** the helpful comments provided by Mark Sudo and three anonymous reviewers.
   我々は，Mark Sudo および他の3名の匿名査読者による有益な論評にも感謝したい。
2. The ecoregions concept has been developed through the effort of many people, too many to **acknowledge** individually here.
   生態地域（エコリージョン）という概念は，大勢の方々の努力によって編み出されてきたので，ここで個々にあげて感謝することができない。

- ~ **acknowledge that** ..... ：~は…を認める

3. The former question **acknowledges that** .....
   前の質問は…を認めている。
4. Komatsu (1992) **has acknowledged that** "the problem of pattern and scale is the central problem in ecotopology."

Acknowledge

Komatsu (1992) は，「パターンとスケールの問題が生態環境学における中心的問題であること」を認めた。

5. USEPA **acknowledges that** site-specific modifications may be appropriate in some circumstances.
   USEPA は，いくつかの状況では地点特定の変更が適切であることを認めている。

- **～ wish to acknowledge .....：～は … に感謝したい**

6. We also **wish to acknowledge** the assistance of Taro Kato for statistical analysis support.
   我々は同じく統計解析を手伝ってくれた Tarou Kato に感謝したい。

- **acknowledgement：感謝**

7. Especially deserving of **acknowledgement** are the people I have learned from and worked with at the Environmental laboratory in Idaho Falls.
   特筆して感謝すべきは，著者が Idaho Falls 市にある環境研究所において多くを学び，共に研究した人たちである。

★感謝の表現としては thank また appreciate などもよく使われるので，それらも参照してください。

★Acknowledgement（謝辞）によく使われる例文をいくつか下に示します。

- **～ is credited for .....：～は … の功績がある**

8. Finally, Isamu Wada and Hitoshi Saigou (deceased) **are credited for** their solid management support for the concept of water quality criteria and monitoring at the Michigan EPA.
   最後に，Isamu Wada と Hitoshi Saigou（故人）は，ミシガン州 EPA での水質クライテリアおよび水質モニタリングの概念に確固たる管理支援をしてくれた。

- **～ contributed to .....** (See **Contribute**)：
  ～は … に貢献した (➡ **Contribute**)
- **～ made contributions to .....** (See **Contribute**)：
  ～は … に貢献してくれた (➡ **Contribute**)
- **～ provided comments .....** (See **Comment**)：
  ～は論評を寄せてくれた (➡ **Comment**)
- **～ would have been possible** (See **Possible**)：
  ～がしえただろう (➡ **Possible**)
- **～ provide support for .....** (See **Support**)：
  ～は … に対する支援をしてくれた (➡ **Support**)
- **support for ～ was provided by .....** (See **Support**)：

- ～について支援してくれたのは…である（➡ **Support**）
  - ～ **has been funded by** ..... ：～は … から資金供給された

9. The research described in this report **has been funded** by the USEPA.
   本章で述べてきた研究は，USEPAから資金提供された。

  - **through contract No.** ..... ：契約 No. … によって

10. This document has been prepared at the ISU Environmental Research Laboratory in Pocatello, Idaho, **through contract** #24-C5-0077, with AOKI Environmental Technology, Inc.
    本章は，Idaho州Pocatello市にあるISU環境研究所において，AOKI Environmental Technology, Inc.との契約 # 24-C5-0077 によって作成された。

  - ～ **is dedicated to** ..... ：～を … に捧げる

11. This work **is dedicated to** the memory of Genki Nakata whose years of service laid the foundation for our efforts.
    本稿を我々の努力を長年支えてくれた Genki Nakata の思い出に捧げる。

  - ～ **are indebted to** ..... ：～は … に恩義を感じている

12. We **are indebted to** the state representatives who provided information for this document, as well as to Shigeru Saito, two anonymous reviewers, and the editors of this book who commented on this chapter.
    我々は，本章のために情報提供してくれた各州代表者に恩義を感じている。また，Shigeru Saito，匿名査読者2名，そして本章にコメントをくださった本書の監修者らにも感謝したい。

  - ★謝辞でなくても Acknowledgement（謝辞）または Disclaimer（但し書き）によく載せられる例文を下に掲げます。

13. It has been subject to the agency's peer and administrative review and approved for publication.
    USEPAの同領域専門家および行政官による評価を経て，公刊が承認された。

14. Mention of trade names or commercial products does not constitute endorsement or recommendation for use.
    商標名や商品への言及は，その利用を必ずしも是認していないし，また推奨もしていない。

Acquire：獲得する；得られる

- **acquire new knowledge about** ..... ：
  … について新しい知識が得られる
- **acquire insight about** ..... ：… について洞察が得られる

1. The method used here may be a new approach not only for gaining better results, but also for **acquiring new knowledge and insights about** toxic behavior.
   ここで使われた方法は，良い結果が得られるだけではなく，毒性挙動について新しい知識と洞察が得られることで，新しいアプローチであるかもしれない。

   ★ for gaining better results ..... と gaining が前に使われているので，..... gaining new knowledge ..... というよりは..... acquiring new knowledge ..... の方が文が生きてきます。

Addition：付加；追加

1. The **addition** of H_2O_2 led to decreased TCE intermediate concentrations and increased DCM intermediate concentrations.
   H_2O_2 の添加は，TCE 中間体の濃度減少と DCM 中間体の濃度上昇に導いた。

   - **In addition**：さらに
   - **in addition to** ..... ：… に加えて
   - **It should also be added that** ..... ：
     … ということも付記しておくべきである

2. **In addition**, assessments of the quality of physical habitat structure are increasingly being incorporated into the biological evaluation of water resource integrity.
   さらに，物理的生息地構造の質アセスメントがますます水資源健全性の生物学的評価に盛り込まれつつある。

3. This framework offers substantial advantages for the interpretation of toxic responses **in addition to** the substantial cost savings.
   この枠組みは，実質的な出費倹約に加え，毒性応答の解釈について実質的な利点をもたらす。

4. The regional reference site framework offers substantial advantages for the interpretation of community responses **in addition to** the derivation of biocriteria.

地域参照地点の枠組みは，生物クライテリア導出に加え，生物群集応答の解釈について実質的な利点をもたらす。

Additional (Additionally) ： 追加の (その上；さらに)

1. **Additional** work is needed to confirm this.
   これを確証するためには，更なる研究が必要である。
2. **Additional** issues concerning level-of-detail are critical to every step of the simulation process.
   細部レベルに関する追加問題は，シミュレーション過程のすべてのステップに重要である。
3. **Additional** information was collected by BTK, Inc.
   追加の情報がBTK社によって集められた。
4. This appendix is subject to revisions based on **additional** information.
   この付録は，追加情報に基づいて修正の適用を受ける。
5. These data provide **additional** information on previously identified Solid Waste Management Units.
   これらのデータは，前に識別された廃棄物処理ユニットについての追加情報を供給する。

   ● **additionally** ：さらに加えて

6. **Additionally**, efforts were made to secure all prior existing data relevant to this subject.
   さらに，この課題に関する既存のデータ全てを安全に保つ努力をした。
7. **Additionally**, some programs are beginning to broaden the scope of sampling from a single type of water body such as a lake or stream to entire watersheds, including land use monitoring.
   さらに，一部のプログラムは，湖沼や河川といった1種類の水域から土地利用モニタリングといった流域全域までサンプリング範囲を拡大している。

Address (Addressing) ： 扱う；取り組む (扱うこと；取り組むこと)

★ Addressは一般に，「何を扱ったか」とか「何をしたか」を強調した言い方と言ってよいでしょう。

   ● ～ address ..... ：～は…を扱う

1. This report **address**es the problem of analyzing the fate of chemicals

Address

A

discharged into receiving waters.
この報告書は，水環境のなかに排出された化学物質の衰退過程を解析するときに生じる問題を扱っている。

2. This report **will address** the possibility of contamination at any of the three sites.
この報告書は，3地点のいずれにおいても汚染可能性があることを強調するであろう。

3. This assessment **addresses** impact rather than only discharger performance.
このアセスメントは，排出者の実績ばかりでなく影響にも取り組んでいる。

4. Water quality standards **must address** chemical, physical, and biological integrity of the nation's waters.
水質基準は国の水域での化学的・物理的・生物学的な健全性に取り組まなければならない。

5. Subsequent calibration of the models **must address** the ability to differentiate between impaired and nonimpaired sites.
その後のモデルの較正は，損傷地点と無損傷地点を差別化できる性能に取り組むべきである。

6. Some programs have been established **to address** specific local or watershed issues or projects.
一部のプログラムは，具体的な局地または流域の問題やプロジェクトに取り組むため設けられてきた。

7. In this work we **address** the volatilization rate of pesticides from water bodies.
この研究では，我々は水域からの殺虫剤の気化速度を強調する。

★ water bodies は河川湖沼（湾，海も含み）などの水環境または水域を意味します。

- **address issues**（See Issue）：問題へ取組む（➡ **Issue**）
- ～ **is addressed**：～は説明される

8. These **are addressed** later.
これらは後に説明される。

9. This issue **can be addressed** only briefly.
この問題にはごく手短に触れる。

10. This paper **is addressed** to answering these questions.
この論文はこれらの質問に答えるのに書かれている。

11. The use of these reactors for potential wastewater treatment applications

has not been addressed to date.
これらの反応器の用途が，下水処理法として利用可能性があるとは，今までは考えられなかった。

- **to address** ..... ：… に対処するため；… を扱うよう

12. **To address** this problem, a three-year project was conducted.
    この問題点に対処するため，3箇年のプロジェクトが実施された。
13. This assessment was designed **to address** the following questions: 1..., 2..., etc.
    このアセスメントは次の質問を扱うよう意図された。1 …，2 …，などである。
14. Further research into sediment transport is needed to develop more effective tools **to better address** these issues.
    これらの問題をより良く扱うためのより効果的なツールを開発するには，堆積物輸送に対するいっそうの研究が必要である。
15. Each of these types of criteria was developed **to address** specific regulatory needs of managements as related to the permitting of point source discharges.
    これらの種類のクライテリアそれぞれは，点源排出の許可に関連した管理の具体的な規制ニーズに取り組むために設定された。

- **addressing**：扱う

16. In addition to **addressing** the requirements of 40CFR270, this report will serve as a basis for RFI work plans.
    40CFR270 の必要条件を扱うことに加えて，この報告書は RFI 仕事計画の基礎となるであろう。
17. Laboratory work **addressing** some aspects of these problems is ongoing at Oregon State University.
    これらの問題のある局面を扱う実験室内での研究がオレゴン州立大学で進行中である。

Adequate (Adequately)：適切な；十分な（適切に；十分に）

1. Data gathering efforts for the Tone River yielded **adequate** stream flow.
   利根川でのデータ収集の努力で適切な流量が得られた。
2. The amount and quality of available information **are** not **adequate** to complete the studies.
   利用可能な情報の量と質は，これらの研究を完成させるのに十分ではない。

- **adequately**：適切に

3. SUIGIN model does not **adequately** represent the mercury behavior in this application.
   SUIGIN モデルは，このアプリケーションでは水銀の挙動を適切に示さない。
4. Short term studies should be undertaken to determine whether constituents are being sampled **adequately**.
   構成要素が適切にサンプルされているかどうかを決めるため，短期の研究がおこなわれるべきである。

Adopt：採用する

1. If this view **is adopted**, ...
   もしこの見解が採用されるなら，…
2. **It should not be adopted** without experimental evidence.
   それは実験的な証拠なしで採用されるべきではない。
3. The initial classification scheme **adopted** in this work appears in Table 3.
   この研究で採用された最初の分類案は，表3に示されている。
4. Colorado and several neighboring states **have adopted** ambitious goals for .....
   コロラドといくつかの隣接する州が…のために意欲的な目標を採用してきた。
5. The most critical cleanup issue facing DOE is the need **to adopt** a more practical policy on the program's objectives.
   DOE が直面している最も重要な汚染除去問題は，プログラムの目的に関して，いっそう実務的な政策を採用する必要性である。

   ★ DOE は Department of Energy（米国エネルギー省）の略語です。

Advance (advancement)：進歩（前進）

1. These played a major role in evaluating water quality management plans and **advanced** treatment justifications.
   これらは，水質管理プランの評価，および高度処理の正当化において重要な役割を演じた。

   - **advances in** .....：…における進歩

2. Research **advances in** these areas will contribute both directly and indirectly to the proposed data base for the state.
   これらの分野での研究の進歩は，直接的にも間接的にも，その州での提案された

データベースに寄与する。
3. Recent **advances in** AOP treatment technology are summarized in the literature.
AOP 処理技術における最近の進歩が文献に要約されている。

- **in advance of .....**：… に先立って

4. Most states are developing bioassessment methods **in advance of** implementing biological criteria programs.
大半の州は，生物クライテリアプログラムの実施に先立って生物アセスメント方法を案出中である。

- **advancement**

5. Much progress and **advancement** has been made in the past decade.
多くの進歩と前進がこれまでの 10 年間になされてきた。
6. It can be determined if refinements and **advancement** in macroinvertebrate taxonomy have been sufficient enough to warrant further adjustments to ICI scoring categories.
大型無脊椎動物分類学における改善と前進が ICI スコアリング区分のさらなる調整を保証するに足るほど十分か否か決定されうる。

★ ICI は，ここでは Index of Community Integrity（生物群集完全性指数）の略語として使われています。

Advantages（Disadvantages）：利点（不利点）

1. The **advantages outweigh the disadvantage that** .....
… という利点が不利点よりもまさっている。
2. **All advantages are** not associated with ....
すべての利点は…と結び付けられない。
3. The **advantages** and **disadvantages** of artificial and natural substrates **have been discussed** extensively.
人造底質および自然底質の利点と欠点について綿密に論じられてきた。
4. The **advantage is** in cost savings by limiting the number of samples sent for laboratory analyses.
利点は，分析のために送られるサンプルの数を制限することによるコスト削減である。
5. **The inherent advantages of** photocatalysis **include** (a) .....; (b) .... ; and (c) .....

Advantages

光触媒反応の固有の利点としては，(a)…，(b)…，そして(c)…，がある。

6. **The advantages of** these increased efforts **are** threefold. First, ... Second, ... Third, ...
   こうした努力増強の利点は，3つある。第一に，… 第二に，… 第三に，…。

   - **One adbantage of** ～ **is that** ......：～の一つの利点は … である
   - **The advantage of** ～ **is that** ......：～の利点は … である

7. **One advantage of** using a database such as RF3 **is that** it enumerates the resource.
   RF3といったデータベース使用の一つの利点は，資源を列挙することである。

8. **An advantage of** examining the trophic status of component populations **is that** it also provides information on functional aspects of the community as well.
   構成要素たる個体群の栄養段階を検討することの利点は，その生物群集の機能面に関する情報も提供することである。

9. **The advantage of** this approach **is that** it would allow land-use planners and managers to establish criteria for the conservation of high-quality watersheds and the restoration of degraded areas.
   この手法の利点は，土地利用の計画者・管理者が高質流域の保全ならびに劣化区域の回復についてクライテリアを定められるようにすることである。

   - **There are advantages with** ......：… に利点がある
   - **There are advantages to** ......：… に利点がある

10. **There are advantages** and **disadvantages with** each.
    それぞれに利点と不利点がある。
11. **There are advantages and disadvantages to** both approaches.
    両方のアプローチに利点と不利点がある。
12. During the application of the impairment criteria protocol, it became apparent **there were several advantages over** an ecoregional method.
    損傷クライテリアのプロトコルの適用中に，生態地域（エコリージョン）方式を上回る利点がいくつか在ることが明らかになった。

    - ～ **has the advantages that** ......：～は … という利点を持つ

13. The model **has the disadvantage that** ....
    このモデルは … という不利な点を持っている。
14. The definition **has the additional advantage** that the IR of mixtures can be calculated by summing the reactivity contributions of the components.
    この定義は，構成要素の反応寄与率を合計することによって，混合物のIR計算

ができるという更なる利点を持つ。
15. Each process **has its own advantages and disadvantages**.
それぞれのプロセスがそれ自身の利点と不利点を持っている。

● ~ **have advantages over** .....：~は … に対して利点を持つ

16. UV disinfection **has several advantages over** the use of traditional halogen disinfectants.
UV 殺菌は，伝統的なハロゲン殺菌剤の使用に対していくつかの利点を持っている。
17. Box plots **have important advantages over** the use of means and standard deviations.
ボックス図は，平均や標準偏差の利用に優る重要な利点を持つ。
18. AO Process **has a number of potential advantages over** other treatment methods including .....
AO プロセスは … といった他の処理方法より多くの潜在的利点を持っている。

● ~ **offer the advantage**：~は利点をもたらす

19. The test **offers the advantage of** being more sensitive and much faster.
このテストには，いっそう敏感で，より迅速であるという利点がある。
20. The use of probabilities **offers certain advantages for** implementation.
確率の利用は，実施上の利点をいくつかもたらす。
21. Performance standards **offer several advantages that** will perpetuate their use in water resource management.
実績基準には，水資源管理においてそれらの利用を永続化させるといった利点がいくつかある。
22. The framework **offers substantial advantages for** the interpretation of fish community responses.
この枠組みは，魚類群集応答の解釈について実質的な利点をもたらす。
23. Field measurements **offer the important advantage of** providing immediate results to the volunteers.
現地測定は，ボランティアに直ちに結果を提示するという重要な利点を持たす。

Affect（Affecting）：影響する（影響を与える）

1. The focus is on the mechanisms that **affect** the distribution of industrial chemicals in the aquatic environment.
焦点は，水生環境で産業化学物質の分布に影響を与える機構（メカニズム）にある。

A Affect

2. Nitrate up to 1 mg/L as N did not **affect** the rate of primary producer during 3 hr of incubation.
   3時間の培養時間では，硝酸態窒素1 mg/Lまでは一次生産者の増殖速度に影響を与えなかった。
3. Japan, together with North America and Europe, is one of the areas in the world most severely **affected by** environmental pollution.
   日本は，北アメリカとヨーロッパと共に，世界中で最もひどく環境汚染の影響を受けている地域の1つである。

   ★ North Americaのように米国の主要な領域は大文字で始まります。例えば，The necessary data were gathered in the North（必要なデータはアメリカ北部で収集された。）となります。

   - **～ is most affected** ：～は最も影響される
   - **～ is affected by .....** ：～は…によって影響がもたらされる

4. This approach leads itself to a better understanding of the nature of the impairment, including which elements or processes of the community **are most affected**.
   この手法は，生物集団のどの要素や過程が最も影響されるかなど，損傷の性質をより深く理解するのに役立つ。
5. Organic suspension **is affected** differently **by** dissolved substances.
   懸濁有機物は，溶解物質によって違った影響がもたらされる。

   - **affecting** ：影響を与える

6. Some chemical changes **affecting** aqueous metal chemistry may be almost immediate.
   水溶性金属化学に影響を与えるある化学変化は，ほとんど瞬間的であるかもしれない。
7. Knowledge of the factors **affecting** the movement and transformations of these substances are of obvious importance in understanding and controlling the hazard.
   これらの物質の移動と変化に影響を与える要因の知識は，その物質の危険性を理解し制御していくために当然重要である。
8. Additional insight into the fate of DDE can be obtained by considering the magnitudes of the parameters **affecting** chemical fate.
   化学物質の衰退に影響を与えるパラメータの規模を考慮することによって，DDEの衰退についての更なる洞察を得ることができる。

Aforementioned：前述の

1. Despite the **aforementioned** impoundments and flow alterations, overall habitat conditions in the Natsui River are good to excellent.
   前述の人工湖と流量改変にもかかわらず，Natsui 川の全般的生息場所状態は，良ないし優である。

Agree (Disagree)：同意する；合意する；一致する（同意しない）

- ～ **agree with** ……：～は…と一致する
- ～ **agree well with** …：～は…とよく合致する
- ～ **agree in principal with** ……：～は…と基本的に一致する
- **agree** …..% **of the time**：の一致率は…％である
- **few** ～ **agree on** ……：…に関して同意している～はほとんどいない

1. Most **agree with** a general definition that …..
   ～という一般的な定義には，ほとんどの人が同意する。
2. The inclusion of Oligochaeta as a toxic tolerant group **agrees with** the findings of Yoshida et al. (1990).
   貧毛綱を毒物耐性集団として含めることは，Yoshida ら (1990) の所見に合致している。
3. The curve shown in this figure **agrees well with** the curve presented recently by Ito et al. (2002).
   この図に示された曲線は，最近 Ito ら (2002) によって発表された曲線とよく一致している。
4. The overall removal of total chromium from the wastewater treatment plant **agreed in principal with** earlier studies.
   この下水処理場からの全クロムの総除去量は以前の研究結果と基本的に一致した。
5. In the results cited above, chemical and biosurvey results **agreed** 58 % **of the time**.
   前記の結果においては，化学的調査結果と生物的調査結果の一致率は，58％であった。

Agreement (Disagreement)：同意；合意；一致（不一致）

1. Some of this **disagreement stems from** differences in individual perceptions of ecosystems.

意見相違の一因は，生態系に対する各自の認識の差である。
2. EPA (2003) studied the level of **agreement between** bioassay **and** instream biosurvey results using data collected near 43 separate facilities.
EPA (2003) は，43 施設の付近での収集データを用いて，バイオアッセイと河川生物調査の結果の一致レベルを研究した。

- **agreement was achieved**：合意が達成された
- **agreement improved**：合致度が上昇した
- **agreement was improved by .....**：… によって合意性が高くなった

3. Reasonable **agreement was achieved**.
妥当な合意が達成された。
4. When the SQM analysis was used, **agreement improved**.
SQM 分析が用いられた場合では，合致度は，上昇した。
5. **Agreement was improved** by adjusting the hydraulic conductivity to a value of 0.01 cm/sec.
水理学的伝導率を 0.01 cm/sec に調整することによって，合意性が高くなった。

- **agreement with ..... is good**：… と良く合意している

6. Quantitatively, **agreement with theory is good** at the lower flow-rates.
数量的に，流速の遅いときに，理論と良く合意する。

- **There is a good agreement between A and B**：
  A と B とは良く合意している
- **There is little agreement on .....**：
  … についての意見はほとんど一致していない
- **There is very little agreement that .....**：
  … ということについては，意見の一致はほとんどみられない
- **There is general agreement that .....**：
  … という一般的な合意がある
- **There is considerable disagreement about .....**：
  … についてかなり大幅な意見の相違がある

7. **There is a good agreement between** the solution of the model **and** the data.
モデル計算結果とこのデータとは良く合意している。
8. Although studies have been made on the cause of nitrification failure, **there is little agreement on** how it was failed.
硝化作用が生じなかった原因についての研究はおこなわれてきたが，その原因についての意見はほとんど一致していない。

9. **There is general agreement that** these ecological regions exist, but there is considerable disagreement about how to define them.
   こうした生態地域(エコリージョン)が存在するという一般的な合意はあるものの，その定義づけについては，かなり大幅な意見の相違がある。

   - **agreement was obtained between A and B**：
     ＡとＢの間に合意が得られた

10. **Good agreement was obtained between** experimentally determined stability constants **and** available literature values.
    実験によって決められた安定定数と入手可能な文献値の間にかなりの一致が得られた。

    - **~ show good agreement between A and B**：
      ～はＡとＢの間で良い合意を示している
    - **~ found good agreement between A and B**：
      ～はＡとＢの間で，かなりの一致を認めた

11. Figure 3 **shows good agreement between** the predicted **and** observed suspended solids concentrations for the first 150 meters downstream.
    図3が，最初の150メートル下流で浮遊物質濃度の予測値と観測値の間で良い合意を示している。

12. Preliminary examinations and comparisons **show good agreement between** the two at most common stations.
    予備的な検討・比較によれば，ほとんどの一般的な調査地点では，これら両者がかなり合致する。

13. EPA (2004) **found** good **agreement between** results in about 20% of cases when effluent toxicity was assessed, and in about 30% of cases when mixing zone toxicity was measured.
    EPAは，流出毒性を評価した際には約20％の事例結果で，また，混合区域毒性を測定した際には約30％の事例結果で，かなりの一致を認めた。

    - **~ is in agreement with .....**：～は…と一致する
    - **~ is in good agreement with .....**：～は…と良く一致する
    - **~ is in general agreement with .....**：
      ～は…と一般的な合意にある
    - **~ are generally in good agreement**：～は一般に良い合意にある
    - **~ is seen to be in agreement with ....**：
      ～は…と合意にあるように見られる

14. These results **are in agreement with** published data.

Agreement

これらの結果は，発表されたデータと一致する。
15. These data **were in agreement with** predictions from the model.
    これらのデータは，モデルの予測と一致した。
16. The value obtained here **is in good agreement with** that of Suzuki et al. (2002).
    ここで得られた数値は Suzuki ら(2002)が報告した値と良く一致する。
17. The two data sets **are generally in good agreement**.
    この2つのデータセットは，一般に良い合意にある。
18. The results **are in general agreement with** the experimental data.
    この結果は，実験データと一般的な合意にある。
19. Quantitatively, the results plotted in Figure 6 **may be seen to be in general agreement with** the theory.
    数量的に，図6にプロットされた結果はその理論と一般的な合意にあるように見られるかもしれない。

- ～ **show agreement** ：～は合意を示す
- ～ **show good agreement with** ……：
  ～は … と良い一致を示している

20. A comparison of the observed data with the computed results **shows good agreement**.
    観察データと計算結果との比較では良い一致を示している。
21. Model predictions **show reasonable agreement** as can be seen from both Figures 2 and 3.
    両方の図2と3から見られるように，モデルによる予測が妥当な合意を示している。
22. These measurements **show** generally **good agreement with** expected concentration values.
    これらの測定値は，予期していた濃度値と一般に良い一致を示している。
23. The relationship between cellular copper content and cupric ion activity **shows good agreement with** the hyperbolic function.
    細胞の銅含有量と銅イオン活性の間の関係は，双曲線関数で示すと良く一致する。

- ～ **indicate a good agreement between A and B**：
  ～はAとBの間で良い合意を示す
- **indicating a reasonable agreement**：妥当な合意を示し
- ～ **have agreement**：～は一致している

24. Measurement of TOC during the treatment **indicated a good agreement**

between the experimentally determined TOC values **and** those calculated from the quantified reaction intermediates.
処理中の TOC の測定が，実験的に決められた TOC の値と定量化した反応中間物から計算された TOC の値の間で良い合意を示した。

25. A linear regression of measurement and predicted values yielded an R^2 of 0.75, **indicating a reasonable agreement**.
測定値と予測値の線形回帰は決定係数（R^2 値）が 0.75 となり，妥当な合意を示した。

26. A comparison of quantitative measures of performance between AMODEL versus the observed residues, and between BMODEL versus the observed residues shows that AMODEL **had** only slightly better **agreement** than BMODEL.
AMODEL 値と観察された残留物，および BMODEL 値と観察された残留物との間の量的測定比較では，AMODEL は BMODEL より少しだけ良く一致していた。

Aid：手助け；助ける

● **aid in .....** ：…するのを助ける

1. This work will **aid in** making predictions about long-term buildup of organic layers on the stone.
この研究は，石への長期にわたる有機層の蓄積を予測するのを助けるであろう。

2. The guidelines are followed step-by-step to demonstrate how the guidelines can **aid in** design of a field study.
この指針が，いかに現地研究の計画の助けになるかを明示するのに，この指針が一歩一歩追髄される。

3. Whether the models we create are good models or poor models depends on the extent to which they **aid** us **in** developing the understanding which we seek.
我々が作るモデルが良いモデルであるか，あるいは貧弱なモデルであるかどうかは，それらのモデルが我々が求める理解・知見を深める手助けする程度による。

● **To aid in .....** ：…するのを助けるために．

4. **To aid in** judging when this condition was attained, ...
この状態がいつ達成されたか判断するのを助けるために，…

● **as an aid to .....** ：…への手助けとして

5. **As an aid to** the user in projecting computer costs, ...

Aim

コンピュータのコストを推定するのにユーザーへの手助けとして，…

6. A brief summary is included below **as an aid to** understanding the results obtained in this project.
   このプロジェクトで得られた結果を理解する手助けとして，短い要約が下に含まれている。

> **Aim** (See also **Goal, Purpose, Objective**)：
> 目的；目的とする；向ける；狙う（➡ **Goal, Purpose**, Objective）

★ Aim, goal, purpose, objective は研究や，本，論文などの目的を示すときに使う表現で，どれも同様な使い方をします。

- **The aim of ～ is to .....**：～の目的は…である
- **The aim of this study is to contribute to .....**：
  この研究の目的は…に貢献することである

1. **The aim of** this book (report, paper) **is to** .....
   この本(報告書，論文)の目的は…である。
2. **The major aim of** this investigation (study) **is to** .....
   この調査(研究)の主要な目的は…である。

★ to .... のところには to examine (検討する)，to evaluate (検討する)，to investigate (研究する)，to test (テストする) といった言葉がよく使われます。しかし，The major aim of this investigation is to investigate ..... といった同様な言葉 (investigation, investigate) が重なる表現はさけましょう。

3. The **aim** of this study **is** also **to** contribute indirectly to the proposed data base for the state.
   この研究の目的は，この州の為に提案されたデータベースに間接的に貢献することでもある。
4. **The major aim** of this investigation **is to** test the hypothesis that copper toxicity to algae and copper content of algal cells are functionally related to free cupricion activity.
   この調査の主要な目的は，藻類への銅毒性そして藻類細胞内の銅蓄積量は機能上遊離銅イオン活量と関係があるという仮説を検証することである。
5. **The aim of** the present study **was to** examine aspects of elimination of PAH by *Mytilus edulis*.
   現在の研究の目的は *Mytilus edulis* による PAH の除去の状況を調べることであった。

- **aim at .....**：…を目的とする

- **aim toward** ..... ： … に狙いを定める

6. All of the above studies **were aimed at** defining the ambient water quality.
   上記の研究のすべてが周辺の水質を定義することに向けられた。
7. The environmental research laboratory contributes to this information through research programs **aimed at** determining the effects of toxic organic pollutants.
   環境調査研究所は，有毒な有機汚染物の影響を決定することを目的とする研究プログラムを通してこの情報に貢献する。
8. What we **are aiming toward** is establishment of the validity of the concept of controlled catalytic biomass.
   我々が狙いを定めているものは制御触媒性バイオマスの概念の正当性を立証することである。

Aliquot ： アリコート；分割量

1. Fifty mL **aliquots** were processed for GC−MS analysis.
   50 mL のアリコートを GC−MS 分析のために処理した。
2. An **aliquot of** the 100 ppm PMPA solution was acidified to pH of 2.0.
   100 ppm PMPA 溶液のアリコートが pH 2.0 に酸化された。
3. **Aliquots** (1.0 mL) were withdrawn with a Hamilton syringe and mixed with 0.5 mL of pentane.
   アリコート (1.0 mL) はハミルトン注射器で抜かれ，0.5 mL のペンタンと混ぜられた。

All ： すべて；全部

1. Decommissioning of these units requires removal of **all** stored explosives.
   これらのユニットを退役させるには，貯蔵されている爆発物を全部撤去することが必要である。
2. This description includes **all** current, reasonably obtainable information on the unit.
   この記述は，このユニットについての現在の，ある程度入手可能な情報すべてを含んでいる。

- **All 〜 are** ...... ：すべての〜は … である
- **All things are** ...... ：あらゆる事物は … である
- **Not all 〜 is** ..... ：すべての〜が … ではない
- **All known** ..... ：すべての周知の…

- **First of all**：まず第一に
- **Above all**：なによりも

3. **All** advantages **are** not associated with ....
   すべての利点は … と結び付けられない。
4. **All** areas storing containerized wastes **are** described.
   コンテナに入れられた廃棄物をしまっておいてあるすべての場所が記述されている。
5. **All things are** somewhat different, yet some things are more similar than others.
   あらゆる事物は，それぞれ若干異なるが，あるものは他のものよりも類似性が強い。
6. **All** our attempts at classification **are** human constructs that help us understand a spatially and temporally varying landscape.
   分類における我々の試みは，すべて人為的構成物であり，空間的・時間的に異なる景観を理解するのに役立つ。
7. **All** corrective actions at Pacific Northwest National Laboratory (PNNL) **will be** managed under the Environmental Restoration (ER) program.
   Pacific Northwest National Laboratory (PNNL) においてのすべての是正処置は，Environmental Restoration (ER) プログラム下で管理されるであろう。
8. **Not all** the information for refining the target population **was** available before field visits.
   現地視察の前には，ターゲット母集団を絞り込むための情報すべてが利用可能ではなかった。
9. **All known** underground storage tanks previously or currently used for the storage of wastes **have been** included.
   以前に，あるいは現在，廃棄物の貯蔵のために使われたすべての周知の地下貯蔵タンクが含まれてきた。

- **In all**：全部で

10. **In all**, at least 24 states can be described as currently using biological criteria to support their water resource management.
    全部で，少なくとも 24 州が水資源管理を支えるために生物学的クライテリアを目下用いているといえる。

- **all of .....**：… のすべて

11. **All of** the pesticides chosen for this study are organophosphorus compounds.
    この研究のために選ばれた殺虫剤のすべてが有機リン化合物である。
12. Virtually **all of** the petroleum industry is based on a series of catalytic

transformations.
石油産業のほとんどすべては，一連の触媒変換に基づいている。
13. The 1116 commercial reactors in the U. S. are **all of** the light-water type.
米国での1116の商業用原子炉のすべては，軽水タイプである。
14. We examined **all** nine **of** the impact types in a two-dimensional framework.
我々は，また，9種類の影響すべてを二次元枠組みで検討した。

- **nearly all of** ..... : ほぼすべての …

15. It is envisioned that by the end of the decade, **nearly all of** the state water resource agencies will have some form of biological monitoring.
この10年代末までに，ほぼすべての州の水資源担当機関が何らかの形の生物モニタリングを有するようになるものと予想される。

- **almost all** ..... (See **Almost**)：ほとんどすべての …（➡ **Almost**）
- **among all** ..... (See **Among**)：
    … のすべてのなかで；すべての … の間で（➡ **Among**）
- **Of all** ..... : … すべてのうち

16. **Of all** the ions removed, several are particularly important.
除去されるイオンすべてのうち，いくつかは特に重要である。
17. **Of all** the BTEX chemicals contained in the various fuels used at Eielson, benzene has the greatest mobility in the groundwater system.
Eielsonで使われている種々の燃料に含まれるBTEX化学物質すべてのうち，ベンゼンが地下水系で最も大きい移動性を持っている。

Allow：許す；させる；可能になる；可能にする

1. Identification of organisms to species **allows** the use of indices.
種のレベルまで生物の識別ができれば，指数の利用が可能になる。
2. The available field data were not sufficiently detailed to **allow** an assessment of spatial variations.
入手可能なフィールドデータは，空間的変移のアセスメントをするためには十分に詳述されていなかった。
3. The advantage of this approach is that it would **allow** land-use planners to establish criteria for the restoration of degraded areas.
この手法の利点は，土地利用の計画家が環境の悪化した区域の回復のためにクライテリアを定められるようにすることであろう。
4. This **has allowed** small towns to begin building treatment plants that will attain the effluent limit specified in the new water quality standards.

A

これにより，小都市は，新規水質基準で明記された放流限度を満たすような処理場の建設を開始できるようになった。

- **～ allow for ..... ：**
  ～は … を許す；～は … を示す；～ は … を可能にする

5. **It allows for** a very small margin of error.
   それは非常に小さい範囲の誤差を許している（それは非常に小さい誤差範囲でできる）。
6. This **allows for** the analysis of incremental changes in aquatic community performance over space and time.
   これは，水生生物群集実績における時間的・空間的な漸増変化の分析を可能にする。
7. Procedures should be developed that **allow for** both regional- and subregional-scale deviations from broadly established water quality criteria.
   広範に定められた水質クライテリアから，地域規模および小地域規模の偏差を示すための手順を設けるべきである。

- **allowing ....：** … を許し；… させ；… を可能にし

8. Regional scale modifications in management practices may be feasible, **allowing for** the significant recovery of impaired aquatic resources.
   管理慣行における地域スケール改善が実行可能であり，これは傷ついた水生資源をかなり回復させる。
9. The planning process can be strengthened by interaction with other programs, **allowing** joint utilization of reference database.
   立案過程は，参照データベースの共同活用を可能にし，他のプログラムとの相互作用で強化されうる。

- **allow comparison ..... (See Comparison)：**
  … との比較を可能にする（➡ **Comparison**）

Almost：ほとんど

- **almost all ....：** ほとんどすべての …
- **almost complete：** ほとんど完全に
- **almost entirely：** ほとんど完全に

1. With increasing pH, **almost all** the cycle esters are hydrolyzed.
   pH の上昇とともに，環エステルのほとんどすべてが加水反応される。
2. DAO Process have a number of potential advantages over other treatment methods including the complete destruction of **almost all** organic

molecules to form water, carbondioxide, and simple organic ions.
DAOプロセスは他の処理方法より多くの可能性がある利点を持っている。それらは，ほとんどすべての有機分子を完全に破壊し水，二酸化炭素と単純な有機物イオンに分解するなどである。

3. The system had approached **almost complete** equilibrium for mercury.
   このシステムは，ほとんど完全に水銀の平衡状態に達していた。
4. The mineralization of organic compounds can be attributed **almost entirely** to biodegradation.
   有機化合物のミネラル化は，ほとんど完全に生物的分解に帰因されるであろう。

- ~ **is almost immediate** : ~は，ほとんど瞬間的である

5. Some chemical changes affecting aqueous metal chemistry may **be almost immediate.**
   水溶性金属化学に影響を与えるある化学変化は，ほとんど瞬間的であるかもしれない。

Along：沿って

1. Eleven dams exist **along** the Columbia River in the United States.
   11のダムが合衆国でコロンビア川に沿って存在する。
2. The company constructed a diversion channel and a small settling pond **along** Rock Creek at its mouth.
   この会社はRockd Creekに沿ってその河口に転換水路と小さい沈降池を建設した。
3. There is no doubt that gypsum is ubiquitous in the tailings deposits **along** Cherry Creek.
   Cherry Creekに沿って選鉱屑堆積層に石膏が遍在していることには疑いがない。
4. The Yakima River runs **along** part of the southern boundary and joins the Columbia River south of the city of Richland.
   ヤキマ川は南境界線の一部に沿って流れ，そしてRichland市の南でコロンビア川と合流する。

   ★ Richlandは町の名前で，the city of Richland (Richland市)のようにthe city of ~と書きますが, Iowa City (アイオワシティー)とかKansas City (カンサスシティー)のようにCity (シティー)が町の名前の一部である場合には，~Cityと大文字でかきます。

- **along with** ..... : …と共に；…と一緒に

5. These studies **along with** our results confirm that ....

これらの研究は，我々の結果と共に…を確証する。
6. Figure 6 is a display of the data plotted **along with** this hypothetical line.
   図6は，この仮説線と一緒に作図したデータの図示である。
7. The computational results **along with** the observed data are shown in Fig. 4.
   計算結果が，観察データとともに図4に示されている。
8. Faunal similarity between midwestern river systems is dependent on river mile distance, **along with** other factors.
   中西部河川系における動物相の類似は，河川距離(マイル)および他の因子に左右される。

Although：だけれども；だが

1. **Although** it is generally believed that .....
   …と一般に信じられているけれども，
2. **Although** factors such as geology and soils are also important, the other factors appear to be the most important in this ecoregion.
   地質や土壌などの因子も重要だが，この生態地域(エコリージョン)では，他の因子が最も重要に思われる。
3. **Although** these indices have been regionally developed, they are typically appropriate over wide geographic areas with minor modification.
   これらの指数は，地域的に設定されてきたものだが，一般に，やや修正すれば広い地理的範囲に適している。
4. **Although** the mechanism of acoustic foam disintegration is not entirely understood, the following acoustic effects are believed to be most important in foam breakage.
   音波によって生じる泡の崩壊機構は完全に理解されていないが，次に示す音波効果は気泡崩壊過程上最も重要であると思われる。
5. Stream habitat modification was identified as the third leading cause of aquatic life impairment in 1992, **although** the database was oriented primarily to the evaluation of point sources.
   河川生息地改変は，1992年に，水生生物損傷の第三の主因と確認された。ただし，このデータベースは，主に点源評価を志向している。

Among：ごとに；において；の間で

1. Opinions differ **among** states.
   諸州で意見の相違がある。

2. The extent and form of biological survey data vary widely **among** states.
生物調査データの程度と形式は，州ごとで大幅に異なる。
3. Cooling water chemistry control varies considerably **among** nuclear power plants.
冷却水の化学的コントロールは，原子力発電所ごとでかなり異なる。
4. Index scores can be compared **among** streams of different sizes.
指数スコアは異なる規模の河川において比較されうる。
5. The approach provides valuable consistency **among** the many different programs.
この手法は，多くの様々なプログラムにおける貴重な整合性をもたらす。
6. Stream classification provides relatively homogeneous classes of streams for which biocriteria may differ **among** the classes.
河川分類は，比較的同質な河川等級を生じさせるが，等級ごとに生物クライテリアが異なることもある。

- among all ......：…のすべてのなかで
- among others：他から

7. **Among all** the techniques of waste management, waste reduction is the common sense solution to the prevention of future hazardous waste problems.
廃棄物管理のすべてのテクニックのなかで，廃棄物削減は将来の危険廃棄物問題の防止に対する共通認識的な解決策である。
8. These indices were selected **among** many **others** on the basis of accuracy, low variability, and simplicity.
これらの指数は，正確性，低い変動性，単純性に基づいて他の指数から選ばれた。

Amount：量

- the amount of ......：…の量
- a large amount of ......：大量の…
- a small amount of ......：小量の…..
- only a limited amount of ......：ほんの限られた量の…
- a considerable amount of .....：かなりの量…..
- the amount and quality of .....：…の量と質
- a certain amount of .....：ある程度の…

1. **A large amount of** circumstantial evidence indicates that .....

多くの付随的な証拠が … のことを指摘している。
2. **The amount and quality of** available information are not adequate to quantify the risk.
利用可能な情報量と質はリスクを数量化するには不十分である。
3. **Only a limited amount of** stream water quality data is available.
ほんの限られた量の河川水質データが利用可能である。
4. **The amount of** data available is limited.
利用可能なデータの量が不十分である。
5. **Small amounts of** $HCOO^-$ were detected.
小量の $HCOO^-$ が検出された。
6. **A considerable amount of** information is available on the effect of inhibitors on culture of *Nitrosomonas sp*.
Nitrosomonas sp. 菌株に対する阻害剤の影響に関するかなりの量の情報が利用可能である。
7. This variation in terminology has caused **a certain amount of** confusion.
用語法におけるこの差異は，ある程度の混乱を起こしてきた。

Analogous：類似している

- ~ **is analogous to** ..... ：~は … に似ている
- ~ **is somewhat analogous to** ..... ：~は … に若干似ている

1. Sample collection, handling, and processing procedures **were analogous to** those previously described for the soil sampling.
サンプル収集，取り扱い，処理の手順は土壌サンプリングのために前に記述された手順に類似していた。
2. The selection of the 25th percentile value **is analogous to** the use of safety factors, which is commonplace in chemical water quality criteria applications.
25パーセンタイル値の選定は，化学的水質クライテリア適用において，ごく一般的な安全因子の利用に似ている。
3. The results can be expressed in a fashion **analogous to** an adsorption isotherm.
この結果は吸着等温線に類似した方法で表現されうる。
4. The use of these two groups **is somewhat analogous to** the use of a fish species and an invertebrate species as standard bioassay test organisms.
これら2集団の利用は，標準的バイオアッセイ試験生物としての魚種および無脊椎動物種の利用にやや似ている。

5. Marui (1993) described the use of rapid assessment approaches in water quality monitoring as **somewhat analogous to** the use of thermometers in assessing human health.
Marui (1993) は，水質モニタリングにおける迅速アセスメント手法の利用について，人間健康を評価する際の体温計に若干似たものと評した．

Analogy ：類似

- **analogy between ..... and .....** ：… と … の間の類似性
- **By analogy with ..... and .....** ：… と … が類似していることから

1. The word is relatively new in mathematics and would seem to have been borrowed from earlier usage because of the **analogy between** the mathematical model **and** the scale model.
この言葉は数学の領域では比較的新しく，そして数学的モデルとスケールモデルの間の類似性のためにより以前の使用法から借りられたように思われる．

2. **By analogy with** the oxidation of formic **and** oxalic acids, it is reasonable to assume a second-order reaction for glyoxylic acid because it has a structure similar to the other two acids.
ギ酸とシュウ酸の酸化反応が類似していることから，グリオキシル酸の反応は第2次反応であると仮定するのが道理的である．なぜなら，それは他の2つの酸に構造が類似しているからである．

Analysis ：分析；解析

1. The **analysis** was run in duplicate.
分析は重複して行なわれた．
2. No particle size **analysis data** is currently available.
粒子径分析データは現在利用可能ではない．
3. **Laboratory analysis** of species composition is labor intensive.
生物種構成の室内分析は，労働集約的である．

- **From an analysis of .....** ：… の解析から
- **Further analyses of .....** ：… 更なる … の解析

4. **From an analysis of** these plots, two general conclusions may be drawn. First, ...
これらのプロットの解析から2つの一般的な結論が引き出されるかもしれない．

Aanalysis

第一に…。

5. **Further analyses of** the data are planned that will employ additional existing geographical classification.
   他の既存地理分類を用いる更なるデータ解析が予定されている。

   - **The data analysis provided** ……：
     データ分析によって … が提示された
   - **The analysis revealed** …… ：この分析によって … が明らかにされた
   - **This analysis demonstrates** …… ：
     この分析によって … が明示される

6. **The data analysis provided** the following conclusions.
   データ分析によって次の結論が提示された。
7. **The analysis revealed** the following: ….., ….., …..
   この分析によって，以下が明らかにされた。それらは…, …, …である。
8. **This analysis demonstrates** the need to interpret fish community information.
   この分析によって，魚類群集情報を解釈する必要があることが明示される。

Analyze ：分析する；解析する

1. The focus of this effort is to **analyze** exhaustively the solutions.
   この努力の焦点は，徹底的に解決策を分析することである。
2. The manner in which biological data **was analyzed** also underwent changes.
   生物学的データのかつての分析方法も変化した。
3. These data **will be analyzed** to assess the status of biological condition and to address the goals as stated above.
   生物状態の現状を評価し，上述の目標に取り組むために，これらのデータが解析されるであろう。

   - **~ was analyzed for** ….. ：～で … の分析がおこなわれた

4. Samples from each of these sites **have been analyzed for** mercury.
   これらの各地点からのサンプルで，水銀分析がおこなわれてきた。
5. Between 16 and 52 samples from each of these sites **have been analyzed for** total arsenic.
   これらの各地点から採取された16から52のサンプルで全ヒ素の分析がされてきた。

Ancillary：補助；補助の；補助的

1. **In ancillary experiments** it was determined that .....
   補助的な実験で … が決められた。
2. **Ancillary to** these assumptions are two others that affect the utility of the theory.
   これらの仮定に対する補助として，他に，理論の有用に影響を与える2つがある。

Another：他の；もう一つの

1. **Another** advantage of this process is that .....
   このプロセスの他の利点は … である。
2. Many facilities will likely have considerable data for one assessment but relatively little for **another**.
   多くの施設で，1つのアセスメントに関してかなりのデータが集められそうだが，他のアセスメントに関するデータは，比較的少ないであろう。
3. The question about what constitutes a sufficiently comprehensive bioassessment is **another** key contemporary issue facing the implementation of bioassessments.
   十分に総合的な生物アセスメントを構成するものが何かに関する疑問は，生物アセスメントの実施が現在直面している，もう一つの重大問題である。

   - **Another reason for .....**：… の別の理由
   - **〜 is another matter**：〜は別問題である
   - **from one time to another**：時間ごとに
   - **from one place to another**：場所ごとに
   - **from one scale to another**：縮尺ごとに
   - **In one way or another**：何らかの方法で

4. The factors that are more or less important vary **from one place to another** at all scales.
   多かれ少なかれ重要な因子は，あらゆる縮尺で場所ごとに異なる。
5. Resource managers and scientists have come to realize that the nature of these resources varies in an infinite number of ways, from one place to another and **from one time to another**.
   資源管理者や科学者は，これらの資源の性格が，場所ごと，時間ごとに無限な様相で異なることを理解するようになっている。

A

6. Without accounting for natural geographic variability it would be difficult to establish numerical indices that were comparable **from one part of State to another**.
自然地理的変動性を説明できなければ，ある州の一地域と別の地域を比較可能な数値指数を定めることは難しいであろう。

7. **Another reason for** using single-purpose frameworks stemmed from the belief that a scientifically rigorous method for defining ecological regions must address the processes that cause ecosystem components to differ **from one place to another** and **from one scale to another**.
単一目的枠組みを用いる別の理由は，生態的地域（エコリレーション）を定義づけるための科学的に厳密な方法が場所ごと，縮尺ごとに異なる生態系構成要素を生じさせる過程に取り組むべきだという思考にあった。

- **Another way to .....** (See **Way**)：…するための別方法（➡ **Way**）
- **at one time or another**：ある時期に

8. All streams relative to watersheds of 30 m^2 or more have been channelized **at one time or another**.
30m^2 以上の流域に関わる河川は，すべてある時期に流路制御を受けてきた。

- **～ is similar to one another**：～は … 互いに似ている

9. Streams within each set should **be similar to one another** regarding "relative disturbance".
各セットにおける河川は，「相対攪乱」に関して似ているべきである。

10. Reference streams tend to **be similar to one another** when compared to reference streams in adjacent regions.
参照河川は，隣接地域における参照河川に比べ，互いに類似している。

Answer (See also **Question**)：回答；答える（➡ Question）

1. **The answer** is that .....
答えは … ということである。
2. **The answer** can be found in .....
答えは … の中にみいだすことができる。
3. Question comes more easily than **answers**.
疑問が回答より容易に生じる。
4. It may raise more questions than it **answers**.
それは回答よりも多くの疑問を提起するかもしれない。

- **obtain an answer** ..... ：回答を導く

5. The protocol provided the means for **obtaining** such **an answer** in a defensible way.
   このプロトコルは，そうした回答を弁護可能な形で導くための手段となった。

    - **The answer to the question is** ..... ：この質問への答えは…である

6. **The answer to those questions lies** in the fact that the stereospecificity of enzymes is not exact.
   それらの質問への答えは，酵素の立体特異性が厳密でないという事実に見いだされる。
7. **The answer to this question is provided by** something called Dynamic Modeling.
   この質問への答えは，ダイナミックモデリングと呼ばれるものによって見いだされる。

    - **answering** ：答えること

8. Site studies will be directed toward **answering** that question.
   現場研究は，その質問に対し答えられる方向にむけられるであろう。
9. This paper is addressed to **answering** these questions.
   この論文は，これらの質問に答えることに取り組んでいる。
10. The modeling of certain constituents will not be nearly as critical to **answering** study question.
    ある特定成分をモデル化することは，研究の疑問に答えることほど重要ではないであろう。

    - **〜 remain unanswered** ：〜は解決されないまま残っている

11. Fundamental questions **remain unanswered**.
    基本的な疑問は解決されないまま残っている。
12. There still **remain** many **unanswered questions** regarding reaction mechanisms under a variety of conditions.
    いろいろな状態下での反応機構に関して解決されてない疑問がまだ多く残っている。

Any ：どんな；どれも；なにも；誰も

1. **Any** noncombustible debris remaining is typically landfilled.
   残っているどんな不燃瓦礫も主として埋め立て地に処分される。

Any

2. Examination of Table 2 provides little evidence **of any** correlation between influent metal concentrations and percentage removal.
表2の考察から，流入水の金属濃度とパーセンテージ除去率との間の相互関係に関する証拠はほとんどなにも出てこなかった。

3. Reviewing the literature on sonochemistry of organochlorine compounds did not lead to **any** publications on the use of ultrasound for chemical monitoring.
有機塩素化合物の超音波化学分解に関する文献レビューは，化学的モニタリングの為の超音波利用に関するどの出版物にもならなかった。

- **at any of .....**：…のいずれにおいても
- **any other .....**：他のいかなる…
- **like any other .....**：他の…と同様に
- **as with any......**：どんな…にもあるように，
- **within any one of .....**：…のいずれかに
- **in any one area**：いずれか1領域での

4. This report will address the possibility of contamination **at any of** the three sites.
この報告書は，3地点のいずれにおいても汚染可能性があることを強調するであろう。

5. No gas evolution was observed, nor were **any other** products detected by GC in either the gas or liquid phase.
ガスの発生が観察されなかったし，また，他のいかなる生成物もGCでは気相でも液相でも検出されなかった。

6. However, **like any other** valued fish species, it does have specific habitat and water quality requirements.
しかしながら，他の貴重な魚種と同様，特有な生息地および水質の要件を有する。

7. **As with any** geochemical model, ...
どんな地質化学モデルにもあるように，

8. Only very small streams have watersheds completely **within any one of** these subregions.
ごく小さな河川がこれらの小地域のいずれかにすべて属する流域を持つ。

9. The implementation of water quality criteria **in any one area** serves to enhance the state's water quality program.
いずれか1領域での水質クライテリアの実施でもその州の水質プログラムを強化できる。

- **In any case**：いずれにせよ
- **In any event**：とにかく

10. **In any case**, there are no continuous records of ecological change over the last 2000 years.
    いずれにしても，過去2000年にわたっての生態変化の継続した記録が存在しない。

11. **In any event**, it was very clear that .....
    とにかく，…のことは非常に明確であった。

- **in any detail**：詳しく：

12. This assumption has not been examined **in any detail**.
    この仮定は，まだ詳しく検討されていない。

- **if any** .....：…か否か
- ....., **if any**, ...：…，仮にあったとしても，…

13. The plots for each metric were examined to determine **if any** visual relationship with drainage area existed.
    各メトリックについての図が検討され，流域面積との目測関係が存在するか否か決定された。

14. This is not an attempt to characterize pristine or totally undisturbed, pre-Columbian environmental conditions ; such conditions exist in only a very few places, **if any**, in the counterminous United States.
    これは無傷の，すなわち全く無攪乱のコロンブス前の環境状態を特性把握しようとすることではない。そのような状態は，米国全土において，仮にあったとしても，ごく少数の場所にしか存在しない。

- **under any conditions**（See Condition）：
  どんな状況であっても（→ **Condition**）

Apparent：明白な；明らかな

1. The **apparent** reduction in the oxygen utilization is most likely the result of soil gas diffusion.
   酸素利用量の明白な減少は，おそらく土壌ガス拡散の結果であろう。

- **It is apparent** .....：…は明白である
- ～ **is apparent**：～は顕著である
- **It became apparent** .....：…が明らかになった

2. **It is apparent that** metal removal efficiencies may be extremely high on ocean.
明らかに，海では金属除去効率が極めて高いかもしれない。
3. North of the line in Wyoming, patterns of logging activity **were apparent**, whereas south of the line they were not.
ワイオミング州境の北では伐採活動パターンが顕著だったのに対し，州境の南ではパターンが認められなかったのである。
4. A decision was made to record this information even though the eventual importance of its use **was** not **apparent**.
たとえその利用の最終的重要性が明らかでなくても，この情報を記録することが決定された。
5. During the application of the impairment criteria protocol, **it became apparent** there were several advantages over an ecoregional method.
損傷クライテリアプロトコルの適用中に，生態地域（エコリージョン）方式を上回る利点がいくつか明らかになった。

Appear：示される；なされる；ように思われる

1. The initial classification scheme adopted in this work **appears** in Table 1.
この研究で採用された最初の分類案は表1に示されている。
2. From time to time, attempts to develop mathematical models of photocatalysis **have appeared**.
時折，光触媒反応の数学モデルを構築する試みがなされてきた。
3. Although the concept **appears** simple on the surface, it can be deceptively complex in application.
この概念は，一見，単純に思えるが，適用時には意外に複雑である。

- **It appears that** .....：…のように思われる
- **There appears to be** .....：…が存在するように思われる

4. **It appears** from the results presented that iron and copper have the highest metal removal efficiencies.
提出された結果から，鉄と銅が最も高い金属除去効率を持つように思われる。
5. **It appears that** there is nothing gained by plotting the data in this fashion.
データをこの様式で作図することによっては，何も得られないように思われる。
6. **There appears to be** considerable difference between the A, B, and C samples.
A，B，Cサンプル間にかなりの相違があるように思われる。
7. **There appears to be** sufficient evidence that water quality will differ for

the upper and middle Natsui River due to geographic factors.
水質が，地理的因子故に Natsui 川の上流と中流で異なることを示すに足る証拠が存在すると思われる。

- ～ **appear to** ..... ：～は…するように思われる
- ～ **appear to be** ..... ：
  ～は…であるように思われる；～は…とみなされる

8. Copper did not **appear** to compete with nickel for the active sites.
   銅が，活性部位をめぐってニッケルと競合するとは考えられなかった。
9. Selenium and sulfate **do not appear to be** related to mine tailings.
   セレンと硫酸塩が，鉱山選鉱廃物と関連があるようには思われない。
10. These lower levels **appear to be** typical of background levels in the area.
    これらの低い濃度はその地域のバックグラウンド濃度の典型であるように思われる。
11. Although factors such as geology and soils are also important, the other factors **appear to be** the most important in this ecoregion.
    地質や土壌などの因子も重要だが，この生態地域（エコリージョン）では，他の因子が最も重要に思われる。
12. This more holistic definition **appears to be** gaining acceptance.
    この総合的な定義は，容認されつつあるようだ。
13. An additional 25 states **appear to be** developing biological criteria for use either in water resource management.
    さらに 25 州は，水資源管理のために生物クライテリアを設定中とみなされる。

Applicable：適用可能

- ～ **is applicable** ：～が適用可能である
- ～ **is most applicable** ：～が最も適用可能である

1. It is important to define the regional extent over which a particular biocriterion **is applicable**.
   ある特定の生物クライテリアが適用可能な地域の範囲を定めることが重要である。
2. A critical issue is to determine the regional extent over which a particular biological attribute **is applicable**.
   重要な課題は，ある特定の生物学的属性が適用可能な地域範囲を決定することである。
3. The steady flow assumption **is most applicable** in this region.
   この地域では，定常流量の仮定が最も適用可能である。

- ~ **is applicable to** ..... ：~は … に適用可能である
4. The fundamental bioassessment approaches **are applicable to** all waterbody types.
基本的な生物アセスメント手法は，すべての水域タイプに適用可能である。
5. The approach in using ultrasound **is** obviously **applicable to** organic compounds which contain halides.
超音波を使うアプローチは，ハロゲンを含む有機化合物に対し明らかに適用できる。

Application ：適用；応用；アプリケーション

1. The distribution of pesticides in reservoirs is established by **application of** the principle of continuity or mass balance.
貯水池における殺虫剤の分布は，連続性の原則あるいは物質収支の適用によって確立される。

- **applications are in progress** ：適用が進行中である

2. While a number of GMODEL **applications are in progress**, few studies are complete.
多数のGMODELアプリケーションが進行中である一方，ほとんどの研究が完成していない。

- ~ **have an application** ..... ：~はアプリケーションがある

3. This analytical instrument **has a wide application**.
この分析機器の適用範囲は広い。
4. Such materials **have many** technological **applications**.
このような材料は技術的アプリケーションが多くある。
5. Ultrasound **has** potentially important **applications** in both homogeneous and heterogeneous catalytic systems.
超音波は，均質触媒システム，そして混成触媒システムの両方で潜在的に重要なアプリケーションである。

- ~ **is put to application** ：~が適用に使われる

6. Models should be tested before they **are put to serious application**.
モデルは，それらが重大な適用に使われる前に，テストされるべきである。

- **in this application** ：このアプリケーションでは
- **in large-scale applications** ：大規模なアプリケーションで
- **in such application** ：このような適用において

- **in an application of .....**：… の応用において
- **in real-life applications of .....**：… の現実の応用において
- **in the application process**：アプリケーション過程での

7. FARM model does not adequately represent the aldicarb behavior **in this application**.
   このアプリケーションでは，FARM モデルは十分に aldicarb 挙動を示さない。
8. It has also become widely accepted **in large-scale applications**.
   それは同じく大規模なアプリケーションで広く受け入れられた。
9. An equation which is of considerable importance **in such application** is .....
   このような適用においてかなりの重要性を持っている式は … である。
10. At each step **in the application process**, we will first explain what need to be done.
    我々は，アプリケーション過程のそれぞれのステップにおいて，何をしなければならないかを最初に説明するであろう。
11. Catalytic reactions are of enormous importance **in both laboratory and industrial applications**.
    触媒反応は，研究と産業的アプリケーションの両方で非常に重要である。

   ★ Application は "technological applications," "engineering applications," "laboratory applications," "industrial applications," "practical applications" のように多様な形で使用されます。

   - **application to .....**：… への応用；… への適用
   - **in the application to .....**：… への適用において

12. Current research is being conducted to test the efficacy of this **application to** lakes, reservoirs, and large rivers.
    目下，この概念を湖沼，貯水池，および大河川へ応用することの効力を試す研究が行われている。
13. The numerical criteria are limited **in their application to** large rivers and streams.
    数値的クライテリアは，大河川および河川，小川に適用が限定される。

Apply：適用する；応用する；当てはめる

1. It is valid to directly **apply** a model.
   モデルを直接応用することが正当である。
2. The water quality criteria **are applied** extensively in both defining use classification and assessing water quality.

水質クライテリアは，利用分類の確定ならびに水質の評価の両方で多用されている。
3. The new method requires a thorough field test before it **can be** safely **applied**.
   この新手法は，安全に適用される前に，綿密な現地試験が必要とされる。

   - **apply to .....** ：…にあてはまる；適用する

4. The present model structure **may apply to** other compounds.
   既存のモデル構造は他の化合物に当てはまるかもしれない。
5. These analytical methods **do not apply to** the river mouth and harbor areas.
   これらの解析方法は，河口および港湾区域には適用されない。

   - **It is difficult to apply A to B**：
     AをBに当てはめることは困難である
   - **~ is applied to .....**：
     ～は…に用いられている；～は…に適用されている

6. It is difficult to **apply** methods of statistical analysis **to** their study.
   統計学的分析法を彼らの研究に当てはめることは困難である。
7. The water quality model **is applied to** Clear Lake.
   この水質モデルはClear湖に用いられている。

Appreciate ：感謝する

★ AppreciateはAcknowledgment(謝辞)によく使われる言葉です。

1. **We appreciated** critical review comments from three anonymous reviewers, which greatly improved an earlier draft of this manuscript.
   筆者は，3人の匿名査読者からの批評的論評に感謝する。それは，本章の草稿を大幅に改善してくれた。
2. The insightful comments of James Morita and Minoru Oki **are appreciated.**
   James MoritaとMinoru Okiの洞察あふれる論評に感謝したい。

Approach ：手法；アプローチ；達する；及ぶ

1. A more detailed explanation of Sakamoto's **approach** is given later in this chapter.
   Sakamotoの手法のより詳細な説明を本章の後半で述べる。

2. The method used here may be a new **approach** not only for gaining better results, but also for acquiring new knowledge and insights about toxic behavior.
   ここで使われた方法は，良い結果を得ることだけではなく，毒性物質の挙動について新しい知識と洞察を得ることに対して，新しい手法であるかもしれない。
   - **This approach is employed**：この手法は使用されている
   - **The approach was used by** ..... ：
     このアプローチが … によって使われた
3. **This approach can be** widely **employed**.
   この手法は広く使用されうる。
4. A two-tiered **approach was used by** Takahashi et al. (2001) to test MMODEL.
   MMODELをテストするために2段階のアプローチがTakahashiら (2001) によって使われた。
   - **The approach has been applied to** ..... ：
     この手法は … に適用されてきた
   - **The approach is applicable to** ..... ：
     このアプローチは … に適用可能である
5. Such **an approach has been** successfully **applied to** .....
   このような手法は，成功のうちに…に適用されてきた。
6. **The approach** chosen **is** obviously **applicable to** other inorganic compounds.
   選ばれたこのアプローチは，明らかに他の無機化合物に適用可能である。
   - **The approach provides**.....：この手法は … をもたらす
7. **The approach provides** valuable consistency among the many different programs.
   この手法は，多くの様々なプログラムにおける貴重な整合性をもたらす。
8. The multivariate **approach provides** a powerful means of maximizing ecological information obtained from periphyton assemblages.
   この多変量手法は，付着生物群集から得られた生態学的情報を最大化するための有効手段となる。
   - **The approach is based on** ..... ：このアプローチは … に基づいている
   - **This approach leads to** ..... ：この手法は … へ導く
   - **This research takes a different approach toward** ..... ：

この研究は … に対して異なったアプローチをとっている

9. **The approach is based on** mass balance and uses the simulation techniques described by Sasaki et al.(2001).
この手法は物質収支に基づいていている，そしてSasakiら(2001)によって記述されたシミュレーション手法を使う。

10. **This approach leads** itself **to** a better understanding of the nature of the impairment.
この手法は，この損傷の性質をより深く理解するのに役立つ。

- **In this approach** ： このアプローチでは

11. An implicit assumption **in this approach** is that the relative condition of reference sites within each region is similar.
この手法における暗黙の仮定は，各地域内での参照地点の相対的な条件状態が似ていることである。

★ Approach が動詞として使用される場合の例文を下に示します。

- **approach** ： 達する；及ぶ

12. The system **had approached** almost complete equilibrium for mercury.
このシステムは，ほとんど完全に水銀の平衡状態に達していた。
13. These models of simple aqueous systems cannot **approach** the complexity of natural sediment−water systems.
これらの単純な水性システムモデルは，自然の沈降堆積物−水システムの複雑さに及ぶことが出来ない。

Appropriate ： 適切な；適切である；適している

1. The **appropriate** control actions can be taken and water quality standards can be achieved.
適切な制御行動がとられ，水質基準が達しうる。

- **It is appropriate for** ..... ： それは … に適切である
- **It is appropriate to** ...... ： … することは適切である
- **It may be appropriate to** ..... ： … することが適切であろう
- **It seemed appropriate to** ..... ： … することが適切だと思われた

2. Just because a model has been found valid for one use does not mean **it is appropriate for** some other use.
ただモデルがある一つの使用で有効であることを見いだされたからといって，そ

3. **It may be appropriate to** reject a model prior to any field testing.
   現場試験以前にはモデルを受け入れないことが適切であろう。
4. **It seemed appropriate to** assume that the same water quality conditions were affecting all the sampling units.
   同一の水質条件が全部のサンプリングユニットに影響していると仮定するのが適切だと思われた。

   - ～ **is appropriate for** ...... ：～が…に適切である
   - ～ **is most appropriate** ：～が最も適している

5. There are certain situations in which each type **is appropriate for** use.
   それぞれのタイプが使用に適切である特定の状態がある。
6. EPA acknowledges, albeit briefly, that site-specific modifications **may be appropriate** in some circumstances.
   EPAは，いくつかの状況ではその地点に特有の修正が適切であることをごく簡単にだが認めている。
7. Some of the functional feeding-group measures in Table 2 **are** also **appropriate for** use.
   表-12.2における摂食機能群の一部も利用に適している。
8. For the broad range of human impacts, comprehensive, multiple metric approaches **is most appropriate**.
   広範な人為的影響については，総合的な複数基準手法が最も適している。
9. It has not yet been determined which level **would be most appropriate** for the water quality standard.
   どのレベルが水質基準に最も適しているかは，まだ未決定である。

a priori ：先験的に；直感的に；アプリオリ

1. The IBI metrics are **a priori** assumed to measure a specific attribute of the community.
   IBIメトリックは，その生物群集の具体的属性を測定すると先験的に仮定されている。
2. The definition of degradation responses **a priori** is justified if clear patterns emerge from specific metrics.
   具体的メトリックから明確なパターンが現れたならば，劣化応答という定義は先験的に正当化される。
3. It is difficult to ensure, **a priori**, that implementing nonpoint source controls will achieve expected load reductions.

Area

面源制御の実施が予想される負荷量の削減を達成するであろうことを先験的に保証することは，難しい。

> **Area** (See also **Region**)：区域；エリア；面積；場所（➡ **Region**）

- **in the area** ：その区域で
- **throughout the study area** ：調査全地域で
- **over wide areas** ：広範な地区に
- **over the whole of the study area** ：調査全地域では
- **among areas** ：各地域での

1. Data were gathered on fish, sediment, water, and aquatic plants **in the area**.
   その区域で魚類，堆積物，水，それに水生植物のデータが収集された。
2. Core samples were collected at selected locations **throughout the study area**.
   調査全地域の選択地点でコア（円筒型）サンプルが収集された。
3. Regional variations in metric details are expected but the general principles used in defining metrics seem consistent **over wide** geographic **areas**.
   メトリックの細部における地域的変動が予想されるが，メトリック確定に用いられる一般原理は，広範な地区に共通なようである。
4. Thickness of tailings deposits varies considerably and cannot be accurately predicted **over the whole of the study area**.
   選鉱屑堆積層の厚さがかなり変化し，調査全地域では正確に予測することができない。
5. By calibrating **over large areas**, decisions about level of protection can be made with a clear understanding of the differences **among areas** over which differing criteria might be established.
   広い地域での較正による保護レベルに関する決定は，異なるクライテリアが設定されうる各地域の差異を明確に理解した上で下されるようになる。

- **This is an area for .....** ：この領域が … である
- **The study area is situated on .....** ：調査地域は … に位置している
- **~ is one of the areas in .....** ：〜は … の地域の1つである
- **The study areas were selected on the basis of .....** ：
  これらの調査地域は，… をベースに選択された
- **Another area of interest is .....** ：
  もう1つの興味あるエリアは … である

6. **This is an area for** further research and an opportunity for interstate

7. The Whitewood Creek **study area is situated on** the northeastern periphery of the Black Hills uplift.
Whitewood Creek 調査地域は Black Hills 隆起地の北東縁部に位置している。
8. Japan, together with North America and Europe, **is one of the areas** in the world most severely affected by environmental pollution.
日本は，北アメリカとヨーロッパと共に，世界中で最もひどく環境汚染による影響を受ける地域の1つである。
9. **The study areas were selected on the basis of** evidence suggesting that they are areas of relatively intense geochemical activity.
これらの調査地域は，比較的激しい地質化学的活動のある地域であることが示唆されている証拠に基づいて選択された。
10. **Another area of interest** in important assessment of dredge spoils discharge operations **is** ....
浚渫物の放流作業に対する重要なアセスメントでの，もう1つの興味あるエリアは，…である。

★ Area（面積）に関する例文を下に示します。

11. The lake has **a total area** of 25,300 km^2, and is divided into three major, distinct subbasins.
この湖は，全面積が 25,300 km^2 あり，三つの主要な水域に明確に分けられる。
12. The central basin is the largest **in terms of area** (16,317 km^2) and has a mean depth of 25 m.
中央水域は面積では最大（16,317 km^2）で，平均湖深が 25m である。
13. In contrast within the unsaturated zone, the saturated zone **is the area** of contamination.
不飽和層内とは対照的に，飽和層は汚染地帯である。
14. Jamaica Bay occupies approximately 52 km^2.
ジャマイカ湾の面積は，およそ 52 km^2 である。

Arise：起こる；生じる；高まる

- **concern arises**（See **Concern**）：関心が高まる（➡ **Concern**）
- **difficulty arises**（See **Difficulty**）：困難が生じる（➡ **Difficulty**）
- **error arises from** …..（See **Error**）…..：からエラーが生じる（➡ **Error**）
- **problem arises from** …..（See **Problem**）：

　　　　　問題は … から起こる（➡ **Problem**）
　　　　● **question arises**（See **Question**）：疑問が起こる（➡ **Question**）

Argue（Argument）：主張する；議論する（議論）

　　　　● **～ has argued that** …… ：～の … という主張によれば
　　　　● **it can be argued that** …… ： … だとの主張もありえる
　　　　● **it has been argued,** …… ： … という主張がなされてきた

1. Koike (1990) **has argued that** ecological regions subsume patterns in the quality and quantity of the space these organisms occupy.
Koike (1990) の主張によれば，生態地域（エコリージョン）は，これらの生物が占有する空間の質・量におけるパターンを包含する。
2. **It can be argued that** full protection of aquatic life requires the use of biological criteria.
水生生物の完全保護には，生物クライテリアが必要だとの主張もありえる。
3. Therefore, **it has been argued**, insufficient attention is directed toward assessing cumulative impact of these policies in terms of costs and outcomes.
したがって，費用と成果の面でこれらの政策の累積影響評価にあまり注意が払われない，という主張がなされてきた。

　　　　● **argument over** …… ： … をめぐる議論:

4. **Arguments over** the merits of narrative vs. numeric biological criteria will likely remain.
記述生物クライテリアと数値生物クライテリアの利点をめぐる議論は，今後も続く見込みが強い。

　　　　● **argument can be made against** …… ：議論が … に対してされうる

5. From a consideration of microbial statistics, **argument can be made against that** ……
微生物統計学の考慮から， … という議論がそれに対してされうる。

As（See also **Above** and **Below**）
：であるように；と同様に（➡ **Above** ならびに **Below**）

1. **As** in all the models, ……
すべてのモデルにあるように，…

　　　　● **as with** …… ： … と同様に

2. **As with** any geochemical model, ...
   どんな地質化学モデルにもあるように，…
3. **As with** all sampling designs, the ideal number of reference sites must be balanced against budget realities.
   あらゆるサンプリング計画と同様に，参照地点の理想的な数は，現実的予算に照らして決定せねばならない。

   - as A as B ： B と比べて A ぐらい

4. Macroinvertebrate changes were not **as great as** in previous years.
   大型無脊椎動物の変化は，それまで数年間ほど大幅でなかった。
5. High-intensity ultrasound dramatically increases the rates of reaction by **as much as** 200-fold.
   強度の超音波は，反応速度を劇的に増やし，それはおよそ 200 倍にも達する。
6. The former can be used just **as much as** the latter as a means of experimentation.
   前者は後者とまったく同じぐらいよく実験手段として使用できる。
7. The groundwater in the alluvium beneath the tailings has a possibility of being comprised of water from **as many as** four sources.
   堆積選鉱屑の下にある沖積層の地下水は，およそ 4 つの水源からの水が混ざっている可能性がある。

   - as predicted by ..... ： … によって予測されるように
   - as can be seen ..... ： … に見られるように．
   - as a consequence of..... ： … の結果として
   - as a follow-up note ： 追記すれば

8. Kimura and Sasaki (2004) reported a linear relationship between the log of Ne/Nin and the applied UV dose **as predicted by** Eqn 1.
   Kimura と Sasaki (2004) は，Ne / Nin と使用した UV 放射量の関係は式 1 によって予測されるように直線関係にあると報告した。
9. Model predictions show reasonable agreement **as can be seen** from both Figures 2 and 3.
   図 2 と図 3 の両方に見られるように，モデルの予測はかなり一致している。
10. Some chemical changes affecting aqueous metal chemistry **as a consequence of** an altered redox potential may be almost immediate.
    酸化還元電位の変化の結果として，水性金属化学に影響を及ぼすある種の化学変化は，ほとんど即時であるかもしれない。
11. **As a follow-up note**, this sewage treatment plant subsequently underwent a major upgrade, and later macroinvertebrate sampling revealed an

improved community downstream.
追記すれば，この下水処理場は，その後大幅な改善がなされ，後の大型無脊椎動物サンプリングで下流生物群集における向上が見られた。

- **as ..... as possible**（See **Possible**）：
  できる限り（➡ **Possible**）
- **as well as .....**（See **Well**）：…と同様に（➡ **Well**）
- **as discussed previously**（See **Discuss**）：
  前に，論じられたように（➡ **Discuss**）
- **as stated previously**（See **State**）：前述のとおり（➡ **State**）
- **as previously mentioned**（See **Mention**）：前述のとおり（➡ **Mention**）
- **as an aid to ......**（See **Aid**）：…への手助けとして（➡ **Aid**）
- **as basis for .....**（See **Basis**）：…の基礎として（➡ **Basis**）
- **as compared to .....**（➡ See **Compare**）：
  …と比較されるように（➡ **Compare**）
- **as follows**（See **Follow**）：次のように（➡ **Follow**）
- **as a result of .....**（See **Result**）：…の結果として（➡ **Result**）
- **as part of .....**（See **Part**）：…の一部として（➡ **Part**）

Ascertain ： 確認する

1. In this study, attempts were made **to ascertain** which parameters were causing toxicity.
   この研究で，どのパラメータが毒性を起こしていたか確認する試みがされた。
2. A number of reference sites were sampled **to ascertain** as accurately as possible the background levels of all parameters under investigation.
   調査中のパラメータすべてのバックグラウンドレベルを可能な限り正確に確認するために，多くの参照地域でサンプルを採集した。
3. Although it was not within the scope of this research **to ascertain** the factors which inhibit nitrite production, the results of the study have incidentally seemed to substantiate the research of others.
   亜硝酸の生成を抑制する要因を確認することはこの研究の範囲外であったけれども，この研究結果は偶然にも他の研究についても実証したようである。

Aspect ： 様相；局面

- **The aspect of ～ is .....**：～の側面は … である

Aspect

- A key aspect of ～ is ..... ：～の重要な点は … である。
- Some aspects of ～ are ..... ：～の一側面は … である
- The most important aspect is ..... ：最も重要なことは … である

1. **That aspect of** the problem **is** being investigated by the Center for Disease Control Office.
   この問題点の様相が，疾病対策センターによって調査されている。
2. **A key aspect of** this approach **is** its iterative nature.
   このアプローチの重要な点は，それの持つ反復特性である。
3. **A specific aspect of** transport and transformation **is** examined from a system's perspective.
   輸送と変化の特定の様相がシステムの見地から検査されている。
4. Catalysis by fixed cell is quite complex and many **basic aspects are** yet to be understood.
   固定細胞による触媒作用は非常に複雑で，多くの基本的な側面がまだ理解されていない。
5. The discussion will emphasize assessments of stream environments, but **aspects of** monitoring ponds and lakes also **will be** mentioned.
   本章の力点は，河川環境アセスメントに置かれているが，湖沼モニタリングについても言及していく。
6. **Some aspects of** these deficiencies **have been** reviewed by Koike(1998).
   これらの欠陥の一側面がKoike(1998)によってレビューされた。
7. **The most important aspect is** the choosing of proper initial concentrations for.....
   最も重要なことは，適切な初期濃度を選択することである。
8. **The potential aspects lie** outside the scope of this paper.
   可能的側面については，この論文の範囲外である。

- ～ **is an important aspect of** ..... ：～は … の重要な側面である
- ～ **explore the aspects of** ..... ：～は … の局面を探究する
- **examine aspects of** ..... ：… の状況を調べる
- ～ **address some aspects of these problems** ：
  ～はこれらの問題の一局面を扱う

9. The reactions are **an important aspect of** a chemical's fate in the environment.
   反応は，環境内で化学物質が衰退する過程で重要な局面である。
10. The project **explored** three important **aspects** of the subjects.
    このプロジェクトでは，3つの重要な局面を探究した。

Assess

11. The aim of the present study was to **examine aspects of** elimination of PAH by Mytilus edulis.
    現在の研究の目的は，Mytilus edulis による PAH の除去の状況を調べることであった。
12. Laboratory work **addressing some aspects of these problems** is ongoing at Washington State University.
    これらの問題の一局面を扱う室内研究がワシントン州立大学で進行中である。

Assess (Assessing)：評価する(評価する際)

1. Fundamental studies are needed **to assess** the possibilities of this method.
   基本的な研究が，この方法の可能性を評価するのに必要とされている。
2. These data will be analyzed **to assess** the status of biological condition and to address the goals as stated above.
   これらのデータが分析されて，生物状態の現状を評価し，上述の目標に取り組むことになろう。
3. Comparisons were made between process operational variables and effluent quality **to assess** the influence of the heavy metals.
   重金属の影響を評価するため，プロセス操作上の変数と排水の水質の比較をおこなった。

   ● **in assessing** .....：… を評価する際.
   ● **for assessing** .....：… を評価するための

4. **In assessing** the potential health hazards of a contaminated marine environment, particular attention should be directed to carcinogens and mutagens.
   汚染された海洋環境の潜在的な健康被害を評価する際，特定の注意が発癌性物質と突然変異誘発性物質に向けられるべきである。
5. Future work is needed, however, to better utilize the ADV as part of resource value assessment such as NRDAs and **in assessing** other types of environmental damage claims.
   しかし，NRDAs といった資源価値アセスメントの一部として，また，他の種類の環境損害請求の評価において ADV をより活用するには，今後の研究が必要である。
6. Approaches **for assessing** the response of stress proteins in organisms when exposed to chemicals are currently under development.
   化学物質に曝された際の生物体内ストレス蛋白質の応答を評価するための手法を目下案出中である。

Assessment：アセスメント

1. This integrated **assessment** was designed to address the following questions: 1....., 2....., etc.
   この総合的なアセスメントは，次の問題を扱うよう意図された。1…，2…，などである。
2. The rapid **assessments** offer standardization of analysis and accuracy of habitat classification.
   迅速なアセスメントは，分析の標準化や，生息地分類の正確性をもたらす。
3. The development and implementation of this **assessment** programs is discusses in detail by Matsui and Etoh (1994).
   このアセスメントプログラムの策定・実施については，Matsui と Etoh (1994) が詳述している。
4. The available field data were not sufficiently detailed to allow an **assessment** of spatial variations.
   利用可能なフィールドデータは，空間的変異のアセスメントをするのには詳細に欠ける。

Assign (Assignment)：割り当てる（指定）

1. Tolerance values **are assigned** on a scale of 0 to 10, with 0 being the least tolerant and 10 being the most tolerant.
   耐性値は，0から10までのスケールで指定され，0が最低耐性を，10が最高耐性を表す。
2. Based on these considerations, a settling velocity of 1.2 m/d **is assigned as** characteristics of the Yoshima River solids.
   これらの考慮に基づいて，1.2m/d の沈降速度が Yoshima 川での固体物質の特徴として割り当てられている。

   - **assigning** …..：… を割り当てること
   - **in assigning** …..：… を示す際に

3. This points out the limitations of **assigning** a single degradation rate coefficient for the entire soil profile.
   これは土壌プロフィール全体に一つの分解速度係数を割り当てることには限界があることを指摘している。
4. These have proved valuable **in assigning** causes and sources to water resource impairments noted by EPA.

これらは，EPAにより指摘された水資源損傷の原因と出所を示す際に有益であると判明している。

5. These played a major role **in assigning** and evaluating use designations, water quality management plans, and advanced treatment justifications.
これらは，使用指定の割当・評価，水質管理プラン，および高度処理正当化において重要な役割を演じた。

> ● **assignment of .....**：…の指定

6. **Assignment of** an appropriate aquatic life use to waterbody is primarily driven by an assessment of biotic integrity.
水域への適切な水生生物利用指定は，主に生物的健全性のアセスメントに左右される。

Associate（Association）：伴う；関係がある（関連性）

> ● **associated with .....**：…と関連する
> ● **～ is associated with .....**：～は…と関係がある
> ● **A and B are closely associated**：AとBは密接に連関している

1. Assumptions **associated with** the assessment exercise are as follows:
アセスメントの実施と関連する仮定は次の通りである。
2. Primary contaminants of concern **associated with** the fuel leakage include BTEX.
燃料漏れと関連する主要な汚染物質としてはBTEXがある。
3. The variability **associated with** seasonal change can be alleviated using species autecologies.
季節変化に伴う変動性は，生物種の個生態学を用いて緩和されうる。
4. There is some empirical evidence that instream biological response and instream ambient toxicity **are closely associated**.
経験的証拠によれば，河川内生物応答と河川内環境毒性は，密接に連関している。
5. Are those in poor condition **associated with** any particular cities?
これらの悪条件は，特定の都市に関連しているのか。
6. If the differences **can be associated with** management practices that have a chance of being altered, it seems wise to combine the regions for the purposes of establishing water criteria.
仮に，その差違が変わる可能性のある管理慣行に関連している場合には，水質クライテリア設定のために諸地域を組み合わせた方が賢明であろう。

- Association：関連性
- ～ have an association with ..... ：～は … と関係がある

7. A measure of how much the quality can be improved can be derived through changing management practices in selected watersheds where **associations** were determined.
水質がどれほど改善されうるかを示す測定は，関連性が決定された流域における管理慣行の変更を通じて得られるだろう。

Assume (Assuming)：仮定する；想定する；推測する（想定して）

1. This approach **assumes that** the control and study site differ only in the presence of pollution.
この手法は，対照地点と研究地点が汚濁の有無においてのみ異なると仮定している。
2. The IBI metrics **are** *a priori* **assumed** to measure a specific attribute of the community.
IBIメトリックは，その生物群集の具体的属性を測定すると先験的に仮定されている。
3. Because the dechlorination reaction rates were relatively slow as compared to chlorophenol (CP) desorption kinetics, solid−liquid partitioning equilibrium **was assumed**, and partitioning rates were not incorporated into the model.
クロロフェノール（CP）の脱着速度と比較して，脱塩素反応速度が比較的遅かったので，固体−液体間の分配平衡状態が仮定され，分配速度はモデルには含まれなかった。

- It is assumed that ..... ： … と想定される
- If it is assumed that ..... ：もし … だとみなすと
- It seems appropriate to assume that ..... ：
  … と仮定するのが適切だと思われる
- It is reasonable to assume that ..... ：
  … であると仮定するのが道理的である
- It is logical to assume that ..... ：
  … であると想定することは論理的である

4. **It is assumed that** 10 % of the total volume of waste material is disposed of as non−hazardous waste.
廃棄物の全容積の10 %が非危険廃棄物として処分されていると仮定される。
5. **It has always been assumed that** the tailings provide a potential to cause

Assumption

degradation of groundwater quality.
選鉱屑が地下水質の低下を起こす可能性があるとは常に想定されてきた。

6. If the contamination were to **be assumed** to reach the screen depth, the concentration would decrease due to dilution up to an order of magnitude.
もし汚染が井戸のスクリーンの深さに達すると仮定されるなら，その濃度は希薄のために一桁減少するであろう。

7. **It seemed appropriate to assume that** the same water quality conditions were affecting all the sampling sites.
同一の水質条件がすべてのサンプリング地点に影響していると仮定するのが適切だと思われた。

8. By analogy with the oxidation of formic and oxalic acids, **it is reasonable to assume** a second-order reaction for glyoxylic acid because it has a structure similar to the other two acids.
ギ酸とシュウ酸の酸化反応が類似していることから，グリオキシル酸の反応は第2次反応であると仮定するのが道理的である。なぜなら，それは他の2つの酸に構造が類似しているからである。

- **Assuming that .....** : … と仮定して，… とすれば，
- **Assuming .....** : … と仮定する

9. **Assuming that** the substance under consideration is increases in accordance with a first-order reaction, the following equation is developed.
考慮中の物質が一次反応に従って増加すると仮定して，次の式がつくられている。

10. We need to examine the use of art in science, rather than **assuming** an either/or scenario.
我々は，二者択一シナリオを仮定する代わりに，科学における芸術の利用を検討する必要があろう。

Assumption：仮定

- **The assumption involved in ～ is that .....** :
 ～に関与する主要な仮定は … である
- **The assumption for ～ is that .....** : ～の仮定は … である

1. The major **assumption involved in** this estimation technique **is that** ....
この概算手法に関与する主要な仮定は … である。
2. A basic **assumption for** the model **was that** biofilm diffusion limitations were not significant.

生物膜拡散限界は重要ではないということが，このモデルの基本的な仮定であった。

- **the assumption is applicable**：この仮定は適用可能である
- **the assumption is valid**：この仮定は有効である
- **this assumption is not realistic**：こうした仮定は現実的ではない
- **the assumption of ～ is misleading**：
  ～についての仮定は誤りを導く
- **the assumption of ～ is appropriate**：～の仮定は適切である

3. The steady flow **assumption is most applicable** in this region.
   定常流の仮定はこの地域で最も適用可能である。
4. Over the time of scale of interest, **this assumption is reasonably valid** and may be used as a basis for a preliminary analysis of the problem.
   我々が関心ある時間のスケールにおいて，この仮定は適度に有効であり，この問題の予備解析の基盤として使用されるかもしれない。
5. This **assumption is not realistic**, as discussed below.
   こうした仮定は，下述のとおり現実的ではない。
6. **The assumption of** neglecting substrate inhibition **is** not particularly **misleading** at low substrate concentrations.
   基質阻害を無視することについての仮定は，低濃度の基質においては特に誤りではない。
7. If the coefficients defining this process are much greater than the other kinetic coefficients, **the assumption of** instantaneous equilibrium **is appropriate**.
   もしこのプロセスを定義している係数が他の速度係数よりかなり大きければ，瞬時平衡の仮定は適切である。

- **assumptions concerning ..... must be made**：
  … に関する仮定が設定されなくてはならない
- **This assumption is supported by .....**：
  この仮定は … によって支持されている
- **This assumption has not been examined**：
  この仮定は，まだ検討されていない

8. Several **assumptions concerning** flow and water quality must be made.
   河川と水質に関するいくつかの仮定が設定されなくてはならない。
9. **This assumption was supported by** transmission electron microscopy studies.
   この仮定は透過型電子顕微鏡(TEM)の研究によって支持された。

10. **This assumption has not been examined** in any detail.
    この仮定は，まだ詳しく検討されていない。

    - **assumption holds true**：仮定が正しい；仮定が有効である
    - **assumption holds for .....**：… に対して仮定が有効である

11. **If this assumption holds true**, the following equation can be used.
    もしこの仮定が正しければ，次の式が使われる。

12. It is important to note that most models represent an idealization of actual field conditions and must be used with caution to ensure that the underlying model **assumptions hold for** the site-specific situation being modeled.
    ほとんどのモデルは，実際の現場の条件を理想化していること，そしてモデルの対象となっている特定な状況に対してモデルの基礎仮定が有効であることを保証するため注意して使われなくてはならないということ，を指摘することは重要である。

    - **Assumptions associated with ..... are as follows**：
      … と関連する仮定は次の通りである
    - **～ rely on the assumption that .....**：～ は … という仮定に左右される
    - **It lends credence to the assumption**：それは仮定に信頼性を与える
    - **Ancillary to these assumptions are .....**：
      これらの仮定に対する補助は … である

13. **Assumptions associated with** the assessment exercise **are as follows**: .....
    アセスメントの実行と関連する仮定は次の通りである。…

14. The bioassay **relies on the assumption that** only free copper ions are toxic.
    このバイオアッセイは遊離銅イオンだけが毒性を持つという仮定に基づく。

15. **It lends credence to the** completely mixed compartment **assumption**.
    それは完全混合区画の仮定に信頼性を与える。

16. **Ancillary to these assumptions are** two others that affect the utility of the theory.
    これらの仮定に対する補助として，他に，理論の有用に影響を与える2つがある。

Attempt：試み；試みる

1. This study is **an attempt** to identify the effect of selected inorganic ions on a fixed-bed catalyst.

この研究は固定層触媒に対する選択された無機イオンの影響を識別する試みである。

2. **All our attempts** at classification are human constructs that help us understand a spatially and temporally varying landscape.
分類における我々の試みは，すべて人為的構成物であり，空間的・時間的に異なる景観を理解するのに役立つ。

- **attempts to extrapolate** ..... : …を推定する試み
- **prior attempts to examine** ..... :
  … を調べようとするこれまでの試み

3. The user should be aware that such effects may complicate **attempts to extrapolate** data for photolysis rates from one aquatic medium to a very different medium.
利用者は，このような影響が光反応速度データを1つの水生媒質からかなり異質の媒質まで外挿法によって推定する試みを複雑にするかもしれないということを認識すべきである。

- **an attempt is made to** ..... : …をするための試みがなされた
- **sufficient attempts were made to** ..... :
  … をするために，十分な試みがなされた
- **few attempts have been made to** ..... :
  … については，ほとんど試みがなされてこなかった
- **greater attempts are made to** ..... :
  … については，より大きい試みがなされている
- **no further attempts were made to** ..... :
  … については，それ以上の試みはなかった

4. In this study, **attempts were made to** ascertain which parameters were causing toxicity.
この研究で，どのパラメータが毒性を起こしていたか確認する試みがなされた。

5. **Sufficient attempts were made** to obtain water samples.
水サンプルを得る十分な試みがなされた。

6. **Few attempts have been made to** validate these models.
これらのモデルを実証する試みがほとんどされなかった。

7. **No further attempts were made to** calibrate the model on the study site.
この研究地域におけるモデル校正のために，それ以上の試みはされなかった。

8. **No attempt is made** here **to** give a quantitative estimate of the error.
ここでは，誤差を定量的に推定をする試みはなにもされない。

- **attempts are being made to .....** ：… する試みがなされている

9. Increasingly, **attempts are being made to** integrate all three types (physical, chemical, and biological) of sampling data.
だんだんと，3種類(物理的，化学的，生物的)のサンプリングデータを統合する試みがなされつつある。
10. Greater **attempts are now being made to** relate metal toxicity to speciation.
金属毒性を化学種に関連づけるために，より大きい試みが今なされている。

- **an attempt to ..... proved unsuccessful** ：
  … の試みは，不成功であることが判明した

11. **An attempt** to use the scheme for classifying aquatic ecosystems **proved unsuccessful**.
水生生態系を分類するための方式を用いる試みは，不成功であることが判明した。
12. **Attempts to** calibrate the model and keep parameters in a reasonable range based on literature data **proved unsuccessful**.
モデルをキャリブレートして，パラメータを文献データに基づく妥当な範囲にとどめる試みは，不成功であることが判明した。

- **There have been some attempet to .....** ：
  … する若干の試みがされてきた
- **There have been many attempts to .....** ：
  … する多くの試みがされてきた
- **This is not an attempt to.....** ：… する試みはされない

13. **There have been some attempts to** estimate copper toxicity in seawater using bioassay.
バイオアッセイによって，海水での銅の毒性を推測する試みが若干されてきた。
14. **There have been many attempts to** explain, describe and quantify these effects.
これらの影響を説明し，記述し，そして数量化する多くの試みがされてきた。
15. **This is not an attempt to** characterize pristine or totally undisturbed, pre-Columbian environmental conditions.
これは無傷の，すなわち全く無攪乱のコロンブス前の環境状態を特性把握しようとすることではない。

- **in an attempt to .....** ：… する試みで

16. **In an attempt to** linearize all of the data taken in the experiments, ...
    実験で得られたデータのすべてを直線化する試みにおいて，…
17. A series of experiments were initiated **in an attempt to** obtain results similar to run #1.
    実験#1に類似する結果を得る試みで一連の実験が始められた。
18. **In an attempt to** classify lakes, streams, and wetlands, Okada et al. (1999) delineated ecoregions of the counterminous United States based on mapped landscape characteristics.
    湖沼，河川，湿地を分類するため，Okadaら(1999)は，地図作製された景観特性に基づき隣接した米国の生態地域(エコリージョン)を表示した。
19. The longitudinal examination of biological sampling results is also performed **in an attempt to** interpret and describe the magnitude and severity of departure from the numerical biological criteria.
    生物サンプリング結果の縦断的検討も，数量的生物クライテリアからの逸脱の規模・程度を解釈し記述するために行われる。
20. Experiments were run changing residence time in the UV chamber **in an attempt to** determine if certain chemicals in the system are more completely destroyed as the residence time is increased.
    滞留時間が増えるにつれて，システム内のある特定の化学物質がいっそう完全に分解されるかどうかを決定する試みで，UV反応器の滞留時間を変えて実験が実施された。

- 〜 attempt to ..... ：〜は…するよう試みる
- We attempted to..... ：我々は…しようと試みた

21. The kinetic model **attempts to** explain the mechanics of sorption.
    この速度論モデルは，吸着のメカニズムを説明するよう試みる。
22. This chapter **has attempted to** describe the status of biological criteria efforts in the United States.
    本章では，米国での生物クライテリアにおける努力の現状を述べる試みをしてきた。
23. The primary purpose of this research work has been **to attempt to** explain the effects of copper toxicity on aerobic biological systems.
    この研究の主要な目的は，好気性生物システムに対する銅の毒性影響を説明しようと試みることであった。
24. **We attempted to** describe the status of groundwater resources in the United States.
    我々は，米国での地下水資源の現状を述べる試みをした。

Attention：注意

1. **The increased attention to** this issue is due to .....
   この問題が注目を集めてきたのは，…のためである。
2. One additional provision of the law **merits attention**.
   同法の一つの追加規定が注目に値する。

   - **much attention has been directed at.....**：
     …にかなりの注意が向けられてきた
   - **insufficient attention is directed toward .....**：
     …にあまり注意が払われない
   - **particular attention should be directed to .....**：
     特定の注意が … に向けられるべきである

3. **Much attention has been directed at** investigating the photcatalytic degradation of organic pollutants.
   有機汚染物質の光触媒分解の研究に多くの注意が向けられてきた。
4. **Insufficient attention is directed toward** assessing cumulative impact of these policies in terms of costs and outcomes.
   費用と成果の面でこれらの政策の累積影響評価にあまり注意が払われていない。
5. In assessing the potential health hazards of a contaminated marine environment, **particular attention should be directed to** carcinogens and mutagens.
   汚染された海洋環境の潜在的な健康被害を評価する際，特定の注意が発癌性物質と突然変異誘発性物質に向けられるべきである。

   - **much attention has been given to.....**：
     …にかなりの注意が向けられてきた
   - **little attention has been given to .....**：
     …についてはほとんど注意が向けられてこなかった
   - **particular attention should be placed on .....**：
     …に特定の注意が置かれるべきである
   - **special attention has been paid to.....**：
     …に特別な注意が向けられてきた
   - **increasing attention has been focused on .....**：
     …にますます多くの注目が集中されてきた

6. **Much attention has been given** recently **to** the development and application of biochemical models.

近年，多くの注意が生化学的モデルの構築と適用に向けられてきた。
7. **Less attention has been paid to** the level of other amino acids.
   他のアミノ酸の濃度には，注意があまり払われていない。
8. **Little** legal **attention has been paid to** this trend.
   この傾向には法律的な注意がほとんど払われていない。
9. **Special attention has been paid to** the molar taurine: glycine ratio.
   タウリン:グリシンのモル比に特別な注意が向けられてきた。
10. **Particular attention should be placed on** the development of treatment methods for removing nitrates from groundwater supplies.
    供給用地下水から硝酸を取り除くための処理方法の開発に特定の注意が置かれるべきである。
11. During the last decade, **increasing attention has been focused on** the ability of aquatic organisms to metabolize chemical contaminants.
    過去15年間に，化学汚染物質を新陳代謝する水生生物の能力にますます多くの注目が集中されてきた。

    ● ～ **pay attention to** ..... : ～は…に注意をむける

12. The first investigator to **pay attention to** the reaction was Yamanaka (1930).
    その反応に最初に注意を払った研究者は Yamanaka (1930) であった。
13. During the first two decades of the CWA, regulatory agencies **paid far more attention to** chemical than to biological integrity of the nations.
    CWA の当初20年間に規制機関は，米国の水域の生物完全度よりも化学完全度に注意を払った。

    ● ～ **draws attention to**..... : ～は…に注意を引いている

14. The widespread public concern about these problems **draws attention to** the fact that .....
    これらの問題点に関しての広範な市民，一般の関心は，…という事実に対し注意を引いている。
15. In recent years, advanced oxidation processes (AOPs) **have drawn considerable attention** because of their ability to decompose biorefractory compounds.
    促進酸化法は，生物処理が難しい化合物を分解する能力があるため，近年，かなりの注目を浴びてきた。

    ● ～ **focused attention on**..... : ～ は…に注意を注いだ
    ● ～ **have focused a great deal of attention on** …:

Attention

～は…に多くの注意を注いできた

16. The USEPA **focused attention on** regions of the United States where acidic or acid-sensitive lakes and streams were expected to be found.
USEPAは，酸性または酸に敏感な湖沼・河川が見出されると予想される米国各地に注意を注いだ。
17. The need for disposal methods **has focused attention on** heavy metals and other toxic metals associated with these sludge.
処分方法の必要性は，これらのスラッジにからむ重金属ならびに他の毒性金属に注意を注いだ。
18. The increasing threat of contamination and the potential hazard **have focused a great deal of attention on** the fate of mercury in the environment.
汚染による増加する脅威と潜在的危険性が，環境中における水銀の衰退に多くの注意を注いできた。
    - ～ **have received attention**.....：～は注目を浴びてきた
    - ～ **have received scanty attention**：
      ～には，ほとんど注意が払われてこなかった
    - ～ **have received insufficient attention**：
      ～には十分な注目が払われてこなかった
    - ～ **have been receiving increasing attention**：
      ～はますます多くの注目を集めてきた
    - ～ **is receiving increased attention**：
      ～は，ますます注目を集めている
19. Periphyton **has received recent attention** related to monitoring program and its use in bioassessment is being evaluated by three states.
付着生物は，最近，モニタリングプログラムに関連して注目を浴びており，生物アセスメントにおける付着生物の利用が3州によって評価中である。
20. Analysis of the lag-phase **has received scanty attention**.
遅滞期段階の解析には，ほとんど注意が払われてこなかった。
21. Very small and intermittent streams **have received insufficient attention** everywhere but Minnesota and Indiana.
小規模な間欠河川には，ミネソタ州およびインディアナ州を除き，十分な注目が払われてこなかった。
22. The problem **is receiving increased attention** throughout the country.
この問題は国全体でさらに注目を集めている。
23. Heavy metal pollutants will continue **to receive increased attention**.
重金属汚染物は注目を集め続けるであろう。

- ～ **have attracted extensive attention** ..... :
  ～は広く注意を引いてきた
24. During the past decade, this process **has attracted extensive attention** as a modern technology for groundwater purification.
    これまでの 10 年間に，このプロセスは，地下水浄化の近代的な技術として広く注意を引いてきた。
  - **It has been brought to attention by** ...... :
    それは…によって注目を浴びるようになった
  - **It has drawn considerable attention** ..... :
    それはかなりの注意を引いた
25. **It has been brought to public attention by** USEPA
    それは USEPA によって大衆の注目を浴びるようになった。
26. **It has drawn considerable attention** in recent years.
    それは近年かなりの注意を引いた。

Attribute (Attributable)：特質；属性；帰する (帰せられる)

1. One of the most important **attributes** is molecular weight.
   最も重要な属性の 1 つが分子量である。
2. While these classifications were based on ecological **attributes**, the criteria associated with each were entirely chemical/physical.
   これらの分類は，生態的属性に基づいていたが，それぞれに関連するクライテリアは全く化学的/物理的なものであった。
   - ～ **is attributable to** ..... : ～ は … に帰せられる
3. This **is** probably **attributable to** the variable nature of activated sludge.
   これはおそらく活性汚泥の可変的性質に帰することが出来る。
4. A robust statistical treatment of these data often can help determine whether or not the biological response **is attributable to** the measured pollutant.
   これらのデータへの確かな統計処理は，生物応答が測定された汚濁物質に帰せられるか否か決定するのに役立つことが多い。
   - **..... is attributed to** ..... : … に 帰因する
5. The eventual mineralization of organic compounds can **be attributed** almost entirely **to** biodegradation.

有機化合物の最終的な無機化は，ほとんど完全に生物的分解に帰因されるであろう。
6. The failure of the model to predict it **may be attributed to** errors in the model input data.
それを予測するモデルの失敗は，モデル入力データのエラーに帰されるかもしれない。
7. The better performance of the QHEI compared to the RBPHQ **is not attributed to** some inherent superiority in the QHEI.
QHEIがRBPHQよりも好成績な原因は，QHEIの本質的優越性ではない。
8. This **can be attributed to** the fact that the optimum catalyst concentration value will be a strong function of the geometry of the system and operating conditions viz., incident intensity of UV light and concentration and type of the pollutant.
これは，触媒の最適濃度が実験装置の形状ならびに操作条件，すなわち，紫外線強度，汚濁物質の濃度と種類，に強く関連するであろうという事実に帰因するかもしれない。

Available：利用可能な；入手可能な

1. The amount of data **available** is limited.
利用可能なデータの量が限られている。
2. **The available** field data were not sufficiently detailed to allow an assessment of spatial variations.
利用可能なフィールドデータは，空間的変位のアセスメントをするのには詳細に欠ける。
3. The amount and quality of **available** information are not adequate to quantify the risk.
利用可能な情報量と質は，リスクを数量化するには不十分である。

● ..... is available .....

4. Only a limited amount of stream water quality data **is available**.
ほんの限られた量の河川の水質データだけが利用可能である。
5. No particle size analysis data is currently available and insufficient laboratory duplicate data **is available**.
利用可能な粒子径分析データが現在得られない，そして利用可能な研究室での重複データが不十分である。
6. A considerable amount of information **is available** on the effect of inhibitors on culture of Nitrosomonas sp.

Nitrosomonas 種の菌株に対する阻害剤の影響に関するかなりの量の情報が利用可能である。
7. To what extent **is** mercury in the form of mercuric sulfide **available for** biological methylation?
どの程度まで，硫化水銀として存在する水銀が生物によってメチル化されるのか。

Average(Averaging)：平均(平均する)

1. Jamaica Bay occupies approximately 52 km^2. **Average** depth is about 4 m at mean low water.
ジャマイカ湾の面積はおよそ 52 km^2 である。平均水深は平均の低水位においておよそ 4 m である。
2. It takes **an average of** about 2 hr per sample to identify the 500 organisms to species.
500 体の生物種識別には，サンプル当り平均約 2 時間要する。
3. To estimate the **average** velocity, use the plot and identify the velocity that corresponds to the average annual discharge.
平均速度を推定するためにプロットを使い，そして平均年度排水量に対応する速度を確認してください。
4. Since the model is intended to compute long-term trends, **average** annual values are simulated and variability within the year is ignored.
このモデルは長期傾向を計算するように意図されるので，年平均の値がシミュレートされ，年内の変動は無視される。

- **average concentration**(See **Concentration**)：
平均濃度(➡ **Concentration**)
- **on average**：平均で
- **on an average**：平均して

5. **On average**, around 5–10 % of the original oil in place can be exploited by means of primary recovery.
第一次回収法では，平均して，原埋蔵量の約 5 %から 10 %のオイルが産出されうる。
6. If the peak flow occurs over a 24-hour period, it would take, **on average**, 6.5-year floods to remove this sediment.
もし最大流量が 24 時間に渡って生ずるとするなら，この堆積物を除くには，平均で 6.5 年に一度の洪水を要するであろう。

- **〜 averaged** .....：〜は平均して … である

Aware

- **..... is averaged**：… が平均された

7. The effluent nickel concentration consistently **averaged** greater than 95 percent of the influent nickel concentration.
排水中のニッケルの濃度は,一貫して平均して流入水中のニッケル濃度の 95 パーセント以上であった。

8. All values were obtained for a time period of 1900 to 1960 except for the Tama River flow which **was averaged** over the period from 1939 to 1960.
すべての値は,1939 から 1960 までの期間にわたって平均された Tama 川の流量以外は,1900 から 1960 までの期間に収集された。

- **..... of averaging**：… を平均すること

9. The dispersion coefficient lump together all the effects **of averaging**.
分散係数はすべての影響を平均し,それをひとまとめにしたものである。

Aware (Awareness)：認識して；知って (認識)

- **〜 be aware of .....**：〜は … を認識する
- **〜 become aware of .....**：〜は … を認識する

1. Industry **should be aware of** the principal statutory provisions pertaining to water quality criteria.
業界は,水質クライテリアに関わる主な法律規定を知っておくべきである。

2. Industry **should become aware of** the process that states are using to derive water quality criteria.
業界は,諸州が水質クライテリアを導出する過程を認識すべきである。

- **there has been an increasing awareness that .....**：
… との認識が強まってきた
- **As we increase our awareness that .....**：
我々は … だと深く認識するにつれて
- **〜 should be aware that .....**：〜は … を認識すべきである

3. In recent years **there has been an increasing awareness that** effective research, inventory, and management of environmental resources must be undertaken with an ecosystem perspective.
近年,環境資源の効果的な研究・調査一覧作成・管理は,生態系を視野に入れて行うべきだ,との認識が強まってきた。

4. **As we increase our awareness that** a holistic ecosystem approach to environmental resource assessment and management is necessary, we must

also develop a clearer understanding of ecosystems and their regional patterns.
我々は，環境資源の評価・管理への総合的生態系手法が必要だと深く認識するにつれて，生態系とその地域パターンも明確に理解しなければならない。

5. The user **should be aware that** such effects may complicate attempts to extrapolate data for photolysis rates from one aquatic medium to a very different medium.
利用者は，このような影響が光反応速度データを1つの水生媒質からかなり異質の媒質まで外挿法によって推定する試みを複雑にするかもしれないということを認識すべきである。

6. This need was part of a larger concern for a framework to structure the management of aquatic resources in general and was coupled with **an increasing awareness that** there was more to water quality management than addressing water chemistry, which had been the primary focus
この必要性は，水生資源全般の管理を構成する枠組みへの大きな関心の一部であり，それまで中心的主題だった水の化学的特性への対処よりも水質管理の方が重要だとの認識と相まっていた。

Axis：軸

1. For purposes of analysis, let these faces be perpendicular to the **z-axis**.
解析するために，これらの面をz軸と垂直に交わらせてください。

2. For example, the proportion of sites with IBI scores <30 is illustrated by following the vertical line originating at 30 on the x axis to its intersection with the CDF, the following the horizontal line until it intersects **the y axis**.
例えば，IBIスコアが30点未満の地点の比率は，x軸の30点で始まる垂直線をCDFとの交差点まで辿った後，それがy軸を横切るまで水平線を辿れば示される。

Base：基本；基部；基点；基礎

- **based on .....**：…に基づいて

1. This report is subject to revisions **based on** additional information.
   この報告書は，追加情報に基づいて修正の対象となる。
2. Data quality can be roughly divided into three categories **based on** program objectives.
   データの質は，プログラム目標に基づいておおよそ以下の3つのカテゴリーに分けられる。
3. **Based on** recent studies of soil ingestion, these values should be considered conservative.
   土壌摂取の最近の研究に基づくと，これらの値は余裕ある値であると考えるべきである。
4. **Based on** the existing data and the conceptual model, the following data needs have been identified: 1) ....., 2) ......, and 3) ....
   既存のデータと概念モデルに基づいて，データの必要性が次のように識別された。1)…，2)…，3)… である。

- ～ **is based on .....**：～は…に基づいている
- ～ **is based upon .....**：～は…に基づいている

5. These descriptions **are based on** information contained in the EPA reports (EPA, 2002).
   これらの記述は，EPA報告書(EPA, 2002)に含まれている情報に基づいている。
6. The approach chosen **is based on** dynamic mass balance and uses the simulation techniques described by Sasaki et al.(2001).
   選択された取り組み方は，ダイナミックな物質収支に基づき，そしてSasakiら(2001)によって記述されたシミュレーション手法を使う。
7. Most filter research work completed to date **has been based on** the measurement of suspended solids(SS).
   今日までに完了したほとんどの濾過の研究は浮遊物質(SS)の測定に基づいていた。
8. Many local and state watershed management decisions **have been based**,

at least in part, **on** data collected by volunteer monitors.
多くの地方および州の分水界管理上の決定は，少なくとも部分的にボランティアモニターの収集データに基づいている。

9. Candidate models are selected **based on** knowledge of aquatic systems, flora and fauna, literature review, and historical data.
候補のモデルは，水生系に関する知見，植物相と動物相，文献査読，史的データに基づいて選ばれる。
10. The formulation **is based upon** a proven concept.
この式はすでに証明されている概念に基づいている。
11. Aoki and Tomoda's (1997) biotic index **was based** directly **upon** the saprobien system.
AokiとTomoda(1997)の生物指数は，腐敗システムに直接基づいていた。

Basic：基本的な

1. The **basic** process is still poorly understood.
その基本的なプロセスはまだ良く理解されていない。
2. The **basic** steps of the risk assessment process are the following: .....
リスクアセスメント過程の基本的な段階は次の様である。それは，…
3. The investigation process consists of six **basic** components: .....
調査プロセスは6つの基本的な構成要素から成る。それらは…
4. A **basic** assumption for the model was that biofilm diffusion limitations were not significant.
生物膜拡散限界は重要ではないということが，このモデルの基本的な仮定であった。
5. Catalysis by fixed cell is quite complex and many **basic** aspects are yet to be understood.
固定細胞による触媒作用は非常に複雑で，多くの基本的な側面がまだ理解されていない。
6. As a **basic** principle of statutory construction, general statements of statutory goals and objectives have no legal force and effect, absent specific operative provisions in the law.
法的解釈の基本原理として，法律上の目的および目標の一般的な文章は法的効力を持たず，法律における具体的な運用既定を欠いている。

Basis：根本原理；根拠；論拠；基礎

1. The opposing concepts of diversity and similarity are **the bases of** most

Basis

classifications.
多様性と類似性という対立的概念が大半の分類の基本にある。

2. While it is recognized that individual waterbodies differ to varying degree, **the basis for** having regional reference sites is the similarity of watersheds within defined geographical regions.
個々の水域が様々に異なることは認識されるものの，地域参照地点を設けることは規定した地理的な地域内の流域が類似していることを根拠としている。

- **as the basis for** ..... ：…の基礎として
- **as the sole basis for** ..... ：…の唯一の根拠として

3. The information could also serve **as a basis for** developing water quality criteria.
この情報は，水質クライテリア設定の基礎にもなる。

4. Equation 3 may be used **as the basis for** a total mass balance.
式3は，総マスバランスの基礎として使えるだろう。

5. This report will serve **as a basis for** RFI work plans developed by ERP.
この報告書は，ERPによって作成されにRFIワークプランの基盤となるであろう。

6. This assumption may be used **as a basis for** a preliminary analysis of the problem.
この仮定は，この問題の予備解析の根拠として使えるだろう。

7. In these cases, biological impairment criteria may be used **as the sole basis for** regulatory action.
これらの場合，生物損傷クライテリアは，規制措置の唯一の根拠となるかもしれない。

- **on a case-by-case basis** ：「ケースバイケース」ベースで
- **on a year-round basis** ：1年中を通してのベースで
- **on a statewide basis** ：全州ベースで
- **on a timely basis** ：タイムリーに

8. Comparison with other technologies **on a case-by-case basis** is limited.
「ケースバイケース」ベースでの他の技術との比較には限りがある。

9. They proposed environmentally sound uses for SMS **on a year-round basis**.
かれらは環境上効果的なSMSの使用を1年中を通してのベースで提案した。

10. The reference site results were pooled **on a statewide basis** prior to constructing the drainage area scatter plots.
流域面積の散点図が作製される前に対象地域の結果が全州ベースでプールされ

た。
11. Remote, difficult to reach areas have also been monitored by citizen volunteers when they were too difficult for state personnel to reach **on a timely basis**.
カバーしづらい遠隔地も，州職員がタイムリーに到着しかねる場合には，市民ボランティアによってモニターされてきた。
12. Usually, expectations have been developed **on a watershed or regional basis**, and have been derived from recent field data.
通常，期待値は，流域または地域ベースで設定されてきており，最近の現地データから導出されてきた。

- **on the basis of ......**：…に基づいて；…をベースに；…によると

13. A relationship was derived **on the basis of** energy considerations.
一つの関係がエネルギーを考慮して導かれた。
14. Impacts may not be totally predictable **on the basis of** habitat assessments alone.
生息場所アセスメントだけではインパクトを完全には予測できないかもしれない。
15. These indices were selected among many others **on the basis of** accuracy, low variability, and simplicity.
これらの指数は，正確性，低い可変性，単純性に基づいて他の多くの指数の中から選ばれた。
16. **On the basis of** estimated rate constants, the oxidation rate of formic acid is faster than oxalic acid over a pH range of 4−8.
推定速度定数を基にすると，4から8のpH範囲では，ギ酸の酸化速度はシュウ酸より速い。
17. The study areas were selected **on the basis of** evidence suggesting that they are areas of relatively intense geochemical activity.
調査地域は，比較的激しい地質化学的活動の地域であることを示唆している証拠をベースに選択された。
18. Several site-specific hydrologic input data required were not collected and had to be estimated **on the basis of** model calibration using historical climatological data.
必要とされる地点特異的な水文入力データのいくつかが集められず，歴史的気候学データを使ったモデルのキャリブレーションをベースに推測しなければならなかった。

- **form a basis of ......**：…の根拠になる
- **form a basis for ......**：…の根拠になる

19. The t-test does not **form the basis of** the determination of impairment, but tightens the conditions for a genuine exceedance.
    t検定は，損傷決定の根拠にならないが，純粋な超過の状態を確かめられる。
20. Success with compounds such as TCE, CHCl₃ and CCl₄ will serve as proof-of-principle and **form a basis for** expanding the research to other pollutant classes.
    TCE，CHCl₃，ならびにCCl₄のような化合物での成功は，「原理の証明」となり，他の種類の汚染物質に研究を拡大するための根拠となるであろう。

    - ~ **provide a basis for** ..... ：~は…の基礎となる
    - ~ **provide a sound basis for** ..... ：
      ~は…のための確かな基盤を提供する
    - ~ **provides a rational basis for** ..... ：
      ~は…のための合理的基礎を提する

21. Mathematical models **provide a basis for** quantifying the inter-relationships among the various toxic chemicals.
    数学モデルは，種々の有毒化学物質間の相互関係を数量化する基礎となる。
22. The TMDL determines the allowable loads and **provides the basis for** establishing or modifying controls on pollutant sources.
    TMDLは許容可能な負荷を決定し，そして汚染源を確立するか，あるいは制御を修正するための基礎を提供する。

    ★ TMDLはTotal Maximum Daily Loadの略語で，一般に「日最大許容負荷量」と訳されています。

23. The use of stability constant **provides a rational basis for** predicting levels of metal accumulation.
    安定定数の使用は，金属の蓄積レベルを予測するための合理的基礎を提する。
24. A three-year ecoregion project **provided a sound basis for** developing realistic water quality standards and beneficial uses within ecoregions.
    3箇年の生態地域プロジェクトは，生態地域内での現実的な水質基準および有益利用の確立のための確かな基盤を提供した。

Because：なぜなら

1. The reaction of gold with hemoglobin is of interest, for example, **because** studies have shown that gold accumulates in the red blood cells of proteins.
   例えば，ヘモグロビンと金の反応は興味深い。なぜなら，金がタンパク質の赤血球に蓄積することが研究されているからである。

2. Only a few rocks need to be sampled, **because** each has a periphyton assemblage with hundreds of thousands of individuals.
   サンプリングに必要なのは，数個の岩石だけである。なぜなら，1個の岩石に個体数が数百ないし数千もの付着生物群があるからである。
3. Some known features of the behavior of iodine are not included in the model, **because** they are not relevant to the problem of interest.
   ヨードの挙動に関してすでに知られている特徴の一部は，モデルに含められていない。なぜなら，それらはここで取り上げている問題点とは直接に関係が無いからである。
4. **Because** the dechlorination reaction rates were relatively slow as compared to CP desorption kinetics, solid−liquid partitioning equilibrium was assumed, and partitioning rates were not incorporated into the model.
   CPの脱着速度と比較して，脱塩素反応速度が比較的遅かったので，固体−液体間の分配平衡が仮定され，分配速度はモデルに取り入れられなかった。

- **This is partly because .....：**
  その理由は一つには … したからである；これは一つには … ということによる

5. Existing multimetric approaches are robust in their ability to measure biological condition. **This is partly because** they incorporate biological and ecological principles that enable an interpretation of exposure/response relations.
   既存のマルチメトリック手法は，生物学的状態の測定能力において確かである。その理由は，一つには，曝露／応答関係の解釈を可能にするような生物学的・生態学的原理を盛り込んでいることによる。
6. The criteria were also applied to multiplate samples. **This is partly because** variability among percent contribution of major groups in multiplate samples from nonimpacted sites was found to be too great to establish a model community.
   このクライテリアは，多プレートサンプルにも適用された。その理由は，一つには，無影響地点での多プレートサンプルの主要グループのパーセンテージの多様性が高すぎて，モデル生物群集を定められないと考えられたからである。

- **Because of .....：** … という理由で；… の故に

7. **Because of** a lack of extensive knowledge on pesticide toxicity, ...
   殺虫剤毒性に関する広い知識が欠如しているという理由で …
8. **Because of** the copious quality of data gathered during the sampling program, ...

サンプリングプログラム期間中に収集された豊富な良質のデータ故に，…
9. **Because of** the lack of data, the methodology was not fully verified.
   データの欠如のために，この方法論は十分に立証されなかった。
10. Perfect charge balance would not be expected **because of** analytical errors in measuring the many parameters used.
    使用された多数のパラメータを測るにあたっての分析エラーのために，完全なチャージバランスは期待されないであろう。

Become：になる

1. All data points in the individual experiments appeared **to become** linear over the range of contact times studied.
   個別の実験でのすべてのデータポイントが，実験した接触時間範囲では直線になるように見えた。
2. The crucial question of whether the water body **becomes** acidic depends on the magnitude of the watershed's acid neutralization capacity.
   水流が酸性になるかどうかという重要な問題は，流域の酸中和容量による。
3. Toxicity assessment of single chemical compounds and of complex industrial effluents **has become** an increasingly difficult task.
   一つの化合物や複合的な産業廃水の毒性査定がますます難しい仕事になった。
4. It **has become** increasingly desirable to use destruction of hazardous wastes as the site remediation technology.
   危険廃棄物の分解をサイト是正技術として使用することがますます望ましくなった。
5. Environmental managers within industry **can become** well acquainted with basic water quality criteria concepts by reviewing the four documents cited above.
   業界内の環境管理者は，上記4つの文書を繙けば，基礎的な水質クライテリア概念を十分に把握することができる。

- ～ **become aware of** ..... ：～は … を認識する

6. Industry personnel **should become aware** of the process that states are using to derive biocriteria and communicate their concerns when technical flaws may result in unrealistic, or invalid, biological expectations outside of reference sites.
   業界人は,，諸州が生物クライテリアを導出する過程を認識して，技術上の欠陥が非現実的もしくは無効な参照地点外の生物学的期待値を生じそうな場合には，懸念を伝えるべきである。

- ~ **become difficult**：～は難しくなる
- **It has become widely accepted in** .....：
  それは … で広く受け入れられてきた
- ~ **have become increasingly evident**：
  ～は，ますます明白になってきた
- ~ **can become increasingly obvious**：～は，ますます明瞭になりうる
- ~ **becomes clear**：… が明確になる

7. The decision **can become difficult** when .....
   …のとき，決定は難しくなる。
8. It **has become widely accepted** in large-scale applications.
   それは大規模なアプリケーションで広く受け入れられてきた。
9. The magnitude and significance of the problem **has become increasingly evident**.
   この問題の規模と重要性はますます明白になってきた。
10. It **has become increasingly evident** that the environmental impact of a particular metal species may be more important than the total metal concentration.
    特定の金属種の環境に対する影響が，金属すべての濃度の影響よりいっそう重要であるかもしれないということがますます明白になってきた。
11. Need for establishment of modified biocriteria due to limitations in habitat quality **will become evident** in applications of habitat assessment routines.
    生息地の質における限界故に，改善された生物クライテリアを定める必要性は，生息地アセスメントルーティンの適用において顕著となろう。
12. Thus, a weight-of-evidence approach **can become increasingly obvious** when hypothesis testing demonstrates the unnecessary usage of independent application
    それ故，証拠加重値手法は，仮説検定により独立適用が不要であることが実証された場合には，ますます明瞭になりうる。

- ~ **is becoming** .....：～は … になりつつある

13. The assessment of biological condition using a suite of metrics to define biocriteria **is** rapidly **becoming** the method of choice among state water resource agencies.
    生物クライテリアを定めるための一連のメトリックを用いた，生物学的状態のアセスメントは，急速に州の水資源担当機関に選択される手法になりつつある。

Begin(Beginning)：始まる（始まり）

1. Three states **have not yet begun to** formulate their bioassessment approach, but have initiated discussion within their own agencies and with the USEPA.
 3州は，まだ生物アセスメント手法の作成を開始していないが，州機関内での討論やUSEPAとの論議に着手した。

   ● ~ **begin with** ..... ：~は … で始まる

2. The efforts that states invest in developing biological criteria **begin with** biological surveys conducted to characterize the condition of the state's water resources.
 諸州が生物クライテリア設定に払う努力は，まず，その州の水資源の状態を把握するため行われる生物調査で開始される。

3. Indeed, the first state biological criteria programs **began with** the effort of state biologists to apply their bioassessment results within a regulatory framework.
 実際，最初の州の生物クライテリアプログラムは，各州の生物アセスメント結果を規制枠組みにおいて適用しようとする州生物学者の努力で始まったのである。

   ● begin ___ing ：___することを開始する

4. This has allowed small towns to **begin building** treatment plants that will attain the effluent limit specified in the new water quality standards.
 これにより，小都市で，新しい水質基準で明記された排出限界を満たすような処理場の建設を開始できるようになった。

   ● **Beginning** ：始まり

5. **Beginning** in 1987 and continuing to the present, annual sampling has documented attainment of the ICI ecoregional biocriterion at this site.
 1987年に始まり現在までずっと，毎年のサンプリングは，この地点でICI生態地域生物クライテリア達成を記録してきた。

6. **Beginning** around 1900 and accelerating greatly in the last 20 years, fish community characteristics have been used to measure relative ecosystem health.
 1990年頃から始まり，過去20年間に大幅に加速したのだが，魚類群集特性を用いて相対的生態健全度が測定されてきた。

7. Some programs **are beginning** to broaden the scope of sampling from a

single type of waterbody such as a lake or stream to entire watersheds, including land use monitoring.
一部のプログラムは，湖沼や水流といった1種類の水域から土地利用モニタリングといった流域全域までサンプリング範囲を拡大し始めている。

- **In the beginning**：まず初めに

8. **In the beginning**, case histories are presented.
まず初めに，事例史を下に述べる。

Behavior：挙動

1. One plausible explanation for this **behavior** is that ......
この挙動についての妥当な説明の1つは…ということである。
2. Not much is known concerning the nature and **behavior of** natural systems.
自然界のシステムの性質と挙動に関してはあまり多くは知られていない。
3. We use mathematics to help understand the **behavior of** some physical process or system.
我々は，物理的プロセス挙動あるいは物理的システム挙動の理解を助けるために数学を使う。
4. The model does not adequately represent the endocrine disruptors' **behavior** in this application.
このモデルは，この応用では内分泌かく乱物質の挙動を十分に表現しない。
5. SOILII model does not adequately represent the aldicarb **behavior** in this application.
このアプリケーションでは，SOILIIモデルは十分にアルジカルブの挙動を示さない。
6. A greater understanding of kinetic **behavior** should lead to better predictions of the **behavior of** PCP during anaerobic bioremediation.
動力学的挙動をより良く理解することは，嫌気的バイオレメディエーションにおけるPCPの挙動をもっと良く予測することにつながる。

Being：であるという；ということ

1. The test offers the advantage of **being** more sensitive and much faster.
このテストには，いっそう敏感で，より迅速であるという利点がある。
2. Lobsters were not fed during their captivity and were kept in the aquarium for at least 10 days before **being** used in the experiment.
ロブスターは飼育中はえさを与えず，少なくとも実験の前10日間は水槽の中で

飼われた。
3. In contrast to the complexity of the processes **being** modeled, the model itself must be as simple to use as possible.
モデル化されているプロセスの複雑さとは対照に，モデルそれ自身の使用ができる限り簡単でなければならない。
4. Perhaps most important to the enhancement of biological criteria efforts are the lessons **being** learned from state experiences.
生物クライテリア努力の増強にとっておそらく最も重要なのは，諸州の経験から学んだ教訓である。
5. The idea **being** brought forward in this research is to measure several of these general parameters.
この研究における前進的考え方は，これらの一般的なパラメータのいくつかを測定するということである。
6. This is especially true in such cases where a waterbody may be perceived as **being** at risk due to new dischargers.
ある水域が，新しい排出のために危険な状態であるものと見なされるかもしれない場合，これは特に真実である。

● ～ **is being ＿ ed**：～が…されている；～が…されつつある

7. That aspect of the problem **is being investigated** by the Center for Disease Control Office.
問題点のこの側面が，病原対策センターによって調査されている。
8. Although freshwater rivers and streams are the focus of this book, the same framework **is being applied** for other resource types.
淡水河川・水流が本書の主題であるが，この枠組みは，他の資源類にも適用されている。
9. Assessments of the quality of physical habitat structure are increasingly **being incorporated** into the biological evaluation of water resource integrity.
物理的生息地構造の質アセスメントがますます水資源健全性の生物学的評価に盛り込まれつつある。
10. Potential uses of biological criteria in the total maximum daily loads (TMDL) process **are** also **being explored** at both the state and federal levels.
日最大負荷量(TMDL)過程における生物クライテリアの潜在的利用も，州レベルおよび連邦レベルで追求されている。
11. Vermont uses biological criteria from ambient stream data to determine whether two different types of biological standards **are being met**.
バーモント州は，周辺水流データからの生物クライテリアを用いて，2種類の生

物基準が満たされているか否か決定している。
12. It **is** currently **being used to** evaluate qualitative data in conjunction with other more traditional sample attributes such as total and EPT richness and overall community composition and balance.
それは，現在，総豊富度と EPT 豊富度，全ての生物群種の構成やバランスといった他のより伝統的なサンプル属性と共に質的データを評価するため用いられている。

★ここでは，EPT は Ephemeroptera/Plecoptera/Trichoptera, カゲロウ目／カワゲラ目／カゲロウ目を意味しています。

- **Attempts are being made to .....** (See **Attempt**)：
  … する試みがなされつつある (➡ **Attempt**)

Believe：思う；信じる

- **..... is believed to .....**：… であると思われる

1. Microorganisms **are believed to** play a significant role in the disappearance of trifluralin in soil.
微生物が土壌でトリフルラリンの消失において重要な役割を演ずると思われいる。
2. Although the mechanisms of acoustic foam disintegration are not entirely understood, the following acoustic effects **are believed to** be most important in foam breakage.
音波によって生じる泡の崩壊機構は完全に理解されていないが，次に示す音波効果は気泡崩壊過程上最も重要であると思われる。
3. An attempt to use the scheme for classifying aquatic ecosystems proved unsuccessful and resulted in the development a different framework **believed to** be more effective.
水生生態系分類方式を用いる試みは，不成功であることが判明して，より効果的と考えられる別の枠組みの案出をもたらした。

- **It is believed that .....**：… と思われる

4. **It is believed that** the decline in dieldrin from the reservoir outlet is due to the decreased aldrin application rates.
貯水池出口でのディルドリンの減少はアルドリンの使用度の減少によると思われる。
5. The mechanism of this decomposition is very complex. Briefly, **it is believed that** when exposed to radiation, water molecules split into ions as shown below.

この分解機構はかなり複雑である。手短かに言えば，放射線にさらされたとき，水の分子は下に示されるようにイオンに分かれると考えられる。

6. Although **it is generally believed that** .....
   一般に … と信じられるけれども

   - **There is a reason to believe that** ..... ：… を信じる理由がある
   - **We believe that** .... ：我々は … を信じる

7. **There is a reason to believe that** it will continue to be so well into the future.
   それが未来に向けて良くあり続けるであろうと確信する理由がある。

8. The question arises as to how **we can believe that** ......
   … ということを我々がいかに信じることができるかという質問が起きる。

9. **We believe** the latter approach may introduce some unintentional bias into the water quality criteria calibration and derivation process.
   我々は，後者の手法が水質クライテリアの較正・導出過程に意図しないバイアスを持ち込みかねないと考える。

Below (See also **As** and **Above**)：下に (➡ As ならびに Above)

1. All other values are at concentrations **well below** this risk level.
   すべての他の値は，このリスクレベルをずっと下まわる濃度にある。
2. A long-standing heavy metals problem in this stream occurred **below** an industrial effluent discharge.
   この水流における長年の重金属問題は，産業廃液排出口の下流で発生した。

   - **as discussed below**：後述のとおり
   - **~ is given below**：~ は下に示される
   - **~ are presented below**：~ を下に述べる
   - **~ is included below**：~ は下に含まれている
   - **Shown below is** .....：下に … が示される

3. This assumption is not realistic, **as discussed below**.
   この仮定は，後述のとおり現実的ではない。
4. Water quality data collected by citizen monitors **are discussed below** by the traditional categories of physical, chemical, and biological conditions.
   市民モニターが収集した水質データについて，物理的状態，化学的状態，生物的状態という従来のカテゴリーによって以下で論じていく。
5. Case histories **are presented below** showing the outcome of impairment criteria testing in a variety of situations.

さまざまな状況での損傷クライテリア試験の結果を示した事例史を下に述べる。
6. A brief summary **is included below** as an aid to understanding the results obtained in this project.
短い要約が，このプロジェクトで得られた結果を理解する助けとして下に含まれている。
7. **Shown below is** a representative breakdown of the application effort into the steps.
このステップへの適用努力の代表的な内訳が下に示されている。

- **~ is discussed below**（See **Discuss**）：
～は後ほど詳述する；～は下に論じられている（➡ **Discuss**）

Benefit：便益；利益

1. The risk and **benefits of** this proactive strategy are discussed in more detail below.
この先取的戦略の持つ危険と便益については，後ほど詳述する。

- **~ provide benefits for .....**：～は … に利益を提供する
- **~ outweighed the benefits**：～が便益をはるかに上回る

2. This wealth of data and experience from the HSPF application on Iowa River **provide major benefits for** the Study.
アイオワ川での豊富なデータとHSPFの適用からの経験は，この研究に主要な利益を提供する。
3. Although every program **provides** some form of educational **benefits**, the extent of educational opportunities varies among the programs。
どのプログラムも何らかの啓発効果を有するものの，啓発機会の度合は，プログラムごとに異なっている。
4. Many other states did not undertake substantial bioassessment programs because they felt the cost of bioassessment **outweighed the benefits**.
他の多くの州は，生物アセスメントの費用が便益をはるかに上回ると考えたため，実質的な生物アセスメントプログラムを行わなかった。

Best：最良の；最善の；最も

1. Carbon provided **the best** treatment overall.
全体として炭素が最良処理であった。
2. If a relationship was observed, a 95％ line of **best fit** was determined.

関係が見られたならば，ベストフィットの95％ラインが決定された。
3. Riffles and runs, with current velocities of 10 to 20 cm/sec, **are best** because biomass is least variable in these habitats.
   流速10〜20cm/sの瀬および急流が最適である。なぜなら，これらの生息場所では，バイオマスの可変性が最も弱いからである。

    - **The best known** .....：最もよく知られている…
    - **The best and most**.....：最善かつ最も…
    - **In the best of**.....：最良の…で

4. **The best known** explanation among those explanations proposed by 〜 is .....
   〜によって提示された説明の中で最もよく知られているのは…である。
5. **The best and most** abundant data for the Iowa River was collected at Marengo, Iowa.
   アイオワ川での最良の，そして最も豊富なデータはアイオワ州のマレンゴで収集された。
6. **The best and most** representative sites for each stream class are selected and represent the best set of reference sites from which the reference condition is established.
   各水流等級について最善かつ最も代表的な地点が選ばれて，最良の対照地点セットを示し，それに基づき参照状態が定められる。
7. **In the best of** situation, it is possible to achieve an annual water balance within five percent.
   最良の状態で，年間の水収支を5パーセント内で達成することが可能である。

    - **one of the best**：最も…の一つは

8. **One of the best** proven uses for biosurvey data is for spatial and temporal trend analysis.
   生物調査データの最も立証された用途の一つは，空間的・時間的傾向分析である。
9. The Scioto River perhaps represents **one of the best** success stories of any river or stream in Ohio.
   Scioto川は，おそらくオハイオ州における河川・水流の最も成功した例の一つである。
10. Periphyton is **one of the best** indicators of disturbances in the catchments as well as of instream chemical alterations.
    付着生物は，集水域における攪乱や水流内での化学的変化に関する最善な指標の一つである。

Better：より優れている

1. The higher the value, **the better** the estimated water quality.
   この値が高いほど推定水質が優れていることになる。
2. A **better** tool was needed to more consistently and accurately characterize the aquatic communities.
   水生生物群集をより一貫性がありかつ正確に特性把握するための優れたツールが必要であった。

   - **It is better to** …… ：…した方が良い
   - **better than** …… ：…より良い

3. **It is better to** identify large areas over which calibrations will be performed rather than small areas.
   キャリブレーションを行うに当たり，狭い地域よりも広い地域を確認した方が良い。
4. A comparison of quantitative measures of performance between AMODEL versus the observed residues, and between BMODEL versus the observed residues shows that AMODEL had only slightly **better** agreement **than** BMODEL.
   AMODELと観察された残留物の間，そしてBMODELと観察された残留物の間の結果の量的比較が，AMODELはBMODELより少しだけ良い一致があったことを示した。

   - ～ **make** …… **a better A than B** ：～は…をBよりも優れたAにする

5. Several new approaches, including multivariate analysis and weighted average metrices, **make** periphyton analysis **a better** indicator **than** it was 5 or 10 years ago.
   多変量分析や加重平均メトリックといった新手法のおかげで，付着生物分析は，5～10年前よりも優れた指標となっている。

   - **to better** …… ：より…する；一層…する

6. Further research into sediment transport is needed to develop more effective tools **to better address** these issues.
   これらの問題をより良く扱うためのより効果的なツールを開発するには，堆積物輸送に対するいっそうの研究が必要である。
7. Future work is needed, however, **to better utilize** the ADV as part of resource value assessment such as NRDAs and in assessing other types of environmental damage claims.

しかし，NRDA のような資源価値アセスメントの一部として，また，他の種類の環境損害請求の評価において ADV をより活用するには，今後の研究が必要である。

- **to better understand .....** (See **Understand**)：
  … をもっと良く理解するために(➡ **Understand**)
- **~ lead to better .....** (See **Lead**)：
  ~は，もっと良く…することにつながる(➡ **Lead**)

Between：間に

1. The model accounts for transport of phosphorus **between** lakes.
   このモデルによって，湖と湖の間のリンの輸送過程が説明される。
2. Due to variability in microbial densities **between** samples, these changes were not statistically significant.
   微生物密度はサンプル間の多様性のため，これらの変動は統計的に有意ではない。
3. Eqn 2 can be integrated **between** limits to yield ...
   式 2 を極限値で積分すると，…を得る。
4. Another way to visualize these trends is to examine changes in the cumulative frequency distribution (CFD) of biological index scores **between** each time period.
   これらの傾向を視覚化するための別方法は，各時期間の生物学的指数スコアの累積度数分布(CFD)における変化を検討することである。
5. Seasonal variability was tested by monthly sampling year-round at two streams, and showed that **between**-month comparisons should not be done, but upstream-downstream sampling on the same date is valid year-round.
   季節による差は，2つの水流での通年サンプリング(毎月)により検証されて，月ごとの比較はすべきでないが，同一日の上流地点-下流地点サンプリングが一年中有効であることを示した。

- **between A and B**：A と B の間

6. During the first "high flow" sampling period **between** June 15 **and** June 24, 2000, ...
   2000 年 6 月 15 日から 6 月 24 日の間の，最初の「流量が多い」時のサンプリング期間中に，…
7. **Between** 19 **and** 46 samples from each of these sites have been analyzed for total arsenic.
   これらのそれぞれのサイトから採取された 19 から 46 のサンプルで全ヒ素の分析

がされてきた。
8. Chloride is in the range of 6 to 41 mg/L and exhibits no significant difference **between** the alluvium **and** the shale.
塩化物が6から41mg/Lの範囲にあり，沖積層と頁岩の間に重要な相違を示さない。
9. Competitive adsorption **between** water vapor **and** a probe contaminant can have a significant influence on the oxidation rate of the contaminant.
水蒸気と調査汚染物質の間の競合性吸着が，汚染物質の酸化速度に重要な影響を与え得る。
10. Source Area SS37 is located approximately 100 m east of Building 43, just east of F Avenue **between** Q Road **and** C Street.
汚染源エリアSS37は，Q道とC通りの間のF大通りのすぐ東にある43号棟の東，およそ100mに位置している。
11. Vertical profiles also exhibited order of magnitude differences **between** the upper **and** lower layer average Lindane concentrations.
縦断プロファイルにおいて，リンデン濃度の上層の平均と下層の平均で桁違いの相違を示した。

- **agreement between A and B**(See **Agreement**)：
  …の間の合意(➡ **Agreement**)
- **competition between** ……(See **Competition**)：
  AとBとの間の競合(➡ **Competition**)
- **correlation between** ……(See **Correlation**)：
  …との間の相互関係(➡ **Correlation**)
- **difference between** ……(See **Difference**)．
  …の間の相違(➡ **Difference**)
- **discrepancy between** ……(See **Discrepancy**)：
  …の間の矛盾(➡ **Discrepancy**)
- **fit between** ……(See **Fit**)：
  …の間で合致に達する(➡ **Fit**)
- **relationship between A and B**(See **Relationship**)：
  AとBの関係(➡ **Relationship**)
- **similarity between** ……(See **Similarity**)：
  …の間の類似性(➡ **Similarity**)
- **boundary between** ……(See **Boundary**)：
  …の境界(➡ **Boundary**)

Bias：偏見

- ～ lend bias to ..... ：～は … に偏りを生じさせる
- ～ introduce bias into ..... ：～は … に偏りを持ち込む

1. This also **lends bias to** the evaluation.
   これもまた，評価に偏りを生じさせる。
2. We believe the latter approach **may introduce** some unintentional **bias into** the biological criteria calibration and derivation process.
   我々は，後者の手法が生物クライテリアの較正・導出過程に無意識な偏りを持ち込みかねないと考える。

Both：両方とも

- both A and B. ：AとBの両方

1. The samples should be taken from **both** the vadose zone **and** the smear zone.
   サンプルは通気帯と汚点帯の両方から採取されるべきだ。
2. These tools are applied extensively in **both** defining use classification **and** assessing water quality.
   これらの手段は，利用分類確定ならびに水質評価において多用されている。
3. Many compounds have a potential for chemical modifications by **both** abiotic **and** biotic systems.
   多くの化合物が，非生物システムと生物システムの両方によって化学変化を起こす可能性を持っている。
4. **Both** the risk-based **and** regulatory-based criteria for benzene were exceeded at groundwater probe FW18.
   ベンゼンについては，地下水調査プローブFW18で危険ベースの基準と規制ベースの基準両方を超えていた。
5. The Big Lost River is **both** a gaining **and** losing stream that fluctuates seasonally and with respect to precipitation and snow melt.
   Big Lost川は浸出河川と侵入河川の両方であり，季節，そして降雨と雪溶け水によって変動する。
6. The emphasis of **both** state **and** federal efforts has been on the development and implementation of bioassessment methods for streams and small rivers, as **both** the scientific **and** experimental databases are greatest for these water bodies.

州および連邦による努力の主眼は，水流・小河川についての生物アセスメント方法の設定・実施に置かれており，また，科学データベースと経験データベースのいずれも，これらの水域について最も大切である。

- ● **In both cases, ...**：両方のケースで；いずれの場合も

7. **In both cases**, good correlation is found between the logarithm of the photochemical rate and the solvent vapor pressure.
両方のケースで，高い相関が光反応速度の対数値と溶媒の蒸気圧の間に見いだされる。

Brief（Briefly）：簡潔な（簡潔に；手短かに）

1. **A brief summary** is included below as an aid to understanding the results obtained in this project.
このプロジェクトで得られた結果を理解する手助けとして短い要約が下に含まれている。

- ● **briefly describe......**：…を簡潔に説明する
- ● **only briefly**：ただ手短かに
- ● **albeit briefly**：ごく簡単にだが
- ● **In brief**：要するに
- ● **Briefly**：手短かに言えば

2. The purpose of this article is to **briefly describe** some critical research and information needs in surface water quality.
この論文の目的は，表流水の水質における重要な研究のいくつかを簡潔に説明することである。

3. Matsuda (1990) **briefly discussed** several studies that have shown biomarker responses correlate with predicted levels of contamination and with site rankings based on community level measures of ecological integrity.
Matsuda (1990) は，予測された汚染レベルや，生態健全性の生物群集レベル測度に基づく地点ランキングと関係する生物標識応答を明らかにした諸研究について簡潔に論じた。

4. Details concerning the experimental procedure are given elsewhere and are described **only briefly** here.
実験手順に関する詳細は，ほかの様々なところで示されているので，ここではただ手短かに記述する。

5. Virginia EPA acknowledges, **albeit briefly**, that site-specific modifications

may be appropriate in some circumstances.
バージニア州EPAは，いくつかの状況では地点特異的な修正が適切であることをごく簡単にだが認めている。

6. The mechanism of this decomposition is very complex. **Briefly**, it is believed that when exposed to radiation, water molecules split into ions as shown below.
この分解機構はかなり複雑である。手短かに言えば，放射線にさらされたとき，水の分子は下に示されるようにイオンに分かれると考えられる。

Bring：もたらす；持ち出す

- ～ **bring up** ..... ：～は…を持ち出す
- **bring forward** ..... ：…を前向きにもたらす
- ～ **bring A into B**：～はBの中にAをもたらす

1. It **brings up** some questions.
   それはいくつかの問題を浮上させる。
2. Another issue the author **brings up** is the intended closure of contaminated facilities.
   この著者が持ち出すもう1つの問題は，汚染された施設の意図的な閉鎖である。
3. The idea **being brought forward** in this research is to measure several of these general parameters.
   この研究において前進的にもたらされているアイデアは，これらの一般的なパラメータのいくつかを測るということである。
4. The phenomena and impacts of acid rain and energy development, the products of our industrialized society, **will bring** often conflicting economic and environmental concerns **into** the public eye.
   酸性雨とエネルギー開発の状況と影響は，我々の工業化された社会の産物であり，公衆の目には，しばしば相いれない経済と環境の問題と写るであろう。

Build (Building)：建設；作成する；構築する；立脚する；蓄積

1. A few chemicals were selected from each group to **build** summary tables for each process.
   それぞれのプロセスに要約表を作成するために，少数の化学物質を各グループから選んだ。

- **build on** ..... ：立脚する
- **built into** ..... ：構築する

2. Insofar as past state biological criteria efforts grew out of biological monitoring efforts, future efforts **will** likely **build on** current monitoring programs.
生物モニタリング努力から発した過去の州の生物クライテリア努力に関する限り，将来の努力は，現行のモニタリングプログラムに立脚するであろう。
3. VanLandingham (1976) claims that over 3000 species have autecological information in the literature, which **can be built into** an autecological database appropriate for the periphyton in a region.
VanLandingham (1976) によれば，文献には3000種以上の個生態学的情報が収められており，1地域における付着生物に適した個生態学的データベースを構築しうる。

- **building.....**：…を建設すること；…を確定すること

4. This has allowed small towns to begin **building** treatment plants that will attain the effluent limit specified in the new water quality standards.
これにより，小都市で，新しい水質基準で明記された排出限界を満たすような処理場の建設を開始できるようになった。
5. These procedures could be based upon **building** a weight of evidence case for modifications.
これらの手順は，修正のための証拠事例の重要性の確定に基づいて行う。

- **buildup of.....**：の蓄積

6. This work will aid in making predictions about long-term **buildup of** organic layers on the stone.
この研究は，石への長期にわたる有機層の蓄積を予測するのを助けるであろう。

Calculate：計算する

1. Scales are collected so that growth rates as well as general size **can be calculated** and compared.
   鱗を採集して生長度や一般的体寸が計算・比較されうる。
2. These are then converted to phi values as in Kudo (1962), and mean particle size **is calculated**.
   これらは，次に Kudo(1962) におけるような phi 値に変換され，粒子サイズ平均が計算される。
3. Using all the estimated rate constants, the **calculated** accumulation of CO_2 is plotted in Figure 2.
   すべての推定速度定数を使って計算された CO_2 の蓄積量を図2に図示する。

   - ～ **is calculated using** ....：～は … を用いて計算される
   - ～ **is calculated by summing** .....：
     … を合計することによって～が計算される
   - ～ **is used to calculate A as a function of B**：～は A を B の関数で計算するのに使われる

4. The SCI **is calculated using** the following steps:
   SCI は，以下のステップを用いて計算される。
5. The IR of mixtures **can be calculated by summing** the reactivity contributions of the components.
   構成要素の反応性貢献度を合計することによって，混合物の IR が計算ができる。
6. The model **can be used to calculate** in-lake concentration of total phosphorus as a function of time.
   モデルは，湖の全リン濃度を時間の関数で計算するのに使うことができる。

   - ～ **is calculated from** .....：～は … から計算される

7. The sedimentation term **is calculated from** the above discussed transport capacity.
   堆積の項は上で論じた輸送容量から計算される。
8. The reactivity of a mixture of VOCs **can be calculated from** the reactivity

of the individual components.
VOCs の混合物の反応性は，それらの個々の成分の反応性から計算することができる。
9. The observed value of the K is definitely higher than **calculated from** Equation 2.
観測されたKの値は式2による計算値より確かに高い。
10. The measured reactivity of a mixture was compared to that **calculated from** the sum of the measured reactivity of the mixture's individual components.
測定された混合物の反応性を，測定された混合物の個々の構成物質の反応性の合計から計算された反応性と比較した。

Calibration：キャリブレーション；較正

1. The Mlwb does not require **a spatial calibration** prior to use.
MIwb は，使用前に空間的較正を必要としない。
2. Subsequent **calibration of the models** must address the ability to differentiate between impaired and nonimpaired sites.
その後のモデルのキャリブレーションは，損傷地点と無損傷地点を差別化できる機能に取り組むべきである。
3. **Calibration procedures** allow normalization of the effects of stream size, so index scores can be compared among streams of different sizes.
較正手順は，水流規模による効果の標準化を可能にするので，異なる規模の水流において指数スコアを比較できる。
4. A 10-mL sample of **calibration standards** was contacted with 1 mL of hexane in conical bottom test tubes followed by shaking for 2–5 minutes by hand.
コニカルボトム試験管で，10 mL の較正標準溶液サンプルに1 mL のヘキセンを加えてから，2-5分間手で振った。

- **～ is calibrated with .....**：～は…を用いて較正された
- **calibrations is performed**：キャリブレーションが行われる

5. The model **has been calibrated with** data collected at Woods Lake.
このモデルは，Woods湖で収集されたデータを用いて校正された。
6. It is better to identify large areas over which **calibrations will be performed** rather than small areas.
キャリブレーションを行うに当たり，狭い地域よりも広い地域を用いた方が良い。

Capable (Incapable)：有能な；できる（できない）

- ~ **is capable of** ..... ：～は…することが出来る
- ~ **is incapable of** ..... ：～は…することができない
- ~ **have been shown capable of** ..... ：
  ～によって…ができることが示されてきた

1. In most aquatic systems these processes **are capable of** reducing free copper levels to very low values.
   ほとんどの水生システムにおいて，これらのプロセスは遊離銅濃度を非常に低い値に下げることができる。
2. The model **is capable of** correlating growth data from many situations in a satisfactory manner.
   このモデルは，多様の条件下からの成長データを満足のいく方法で関連づけることができる。
3. The ecoregion of Oregon has been so extensively ditched and drained that many of the small streams in this region **are incapable of** supporting a WWH use.
   オレゴン州の生態地域は，徹底的に溝掘りと排水が行われてきたので，この地域の小河川の多くがWWH使用に準じない。

   ★ここでは，WWH は Warm Water Habitat の短縮語として使われています。

4. The resultant interparticle collisions **are capable of** including dramatic changes in surface morphology, composition, and reactivity.
   結果として生じる粒子間の衝突は，表面形態，成分，反応性などに劇的な変化を生じさせることができる。
5. Photocatalytic oxidation processes **have been shown capable of** mineralizing a wide range of naturally occurring and anthropogenic organic compounds.
   触媒光反応による酸化プロセスによって，広範囲の自然態物質や人工有機化合物を無機物に変化させることができることが示されてきた。
6. The model has been judged potentially **capable of** meeting the user's needs.
   このモデルは，潜在的に使用者の必要性を満たすことが出来ると評価されてきた。

Capacity：容量

1. The sedimentation term is calculated from the above discussed transport

capacity.
   堆積の項は上で論じた輸送容量から計算される。
2. The crucial question of whether the water body becomes acidic depends on the magnitude of the watershed's acid neutralization **capacity**.
   水流が酸性になるかどうかという重要な問題は、流域の酸中和容量による。
3. Patterns in human activities often vary as a function of ownership or political unit, as well as ecoregion, which reflect differences in potential and **capacity**.
   人間の活動は、しばしば所有単位や政治単位、そして生態地域に相関して異なる。生態地域は、可能性と容量における差を反映している。

- **with a capacity of** ..... ：容量が…の
- **～ have the capacity of** ..... ：～は…の能力を持っている

4. Cell 4 contains a 4 m wide by 8 m long pool **with a capacity of** about 200,000 liters.
   セル4は幅4メートルに長さ8m、容量が約200,000リットルのプールを内蔵している。
5. In the event of a spill emergency, the gate valves at the B-1 bypass **have the capacity of** diverting the creek flows to Pond B-1.
   漏出緊急時の場合、B-1バイパスにある水門弁は、小川の流れをB-1池にそらす能力を持っている。

Care：配慮；入念に

- **care must be taken to** ..... ：
  …するよう配慮すべきである；…するよう注意を払うべきである
- **care was taken to** ..... ：…するよう注意が払われた

1. **Special care must be taken to** exclude anomalous sites.
   異常地点を排するよう特に配慮すべきである。
2. In selecting reference sites, **care must be taken to** avoid including anomalous stream sites and watersheds.
   対照地点の選定に際して、異常な水流地点および流域を含むことを避けるよう配慮しなければならない。
3. Critical to the process of interpreting and integrating the source material is the **care that must be taken to** avoid defining regionalities of particular ecoregion components.
   資料の解釈・統合過程できわめて重要なのは、特定の生態系構成要素で地域性を

画定するのを避けるよう配慮すべきことである。
4. It is unlikely that extreme **care was taken to** prevent deposition of DDT into the Silver River during the 27 years of plant operation.
27年の工場操業期間中，Silver川へのDDTの堆積を妨ぐために過度の注意が払われたという事は，ありそうもない。

- **with considerable care**：入念に

5. These choices must be made **with considerable care** and documented so that they do not include fundamentally different communities.
こうした選択をする際，根本的に異なる群集が含まれないようにするため，かなり入念に行って文書化しなければならない。

Carry out (See also **Investigation**, **Study**)
：実施される；行われる（→ **Investigation**, **Study**）

1. Koike and Ikeda (2002) **carried out** an investigation to identify industrially significant nitrogen.
KoikeとIkeda (2002)は工業的に重要な窒素を識別するための調査を実施した。
2. Cheng et al. (1998) **carried out** experiments to compare the metal uptake in active and non-active sludge.
Chengら(1998)は，活性汚泥および非活性汚泥での金属取り込みを比較する実験を実施した。

- **～ is carried out**：～が行われる

3. Relatively few statistical sample surveys **have been carried out** on lake and streams.
湖沼・水流について行われてきた統計的サンプル調査は，相対的にほとんどない。
4. Biological monitoring **should be carried out** in concert with chemical/physical monitoring.
生物モニタリングは化学的/物理的モニタリングと調整して実施されるべきである。
5. Efforts to assess, research, and manage the ecosystems **are** normally **carried out** via extrapolation from data gathered from single-medium/single-purpose research.
生態系を評価・研究・管理する努力は，通常，単一媒体／単一目的の研究で収集されたデータからの推定によって行われる。
6. A series of batch tests **was carried out** to test the hypothesis that the same enzymes are used to attack chlorophenols with the same position-specific dechlorination reactions, resulting in competition between these

compounds.
位置特異的な同じ脱塩反応によって，同じ酵素が700フェノールを攻撃し，これらの化合物の間の競合反応をもたらす，という仮説をテストするために一連のバッチ試験が行われた。

Case：ケース；事例；場合

1. It likely represents **a worst case** under low flow conditions.
   それは多分，低流量状態下で最悪のケースを代表する。
2. We found good agreement between results in about 20 % **of cases** when effluent toxicity was assessed, and in about 30 % **of cases** when mixing zone toxicity was measured.
   流出毒性を評価した際には約20％の事例と，混合区域毒性を測定した際には約30％の事例の結果でかなりの一致が見られた。

   ● **on a case-by-case basis**：ケースバイケース；状況によって

3. The latter can be dealt with **on a case-by-case** or site-specific **basis** if necessary.
   後者は，必要ならケースバイケースもしくは地点特異的ベースで対処されうる。
4. Comparison with other technologies **on a case-by-case basis** is limited.
   「ケースバイケース」ベースでの他の技術との比較には限りがある。
5. Persons involved in possible field tests must make this determination **on a case-by-case basis**.
   可能な現地テストに関係している人々は，状況によってこの決定をしなくてはならない。
6. The Division of Water Pollution Control (DPC) is using upstream reference sites to assess stream impacts **on a case-by-case basis**.
   水質汚濁防止局(Division of Water Pollution Control: DPC)は，上流の参照地点を用いて水流の影響をケースバイケースで評価している。

   ● **in the case ......**：…の場合；…の事例において
   ● **in the case of .....**：…の場合

7. **In the case** illustrated here, about 82 % of the sites do not achieve their respective goals.
   ここで示した事例において，地点の約82％がそれぞれの目標に達成していない。
8. **In the case of** Iowa, these regulatory water quality criteria are supported by the most extensive sampling, assessment, and implementation program in the nation.

Case

アイオワ州の場合，規制的な水質クライテリアは，米国において最も綿密なサンプリング，アセスメントおよび実施プログラムに裏づけられている。

- **in every case**：事例すべてにおいて
- **in nearly every case**：ほぼ決まって；ほとんどの場合

9. **In every case** where impairment was indicated, we would not have been able to assign the source without bracketing the discharge with upstream–downstream sites.
損傷が示された事例すべてにおいて，我々は，上流地点−下流地点での排出を一括せずには損傷源を決められなかっただろう。

10. Response to the question **in nearly every case** has been that ecoregions are the desired regional framework.
この質問への返答は，ほとんどの場合，生態地域が望ましい地域枠組みだというものだった。

- **in most cases**：ほとんどの場合
- **in most other cases**：他の大半の事例では

11. **In most cases**, only a limited amount of stream water quality data is available.
ほとんどの場合，ほんの限られた数の水流水質データが利用可能である。

12. **In most cases**, both anions and cation exchange is required to achieve the desired degree of water purification. **In some** installations, the anion and cation resins are contained in separate tanks. **In others**, they are mixed together.
ほとんどの場合，望まれる度合いの水の浄化を達成するには，アニオン交換とカチオン交換の両方が必要とされる。一部の施設では，アニオン・カチオン交換樹脂が別々のタンクに詰められており，また他の施設では，それらが混合されている。

13. **In most other cases**, aquatic life use decisions are made with biosurvey and habitat data.
他の大半の事例では，水生生物使用の決定が生物調査および生息場所データでもって下される。

- **In any case**：いずれにせよ；いずれにしても
- **In many cases**：多くの場合
- **In particular case**：特定の場合
- **In such case**：このような場合
- **In some cases**：ある場合には；一部の場合
- **In other cases.**：他のケースで，

14. **In any cas**e, there are no continuous records of ecological change over the last 2000 years.
いずれにしても，過去2000年にわたっての生態変化の継続した記録が存在しない。
15. **In many cases** they can be avoided by very careful geographic analyses and reconnaissance.
多くの場合，こうした事態は，非常に入念な地理的分析と探査によって避けることができる。
16. **In there particular case**, the exposed population grew slower.
それら特定の場合，(毒物に)さらされた個体数は繁殖が遅かった。
17. This is especially true **in such cases** where a water body may be perceived as being at risk due to new dischargers.
ある水域が新排出者のせいで危険な状態にさらされてると見なされるかもしれない場合，これは特に事実である。
18. **In some cases** scales are collected so that growth rates as well as general size can be calculated and compared.
一部の場合，鱗を採集して生長度や一般的体寸が計算・比較される。
19. **In some cases**, TMDL development can be straight-forward and relatively simple. In other cases, a phase approach may be more appropriate.
ある場合には，TMDL開発は単純で，そして比較的簡単であり得る。他のケースでは，段階的アプローチがいっそう適切であるかもしれない。
20. **In some cases**, biological impairment criteria may detect water quality problems that are undetected or underestimated by other methods. In these cases, biological impairment criteria may be used as the sole basis for regulatory action.
一部の場合には，生物損傷クライテリアが他の方法で探知されなかった水質問題，もしくは過小評価された水質問題を探知できるかもしれない。これらの場合，生物損傷クライテリアは，規制措置の唯一の根拠となろう。

● **as is the case.....** : …のように

21. Where subregions represent bands of different mosaics of conditions, **as is the case** in some western mountainous ecoregions, it may be necessary to choose reference sites that comprise watersheds containing similar proportions of different subregions.
一部の西部山脈生態地域のように小地域が異なる状態モザイクのバンドを表した場合には，異なる小地域の類似部分を含んだ流域からなる参照地点を選ぶ必要があろう。

Categorize (Category)：類別する；列挙する（カテゴリー；範疇）

1. These would **be categorized as** diseases, anomalies, or metabolic processes.
   これらは，疾病，異常，または代謝過程に類別される。
2. Metrics **are categorized in** broad classes of community structure, taxonomic composition, individual condition, and biological processes。
   メトリックは，生物の群集構造，分類群構成，個体状態，ならびに生物学的過程の広範な等級において類別される。

   - ～ **fit into** ..... **categories** ：～は … の範疇に収まる
   - ～ **are divided into** ..... **categories** ：～は … のカテゴリーに分けられる

3. Objectives of volunteer monitoring programs **fit into** three general **categories**.
   ボランティアモニタリングプログラムの目標は，一般的な3つの範疇に収まる。
4. Data quality **can be** roughly **divided into** three **categories** based on program objectives:
   データの質は，大体プログラム目標に基づいて3つのカテゴリーに分けられる。
5. Biological processes **can be divided into** several **categories** for consideration as potential metrics.
   生物学的過程は，潜在的メトリックとして考慮するに当たり，いくつかの範疇に分けることができる。

Cause：原因

1. The overall objective of this research is to conduct basic studies into possible **causes of** biological nitrification process instability.
   この研究の総合目的は，生物硝化プロセス不安定性を起こしうる原因に関する基礎研究を行うことである。

   - ～ **is** ..... **caused by** ..... ：～は … によって引き起こされる
   - ～ **is the cause of** ..... ：～は … の原因である
   - ～ **cause** ..... ：～は … をひきおこす
   - **as a cause for** ..... ：… の原因として
   - **the cause of** ～ **is** ..... ：～の原因は… である

2. The odor in the untreated air **is caused** to a large extent by reactive chemical species.

浄化されていない空気の臭気は，主に反応性のある化学種によって引き起こされる。
3. The excess nutrients **are the primary cause of** the reduced oxygen levels observed in the lake.
この湖で観察される酸素レベルの減少は，過剰栄養が主要な原因である。
4. It has always been assumed that the tailings provide a potential **to cause** degradation of groundwater quality.
選鉱くずが地下水質の劣化を起こす可能性があるとは常に想定されてきた。
5. Thus, loss of membrane integrity was not proven **as a cause for** the observed difference in removal between Tests 1 and 2.
それゆえ，膜質の低下が，試験1と試験2の間で観察されたの除去率の相違の原因であるとは証明されなかった。
6. Stream habitat modification was identified **as the** third leading **cause of** aquatic life impairment in the 1992 Michigan Water Resource Inventory.
水流生息地改変は，1992年のミシガン州水資源調査一覧において，水生生物損傷の第三の要因と確認された。

Center (Central) ：集中する (中心的)

1. Much of this speculation **has centered on** arsenic.
かなりの疑いがヒ素に集中してた。
2. Our interest **centers on** the land-based portion of the N cycle.
我々の関心は窒素循環の地上ベースの部分に集中している。
3. Although the effort **centered on** surface water quality, limited sampling was conducted on tailings deposits.
努力が表流水の水質に集中したけれども，サンプリングが選鉱屑堆積層でわずかではあるが行なわれた。

- ～ **is of central importance** ：～は中心的重要性を持つ

4. The origin and flow of groundwater in the alluvium **is of central importance** in this study.
この研究では，沖積層での地下水の水源と流量が中心的な重要性を持っている。

Century：世紀

- **over a century** ：1世紀にわたり

1. Autecological information for many algae, particularly diatom, has been recorded in the literature for **over a century**.

多くの藻類，特に珪藻についての個生態学的情報は，1世紀にわたり文献に記録されてきた。

2. **Over the past two centuries**, vast quantities of energy have been used.
   これまでの2世紀にわたって，膨大な量のエネルギーが使われてきた。

   - **by the turn of this century**：この世紀の到来時までに
   - **since the turn of the century**：今世紀初頭以降
   - **after the turn of the century**：この世紀の到来時の後に

3. **By the turn of this century**, it is expected that ......
   この世紀の到来時までに，…が期待される。

4. The improvement is even more remarkable when conditions **since the turn of the century**, when the river lacked any fish life for a distance of nearly 40 mile, are considered.
   この川で約40mileにわたり魚類が生息していなかった今世紀初頭移行の状態を考慮すれば，この改善は，一層明らかである。

   - **Early this century**：今世紀初期
   - **to the early part of this century**：今世紀前半へ
   - **From ..... to the end of the century**：…から，その世紀の終わりまで
   - **within a century**：1世紀以内に
   - **dating from the 19th century**：19世紀から

5. **Early this century**, before the scientific community conceived of polymer synthesis, ...
   今世紀初期，ポリマー合成が科学社会のなかで考えだされる以前に，.....

6. Some of the roots of the sample survey approach can be traced **to the early part of this century**.
   サンプル調査手法のルーツの一部は，今世紀前半へ遡れる。

7. **From** the 1960's **to the end of the century**, many gold mining companies discharged tailings to Whitewood Creek or its tributaries.
   1960年代からその世紀の終わりまで，多くの金採鉱企業が選鉱廃棄物をWhitewood Creekあるいはその支流に排出した。

8. High iron low-pH water will eventually occupy the entire tailings mass, probably **within** several decades or a **century**.
   おそらく数十年あるいは1世紀以内に，高濃度の鉄分を含有しpHが低い水が結局は選鉱廃棄物の全てを占めることになるでしょう。

9. The ecoregion is affected by significant and widespread historical land use and stream channel modifications **dating from the 19th century**.
   この生態地域は，19世紀から，甚大かつ広範な史的土地利用および水流路改変

の影響を受けている。

Certain (Certainly)：特定の；ある；一部の；確かな（確かに）

1. It enhances the importance of **certain** model parameters.
   それはある特定のモデルパラメータの重要性を高める。
2. There are **certain** situations in which each type is appropriate for use.
   類型それぞれが使用上適切なある特定の状態がある。
3. The modeling of **certain** constituents will not be nearly as critical to answering study question
   ある特定成分をモデル化することは，研究課題に答えることほど重要ではないであろう。
4. It has long been known that **certain** groups of macroinvertebrates are more pollution-tolerant than others.
   大型無脊椎動物の一部のグループが他集団よりも汚染物質耐性が強いことは，かなり以前から知られている。
5. This variation in terminology has caused **a certain amount of** confusion.
   術語学におけるこのずれは，ある程度の混乱を起こしてきた。

   - **~ is somewhat uncertain**：～は，いくぶん不確実である
   - **it is almost certain that** .....：…は，ほとんど確実である

6. Certain Results from these calculations **are somewhat uncertain**.
   これらの計算からのある結果はいくぶん不確実である。
7. **It is almost certain that** the volatilization characteristics of the pesticide can be predicted.
   ほとんど確実に殺虫剤の気化作用の特徴が予測できる。
8. However well founded the legal basis for water quality criteria and their implementation, **it is almost certain that** they will be challenged in various ways.
   水質クライテリアとその実施についての法的根拠がいかに十分に確立していても，様々な形で異議申立てされることは，ほぼ確実である。

   - **certainly**：確かに

9. His article **certainly** fell into that category.
   彼の論文は確かにその部門に分類された。
10. The fit between the model and data is **certainly** good.
    モデルとデータは確かに一致している。
11. **Certainly** streams such as this one should be protected and not be allowed

to degrade to standards and expectations set for streams typical of most of the region.
確かにこのような水流は，保護されるべきであり，その地域の大半の典型となる一連の水流についての基準や期待値を下げることは許されない。

Certainty：確かさ；確実性；確実度

- with a known degree of certainty：既知の確実度で
- with a reasonable degree of certainty：妥当な確実度で

1. Quantifiable estimates of land use can be made **with a known degree of certainty**.
   土地利用の計量可能な推定は，既知の確実度でもって得られる。
2. Rankin (1989) developed a relationship between the IBI and the QHEI sufficient to forecast nonattainment due to habitat degradation **with a reasonable degree of certainty**.
   Rankin (1989) は，生息場所劣化による未達成を妥当な確実度で予測しうる IBI と QHEI との関係を定めた。

Challenge (challenging)：課題；挑戦（課題）

1. There are technical **challenges** for industry during the development of state-specific water quality criteria.
   州別の水質クライテリアの設定において，業界にとって技術的課題が存在する。
2. The real **challenge** for industry and state agencies is that water quality criteria will vary tremendously both among states and within states.
   業界と州機関にとって現実的な課題は，水質クライテリアが州間で，また州内でそれぞれ大幅に異なることである。

- ～ present a challenge：～は課題をもたらす
- ～ poses a challenge：～は課題をもたらす

3. They reflect practical situation and as a result **present** the greatest **challenge** with respect to modeling.
   それらは実際的な状況を反映し，その結果としてモデリングに関して最大の挑戦となる。
4. The development and adoption of biocriteria into state water quality standards **poses challenges** as well as opportunities.
   州水質基準における生物クライテリアの設定・採択は，課題ばかりでなく，機会

ももたらす。
> ● **challenging**：課題

5. Understanding the underlying concepts and theoretical assumptions of biocriteria may be **a challenging** to environmental managers that have little or no experience with biological assessments.
生物クライテリアの根本的概念と理論的仮定を理解することは，生物アセスメントの経験に乏しい，もしくは経験のない環境管理者にとって課題になろう。

Chance：確率；可能

1. By offering a more robust framework, the **chance** for deriving an inappropriate criterion is greatly reduced.
より確かな枠組みを提供することによって，不適切なクライテリアが導出される確率が大幅に低下する。
2. If the differences can be associated with management practices that have a **chance** of being altered, it seems wise to combine the regions for the purposes of establishing water quality criteria.
仮に，その差違が変化可能である管理慣行に関連している場合には，水質クライテリア設定のために諸地域を統合した方が賢明であろう。

Change (Changing)：変更（変化すること）

1. Their activities may relate to thermal **changes**, flow **changes**, sedimentation, and other impacts on the aquatic environment.
彼らの活動は熱変化，流量変化，堆積，それに水生環境への影響と関連づけられるかもしれない。
2. There is also a need to develop models which can be used to predict water quality **changes** within distribution systems.
また，給水システム内での水質変化を予測するために使うことができるモデル構築の必要性もある。
3. Due to variability in microbial densities between samples, these **changes** were not statistically significant.
微生物の密度はサンプル間で多様であるため，これらの変動は統計的に有意ではない。
> ● **changes in .....**：…における変化

4. When used in a monitoring program, these methods can be used to help

diagnose the cause **of changes in** the aquatic system.
モニタリングプログラムで用いた場合，これらの手法は，水生系における変化原因を判定するのに役立つ。

5. Such information has proven very useful for assessing **changes in** climate, salinity, pH, and nutrients.
こうした情報は，気候，塩度，pH，養分における変化を評価するため非常に有用だと判明している。

6. This allows for the analysis of **incremental changes in** aquatic community performance over space and time.
これは，水生生物群集実績における時間的・空間的な漸増変化の分析を可能にする。

- **This is a change .....** ：これは … の変化である
- **There has been a change** ：変化がみられた

7. In any case, **there are no** continuous records of ecological **change** over the last 2000 years.
いずれにしても，これまでの 2000 年にわたって生態系変化の継続した記録がない。

8. Increase in concentrations was observed for TCE solution with the probe, but **there were no measurable changes** for TCA solutions.
濃度の上昇が TCE 溶液で観察されたが，TCA 溶液では測定可能な変化がなかった。

9. Although **this is a substantial change in** species, the assessment of stream condition based on indices is much less variable.
これは，相当大幅な生物種変化だか，指標に基づく水流状態アセスメントは，可変性がはるかに弱まる。

10. For each index **there has been a** significant positive **change** over time at most sites.
各指標について，大半の地点で，経年的に有意なプラス変化が見られた。

- ～ **underwent changes** ：～は変化した

11. The manner in which biological data was analyzed also **underwent changes**.
生物学的データの分析方法も変化した。

- ～ **change** ：～は変化する

12. The arsenic concentrations **will not change** very much in the near future regardless of any channel shifting.
水路が移動したとしても，ヒ素の濃度は近い将来ではそれほど変化しないであろう。

13. The periphyton assemblage **may change** up to 30 % in taxa present from year to year within the spring−summer period.

付着生物群は，春・夏期には年ごとに 30 ％もの群集変化が生じうる。
14. In response to these findings, the Commission **changed** the manner in which it assigned aquatic life uses to unclassified water bodies
これらの所見に応えて委員会は，未分類水域への水生生物利用の指定方法を変更した。

- **changing** ...... : … を変化させて

15. Experiments were run **changing** residence time in the UV chamber in an attempt to determine if certain chemicals in the system are more completely destroyed as the residence time is increased.
滞留時間が増えるにつれて，システム内のある特定の化学物質がより完全に分解されるかどうかを明らかにするために，UV 反応器の滞留時間を変えて実験が実施された。
16. There is general acceptance of the notion that organic compounds may play a major role in **changing** speciation of trace metals.
有機化合物が，微量金属の種の変化に主要な役割をしているであろうという概念が一般的に受け入れられている。
17. These methods provide an indication of what specific environmental characteristics are **changing** over time.
これらの手法は，どの環境特性が経時的に変化しているか示す指標となる。

Characteristics : 特徴

1. The primary concern in selecting the two sites is to assure that the physical **characteristics** are as similar as possible.
2 地点の選定における主な関心事は，物理的特性ができる限り似るよう努めることである。
2. The Forest Inventory Assessment also uses a survey approach and focuses on **characteristics** related to timber production.
森林資源一覧アセスメントもサンプル調査を用いており，材木生産に関わる特性に主眼が置かれている。
3. Physical **characteristics** of individuals that may be useful for assessing chemical contaminants would result from microbial or viral infection, some sort of tetragenic or carcinogenic effects during development of that individual.
化学的汚染アセスメントに有用だと思われる個体の身体特性は，細菌・ウイルス感染や，その個体の発生中における何らかの催奇形影響または発癌影響に起因する。

- **..... as characteristics of .....**：…の特徴として
4. A settling velocity of 1.2 m/d is assigned **as characteristics of** the Gold River solids.
   1.2m/d といった沈降速度が Gold 川の固形物の特徴として使われている。

Characterize（Characterization）：
特徴づける；みなす；特色とする；特性把握する（特性把握）

1. At present, few states **can characterize** even a small fraction of their water resources with biological survey data.
   現在，生物調査データで水資源のごく一部でも特性把握できる州は，きわめて少数である。
2. Regions **characterized** by karst topography, extensive sandy soils, or excessive aridity are examples of areas where watersheds are less important.
   カルスト地形，広大な砂質土壌，極度の乾燥を特色とする地域は，流域があまり重要でない区域の例である。
   - **~ is characterized by .....**：
     ～の特徴は … である；… によって特徴づけられる
   - **~ can be characterized as .....**：～は … と特性把握されうる
3. The most heavily polluted zone **was characterized by** only a few highly tolerant species.
   最もひどく汚染された地域は，少数の強い抵抗力のある種だけによって特徴づけられた。
4. A number of these attributes **can be characterized by** metrics within four general classes.
   これら多数の属性は，4つの一般的的等級におけるメトリックによって特性把握されうる。
   - **to characterize .....**：… を特性把握するため
   - **to more accurately characterize .....**：
     … をより正確に特性把握するため
5. Samples taken from riffles or runs are sufficient **to characterize** the stream.
   瀬または急流から採取されたサンプルで，十分にその水流を特性把握できる。
6. This is not an attempt **to characterize** pristine or totally undisturbed

environmental conditions.
これは無傷の,すなわち全く無攪乱の環境状態を特性把握しようとすることではない。
7. In 1970–1971, a series of studies by the U.S. EPA (1973) were undertaken **to characterize** the discharge of tailings to Wood Creek.
1970 から 1971 までに,Wood Creek への選鉱廃物の排出の特性把握するために一連の研究が U.S. EPA (1973) によって着手された。
8. A better tool was needed **to more consistently and accurately characterize** the aquatic communities.
水生生物群集をより整合的かつ正確に特性把握するためには,より優れたツールが必要であった。

- **Characterization** ：特性評価；特性把握

9. Our unpublished data suggest the methods to be of promise for the **characterization of** sediments.
我々の未発表のデータは,堆積物の特性評価のために有望になるであろう方法を示唆している。

Choice ： 選択

1. **The choice of** groups to be identified and analyzed is related to the measures that will be used.
識別・分析されるべき集団の選択は用いられるであろう測定方法に関連している。
2. The assessment of biological condition using a suite of metrics to define biocriteria is rapidly becoming **the method of choice** among state water resource agencies.
生物クライテリアを定めるための一連のメトリックを用いる生物学的状態アセスメントは,急速に州の水資源担当機関にとって選択手法になりつつある。

- **choices is made** ：選択が行われる

3. Obviously, **these choices must be made** with considerable care and documented so that they do not include fundamentally different communities.
明らかに,こうした選択は,根本的に異なる群集が含まれないようにするため,かなり入念に行われ,文書化されなければならない。

Choose (Choosing) ： 選ぶ；選択する (選択すること)

1. States **may choose** to concentrate on one or more of these uses of

biological criteria when developing their programs.
諸州は，プログラム策定時に，生物クライテリアの上記用途のうち1つないし複数に専念することを選ぶかもしれない。
2. It may be necessary **to choose** reference sites that comprise watersheds containing similar proportions of different subregions.
異なる小地域の類似部分を含んだ流域からなる参照地点を選ぶ必要があろう。
3. The range of inhibition concentrations tested **was chosen** arbitrarily, depending upon the effect of each.
それぞれの影響に基づいて，この試験での阻害濃度の範囲が恣意的に選択された。
4. All of the pesticides **chosen** for this study are organophosphorus compounds.
この研究のために選ばれた殺虫剤のすべてが有機りん化合物である。
5. The approach **chosen** is based on mass balance and uses the simulation techniques described by Sasaki et al.(2001).
選択された手法は，物質収支に基づいていて，そして Sasaki ら (2001) によって記述されたシミュレーション技巧を使う。

- **choosing of** ..... ：…を選択すること

6. The most important aspect is the **choosing of** proper initial concentrations for.....
.....のために最も重要なことは，適切な初期濃度を選択することにある。

Cite：引用する；記する

1. In the results **cited** above, chemical and biosurvey results agreed 58％ of the time.
前記の結果においては，化学的調査結果と生物的調査結果の一致率は，58％であった。
2. Environmental managers within industry can become well acquainted with basic design concepts by reviewing the four documents **cited above**.
業界内の環境管理者は，上記4つの文書を繙けば，基礎的な設計概念を十分に把握することができる。

Clarity：明確さ；明白

- **For the sake of clarity** ：明快にするために

1. **For the sake of clarity** in presenting the test results, most of the details

concerning the theoretical background of the tests and the analytical procedures are provided in Appendix B.
テスト結果の発表にあたってそれを明快にするために，試験と分析手法の理論的な背景の詳細の多くを付録Bに示す。

Classify (Classification)：分類する(分類)

1. Protocol III is used **to classify** streams more finely as unimpaired, or as slightly, moderately, or severely impaired.
   プロトコルIIIは，水流を無損傷，軽度損傷，中度損傷，重度損傷と，より細かく分類するため用いられている。
2. In an attempt **to classify** lakes, streams, and wetlands, Omernik et al. (1987) delineated ecoregions of the counterminous United States based on mapped landscape characteristics.
   湖沼，水流，湿地を分類するにあたり，Omernikら(1987)は，地図作製された景観特性に基づき隣接した米国諸州の生態地域を表示した。

   - ～ **is classified by** ……：～は…によって分類される
   - ～ **can be classified into** ……：～は…に分類される
   - **classifying**……：…の分類

3. Samples **were classified by** location and cooled.
   サンプルは場所によって分類されて，そして冷やされた。
4. The contaminated areas **can be classified into** three main groups.
   汚染地域は3つの主なグループに分類することができる。
5. Existing herbicides transport models **can be classified into** physical models.
   既存の除草剤輸送モデルは物理的モデルに分類される。
6. An attempt to use the scheme for **classifying** aquatic ecosystems proved unsuccessful.
   水生生態系分類方式を用いる試みは，失敗に終わった。

   - **classification**：分類

7. The initial **classification** scheme adopted in this work appears in Table 2.
   この研究で採用された最初の分類案は表2に示されている。
8. All our attempts at **classification** are human constructs that help us understand a spatially and temporally varying landscape.
   分類における我々の試みは，すべて人為的構成物であり，空間的・時間的に異なる景観を理解するのに役立つ。

9. The opposing concepts of diversity and similarity are the bases of most **classifications**.
   多様性と類似性という対立的概念が大半の分類の基本にある。
10. While these **classifications** were based on ecological attributes, the criteria associated with each were entirely chemical/physical.
    これらの分類は，生態的属性に基づいていたが，それぞれに関連するクライテリアは全く化学的/物理的なものであった。

Clear（Clearly）：明確な；明らかな；明白な（明確に；明らかに；明白に）

- **It is clear that** ..... ：… ということは明白である
- **It is clear from ～ that** ..... ：～ から… ということは明らかである
- **It is not clear whether** ..... ：… かどうか明らかでない

1. From the above, **it will be clear that** ......
   上記から，… は明確になるであろう。
2. In any event, **it was very clear that** .....
   とにかく，… のことは非常に明確であった。
3. **It was clear that** a better tool was needed to more consistently and accurately characterize the aquatic communities.
   水生生物群集をより整合的かつ正確に特性把握するための優れたツールが必要なことは明らかであった。
4. **It is not clear from** their results **whether** the effect of ～ is due to .....
   ～の影響が … によるものかどうかは，彼らの結果からは明らかではない。

- **～ become clear** ：～が明確になる
- **～ make it clear that** ..... ：～は … ということを明らかにする
- **What makes it clear is that** ..... ：明らかなことは … ということである

5. The ecoregion patterns **became clear**.
   生態地域パターンが明確になった。
6. The model **makes it clear that** .....
   このモデルは … を明らかにする。
7. **What** this example **makes clear is that** PCBs can be degraded by microorganisms.
   この例から明らかになることは，PCBsが微生物によって分解可能だということである。

- **clear evidence**（See **Evidence**）：明確な証拠（➡ **Evidence**）
- **clear understanding**（See **Understanding**）：

明確な理解（➡ **Understanding**）：
- **clearly**：明らかに

8. Figure 6 shows **clearly** that .....
   図6が明らかに…を示している。
9. The result **clearly** shows that .....
   その結果は…ということを明らかに示している。
10. The fact provides **clearly** that .....
    この事実は…を明らかに提示している。
11. **Clearly**, much work needs to be done to evaluate and polish the proposed technique.
    明らかに，かなりの研究が，提案された技巧を評価して，そしてそれに磨きをかけるためには必要である。
12. **Clearly**, other chemical, toxicological, physical, and source information must be used as part of the overall assessment process.
    明らかに，他の化学的，毒物学的，物理学的および由来の情報は，アセスメント過程全体の一部として用いられなければならない。

- **unclear**：不明確な
- **It is unclear if .....**：…であるかどうかは不明確である

13. **It is likewise unclear** how well limestone reactors will neutralize more highly concentrated solutions.
    石灰岩を使った反応器が，より濃縮された溶液をどれほどよく中和するかは同じく不明確である。
14. **It is unclear if** such organic buildup will increase coliform concentrations in treated waters.
    このような有機物の蓄積が，処理水中の大腸菌の濃度を増やすかどうかは不明確である。

Clearance (Clearing)：排出；清掃（清掃すること）

1. Knowledge of the rate constants governing the uptake and **clearance** of chemicals in fish is required.
   魚体内で，化学物質の吸収・排出を左右している速度定数の知識が必要とされる。
2. Stream channelized under the auspices of the Kentucky Drainage Law are subject to routine maintenance activities, which include herbicide application, tree removal, sand bar removal, and the snagging and **clearing** of accumulated woody debris.
   ケンタッキー州排水法のもとで流路制御された水流は，定期保守活動の対象とな

る。これに含まれるのは，除草剤散布，高木除去，砂洲除去，木質堆積物の除去・清掃などである。

Close（Closely）：近接した（密接に；綿密に）

- **close to** ..... ：…に近い
- **closer to** ..... ：…により近い

1. Mercury was detected at levels **close to** the recommended drinking water criteria.
水銀が飲料水推奨クライテリアに近いレベルで検出された。
2. The year 1988 and 1991 were abnormally dry and both had extremely low flows. Conversely, rainfall and river flows during 1981 and 1986 were **closer to** normal.
1988年と91年は，異例に乾天で，いずれも流量がきわめて少なかった。対照的に，1981年から86年までは，雨量と河川流量が平年に近かった。

- **close proximity to** ..... ：…の近く
- **within or in close proximity to** ..... ：…の内部または近隣で

3. The Big Lost River travels the INEEL and **is close proximity to** facilities and waste disposal / storage areas.
Big Lost川はINEEL地区に達し，そしてその施設と廃棄物処理/貯蔵地域の近くを流れている。
4. The extreme range of outliers in the Municipal Conventional impact type for percent DELT anomalies greater than 10％ was observed mostly **within or in close proximity to** WWTP mixing zones.
DELT異常パーセントについて，都市通常影響における外れ値の極端な値域（10％以上）が観察されたのは，主に下水処理場混合地帯の内部または近隣であった。

- **closely** ：密接に；綿密に
- **closely to** ..... ：…に近い

5. There is some empirical evidence that instream biological response and instream ambient toxicity are **closely** associated.
経験的証拠によれば，水流内生物応答と水流内環境毒性は，密接に関連している。
6. Chemical transport and transformation processes are examined more **closely**.
化学物質の輸送・変化プロセスがいっそう綿密に検討される。

7. As can be seen from Table 4, the BIOMODEL estimates of biotic DDTR concentration correspond fairly **closely to** measured concentrations.
表4から見られるように，BIOMODELによって推定された生体内のDDTRの濃度は測定された濃度にかなり近い。

Coefficient (See also **Constant**)：係数（→ **Constant**）

1. The **coefficient** is a function of the size, shape, and density of the suspended particles.
この係数は浮遊粒子の大きさ，形，密度の関数で示される。
2. The dispersion **coefficient** lump together all the effects of averaging.
分散係数はすべての影響を平均し，それをひとまとめにしたものである。
3. Table 2 gives some literature values for the vertical dispersion **coefficient**.
表2に，鉛直拡散係数の文献値をいくつか示す。
4. The desorption and adsorption **coefficients** are orders of magnitude greater than the settling coefficient.
脱着・吸着係数は沈降係数より桁ちがいに大きい。
5. Kinetic **coefficients** from the single-Cp tests were incorporated into a Michaelis-Menten competitive inhibition model.
単一Cpテストからの速度係数がMichaelis-Menten競合抑制モデルに取り入れられた。
6. If the **coefficients** defining this process are much greater than the other kinetic coefficients, the assumption of instantaneous equilibrium is appropriate.
もしこのプロセスを定義している係数が他の速度係数よりかなり大きければ，瞬時平衡の仮定は適切である。

● correlation coefficient：相関係数

7. The value for **the correlation coefficient** for each lines was greater than 0.96.
線それぞれの相関係数の値は0.96より大きかった。
8. The data were fitted by linear regression and generally the **correlation coefficients** were greater than 0.90.
このデータは線形回帰によってフィットされ，相関係数は一般に0.90以上であった。
9. This **correlation coefficient** was 0.99 indicating a surprising good fit to the data points employed.
この相関係数は0.99で，使用したデータポイントに驚くほど一致することを示唆している。

Combine (Combination)：組み合わせる（組み合わせ）

1. If conditions are similar among regions, or if the differences can be associated with management practices that have a chance of being altered, it seems wise to **combine** the regions for the purposes of establishing water quality criteria.
   仮に，各地域における状態が似ている場合や，その差違が変化する可能性のある管理慣行に関連している場合には，水質クライテリア設定のために諸地域を組み合わせた方が賢明であろう。

   ● ～ **combine A with B**：～はAをBと組み合わせる

2. The present research **combines** sonication **with** commercially available probes and offers a simple approach toward field monitoring.
   現在の研究は超音波破砕を商業的に利用可能な測定器と組み合わせ，現地モニタリング用の単純な手法を提供する。

   ● **combination of A and B**：AとBの組み合わせ

3. The **combination of** UV light **and** hydrogen peroxide has been proven effective at treating aqueous solutions contaminated with chlorinated organics.
   紫外線と過酸化水素の組み合わせは，有機塩化物で汚染された水溶液を処理するのに効果があると証明されている。

4. Subregions of the Southern Rocky Mountains Ecoregion are characterized by **different combinations of** vegetation, elevation, land use, **and** climate characteristics.
   Southern Rocky 山脈生態地域の中の小地域を，異なる植生，高度，土地利用，および気候特性の組み合わせによって特性把握する。

5. These strategies must rely on assessing **various combinations of** chemical-specific **and** WET criteria in the absence of instream biosurvey data.
   これらの戦略は，水流内生物調査データがない場合，化学特異的なクライテリアとWETクライテリアの様々な組み合わせを評価することに頼らなければならない。

   ● **in combination with** ……：…との組み合わせ；…との併用

6. Relying exclusively on this measurement of standing crop is not recommended；however, **in combination with** other indicators, such as DBI, AFDM can be useful.

生物体量の測定に過度に頼ることは推奨できない。しかし，DBI，AFDMなど他の指標との組み合わせは有用であろう。
7. The Biological Response Signatures **in combination with** water quality, effluent, and habitat data was useful in demonstrating that the continuing impairment in the Ohtsu River was due to toxic impacts as opposed to habitat.
生物応答サインと，水質，流出および生息場所データの併用は，Ohtsu川の継続的損傷が生息場所でなく毒性影響に起因することを実証するうえで有用であった。

Come：くる；なる

1. Question **comes** more easily than answers.
答えをだすより質問をだすほうがいっそう容易である。
2. In the decade of the 1970's water quality modeling **has come of age**.
1970年代の10年間に水質モデリングが成熟してきた。
3. Resource managers and scientists **have come to realize** that the nature of this resource varies in an infinite number of ways, from one place to another and from one time to another.
資源管理者や科学者は，これらの資源の性格が，場所ごと，時間ごとに無限な多様性で異なることを理解するようになってきた。

- ～ **come from** ……：～は…からのものである

4. Data for this study **come from** …..
この研究に用いられたデータの出所は…である。
5. Mercury found in the environment **comes from** two major sources.
環境において見いだされた水銀には2つの主要な出所がある。
6. The difficulty **comes from** trying to model a biological process chemically.
生物プロセスを化学的にモデル化しようとすることから難しさが生じる。

- ～ **come into play**：～は有用になる

7. The hydraulic conductivity **comes into play** only when quantitative discharge calculations are made.
流体の伝達速度は，排出量を計算するときにだけ，有用になる。

Comment：論評

1. Toru Takeshima **provided comments on** earlier versions of the manuscript.

Toru Takeshima が初版の草稿を論評してくれた。
2. Kenichi Yoshioka **provided insightful comments on** Yellow River integrity.
Kenichi Yoshioka は，Yellow 川の健全性について洞察あふれる論評を寄せてくれた。

Common (Commonly)：通常；一般的な（一般に）

1. Rate enhancements of more than tenfold **are common**.
10倍以上の反応速度の増速は通常である。
2. One of the most **common** spatial frameworks used for water quality management has been that of watersheds
水質管理に用いられる最も一般的な空間枠組みは，流域であった。
3. One **common** source of error arises from the difficulty of obtaining a representative soil sample and the lack of reproducibility of organics analysis of soils.
エラーを生じさせる一般的原因1つは，代表的な土壌サンプル採集の困難さ，それに土壌の有機分析の 再現性の欠如である。
4. Among all the techniques of waste management, waste reduction is the **common sense** solution to the prevention of future hazardous waste problems.
廃棄物管理のすべてのテクニックのなかで，廃棄物削減は未来の有害廃棄物問題の防止に対する一般的解決法である。

● **commonly**：一般に

5. Chromium was detected at higher than **commonly** expected values throughout the study area.
クロミウムが一般的予想値より高い値で調査地域全体で検出された。
6. Arsenic is well above the upper range **commonly** found in soils.
ヒ素（の濃度）は一般に土壌に見られる（濃度）範囲の上限の更に上にある。

Comparable：相当；類似の；比較の

1. This figure shows that A, B, and C give **comparable** results.
この図はA，B，Cが類似の結果を出すことを示す。
2. These models, PESTAN, PISTON, and PRZM give **comparable** results.
これら PESTAN, PISTON, PRZM のモデルは類似の結果をだす。
3. Few **comparative** studies have been done to investigate the distribution of

aromatic hydrocarbons and PCB in marine organisms.
海洋生物体内における芳香炭化水素とPCBの分布を調査する比較研究はほとんどされていない。

4. Streams draining **comparable** watersheds within the same region are more likely to have similar biological, chemical, and physical attributes than from those located in different regions.
同じ地域内の類比可能な流域を流れる水流は，異なる地域に所在する水流に比べ，類似の生物的・化学的・物理的属性を持つ見込みが高い。

5. One relevant question is: how **comparable** are reference site to the mainstream Muskingum River near Connersville Station?
一つの関連する問題は，「参照地点がConnersville駅付近のMuskingum川本流とどの程度類比可能か？」ということである。

- ～ **is comparable with** ..... ：～は … に相当する
- ～ **is comparable to** ..... ：～は … と比較できる
- ～ **is comparable from** ..... **to another** ：
  ～は … と別の … を比較可能である

7. This **is comparable with** the value found by Morita.
これはMoritaによって見いだされた数値に相当する。

8. The low activation energy obtained in this study **is comparable to** the results from many other studies in alkene oxidation.
この研究で得られた低活性エネルギーは，アルケン酸化反応における他の多くの研究の結果と比較できる。

9. The results of chemical analyses of water samples from the eight water-supply wells **are** not directly **comparable to** the results of analyses of piezometer samples.
8つの給水用井戸から採集した水サンプルの化学分析結果は，ピエゾメーターからのサンプルの分析結果とは直接には比較できない。

10. Without accounting for natural geographic variability it would be difficult to establish numerical indices that **were comparable from** one part of State **to another**.
自然の地理的多様性を説明できなければ，ある州の一地域と別の地域を比較可能な数値指標を定めることは難しいであろう。

Compare (Comparing)：比較する（比較）

1. In this paper, we **compare** laboratory results to previously reported field results.

この論文では，我々は研究室での結果を既報の野外調査の結果と比較している。
2. Cheng et al. (1998) carried out experiments **to compare** the metal uptake in active and non-active sludge.
Cheng ら (1998) は，活性汚泥および非活性汚泥における金属取り込みを比較する実験を実施した。
3. Few studies have undertaken **to compare** data collected by volunteers **with** that of data collected by professionals.
ボランティアの収集データと専門家の収集データを比較する研究は，ほとんどない。
4. In some cases, scales are collected so that growth rates as well as general size **can be** calculated and **compared**.
一部の場合，鱗を採集して生長度や一般的体寸が計算・比較されうる。

- 〜 **is compared with** ..... ：〜 を … と比較する
- 〜 **is compared to** ..... ：〜 を … と比較する；〜 は … と比較される
- 〜 **is compared among** ：〜 を … において比較する

5. The simulation results **were compared with** the observed data.
シミュレーションの結果を観測データと比較した。
6. Results of water quality analyses **are compared to** recommended standards for drinking water.
水質分析の結果が飲料水の推奨基準と比較されている。
7. The measured reactivity of a mixture **was compared to** that calculated from the sum of the measured reactivity of the mixture's individual components.
測定された混合物の反応性と，測定された混合物の個々の構成物質の反応性の合計から計算された反応性を比較した。
8. Calibration procedures allow normalization of the effects of stream size, so index scores **can be compared among** streams of different sizes.
キャリブレーション手順は，水流規模による効果の標準化を可能にするので，異なる規模の水流において指数スコアを比較できる。

- **compared to** ..... ：… に比べ
- **as compared to** ..... ：… と比較されるように
- **when compared to** ..... ：… に比べる場合
- **when comparing A with B** ：AとBとの比較では

9. **Compared to** the Scioto River, the Ottawa River exhibits evidence that, despite some improvements, makes it one of the most severely impaired rivers in the state.

Scioto 川に比べ，Ottawa 川は，若干の改善にもかかわらず，同州で最もはなはだしく悪化した河川の一つという証拠を示している。
10. The Hocking River presents a stark contrast in attainment status **compared to** the Kokosing River.
    Hocking 川は，Kokosing 川に比べ，達成状況において明確な対比を示している。
11. Relative abundance of taxa refers to the number of individuals of one taxon **as compared to** that of the whole community.
    分類群の相対存在量とは，生物群集全体の個体数に比した一分類群の個体数を意味する。
12. Because the dechlorination reaction rates were relatively slow **as compared to** chlorophenol (CP) desorption kinetics, solid−liquid partitioning equilibrium was assumed.
    クロロフェノール (CP) の脱着速度と比較して，脱塩素反応速度が比較的遅かったので，固体−液体間の分配平衡状態が仮定された。
13. Reference streams draining watersheds that are within a particular region tend to be similar to one another **when compared to** reference streams in adjacent regions.
    特定地域内に存在する参照水流流域は，隣接地域における参照水流に比べ，お互い類似傾向にある。
14. **When comparing** all of the stations with Station 6, only Station 7 was not significantly different.
    観測地点のすべてと観測地点6との比較では，観測地点7だけが有意に異なっていなかった。

Comparison：比較

1. Table 7 presents a summary of this **comparison**.
   この比較の要約を表7に提示する。
2. Seasonal variability was tested by monthly sampling year−round at two streams, and showed that between−month **comparisons** should not be done, but upstream−downstream sampling on the same date is valid year−round.
   季節による差は，2つの水流での通年サンプリング(毎月)により検証されて，月ごとの比較はすべきでないが，同一日の上流地点−下流地点サンプリングが一年中有効であることを示した。

   ● **A comparison of ....**：…の比較

3. **A comparison of** the two groups showed two prominent features.

この2つのグループの比較によって，2つの顕著な特徴が示された。
4. **A comparison of** predicted **and** measured Alachlor residues in the soil is given in Figure 3.
土壌におけるアラクロールの残余の予測値と測定値の比較を図3に示す。
5. **The comparison of** the two different time period showed that .....
これら2つの時期の比較は，.....を示した。

- **A comparison between A and B** ： AとBの間の比較
- **A comparison between A versus B** ： AとBとの比較

6. **A comparison between** SEASON results **and** those of Imanaka et al.(1983) was obtained.
Imanakaら(1983)の結果とSEASONの結果との比較が得られた。
7. **Comparisons of** toxicity data **between** tests conducted on-site to tests conducted off-site on samples have shown little variation.
オンサイトとオフサイトで実施した試験の毒性データを比較したところ，サンプル間にほとんど変移が見られなかった。
8. **A comparison** of quantitative measures of performance **between** AMODEL **versus** the observed residues, and **between** BMODEL **versus** the observed residues shows that AMODEL had only slightly better agreement than BMODEL.
AMODELと観察された残留物の間，そしてBMODELと観察された残留物の間の成果の量的手法の比較から，AMODELはBMODELより少しだけ良い一致があったことが示された。

- **comparison with** ..... ： … と比較して
- **in comparison with** ..... ： … と比較して… ；….に比べると
- **for the purposes of comparison of A with B** ： AとBの比較のために
- **comparison to** ..... ： … との比較

9. **Comparison with** other technologies on a case-by-case basis is limited.
他の技術とのケースバイケースでの比較が乏しい。
10. **A comparison of** the observed data **with** the computed results shows good agreement.
観察データと計算結果との比較が良い一致を示している。
11. **Comparison of** surface water analyses **with** previous sample are presented in Table 1.
表流水の分析値と以前のサンプルの分析値との比較を表1に示す。
12. Protocol II is based on **comparisons with** reference conditions.
プロトコルIIは，参照状態との比較に基づいている。

13. To put such a member into perspective **in comparison with** other risks, we need to examine data on facilities from other causes.
他のリスクとの比較において，このようなメンバーを総体的に考えるために，我々は他の原因からの施設データを調べる必要がある。
14. **For the purposes of comparison of** output predictions **with** field data., ...
出力予測値と現地データの比較のために，
15. **Comparison to** reference conditions is essential to evaluate the extent to which study sites are influenced by human actions.
研究対象地点が人間の活動の影響を受ける度合を評価するには，参照状態との比較が不可欠である。

- **Comparison was made with** ..... ：…で比較を行った
- **Comparisons were made by** ..... ：…によって比較を行った
- **Comparisons were made between A and B**：AとBとの比較を行った

16. **Comparisons were made with** two types of data sets.
2種のデータセットで比較した。
17. **Comparisons were made by** site on the percentage of reference condition score.
地点ごとに，参照状態スコアのパーセンテージによって比較が行われた。
18. Fish are weighed and measured and **comparisons are made between** fish in contaminated **and** "clean" areas.
魚類を計量・測定し，汚染区域と「清浄」区域における魚類の比較を行う。
19. **Comparisons were made between** process operational variables **and** effluent quality to assess the influence of the heavy metals.
重金属の影響を評価するため，プロセス操作上の変数と排水の水質の比較をおこなった。
20. The cupric ion concentration was measured and **a comparison was made with** cupric ion concentrations predicted by a chemical equilibrium model.
銅イオン濃度を測り，そして化学平衡モデルによって予測した銅イオン濃度と比較した。
21. To date no formal **comparison has been made between** data collected by volunteers in Maryland **and** data collected by biologists. However, preliminary examinations and comparisons show good agreement between the two at most common stations.
現在までのところ，メリーランド州におけるボランティアの収集データと生物学者の収集データについて，公式な比較はなされていない。しかし，予備的な検討・比較によれば，大半の一般的な観察所では，これら両者がかなり合致する。

- **make comparison between A and B** ：AとBを比較する
- **In making comparison between A and B** ：
  AとBを比較するにあったては

22. The first tier involved **making comparisons between** RBCA predictions **and** MEPAS predictions. The second tier of testing involved comparing MEPAS predictions with field data.
最初の段階では，RBCAによる予測値をMEPASによる予測値と比較した。試験の第2段階では，MEPASによる予測値を現地データと比較した。

23. **In making any comparison between** the model estimates **and** the measured data, it must be noted that....
モデルからの予測値と測定データを比較するにあったては，....が指摘されなくてはならない。

- **allow comparison** ：比較を可能にする

24. Numerous pollutant studies have already been performed on the $< 63-\mu$ m fractions, **allowing better comparison of** results.
多くの汚染の研究が$< 63-\mu$mの部分でなされてきたので，結果との比較をより可能にしている。

25. Results are presented in several forms, but are normalized to **allow direct comparison with** other published work.
結果がいくつかの形式で提出されているが，標準化されているために発表されている他の研究結果との直接比較を可能にしている。

- **In a comparison of** ..... ：…の比較では

26. **In the comparison of** the Toxic and CSO/Urban impact types, the index yielded similarly low results of each.
毒性影響とCSO／都市部影響の比較において，この指標は，それぞれについて同じく低い結果となった。

27. **In a comparison of** the photocatalytic degradation of acetic acid with that of the chloroacetic acids in oxygenated aqueous dispersions of TiO_2, the order of the initial rates of CO_2 release was dichloroacetic acid > chloroacetic acid > acetic acid.
TiO_2の拡散した酸化的水において，酢酸の光触媒分解反応とクロロ酢酸の光触媒分解反応の比較では，CO_2放出の初期速度はジクロロ酢酸＞クロロ酢酸＞酢酸の順だった。

Compile：まとめる；コンパイルする

1. The information provided in each SWMU description **is compiled**.
   SWMUの記述それぞれに示されている情報がまとめられた。
2. The most detailed and current consensus on the tolerance of diatoms **was compiled by** Imai and Nakata (1997).
   珪藻の耐性に関する最も詳細かつ最新のコンセンサスをImaiとNakata (1997) がまとめた。
3. Extensive information on solid waste management units located at LAN Laboratory **has been compiled** by EPA.
   LAN研究所にある廃棄物管理ユニットについての多くの情報がEPAによってまとめられてきた。

Complicate (Complication)：複雑にする（複雑）

1. The user should be aware that such effects may **complicate** attempts to extrapolate data for photolysis rates from one aquatic medium to a very different medium.
   利用者は，このような影響が光反応速度データを1つの水媒質からかなり異質の媒質外挿法によって推定する試みを複雑にするかもしれないということを認識すべきである。
2. There is a further **complication to** the analysis of groundwater flow in the unsaturated zone.
   不飽和帯での地下水流の解析にはそれ以上の複雑さがある。
3. In doing so, the **complicated** pattern may be avoided.
   そうすることにおいて，複雑なパターンが避けられるかもしれない。

Comply (compliant)：従う（従順な）

- **to comply with**……：…に従って
- **～ is compliant with**……：～は…に従う

1. **To comply with** the standards, industrial effluents could be diverted from direct discharge to sewers.
   この基準に従って，産業廃水が直接放流から下水道への放流に変えられるかもしれない。
2. This relationship **is compliant with** the "fixed demand" theory proposed by Wu and Yao (2001).

この関係は，Wu and Yao (2001) によって提案された「不変需要」の理論に従う。

Comprise：からなる

1. These types of streams and watersheds **would comprise** relatively undisturbed references for the region.
   この種の水流および流域は，その地域の比較的影響されていない対照地点からなるであろう。
2. It may be necessary to choose reference sites that **comprise** watersheds containing similar proportions of different subregions.
   異なる小地域の類似部分を含んだ流域からなる参照地点を選ぶ必要があろう。
3. Ecoregions **comprise** regions of relative homogeneity with respect to ecological systems involving interrelationship among organisms and their environment.
   生態地域は，生物およびその環境の相互関係を伴う生態系に関して比較的同質な地域からなる。

- ～ **is comprised of** ..... ：～は…からなる

4. The numeric indices used to help define the narrative classification system **were comprised of** single-dimension measures.
   記述的分類システムを定義するのに用いられた数量的指標は，単一次元の測定値からなっていた。
5. The groundwater in the alluvium beneath the tailings has a possibility of **being comprised of** water from as many as four sources.
   堆積選鉱屑の下にある沖積層の地下水は，およそ4つの水源からの水が混ざっている可能性がある。

Concentration：濃度

1. **The concentrations of** TCA in the effluent of Port-1 and Port-2 were 0.25 mg/L and 0.40 mg/L, respectively.
   ポート-1とポート-2からの流出水中のTCAの濃度は，それぞれ0.25 mg/Lと0.40 mg/Lであった。
2. Measured **concentrations of** arsenic in Redwood Creek during low flow were 0.025 ± 0.003mg/L.
   Redwood Creekで測定されたヒ素の濃度は低流量時で0.025 ± 0.003 mg/Lであった。
3. Similar **concentration** profiles were also obtained for the other soils that

were tested.
類似の濃度プロフィールが，試験した他の土壌でも得られた。

4. If the contamination were to be assumed to reach the screen depth, the **concentration** would decrease due to dilution up to an order of magnitude.
もし汚染がスクリーンの深さに達すると想定されるなら，その濃度は希釈され一桁減少するであろう。

- **average concentration**：平均濃度

5. **The average concentration** in the reference soils was 1.9 mg/kg.
対照土壌での平均濃度は 1.9 mg/kg であった。

6. During period 1, **average** influent and effluent CF **concentrations** were 0.30 and 0.11 mg/L, respectively.
期間 1 では，流入水および流出水中での平均の CF 濃度は，それぞれ 0.30 および 0.11 mg/L であった。

- **concentration above (below)** .....：… 以下（以上）の濃度

7. Soil metal **concentrations** do not appear to be elevated **above** expected natural background levels.
土壌での金属濃度は，自然に存在するとされるバックグラウンドレベル以上に上昇しそうもない。

8. The study by EPA confirmed the existence of arsenic **concentrations** elevated **above** background concentration.
EPA による研究で，バックグラウンド濃度より高濃度のヒ素の存在が確認された。

9. A background soil sample collected at a depth of 40 to 41 ft had a TPH **concentration below** the 3 mg/kg detection limit.
40 から 41 フィートの深さから収集されたバックグランド土壌サンプルでは，TPH の濃度が検出限界の 3 mg/kg 以下であった。

- **concentration up to** .....：… までの濃度

10. **The concentration of** nitrate **up to** 1 mg/L as N did not affect the rate of primary producer during 3 hr of incubation.
硝酸塩の濃度が最高 N に換算して 1 ミリグラムまででは，培養時間 3 時間の間に一次生産者の成長速度に影響を与えなかった。

- **concentration is higher than** .....：濃度は … より高い
- **higher the concentrations are, the higher the B is**：
濃度が高ければ高いほど，B が高い

11. The predicted CO_2 **concentration is higher than** the observed concentration at later time.
    予測された CO_2 濃度は，後の時間に観察された濃度より高い。
12. The time course data set shows that **higher** the MIBK inlet **concentrations are, the higher** the MIBK degradation rates **are**.
    経時的なデータセットは，MIBK の入口での濃度が高ければ高いほど，MIBK の分解速度が速いことを示している。

    - **concentration greater(less) than** ..... ：… 以上(以下)の濃度

13. The effluent nickel **concentration** averaged **greater than** 95 percent of the influent nickel concentration.
    排水中のニッケルの濃度は，平均して流入水中のニッケル濃度の 95 パーセント以上であった。

    - **at concentration of** ..... ：… の濃度で

14. The toxicity was most evident **at** cadmium **concentration of** 0.4 mg/L.
    毒性はカドミウムの濃度が 0.4 mg/L で最も明白であった。

    - **variation in(of) concentration** ：濃度の変化

15. Figure 3 shows **the variation of** oxalic acid **concentration** with irradiation time.
    照射時間によるシュウ酸の濃度変化を図 3 に示す。
16. Figure 4 shows **the variation in** parathion concentration in ppm with the irradiation time at various pHs.
    様々な pH 値での照射時間に対するオアラチオンの濃度変化を ppm で図 4 に示す。

    - **increase(decrease) in concentration** ：濃度の増加(減少)
    - **increase(decrease) of concentration** ：濃度の増加(減少)

17. The pH dramatically increases with depth with a corresponding **increase in** calcium and magnesium **concentrations**.
    pH は，カルシウムとマグネシウム濃度の増加に対応し，水深が増すと共に劇的に増加する。
18. The H_2O_2 formation rate appeared to decrease with an **increase in** the initial **concentration** of MTBE.
    H_2O_2 の生成速度は，MTBE の初期濃度の増加とともに減少するように思われた。
19. Ordinate values in Figure 2 are plotted logarithmically to show linear correspondence to logarithmic **increase of** bacterial **concentration**.

図2の縦座標には，バクテリア濃度の増加が対数直線関係にあることを示すために対数値がプロットされている。

20. It has been concluded from the chemical form of the heavy metals that atmospheric contributions play the most important role and that a 3.5-fold **increase of** the lead **concentration** and a 2.5-fold **increase of** the cadmium **concentration** are from this atmospheric source.
大気中からの寄与が重要な役割を果たしていること，そして，大気中からの寄与が鉛の3.5倍上昇やカドミウムの2.5倍上昇の原因であることが重金属の化学的な形体から結論づけられている。

- **in the range of** ..... (See **Range**)：… の範囲で(➡ **Range**)
- ..... **order of magnitude** (See **Magnitude**)： … 倍；… の規模(➡ **Magnitude**)

Concentrate：的をしぼる；専念する；集中する

- **concentrate on** .....：… に的をしぼる；… に専念する

1. The experiments described in this article **concentrated on** the leaching characteristics of selected arsenic-bearing wastes.
この論文に述べられている実験は，選択されたヒ素含有廃棄物の浸出特性に的をしぼっていた。

2. It will **concentrate on** the following issues: a) ....., b) ....., and c) .....
次の事に的がしぼられるであろう。それらはa)…，b)… およびc)… である。

3. Although a great deal of effort has been devoted to studying the problem of pollution, most work has **concentrated on** freshwater and terrestrial systems.
多くの努力が汚染問題の研究に捧げられてきたが，ほとんどの研究は真水と地生の系に専念されてきた。

4. States may choose to **concentrate on** one or more of these uses of water quality criteria when developing their programs.
諸州は，プログラム策定時に，水質クライテリアの上記用途のうち1つないし複数に専念することを選ぶ場合もある。

Concept (Conceptual)：概念(概念の)

1. The formulation is based upon **a proven concept**.
この式はすでに証明されている概念に基づいている。

2. Seasonality is **a well-understood concept**.

季節性は，十分に理解されている概念である。
3. **The opposing concepts of** diversity and similarity are the bases of most classifications.
多様性と類似性という対立的概念が大半の分類の基本にある。
4. **The central concept** is that two independent processes describe the basic motion of a particle.
この中心的な概念は，2つの独立したプロセスが粒子の基本運動を説明するということである。
5. Understanding the **underlying concepts** may be a challenging to environmental managers that have little or no experience with biological assessments.
根本的概念を理解することは，生物アセスメントの経験に乏しい，もしくは経験のない環境管理者にとって課題になろう。
6. The ecoregions **concept has been developed through** the effort of many people.
生態地域という概念は，大勢の方々の努力によって編み出されてきた。
7. Kimoto et al. (2004) **have strengthened the concept through** their interactions and testing.
Kimotoら(2004)は，対話や試験を通じてこの概念を補強してくれた。

- **In concept** ：概念上

8. **In concept**, there is no legal reason why NPDS permit cannot be used to implement biocriteria.
概念上，NPDES許可を用いて生物クライテリアを実施してはならないという法的理由は，存在しない。

- **Conceptual** ：概念の

9. Metrics that are poorly defined or based on a flawed **conceptual** basis provide erroneous judgments with the potential for erroneous management decisions.
きちんと定められていないメトリックや，欠陥ある概念ベースに基づいたメトリックは，誤った判断を生じさせ，管理上の判断ミスを招く可能性がある。

Concern (Concerning) ：懸念；関心事（関して）

1. Considerable environmental **concern has focused on** these nonbiodegradable and toxic heavy metals.
かなりの環境に関する懸念が，これらの生分解不可能な，および毒性のある重金

属に集中されてきた。
2. Improved detection techniques have now **shifted concern to** the threat posed by toxic chemicals.
改善された検出技術により，今は関心は毒性化学物質による脅威に移行した。
3. Metals and metalloids that form alkyl compounds **deserve** special **concern** because these compounds are volatile and accumulate in cells.
アルキル化合物を生成する金属と半金属は，特別な懸念に値する。なぜなら，これらの化合物は揮発性であり，細胞に蓄積するからである。

- **there is concern**：懸念がある

4. While the quality of some regions has improved during recent years, **there is still great concern** regarding the marine environment.
近年環境が良くなってきた地域がある一方，海洋環境に関してまだ大きい懸念がある。
5. **There has been concern for** a mechanism to structure the assessment and management of nonpoint source pollution.
面源汚濁の評価・管理を構成するメカニズムへの懸念があった。

- **The concern is ......**：関心事は…である

6. Many of Sakai's **concerns are** based on the premise that they are heterogeneous.
Sakaiの関心事の多くは，それらが異質だという前提に基づいている。
7. **One of the major concerns with** regional reference sites **is** their acceptable level of disturbance.
地域参照地点に関わる重要事の一つは，撹乱の許容可能レベルである。
8. **The primary concern** in selecting the two sites **is** to assure that the physical characteristics are as similar as possible.
2地点の選定における主な関心事は，物理的特性ができる限り似るよう努めることである。
9. **Much of the concern about** method **is** with the potential for misuse and abuse by attempts to justify increasing loadings of pollutants.
この方法に関する懸念の大半は，汚濁物質の付加増大を正当化する試みによって誤用・濫用される可能性をめぐるものである。

- **concern is expressed**：懸念が表わされる

10. **Concern is** frequently **expressed about** the potential for biological criteria to be underprotective.
生物クライテリアが保護過少な可能性について，懸念がしばしば表明される。

11. **Some concern was expressed** by the author of the report **over** the accuracy of the majority of the data.
    大多数のデータの正確さについて，若干の懸念がこの報告の著者によって表わされた。

    - **concern arises** ：関心が高まる

12. **Concern arose about** the possibility of an explosion.
    爆発の可能性について関心が高まった。
13. **This concern has arisen** lately as analytical techniques for detecting water contaminants have improved to the point where it is possible to detect contaminants at the part per-trillion level.
    水汚染物質を検出するための分析技術が汚染物質を ppt レベルで検出可能なまでに向上したため，最近この関心が高まった。

    - **concern with ..... has been growing** ：…の関心が高まってきている

14. The environmental **concern with** endocrine disruptors **has been growing** for the last several years.
    ここ数年の間，内分泌かく乱物質の環境に対する関心が高まってきている。

    - **～ is of concern** ：～は関心事である
    - **～ can be of significant concern in .....** ：
      ～が…においてかなりの懸念となりうる
    - **～ is of utmost concern** ：～は極めて重要な関心事である
    - **～ of primary concern is .....** ：関心の高い ～ は … である

15. The source of contaminated fish in the Seta River **is of** significant **concern**.
    Seta 川の汚染された魚の原因は重要な関心事である。
16. 1, 4-Dioxane **is of** environmental **concern** for assorted reasons.
    1,4-Dioxane はさまざまな理由で環境の関心事である。
17. Heavy metals **have been** the cause **of** particular environmental **concern**.
    重金属は特定の環境に関する懸念の原因となってきた。
18. The type of error **can be of significant concern in** a regulatory environment.
    誤差の種類が，規制する環境においてかなりの懸念となりうる。
19. The potential danger these releases present to human and the environment **is of** utmost **concern**.
    これらの排出が人間と環境に与える潜在的危険性は極めて重要な関心事である。
20. The chemicals **of primary concern** were: naphthalene, ..., ...
    関心の高い化学物質はナフタリン，…，…であった。

21. Primary **contaminants of concern** associated with the fuel leakage include BTEX constituents.
燃料漏れと関連し懸念される主汚染物質は BTEX 成分を含む。

- ~ **are matter of growing concern**：
~は，これからますます懸念される事態である

22. The possible long-range effects of the disposal of trace metals in the coastal environment **are matter of growing concern**.
沿岸環境において微量金属の処分による長期にわたる潜在的影響は，これからますます懸念される事柄である。

- **concern over** ..... ：… についての懸念

23. Recently, the **concern over** formaldehyde as a pollutant has been increased.
最近，ホルムアルデヒドの，汚染物としての懸念が増えてきている。
24. In recent years, **concern over** the toxic effects of chemical substance in the environment has increased dramatically.
近年，環境中の，化学物質による毒性の影響についての関心がめざましく増加してきた。
25. **Some concern** was expressed by the author of the report **over** the accuracy of the majority of the data.
大多数のデータの正確さに関するいくつかの懸念がこの報告書の著者によって明示された。
26. This finding led to **concern over** potential impacts of PCB contamination on recreational fisheries in the Wood River.
この結果によって，Wood 川のレクリエーション用漁場に及ぼす PCB 汚染の潜在的影響が懸念されるようになった。

- ~ **is concerned with** ..... ：~は … に関する；~は … を対象にする

27. This study **is concerned with** the fate and persistence of DDT.
これは DDT の行方と残存性に関する研究である。
28. The purification of air in residential and commercial buildings **is** primarily **concerned with** gaseous, particulate, and microbial contaminants.
住宅および商用建物内での空気の清浄化は主にガス，微粒子，微生物からなる汚染物質を対象にしている。

- ~ **is concerned about** ..... ：~は … について懸念する

29. The regulated community **is concerned about** the potential for more stringent permits and other restrictions.

Concert

被規制業界は，より厳しい許可限度および他の制限の可能性について懸念している。

30. The regulated community **is** also **concerned about** taking on responsibility for conducting the ambient monitoring required to implement biocriteria.
被規制業界は，生物クライテリア実施に必要な環境モニタリングの実行責任を負うことについても懸念している。

- ～ **caused a great deal of concern regarding** ……：
～は … について強い関心を生じさせた

31. This language was not trivial and **caused a great deal of concern regarding** how to define "integrity" (especially biological integrity) and the measurements to be applied.
この文言は，些細なものでなく，「健全性」(特に生物健全性)の定義づけと適用すべき測定値について強い関心を生じさせた。

- **concerning** ……：… に関して

32. Several assumptions **concerning** flow and water quality must be made.
流量と水質に関し，いくつかの仮定が設定されなくてはならない。
33. Not much is known **concerning** the nature and behavior of natural systems.
自然系の性質と挙動に関してはあまり多くは知られていない。
The questions **concerning** drinking water are addressed in this article.
飲料水に関する質問が，この論文で取り扱われている。
34. Details **concerning** the experimental facilities are described only briefly here.
実験施設に関する詳細については，ここではただ手短かに記述する。
35. Little is known **concerning** the fate of discharged hydrocarbon in estuarine ecosystems.
河口生態系において，放出された炭化水素の衰退に関してはほとんど知られていない。
36. For the sake of clarity in presenting the test results, most of the details **concerning** the theoretical background of the tests and the analytical procedures are provided in Appendix B.
実験結果の提出にあたってそれを明快にするために，試験の理論的な背景と分析手法の細部の多くを付録Bに示す。

Concert：調整する；協力する

- **in concert with** ……：… と調整して

1. Biological monitoring should be carried out **in concert with** chemical/physical monitoring.
生物モニタリングは化学的・物理的モニタリングと調整して実施されるべきである。

Conclude：結論とする

- ~ **conclude that** ..... : ~は … と結論づける

1. From what has been discussed above, **we can conclude that** .....
上に論じられたことにより … と結論づけることができる。
2. Kurihara (2004) **concluded that** seasonal and habitat changes were responsible for much of the variability of species abundance at a given site.
Kurihara (2004) は，結論として，ある特定地点の生物種多様性の変化を大きく左右するのは季節的変化や生息地変化である，と述べた。

- **it is concluded that** ..... : … と結論づけられる
- **it cannot be definitely concluded that** ...... : … とは確定的には結論づけられない
- **it seems reasonable to conclude that** ..... : … と結論つけるのが妥当のように思われる

3. **It was concluded that** ICI scores are consistent at locations where little man-induced change has occurred.
人為的変化があまり生じていない場所では，ICIスコアは，一定であると結論づけられた。
4. **It is concluded that** in most aquatic systems these processes are capable of reducing free copper levels to very low values.
結論として，ほとんどの水生システムにおいて，これらのプロセスは遊離銅レベルを非常に低い値に下げることができる。
5. **It has been concluded** from the chemical form of the heavy metals **that** atmospheric contributions play the most important role and **that** a 3.5-fold increase of the lead concentration and a 2.5-fold increase of the cadmium concentration are from this source.
大気中からの寄与が重要な役割を果たしていること，そして，大気中からの寄与が鉛の3.5倍上昇やカドミウムの2.5倍上昇の原因であることが重金属の化学的な形体から結論づけられている。
6. Since many of the species were migratory, **it cannot be definitely concluded that** this contamination caused the decline.

多くの種が移住性であるため，この汚染が種の減退をもたらしたとは確定的には結論づけられない。

Conclusion：結論

1. These **conclusions** may be incorrect.
   これらの結論は誤っているかもしれない。
2. **It was the author's conclusion that** seasonal and habitat changes were responsible for much of the variability of species abundance at a given site.
   ある特定地点の生物種多様性の変化を大きく左右するのは季節的変化や生息地変化である，というのが著者の結論である。

   ● conclusion is that ……：結論は … ということである

3. One possible **conclusion** from our analyses and the biocriteria framework in general **is that** the increased "data dimensions" afforded by the more detailed taxonomy translated into more powerful and sensitive analytical tools such as the biological response signatures.
   筆者の分析や生物クライテリア枠組み全般からの一つの結論として可能なのは，より詳細な分類群による「データ次元」増大が生物応答サインといったより強力かつ敏感な分析ツールを生じるということである。

   ● draw a conclusion：結論を引き出す
   ● We can extract conclusions that ……：
     我々は，… という結論を引き出すことができる

4. The field investigation has been conducted to provide adequate information **to draw a conclusion**.
   現地調査が，適切な情報を提供し，結論を引き出すために実施されてきた。
5. At this point there is insufficient information on source area ST5 **to draw a conclusion** on probable risk.
   この時点では，起こりうるリスクについて結論を引き出すには，汚染源ST5についての情報が不十分である。
6. It would be a serious mistake **to draw the conclusion that** the only important function of stream flow is to dilute pollutant concentrations.
   水流量の唯一の重要な機能は汚染物質濃度を希釈することだという結論を下すのは，重大な誤りである。

   ● conclusions is drawn：結論が引き出される

7. From an analysis of these plots, two general **conclusions may be drawn**.

First, ...
これらのプロットの解析から2つの一般的な結論が引き出されるかもしれない。第一に…

8. From the results of this study, the following **conclusions can be drawn**: ......
この研究の結果から，次の結論が引き出されうる。
9. From the results and observations obtained, the following **conclusions were drawn**: .....
得られた結果と観察から，次の結論が引き出された。
10. Two main **conclusions can be drawn from** these profiles. First, ...
2つの主な結論がこれらのプロファイルから引き出されうる。第一に…。
11. **The following conclusions can be drawn from** the data given in Fig. 5.
図5に示されたデータから次の結論が引き出されうる。
12. Basic laboratory studies are being conducted so that from the field data, general **conclusions may be drawn** and applied elsewhere.
基礎的な実験室での研究が行なわれている。そうすることによってフィールドデータから，一般的な結論が引き出され，ほかのところにも応用されるであろう。

- conclusion is reached ：結論に達する
- conclusion is established from ..... ：… から結論が確立される

13. Incorrect **conclusions can be reached**.
誤った結論に達する可能性がある。
14. From the results obtained in these investigations, the following **conclusions were reached**.
これらの調査で得られた結果から，次の結論に達した。
15. After evaluation of the study results, the following general **conclusions were reached**: .....
研究結果の評価後，次の一般的な結論に達した。それらは.....
16. Specific **conclusions are reached for** each metal.
金属それぞれに特定の結論が得られた。
17. Similar **conclusions had been reached** by state and independent researchers even earlier.
同様の結論は，さらに以前にも州および独立の研究者によって述べられてきた。
18. The following **conclusions were established from** the bench-scale testing: 1) …, 2) …
ベンチスケールの試験から次の結論が確立された。それらは1)… 2) … である。

- **~ provide conclusion**：～は結論を提供する
- **~ lead to conclusion**：～は結論に導く
- **~ lead us to the conclusion that .....**：
  ～は我々を…という結論へ導く

19. Extensive laboratory investigations and data analysis **provided** the following **conclusions**.
    大規模な室内調査とデータ解析が次の結論を提供した。
20. The results may **lead to** incorrect **conclusions**.
    この結果は間違った結論に至る可能性がある。
21. The data **lead to** some direct **conclusion about** the usefulness of the three models in safety assessment of new chemicals.
    このデータは，新化学物質の安全性アセスメントでの3つのモデルの有用性について，ある直接的結論に導く。
22. The results **lead us to the conclusion that** the arsenic was derived from bacteria action solubilizing arsenic from the arsenopyrite in the waste deposits.
    この結果は我々を，このヒ素は堆積廃棄物中の硫ヒ鉄鉱からヒ素を溶かしだすバクテリアの働きから発生したという結論へ導く。

- **~ support the conclusion that .....**：～は…という結論を支持する

23. These results **support the conclusion that** chloride is the major ionic product.
    これらの結果は，塩化物が主要なイオン生成物であるという結論を支持する。
24. Data exist from several years of studies that **support** these **conclusions**.
    この結論を支持するような数年にわたる研究によるデータが存在している。

- **In conclusion,**：結論として，

25. **In conclusion**, the method described in this paper can improve the interpretation of the results.
    結論として，この論文に記述された方法によって，この結果の解釈を改善することができる。

Conclusive (Inconclusive) (Conclusively)：
決定的（非決定的）（確実に；断定的に；結論的に）

1. There is no **conclusive** proof that .....
   … という確証はない。

2. Even with a record of that length, results **are inconclusive**.
   こんなに長い記録でさえ，結果はまだ確実ではない。
3. This was not proved **conclusively**, however.
   しかしながら，これは断定的には証明されなかった。
4. Based on their analysis, they **conclusively** stated the following.
   彼らの分析に基づいて，彼らは結論的に次のことを述べた。

Condition：条件

1. Eventually a steady-state **condition** will be achieved in the reactor.
   最終的に，この反応器において定常状態が達成されるであろう。
2. At this rate, low-pH **conditions** will reach the bottom of the Nordic Main tailings within several decays.
   この速度では，pHが低い条件が数十年以内にNordic Main選鉱屑層の底に達するであろう。
3. For such extreme operating **conditions**, deviations between model prediction and experiments can be expected.
   このような過度の操作条件では，モデル予測値と実験値の間のずれが予想できる。

   - **In the steady-state condition**：定常状態で
   - **under the conditions of** .....：…の条件下で
   - **under a variety of conditions**：いろいろな条件下で
   - **under any conditions**：どんな状況であっても

4. **In the steady-state condition**, only about 5 % of the VOC has been converted to product.
   定常状態で，ほんのおよそ5％のVOCが生成物に変換された。
5. **Under the conditions of** the present experiments, chloride was the major ionic product, and small amounts of $HCOO^-$ were detected as well.
   今回の実験条件下では，塩化物が主要なイオン生成物であった，そして同様に小量の$HCOO^-$も検出された。
6. There still remain many unanswered questions regarding reaction mechanisms **under a variety of** ultrasonic **conditions**.
   いろいろな超音波状態下での反応機構に関して数多くの質問が応えられずに残っている。
7. **Under steady-state conditions**, CTEX removal rates were greater than 98 % in both reactors.
   定常状態の条件下では，CTEXの除去速度は両方の反応器で98％以上であった。

8. **Under these conditions**, several thousand years would be required for removal of entire gypsum content of the tailings.
これらの条件の下では，選鉱くずに含まれる石膏の全部を除去するには，数千年が必要とされるであろう。
9. The same organism may have an order of magnitude more chlorophyll **under** low-light and nutrient-rich **conditions** than **under** high-light and nutrient-poor **conditions**.
同生物が，多光量・低養分状態よりも少光量・高養分状態の方が十倍ほど多くのクロロフィルを有するかもしれない。
10. Examination of local gaseous and liquid concentrations permitted an explanation of the complex events occurring **under** transient state **operating conditions**.
局地的なガスと液体の濃度の検討から，過渡状態の運転条件下で生じる複雑なイベントの説明ができる。
11. These two principles hold in all cases **under any conditions**.
これらの2つの原則は，どんな状況であってもすべての場合に適用される。

Conduct：実施する；行われる

1. California EPA **has conducted** biosurveys throughout the state for nearly 20 years.
カリフォルニア州 EPA は，約20年間にわたり州全域で生物調査を行ってきた。
2. The EPA **conducted** a more extensive review of the contaminant content of the existing deposits.
EPA は，既存の堆積物での汚染物質含有量のいっそう広範なレビューを実施した。
3. The experiments **were conducted** in glass bottles at room temperature.
この実験は，室温でガラスのビンの中で行われた。
4. In conjunction with these toxicity tests, biological field surveys **were conducted**.
これらの毒性試験と合同して，生物学的なフィールド調査が行われた。
5. Over the past few decades, a considerable number of studies **have been conducted** on the effects of temperature on the growth of nitrifying bacteria.
ここ数十年にわたって，温度が硝化菌の成長に及ぼす影響に関する研究がかなりの数行われてきた。

Confidence：信頼

1. The generalized equations to be used for design procedures must be developed considering the entire range of operating parameters and must contain a large number of data points to enhance the **confidence** in the design.
   一般化されたこの式が手法の設計に使われるためには，操作パラメータの全範囲が考慮されなければならない。また，設計の信頼度を上げるための多くのデータがなくてはならない。
2. The 95％ **confidence** interval is included to display the uncertainty of the sample estimates.
   サンプル推定の不確実性を示すため，95％信頼区間が含まれている。

Confirm：確証する

1. These studies along with our results **confirm** that .....
   これらの研究は，我々の結果と共に … を確証する。
2. Calculation of the t–test **confirmed that** the impairment was significant.
   この損傷は有意であったことがt検定の計算によって確証された。
3. Further investigations are needed to **confirm** this hypothesis.
   この仮説を確証するために，よりいっそうの調査が必要とされる。
4. The results **confirm** the hypothesis that copper uptake by algae is related to the free cupric ion activity and is independent of the total copper concentration.
   この結果は，藻類による銅の摂取は自由な銅イオン活性と関係があり，全銅濃度とは関係がないという仮説を確証する。
5. Resampling of these wells by EPA **confirmed** the existence of arsenic concentrations elevated above background concentration.
   これらの井戸でのEPAによる再度のサンプリングによって，ヒ素の濃度がバックグランド濃度以上であることが確認された。
6. It is regrettable that no substantive body of evidence **confirming** ...... has been generated.
   … を確証する実質的な一連の証拠をつくれなかったことは残念である。

Conflict (Conflicting)：抵触（相いれない）

1. In our experience the following are the situations where **conflicts** have arisen in Kansas.
   筆者の経験によれば，カンサス州で抵触が生じた事態は以下のとおりである。
2. The phenomena and impacts of acid rain and energy development, the

products of our industrialized society, will bring often **conflicting** economic and environmental concerns into the public eye.
酸性雨とエネルギー開発，我々の工業化された社会の産物，の現象と影響は公衆の目にはしばしば相いれない経済の，そして環境の問題と写るであろう。

Confusion：混乱；混同

1. Any discrepancies should be resolved before actually drawing a sample to prevent as much **confusion** as possible.
   いかなる不一致も，混乱をできる限り防ぐため，実際のサンプリング前に解決されるべきである。

   - **a certain amount of confusion**：ある程度の混乱
   - **a great deal of confusion**：かなりの混同

2. This variation in terminology has caused **a certain amount of confusion**.
   術語学におけるこのずれは，ある程度の混乱を起こしてきた。
3. Examination of the literature on sorption reveals **a great deal of confusion** regarding the time to reach the sorption equilibrium.
   吸着についての文献調査によれば，吸着平衡に達するまでの時間に関してかなりの混同がある。

Conjunction：共に；合同

   - **in conjunction with .....**：…と合同して；…と共に

1. **In conjunction with** these toxicity tests, biological field surveys were conducted.
   これらの毒性試験と並行して，生物学的なフィールド調査が行なわれた。
2. These results, **in conjunction with** the previous information on population structure, support the conclusion that ......
   これらの結果は，人口構造についての前の情報と共に，.....という結論を支持する。
3. It is currently being used to evaluate qualitative data **in conjunction with** other more traditional sample attributes such as total and EPT richness and overall community composition and balance.
   それは，現在，総豊富度とEPT豊富度，全ての生物群種の構成やバランスといった他のより伝統的なサンプル属性と共に質的データを評価するため用いられている。

Connection：関係

- ~ has a connection with ..... ：~ は … と関係がある

1. The elevated nitrate concentration in groundwater **has a** deep **connection with** the usage of fertilizer.
地下水での硝酸塩濃度の上昇は肥料の使用量と深い関係がある。

Consequence (Consequently)：結果（結果として；その結果）

1. Levin (1992) stressed that to gain an understanding of the patterns of ecosystems in time and space and the causes and **consequences of** patterns, we must develop the appropriate measures and quantify these patterns.
Levin (1992) は，時間・空間における生態系パターンや，パターンの原因・結果を理解するために，我々が適切な測度を定め，これらのパターンを数量化しなければならないことを強調した。

- **as a consequence of** ..... ：… の結果として

2. Some chemical changes affecting aqueous metal chemistry **as a consequence of** an altered redox potential may be almost immediate.
酸化還元電位の変化の結果として，水性金属化学に影響を及ぼすある種の化学変化は，ほとんど即時であるかもしれない。

- **In consequence**：その結果として

3. **In consequence**, there are substantial data available to aid in quantifying the fate of contaminants entering the soil zone through surface deposition.
その結果として，地表面堆積を通して土壌層へ入る汚染物質の行方を定量化するのに十分なデータが利用可能である。

- **Consequently**：結果として；その結果，

4. **Consequently**, these compounds were not decomposed in the laboratory.
結果として，実験室ではこれらの化合物は分解しなかった。
5. **Consequently**, Arkansas determined that the water quality standard driving this process needed revision.
その結果，アーカンソー州は，この過程を導く水質基準の改正が必要だと決定したのである。

Consensus：意見の一致

1. **There is a growing consensus** among researchers that excess nutrients are the primary cause of the reduced oxygen levels observed in the lake.
栄養分の過剰がこの湖で観察される酸素レベル減少の主要な原因であるというのが研究者の間で増えている一致した意見である。

Conservative：余裕ある；保守的な

1. The assumption will result in very **conservative** estimates for nutrient concentrations.
この仮定は栄養塩濃度に対しかなり余裕ある推定値をもたらすであろう。
2. It should be noted that our soil data comparisons are an extremely **conservative** view of the data.
我々の土壌データの比較は，極めて保守的なデータの考察であることが指摘されるべきである。
3. Based on recent studies of soil ingestion, these values should **be considered conservative**.
土壌摂取の最近の研究に基づくと，これらの値は余裕ある値であると考えるべきである。
4. The exposure periods are most likely overestimated and therefore the corresponding soil cleanup guidance should **be considered conservative**.
暴露期間が過大に見積もられた見込みが最も高く，そのため，対応する土壌浄化指導は余裕があると考えられるべきである。
5. Assuming that the substance under consideration **is nonconservative** and decay or increases in accordance with a first-order reaction, the following equation is developed.
考慮中の物質は反応性があり，一次反応に従って分解または生成するとの仮定で，次の式が構築される。

Consider：考慮する

1. Definition of these regions must **consider** the regional tolerance, resilience, and attainable quality of ecosystems.
これらの地域の定義は，生態系の地域的な耐性，弾性，到達可能な質を考慮せねばならない。
2. Factors influencing the partition of heavy metals **are considered** in detail.

重金属の分配に影響を与える要因が詳細に考慮される。
3. Based on recent studies of soil ingestion, these values **should be considered** conservative.
   土壌摂取の最近の研究に基づくと，これらの値は余裕ある値であると考えるべきである。
4. The results of the arsenic analyses **can be considered** relative to the flow conditions in the creek.
   ヒ素分析の結果は，小川の流れの状態と比較して考慮されえる。
5. Nonexceedance of the criteria for all indices **would be considered** protective of the community.
   指数すべてについてクライテリアの超過になしには，その生物群集が保護されているとみなされるであろう。
6. If production of a site **is considered** high based on organism abundance and/or biomass, biological condition **would be considered** good.
   生物の存在量および／または生物体量に基づきある地点での生産が高いとみなされるならば，生物学的状態は良好だと考えられよう。
7. **It was considered** appropriate for this study for several reasons.
   それはいくつかの理由でこの研究に適切であると思われた。
8. **When considered** on the basis of agencywide water programs, this percentage is approximately 6%.
   機関全体の水関連プログラムベースで考えた場合，この割合は，約6％である。

- considering ..... ：…を考慮すると

9. **Considering** all the data obtained in the study, the reaction appears to be zero-order, i.e., the rate is independent of the ammonia concentration.
   この研究で得られたすべてのデータを考慮すると，この反応はゼロ次であるように思われる。すなわち，その速度はアンモニアの濃度の影響を受けない。

- **Consideration**(See **Consideration**)：考慮(➡ **Consideration**)

Considerable：かなりの

1. The discovery has **considerable** environmental significance.
   この発見は，かなり環境的な重要性をもつ。
2. The value obtained may be subject to **considerable** error.
   得られた値は，重大なエラーに左右されるかもしれない。
3. It has drawn **considerable** attention in recent years.
   それは近年かなりの注目を集めた。

4. There is **considerable** validity in this theory.
   この理論には，かなりの妥当性がある。
5. During the past few weeks, the project has made **considerable** progress.
   これまでの数週の間に，このプロジェクトはかなりの進歩を成し遂げた。
6. There appears to be **considerable** difference between the A, B, and C samples.
   A，B，Cサンプル間にかなりの相違があるように思われる。
7. There is general agreement that these ecological regions exist, but there is **considerable** disagreement about how to define them.
   こうした生態地域（エコリージョン）が存在するという一般的な合意はあるものの，その定義づけについては，かなり大幅な意見の相違がある。
8. **Considerable** environmental concern has focused on these nonbiodegradable and toxic heavy metals
   かなりの環境に関する懸念が，これらの生物的分解不可能な，および毒性のある重金属に集中されてきた。
9. Many facilities will likely have **considerable** data for one assessment but relatively little for another.
   多くの施設で，1つのアセスメントに関するデータベースは，豊かだが，他のアセスメントに関するデータベースは，貧弱である。

- **a considerable number of** ..... : かなりの数の…
- **a considerable amount of** ..... : かなりの量の…

10. Over the past few decades, **a considerable number of** studies have been conducted on the effects of temperature on the growth of nitrifying bacteria.
    ここ数十年にわたり，温度が硝化菌の成長に及ぼす影響に関する研究がかなり行なわれてきた。
11. **A considerable amount of** information is available on the effect of inhibitors on culture of *Nitrosomonas sp*.
    Nitrosomonas種の菌株に対する阻害剤の影響に関するかなりの量の情報が利用可能である。

Consideration：考慮

1. A relationship was derived on the basis of energy **considerations**.
   一つの関係がエネルギー考慮に基づいて導かれた。
2. **Considerations of** statistical power should be integrated into the risk assessment process.
   統計の利用性への検討をリスクアセスメント（危険評価）過程に盛り込むべきであ

る。
3. Biological processes can be divided into several categories **for consideration** as potential metrics.
生物学的プロセスは，潜在的メトリックとして考慮するに当たり，いくつかの範疇に分けることができる。

- **Based on these considerations**：これらの考慮に基づいて
- **Due to the above considerations** .....：上記の考慮から
- **From a consideration of** .....：…の考慮から

4. **Based on these considerations**, a settling velocity of 1.2 m/d is assigned as characteristics of the James River solids.
これらの考慮に基づいて，1.2m/d の沈降速度が James River の固体の特徴として与えられている。

5. **Due to the above considerations**, it is felt that.....
上記の考慮から，…かと思われる。

6. **From a consideration of** microbial kinetics, argument can be made against that .....
微生物生成速度の考慮から，…対して議論されうる。

- **Little consideration has been given to** .....：
…はほとんど考慮されてこなかった

7. **Little consideration has been given to** the effects of salt water on the nitrification process in an activated sludge treatment system.
活性汚泥処理システムにおいて，塩水が硝化過程に及ぼす影響はほとんど考慮されてこなかった。

- **~ take into consideration** .....：
～は…を考慮する；～は…を考慮に入れる
- **taking ..... into consideration**：…を考慮して
- **taking into consideration** .....：…を考慮して

8. Ecosystem should also **be taken into consideration**.
生態系も考慮されるべきである。

9. One must **take into consideration** the season, as well as precipitation and temperature deviations (both long and short term).
季節，並びに降水量・温度の偏差（長期的／短期的）を考慮しなければならない。

10. The analysis is based solely on the presence or absence of a group and does not **take into consideration** the abundance of that group.
この分析は，ある集団の有無のみに基づいており，その集団の量を考慮していな

い。
11. **Taking this into consideration**, the federal government enacted the following legislation.
これを考慮に入れて，連邦政府は次の法律を制定した。
12. The means to achieve the standard may be flexible, **taking into consideration** site-specific factor.
基準を達成する手段には，地点特定因子を考慮したうえでの柔軟性をとりうる。

- **under consideration**：考慮中の

13. Assuming that the substance **under consideration** increases in accordance with a first-order reaction, the following equation is developed.
考慮中の物質が一次反応にしたがって増加するという仮定のもとで，次の式が求められた。

Consist：からなる

- ~ **consists of** .....：~は…からなる

1. Development and implementation of water quality criteria **consists of** four primary steps: ....., ...
水質クライテリアの設定・実施は，以下の4つの主要ステップからなる。それらは…
2. One subregion **consists of** disjunct areas at or above timberline with heavy snow pack and most of the alpine glacial lakes in the ecoregion.
ある小地域は，高木限界以上の豪雪地帯を持つ離散区域からなり，アルプス型氷河湖の大半がこの生態地域（エコリージョン）にある。
3. For natural substrates, sampling gear **consists of** a small knife, a tooth brush, and aluminum foil for rocks that can be removed from the stream.
自然底質の場合，サンプリング用具は，小型ナイフ，歯ブラシ，アルミ箔であり，水流から取り除ける岩石に使用する。

Consistency：整合性

1. These criteria will insure **consistency in** water quality assessment, licensing and certification, and enforcement of water quality standards.
これらのクライテリアは，水質評価，認可・認定，ならびに水質基準の執行における整合性を保証するであろう。
2. An approach that can use the same framework and information provides

valuable **consistency among** the many different programs in which the protection of aquatic life is a goal.
同じ枠組みおよび情報を用いうる手法は，水生生物保護を目標とする多くの様々なプログラムにおける貴重な整合性をもたらす。

Consistent (Inconsistent) (Consistently)：
整合的；一貫した（つじつまが合わない）（一貫して）

1. It is of critical importance in biological monitoring to collect a **consistent** and reproducible sample.
   整合的かつ再現可能なサンプルを採集することは，生物モニタリングにおいてきわめて重要である。
2. There was no **consistent** relationship between the specific oxidation rate and the concentration of organic matter in the reactor.
   この反応槽で，特定な酸化速度と有機物の濃度の間には一貫した関係がなかった。
3. Whatever the explanation, the weak and **inconsistent** detection of naphthalene did not allow estimation of a threshold concentration by the method used here and elsewhere.
   説明が何であるとしても，微弱，かつ不規則なナフタリンの検出は，ここで使われた方法，そしてほかのところでも使われた方法による濃度閾の推測を難しくした。
4. The general principles used in defining sites seem **consistent** over wide geographic areas.
   地域の確定に用いられる一般原理は，広範な地理的区域に共通なようである。

    - **consistent with .....** ：…と一貫して
    - **～ is consistent with .....** ：
      ～は…と一致する；～は…と一貫している

5. **Consistent with** the results of Wu and Yao (2001), ...
   Wu and Yao (2001)の結果と一貫して，…
6. These results **are consistent with** the fact that .....
   これらの結果は…という事実と一貫している。
7. This result **is consistent with** other studies that .....
   この結果は…する他の研究と一致している。
8. The biocriteria **are consistent with** a good level of community performance.
   生物クライテリアは，高水準の生物学的能力と一貫している。

- **consistently**：一貫して

9. The effluent nickel concentration **consistently** averaged greater than 95 percent of the influent nickel concentration.
   排水中のニッケルの濃度は，一貫して平均して流入水中のニッケル濃度の 95 パーセント以上であった。
10. Effluent TCA concentrations from the reactor CF-2 were **consistently** lower than influent TCA concentrations.
    反応槽 CF-2 からの排水中の TCA 濃度は，一貫して流入水の TCA 濃度より低かった。
11. A better tool was needed to more **consistently** and accurately characterize the aquatic communities.
    水生生物群集をより整合的かつ正確に特性把握するための優れたツールが必要であった。
12. A **consistently** high percentage of agreement occurred only when both the volunteer and professional analyses rated the site in the fair/poor ranges.
    高い一致が一貫して生じる事例は，ボランティアと専門家の両方が可／不可の範囲にある地点を対象とした場合に限られていた。

Constant (See also **Coefficient**)：一定の；定数（→ **Coefficient**）

1. It may be valid to argue that the solar spectrum is sufficiently **constant**.
   太陽光線スペクトルが（条件を満たすのに）十分一定であるという議論は妥当であろう。
2. The estimated **rate constants** are listed in Table 2.
   推定される速度定数を表 2 にリストする。
3. These **constant values** were found to have a satisfactory fit to all the experimental data.
   これらの定数は，満足のいく程度で，実験データをすべて満足することが判明した。
4. The **rate constants** for oxidation of carboxylic acids are strongly correlated with pH over a range of 4–8.
   カルボン酸の酸化速度定数は 4 から 8 の範囲で pH と強い相関関係がある。
5. The **stability constant** was determined by the method of Miyamoto and Kobayashi (1986) with minor modifications.
   安定度定数は，Miyamoto と Kobayashi (1986) の方法を少しの修正した方法で決定された。
6. Knowledge of the **rate constants** governing the uptake and clearance of

chemical in fish is required.
魚体内での化学物質の摂取・除去を左右する速度定数の知識が必要とされる。

Contact：接触する

1. A 10–mL sample of calibration standards **was contacted with** 1 mL of hexane in conical bottom test tubes followed by shaking for 2–5 minutes by hand.
   コニカルボトム試験管で，10 mL の較正用標準溶液サンプルに 1 mL のヘキセンを加えてから，2–5 分間手で振った。
2. All data points in the individual experiments appeared to become linear over the range of **contact times** studied.
   個別の実験でのすべてのデータポイントが，検査された接触時間範囲では直線になるように見えた。

- ～ **is in direct contact with** .....：～は … と直接接触する

3. They **are in direct contact with** the water and are directly affected by water quality.
   それらは，水と直接接触し，水質の影響を直接受ける。

Contain：含む

1. These descriptions are based on information **contained** in the EPA reports (EPA, 2002).
   これらの記述は，EPA の報告 (EPA，2002) に含まれている情報に基づいている。
2. The approach in using ultrasound is obviously applicable to organic compounds which **contain** halides.
   超音波を使うアプローチは，ハロゲンを含む有機化合物に対し明らかに適用できる。
3. Cell 4 **contains** a 4 m wide by 8 m long pool with a capacity of about 200,000 liters.
   セル 4 は幅 4 メートルに長さ 8 m，容量が約 200,000 リットルのプールを内蔵している。
4. These plots **contain** sample size, medians, ranges with outliers, and 25th and 75th percentiles.
   これらの図は，サンプル規模，中央値，外れ値の範囲，25 パーセンタイルおよび 75 パーセンタイルを含んでいる。
5. These simplified models **contain** two types of terms describing the

- containing ..... : … を含んでる

6. Lobsters were maintained in an aquarium **containing** artificial seawater at 10 ℃.
ロブスターは，人口海水を入れた水槽の中で 10 ℃で飼育された。
7. The DDT plant discharged wastes **containing** DDT residues into a drainage ditch which flowed into the river.
DDT 製造プラントが DDT 残余を含んでいる廃液を川に流れ出る排水溝に放流した。
8. The company deposited 1,000 tons of wastes **containing** chromium around Tokyo and in the neighboring Chiba Prefecture.
この会社は，東京周辺と隣接する千葉県に 1,000 トンのクロムを含んでいる廃棄物を捨てた。

Content : 含有量；含有物

1. The error bars show the minimum and maximum average measured soil-water **contents**.
エラーバーは，測定された土壌の水分含有量の最小・最大平均値を示している。
2. The relationship between cellular copper **content** and cupric ion activity shows good agreement with the hyperbolic function.
細胞の銅含有量と銅イオン活性の間の関係は双曲線関数で示すと良く一致する。
3. The EPA conducted a more extensive review of the contaminant **content of** the existing deposits.
EPA は，既存の堆積物での汚染物質含有量のいっそう広範なレビューを実施した。
4. Under these conditions, several thousand years would be required for removal of entire gypsum **content of** the tailings.
これらの条件下では，選鉱廃物の含有物である石膏全部の撤去には数千年が必要とされるであろう。

Context : 文脈；状況；関連

- in the context of ..... : … という文脈において

1. Migration of contaminants from the subsurface occurs **in a** variety of waste

management **contexts**.
地表面下からの汚濁物質の移動は，さまざまな廃物管理状況で生じる。
2. At least **in the context of** antidegradation, USEPA recognized that human and ecological use protection involves more than maintenance of chemical water quality.
少なくとも劣化防止の文脈において，USEPAは，人為的・生態的な使用保護が化学的水質維持以上のものを伴うことを認識していた。
3. The question of chemical speciation poses one of the most difficult problems to be resolved by the chemist, especially **in the context of** synergistic effects as encountered in natural water.
化学的種分化への問いは，特に自然水域で見られる相乗効果という文脈において，化学者によって解決されるべき最も難しい問題の1つを提示している。

- **within ..... context** ：…文脈において

4. Data are evaluated **within** the ecological **context** (water body type and size, season, geographic location, and other elements) that defines what is expected for similar water bodies.
データは，生態学的文脈（水域の種類・規模，季節，地理的位置，および他の要素）において評価され，これによって類似の水域について何が予想しうるかが決まる。

Continue ：続ける

1. Not only with this result in a loss of valuable information, it will likely result in the **continued** degradation of the aquatic resource.
これは，貴重な情報の損失を招くばかりか，水生資源の継続的な悪化を招く見込みが強い。
2. The condition of the pavement in the area needs to be evaluated to determine the potential for **continued** infiltration and leaching.
その区域の舗装面の状態を評価し，継続的な浸入・浸出の可能性が判断されるべきである。

- **continues to .....** ：…し続ける

3. as evidence **continues to** mount that .....,
.....の証拠が増え続けるにつれて
4. Heavy metal pollutants **will continue to** receive increased attention.
重金属汚染物は注目を集め続けるであろう。
5. There is a reason to believe that it **will continue to** be so well into the

future.
　それが未来に向けて良くあり続けるであろうと確信する理由がある。
6. Many thousands of miles of United States streams have been and **continued to** be degraded each year.
　米国の河川は，年に数千 mile も悪化してきたし，今も悪化し続けている。
7. The development of ecological regions has been, and **will** probably **continue to** be, challenging and controversial.
　生態地域（エコリージョン）の設定は，従来，そして今後もおそらく，課題多く，論争の的となり続けることであろう。
8. There is little question that aquatic resources have been and **continue to** be degraded by a myriad of land use and resource use activities.
　水生資源が無数の土地利用・資源利用活動によって劣化してきたこと，また劣化し続けていることには，ほとんど疑問の余地がない。

Contrary：逆に；正反対に

- **contrary to** ..... ：… に反して；… とは逆に
- ～ **is contrary to** ..... ：～ は … と反対である

1. **Contrary to** this common assumptions, ...
　この一般的な仮定とは逆に，…
2. **Contrary to** this hypothesis, the results indicated that .....
　この仮説とは逆に，結果は … を示した。
3. Their results **are contrary to** ours.
　彼らの結果は我々の（結果）と正反対である。
4. The preceding discussion of the analysis of the capabilities of biological survey data to discriminate different types of impacts **is contrary to** several of the assertions of Suzuki (1999) in a critique of community indices.
　生物学的調査データが様々な影響の種類を識別できる能力への分析に関する以前の考察は，生物群種指数を批判した Suzuki (1999) の主張の一部に反している。

Contrast：対照的

- **By contrast**：それと対照して

1. Methylmercury is almost completely absorbed. **By contrast**, inorganic mercury is only poorly absorbed.
　メチル水銀はほとんど完全に吸収される。それと対照して，無機水銀はわずかに吸収されるだけである。

- **~ contrast with .....** ：～は…と対照をなす；～は…と反する
2. The inclusion of Oligochaeta as a toxic tolerant group **contrasts with** the original findings of Brinkhurst (1965).
貧毛綱を毒物耐性集団として含めることは，Brinkhurst (1965)の元の成果に反している。

    - **in contrast** ：対照的に；逆に
3. **In contrast**, cavitations and the shock waves can accelerate solid particles to high velocities.
それと対照的に，キャビテーションと衝撃波は固体の微片を高速度に加速することがでる。
4. **In contrast**, areas of Main with relatively intact landscapes and stream habitat often have high biological integrity.
逆に，メイン州で景観および水流生息場所が比較的損なわれていない区域は，生物学的健全性が高いことが多い。
5. Impact standards, **in contrast**, require that a certain result must be achieved.
対照的に，影響基準は，ある結果が達成されることを要求している。

    - **in contrast to .....** ：…と対照的に
    - **~ is in contrast to .....** ：～は…と対照的である
6. **In contrast to** the complexity of the processes being modeled, the model itself must be as simple to use as possible.
モデル化されているプロセスの複雑さとは対照に，モデルそれ自身の使用ができる限り簡単でなければならない。
7. **In contrast to** many other human impacts, habitat loss can be essentially irretrievable over a human time frame.
他の多くの人為的影響と異なり，生息場所消失は，基本的に人間の時間枠では取り返しがつかない。
8. **In contrast to** the relatively low public concern about periphyton, they have been used extensively in the analysis of water quality for several decades.
付着生物は，一般市民からの関心が比較的低いのに比して，数十年にわたり水質分析で盛んに用いられてきた。
9. Trace metals are not usually eliminated from the aquatic ecosystems by natural processes, **in contrast to** most organic pollutant.
微量金属は，ほとんどの有機汚染物質と対照的に，通常自然のプロセスによって水生生態系から排除されない。

10. These conclusions **are in contrast to** those of previous researchers.
    これらの結論は前の研究者たち（の結論）と対照的である。

    ★ Conversely を使って「対照的に」と表現することもあります。

11. The year 1988 and 1991 were abnormally dry and both had extremely low flows. **Conversely**, rainfall and river flows during 1981 and 1986 were closer to normal.
    1988年と91年は，異例に乾天で，いずれも流量がきわめて少なかった。対照的に，1981年から86年までは，雨量と河川流量が平年に近かった。

Contribute (Contribution) ：寄与する；貢献する（貢献）

● ~ **contribute to**..... ：～は … に寄与する

1. Natural variability also **contributes to** uncertainty.
   自然的可変性も不確実性に寄与する。
2. A number of factors **contribute to** the differences in measured values of DDTR concentrations in biota.
   多くの要因が，生物体内DDTRの測定濃度の相違に起因している。
3. Mathematical models can **contribute to** lake water quality control in two ways. First, ...and secondly, ...
   数学モデルは，2つの点で湖の水質管理に貢献する。第一に，......。そして第二に，......。
4. The environmental research laboratory **contributes to** this information through research programs aimed at determining the effects of toxic organic pollutants.
   環境調査研究所は，有毒な有機汚染物の影響を決定することに向けられた研究プログラムを通してこの情報に貢献する。
5. Hirai and Maruyama **contributed** extensively **to** the early development of biological integrity, ecoregions, reference sites, and biological monitoring.
   HiraiとMaruyamaは，生物学的健全性，生態地域（エコリージョン），参照地点，生物モニタリングの初期の発展に大いに貢献した。

● **contribution from** ..... ：… による貢献

6. **The contributions from** Takahiro Konishi, Yuji Ikeda, and Yoshiaki Okuda at this workshop and their subsequent discussions were invaluable in putting together this chapter.
   同ワークショップでのTakahiro Konishi, Yuji Ikeda, およびYoshiaki Okudaに

よる貢献と，その後の討論は，本章の執筆に大いに役立った。

- ~ **made contributions to** ..... ：～は…に貢献してくれた
- **contributions were made by** ..... ：貢献してくれたのは…であった

7. Other staffs who also **made contributions to** the process include Hiroaki Nakajima, Hideki Morita, and Hiromasa Sakamoto.
   本章に貢献してくれた他職員は，Hiroaki Nakajima, Hideki Morita, そして Hiromasa Sakamoto である。
8. **Extensive contributions were made by** Makoto Fukuda, and Masahiro Miyata.
   特に貢献してくれたのは，Makoto Fukuda, および Masahiro Miyata であった。

Control (Controlling)：
制御；管理；対照（コントロールすること；制御すること）

1. These animals were introduced to the caldera as a means of mosquito **control**.
   これらの動物は，カルデラに蚊の制御手段として導入された。
2. Such information can influence decisions **to control** certain substances or processes that might have been overlooked or underrated in an evaluation based on only one group.
   そのような情報は，1集団のみに基づく評価において看過または過小評価されてきた物質や過程を制御する決定に影響を及ぼしうる。
3. This approach assumes that the **control** and study site differ only in the presence of pollution.
   この手法は，対照地点と研究地点が汚濁の有無においてのみ異なると仮定している。
4. To help us understand the earlier experiments, we are now running more **controlled** experiments.
   以前の実験を理解するために，我々は今いっそう制御された実験を行なっている。

- **controlling** ：コントロールすること；制御すること

5. Insights into the factors **controlling** the chemistry of arsenic in the tailings are provided by the results of the laboratory experiments.
   選鉱廃物中のヒ素の化学的性質をコントロールしている要因が，室内実験の結果によっての洞察される。
6. Knowledge of the factors affecting the movement and transformations of these substances are of obvious importance in understanding and **controlling** the hazard.

これらの物質の移動と変化に影響を与える要因の知識は，その物質の危険性を理解しコントロールしていくために当然重要である。

Core：コア

1. One or two **cores were taken from** each lysimeter.
   一つまたは二つのコアをライシメターから採集した。
2. Soil concentrations for the **coring** periods are shown in Figures 4.
   コア収集期間での土壌における濃度を図4に表わす。
3. The soil **core** analyses correspond to values averaged over 10 cm intervals from the surface to 60 cm depth.
   この土壌コア分析結果は，地表から深さ60 cmまで10 cm間隔の平均値である。

★ Core に関連する例文を下に示します。

4. The bottom 10 cm of soil was analyzed in 2 cm layers.
   下から10 cmの土壌を2 cmごとに分析した。

Correlate（Correlation）：相関している（相互関係）

- ～ **is correlated with** ……：
   ～は…に相関している；…は…と関連している

1. The condition of the immediate riparian zone **was correlated with** the degree of impairment.
   隣接水辺地帯の状態は，損傷度に相関していた。
2. The rate constants for oxidation of carboxylic acids **are strongly correlated with** pH over a range of 4–8.
   カルボン酸の酸化速度定数は4から8の範囲でpHと強い相関関係がある。
3. Matsuda (1990) briefly discussed several studies that have shown biomarker responses **correlate with** predicted levels of contamination and with site rankings based on community level measures of ecological integrity.
   Matsuda (1990) は，予測された汚染レベルや，生態学的健全性の生物群集レベル測度に基づく地点順位と相関するバイオマーカーを明らかにした諸研究について簡潔に論じた。

- ～ **show strong correlation with** ……：～は…と強い相互関係を示す
- ～ **have a significant correlation with** ……：
   ～は…と有意な関係がある

4. During the first step in TCE oxidation, the rate constant k does not **show strong correlation with** its concentration.
   TCE 酸化反応の初期のステップにおいては，反応速度定数 k は TCE の濃度と強い相互関係を示さない。

   ● **correlation between A and B**：A と B の間の相互関係

5. Examination of Table 2 provides little evidence of any **correlation between** influent metal concentrations **and** percentage removal.
   表2の考察から，流入水の金属濃度と除去率との間の相互関係に関する証拠は，ほとんどなにも出てこなかった。

6. In both cases, good **correlation** is found **between** the logarithm of the photochemical rate **and** the solvent vapor pressure.
   両方のケースで，良好な相互関係が光反応速度の対数値と溶媒の蒸気圧の間に見いだされる。

7. There was a significant **correlation between** the calculated zinc concentration **and** the number of cells present.
   計算された亜鉛濃度と存在しているバクテリア数の間に有意な相互関係があった。

   ● **Correlation coefficient**(See **Coefficient**)：相関係数（➡ **Coefficient**）

Correspond：対応する；一致する；符合する

   ● **~ correspond to** .....：
   ~は … に対応する；~は … と一致する；~は … と符合する

1. As can be seen from Table 4, the BIOMODEL estimates of biotic DDTR concentration **correspond** fairly closely **to** measured concentrations.
   表4から見られるように，BIOMODEL によって推定された生体内の DDTR の濃度は測定された濃度にかなり近い。

2. These streams were all sampled during the summer period (June through September), to **correspond to** critical low flow and elevated temperature conditions.
   これらの河川は，限界低流量および高温状態に対応するため，すべて夏季(6～9月)にサンプリングされた。

3. The problem with using this type of framework for geographic assessment and targeting is that it does not depict areas that **correspond to** regions of similar ecosystems or even regions of similarity in the quality and quantity of water resources.
   この種の枠組みを地理アセスメントおよびターゲット設定に用いるうえでの問題

点は，類似生態系の地域や，水資源の質・量において類似な均一地域に符合する区域を描かないことである。

- **correspond with** ...... ：…と一致する

4. The model values **correspond** reasonably well **with** the experimental data.
モデル値は実験データとかなりよく一致する。

- **corresponding** ...... ：対応する…

5. The pH dramatically increases with depth with a seemingly **corresponding** increase in calcium and magnesium.
pHは深さとともに劇的に上昇し，うわべはカルシウムとマグネシウムの上昇と対応するかのようである。
6. The exposure periods are most likely overestimated and therefore the **corresponding** soil cleanup guidance should be considered conservative.
暴露期間が過大に見積もられた見込みが最も高く，そのため，対応する土壌浄化指導は保守的であると考えられるべきである。

Correspondence：通信；関係

1. Other sources of information include interviews and review of internal **correspondence**.
他の情報源はインタビュー，そして内部通信のレビューである。
2. Ordinate values in Figure 2 are plotted logarithmically to show linear **correspondence** to logarithmic increase of bacterial concentration.
図2の縦座標には，バクテリアの濃度の対数増殖が直線的であることを示すために対数値がプロットされている。

Cost：コスト；費用

1. The advantage is in **cost savings** by limiting the number of samples sent for laboratory analyses.
利点は，分析のために送られるサンプルの数を制限することによるコスト削減である。
2. Possibilities exist for benthic macroinvertebrates, such as insect larval head capsule abnormalities or aberrant net-spinning activities of certain caddis flies, but these metrics **are** currently **cost-prohibitive**.
底生大型無脊椎動物について，昆虫幼生の頭部膜異常や，トビケラの異常な造網行動といった可能性が存在するものの，これらのメトリックは，目下，高費用すぎる。

3. Approaches for assessing the response of stress proteins in organisms when exposed to chemicals are currently under development, but they must **be cost-effective** before they are likely to be widely used in biological assessment and criteria program.
化学物質に曝された際の生物体内のストレス蛋白質の応答を評価するための手法を目下案出中である。だが，費用効果的でなければ，これらが生物アセスメントおよび生物クライテリアプログラムにおいて多用される見込みは，低いと思われる。

Couple：いくつかの；連結する

1. The idea that conditions were pristine in North America prior to Europian settlement has been convincingly challenged in the past **couple of** decades.
欧州人の入植前の北米では自然（無傷）の状態だったという見解にたいして，ここ数十年，説得力ある異議が申し立てられてきた。
2. This need was part of a larger concern for a framework to structure the management of aquatic resources in general and **was coupled with** an increasing awareness that there was more to water quality management than addressing water chemistry, which had been the primary focus.
この必要性は，水生資源全般の管理を構成する枠組みへの大きな関心の一部であり，それまで中心的主題だった水の化学的特性への対処よりも水質管理の方が重要だとの認識と相まっていた。
3. EXAMS **couples** data describing the behavior of a chemical toxicant **with** data describing relevant transport, physical, chemical, and biological characteristics of any given aquatic system to predict the steady-state distribution of the chemical in the system.
EXAMSモデルは，毒性化学物質の挙動を説明するデータと関連する輸送，物理学的・化学的・生物学的特性を説明するデータと連結し，このシステムにおいてこの物質の定常状態での分布を予測する。

Course：通る；コース

1. The Big Lost River **courses through** the Idaho National Laboratory.
Big Lost 川はアイダホ国立研究所地区を通り流れている。
2. In large part, the properties of a specific energy source determine the **course of** a chemical reaction.
多くの場合，特定のエネルギー源の特性が化学反応のコースを決定する。

- **time course**：時間経過

3. The **time course** data set shows that higher the MIBK inlet concentrations are, the higher the MIBK degradation rates are.
   時間経過のデータセットは，MIBKの入口での濃度が高ければ高いほど，MIBKの分解速度が速いことを示している。
4. Further investigations into the **time course** of metal uptake in activated sludge have revealed that this process occurs in two stages.
   活性汚泥中で時間とともに変わる金属摂取をさらに調査し，このプロセスが2つの段階で起こることを明らかにした。

- **during the course of .....** ：… の間に … の過程で

5. **During the course of** this research it was observed that .....
   この研究過程で … が観察された。
6. **During the course of** testing, several modifications were made to the criteria.
   検定中にクライテリアに対しいくつかの修正が行われた。
7. A number of these documents were reviewed **during the course of** this study and the review document is appended as Appendix A.
   この研究過程で多くのこれらの証拠書類がレビューされ，レビュー書類が付録Aとして付加されている。
8. To study the kinetics of destruction, 10−mL samples were periodically withdrawn from the reactor for analysis **during the course of** experiments.
   分解速度を研究するために，実験中にサンプル 10−mL を反応器から定期的に抜いた。

Court (See also Law)：法廷（→ Law）

1. **Courts** have noted that the implementation of water quality criteria that are not based on chemical−specific numeric criteria can be difficult as a matter of practice.
   それでも法廷は，化学数値クライテリアに基づいていない水質クライテリアは，実際問題として難しいと指摘してきた。
2. **State courts** have not yet ruled on the validity of biocriteria per se.
   同州の裁判所は，生物クライテリア自体の適用の有効性をまだ裁定していない。
3. Traditionally, **federal courts** defer to reasonable administrative agency interpretations of federal statutes, so long as those readings are not foreclosed by the language or legislative history of the statute.
   伝統的に連邦裁判所は，連邦法への行政機関による合理的な解釈が法律の文言または立法史によってあらかじめ排除されていない限り，それらに従っている。

4. The redesignation to WWH was upheld throughout the appeals process including the Illinois **Supreme Court**.
   WWHへの指定変更は,イリノイ州最高裁など上訴過程全体で是認された。
5. The U.S. **Supreme Court** and lower **courts** have indicated repeatedly that water quality standards must be enforced through NPDES permits.
   米国最高裁判所およびその下位裁判所は,これまで再三,水質基準がNPDES許可を通じて実施されねばならないことを主張してきた。
6. In fact, in finding that dams were not required to have NPDES permits, one **court** rule that the fact that dams may cause "pollution" does not necessitate an NPDES permit where there is no addition of "pollutants."
   実のところ,ダムがNPDES許可を必要としないという所見において,ある裁判所は,「汚濁物質」の追加がない場合,ダムが「汚濁」を生じうるという事実は,NPDES許可を必要とせしめない,と裁定した。

● **in the courts**：法廷で

7. So long as the science is sound and well supported, the findings likely will be upheld **in the courts**.
   科学が堅実で十分に裏づけられる限り,その所見は,法廷で是認されるだろう。
8. Two of the proposed use designation changes were challenged **in court** because upgrades in designated use could possibly result in more stringent permit limitations.
   使用指定における昇格は,より厳しい許可限度をもたらす可能性があるため,指定変更案のうち2つは法廷で異議申立てされた。

★法廷に関連する例文を下に掲載します。

9. The appeal of the revisions of the existing LWH use assigned to the Cuyahoga River in northeastern Ohio to the WWH use was heard before the Ohio Environmental Board Review (EBR) in 1988.
   オハイオ州北東部のCuyahoga川下流へ割り振られた既存のLWH使用からWWH使用への改正に対する上訴は,1988年にオハイオ州環境審査委員会(EBR)で審理された。
10. The revision of the LWH use for the Ottawa River in northwestern Ohio to WWH was settled in favor of the revised use prior to a decision by the Ohio Environmental Board Review (EBR).
    オハイオ州北西部のOttawa川についてのLWH使用からWWH使用への改正は,オハイオ州環境審査委員会(EBR)の決定に先立つ使用改正に有利な形で和解となった。

Critical：重大な；欠かせない

1. Accurate representation of peak flow may **be critical to** a design study for a flood control structure.
正確なピーク流量の表記は，洪水防止構造物の設計研究に欠かせないかもしれない。
2. Additional issues concerning level-of-detail **are critical to** every step of the simulation process.
「細部のレベル」に関する追加の問題は，シミュレーション過程のすべてのステップにて重要である。
3. **Critical to** the process of interpreting and integrating the source material is the care that must be taken to avoid defining regionalities of particular ecoregion components such as fish or macroinvertebrate characteristics, or patterns in a single, or a set of, chemical parameters.
資料の解釈・統合過程できわめて重要なのは，魚類または大型無脊椎動物の特性や，単一または一連の化学的パラメータといった特定の生態系構成要素で地域性を画定するのを避けるよう配慮すべきことである。
4. Definition of these regions **is critical for** effectively structuring biological risk assessment.
これらの地域の定義は，生物リスクアセスメントの効果的な構築にとってきわめて重要である。
5. Although this was not part of the original experimental design, the response of the process to an original load **is critical in** determining its utility for this application.
これは本来の実験計画法の一部ではなかったけれども，本来の負荷量に対するプロセスの反応は，このアプリケーションではその利用を決定することにおいてきわめて重要である。

Crucial：決定的な；欠くことのできない；不可欠な

1. **It may be crucial** in determining the validity of soil samples.
土壌サンプルの有効性を決定することにおいて，それは欠くことができないかもしれない。
2. The **crucial question** of whether the water body becomes acidic depends on the magnitude of the watershed's acid neutralization capacity.
水成が酸性になるかどうかについてという重要な問い，流域の酸中和容量によって決まる。

- **~ is crucial to .....** : ~が…するのに不可欠である

3. Understanding the reactions of PCBs with OHC **is crucial to** predicting the future persistence of PCBs and the hazards associated with their presence.
PCBの将来の持続性およびPCBの存在による危険性を予測するのに，PCBとOHCの反応を理解することが不可欠である。

Data：データ

1. **Data** for this study come from .....
   この研究に用いられたデータの出所は … である。
2. **Data** for this research are based on .....
   この研究に用いられたデータは … に基づく。
3. The model has been calibrated with **data** collected at Woods Lake.
   このモデルは，Woods 湖で収集されたデータを用いて較正された。

   - **available data**：入手可能なデータ
   - **data available for .....**：… のために利用可能なデータ
   - **data available to.....**：… するのに利用可能なデータ

4. **The available field data** were not sufficiently detailed to allow an assessment of spatial variations.
   入手可能なフィールドデータは，空間的変動のアセスメントができるほど詳細ではなかった。
5. A modeling strategy which makes full use of **available data** must be devised.
   利用可能なデータを十分に活用するモデリング戦略が考案されなくてはならない。
6. **Data available for** assessing the effects of acid precipitation do not measure the worst-case condition.
   酸性雨の影響を評価するために利用可能なデータは，最悪ケースの状態での測定ではない。
7. There are very few experimental **data available to** support .....
   … を支持する実験データがほとんどない。

   - **data are available**：データが入手可能である

8. Only limited **data are available**.
   ほんの限られたデータが入手可能である。
9. **Data are available from** the late 1800s to the present.
   1800 年代後期から現在までのデータが入手可能である。
10. **Data are not yet available** to examine this possibility quantitatively.

この可能性を量的に検討するためのデータがまだ得られていない。
11. ....., almost **no reliable data** pertaining to the subject were available.
 …，このテーマに関して信頼できるデータがほとんどなかった。
12. **Very few measured data were available** for the instream pesticide modeling.
 水流中の殺虫剤モデリングのための測定データが極めて少ない。

 ● **data on** ..... ：…のデータ

13. Quantitative **data on** byproducts analysis is lacking.
 副性成物分析の量的データが欠けている。
14. **Data on** the kinetics of bioaccumulation of mercury by zooplankton will be presented.
 動物プランクトンによる水銀の生物濃縮速度に関するデータが発表されるであろう。
15. There is very limited published **data on** the presence of mercury in the Sakura River.
 Sakura川において，水銀の存在を発表したデータがほとんどない。
16. Table 1 summarizes quantitative **data on** .....
 表1は…の量的なデータを要約する。

 ● **data were gathered from** ..... ：…からデータが収集された
 ● **data were gathered on** ..... ：…のデータが収集された
 ● **data are collected from** ..... ：データは…から収集されている

17. **Data were gathered from** a number of experiments using a pilot-scale reactor.
 試験規模の反応装置（リマクター）を使った多数の実験から，データが収集された。
18. **Data gathered from** the piezometers indicate that.....
 ピエゾメーターから集められたデータが…のことを示している。
19. Because of the **copious quality of data gathered** during the sampling program, ...
 サンプリングプログラム期間中に収集された豊富な良質のデータによって，
20. **Data were gathered on** fish, sediment, water, and aquatic plants in the area.
 その地域で魚，堆積物，水，水生植物のデータが収集された。
21. Few studies have undertaken to compare **data collected by** volunteers with that of data collected by professionals.
 ボランティアの収集データと専門家の収集データを比較する研究は，ごく少数である。
22. Once field **data are collected**, processes, and finalized, the next step is to

reduce the **data** to scientifically and managerially useful information.
現地データが収集・処理・最終確定されたならば，次のステップは，データをまとめて科学的・管理的に有用な情報にすることである。

23. Many local and state watershed management decisions have been based, at least in part, on **data collected** by volunteer monitors.
多くの地方および州の流域管理上の決定は，少なくとも部分的にボランティアモニターの収集データに基づいている。

    - **collection of this type of data** ：この種のデータ収集
    - **data gathering** ：データ収集
    - **data collection** ：データ収集

24. **Collection of this type of data** will likely increase over the next few years as biological criteria are developed in each state and as USEPA refines the guidance on how to incorporate biological community parameters into state standards.
各州で生物クライテリアが設定され，USEPAが生物群集パラメータを州基準に盛り込む方法に関する指針を調整するにつれて，この種のデータ収集は，今後数年間に増える見込みが強い。

25. **Data gathering efforts** for the Boise River yielded adequate stream flow.
Boise川でのデータ収集の努力が，適切な流量をもたらした。

    - **data exist** ：データが存在する
    - **have data** ：データがある

26. **Data exist** from several years of studies that support these conclusions.
数年間の研究からなるデータが存在し，この結論を支持している。

27. Many facilities will likely **have considerable data** for one assessment but relatively little for another.
多くの施設で，1つのアセスメントに関するデータベースは，豊かだが，他のアセスメントに関するデータベースは，貧弱である。

    - **data is entered into .....** ：データは…へ入力される
    - **～ enter data into .....** ：～は…へデータを入力する

28. **The data is entered** directly into the electronic database.
データは，直接，電子データベースへ入力される。

29. Laboratory technicians must have the ability to use a microscope and **enter data into** a personal computer.
実験技師は，顕微鏡使用ならびにパソコンへのデータ入力の能力を備えていなければならない。

- **data are tabulated** ：データが表にされる
- **data are evaluated** ：データが評価される
- **data are translated into** ……：データが…へ変換される

30. **Data are tabulated** in the field (fish and habitat) and laboratory (macroinvertebrates), and documented via chain-of-custody procedures.
データは，現地(魚類と生息地)および実験室(大型無脊椎動物)において表にされ，過程管理手順を通じて文書化される。

31. **Data are evaluated** within the ecological context (waterbody type and size, season, geographic location, and other elements) that defines what is expected for similar water bodies.
データは，生態学的状況(水域の種類・規模，季節，地理的位置，および他の要素)において評価され，これによって類似の水域について何が期待しうるかが決まる。

32. The manual provides a good overview of how reference site **data were translated into** regional water quality criteria.
このマニュアルは，参照地点データが地域的水質クライテリアへ変換される様相に関する適切な概観を提供している。

- **data demonstrate** ……：データは…を明示する
- **data suggest** ……：データは…を示唆する
- **data provide** ……：データは…を提供する

33. **The data** summarized in Fig. 3 **demonstrate** that ……
図3に要約されているデータは…を明示している。

34. The body of **data suggests** a possible and effective approach toward the degradation of a wide variety of harmful toxic organic pollutants in wastewaters and toward the purification of drinking water.
多くのデータが，排水中の多様な有害毒性有機物の分解や飲料水の浄化に対して可能かつ効果的なアプローチを示唆している。

35. These **data provide** additional information on previously identified Solid Waste Management Units.
これらのデータは，前に確認された廃棄物処理施設についての追加情報を提供する。

36. **These data provided** some insight into the mechanisms responsible for the transport and transfer of the chemical and the necessary information to develop and calibrate a model of this system.
これらのデータは，この化学物質の輸送と転移の原因であるメカニズムに関する洞察を提供し，またこのシステムのモデルを構築し較正するために必要な情報を提供した。

Data

- **analysis of data**：データ分析
- **interpretation of data**：データの解釈
- **treatment of data**：データ処理
- **in examining data**：データを調べると
- **examination of data**：データの検討

37. Further **analyses of the data** are planned that will employ additional existing geographical classification.
他の既存の地理分類を用いるいっそうのデータ分析が予定されている。
38. Extensive laboratory investigations and **data analysis** provided the following conclusions.
大規模な研究室での調査とデータ分析が次の結論を提示した。
39. **Interpretation of** sampling **data** will require considerations of zoogeography, historical abundance and distribution, and historical ranges of variability.
サンプルからのデータの解釈には，動物地理，歴史的な数量・分布，歴史的な可変値域を考慮する必要があろう。
40. A robust statistical **treatment of these data** often can help determine whether or not the biological response is attributable to the measured pollutant.
これらのデータへの確かな統計処理は，生物応答が測定された汚濁物質に帰因するか否か決定するのに役立つことが多い。
41. **In examining the measured data** further, it was discovered that ......
観測されたデータをさらに調べると，… が発見された。

- **unpublished data**：未発表のデータ

42. Our **unpublished data** suggest the methods to be of promise for the characterization of sediments.
我々の未発表のデータは，堆積物の特性評価のために有望になるであろう方法を示唆している。

★データの量と質に関係する表現例を下に掲げます。

43. The project has developed **a large store of data**.
このプロジェクトは多くのデータをもたらしてきた。
44. The **data** on ..... are insufficient.
… についてのデータは不十分である。
45. Because of **the lack of data**, the methodology was not fully verified.
データの欠如のために，この方法論は十分に立証されなかった。

46. The best and most **abundant data** for the Iowa River was collected at Marengo, Iowa.
アイオワ川での最良の，そして最も豊富なデータはアイオワ州のマレンゴで収集された。
47. **Much of the data** initially entered into the data base was taken from a compilation provided by Itec, Inc.
最初にデータベースに入力されたデータの多くはItec社によって提供された編集物からのものであった。
48. This **wealth of data** and experience from the HSPF application on Kansas River provide major benefits for the Study.
Kansas川での豊富なデータとHSPFアプリケーションからの経験は，この研究に重大な利益をもたらす。
49. Some concern was expressed by the author of the report over the accuracy of the **majority of the data**.
大多数のデータの正確さに関する懸念が若干この報告の著者によって明示された。
50. **The amount of data** available is limited.
利用可能なデータの量が不十分である。
51. However, **quantitative data** on byproduct analysis is lacking.
しかしながら，副生成物の分析に関する量的なデータが不十分である。
52. In most case, **only a limited amount of** stream water quality **data** is available.
ほとんどの場合，ほんの限られた量の河川水質データが利用できる。
53. **Limited data on** pesticide residues are presented in Figure 9.
数少ない殺虫剤残余のデータを図9に示した。
54. Many facilities will likely have **considerable data** for one assessment but relatively little for another.
多くの施設で，1つのアセスメントに関するデータベースは，豊かだが，他のアセスメントに関するデータベースは，貧弱である。
55. The **extent and form** of biological survey **data** vary widely among states.
生物調査データの程度と形式は，州ごとで大幅に異なる。

Date：日付

- **after this date**：この日付後に
- **at the later date**：後に
- **on the same date**：同一日の

1. New SWMUs identified **after this date** will be included in the next revision

to this report.
この日付後に認められた新しいSWMUsは，この報告書の次の修正版に入れられる。

2. The axiom follows "When in doubt choose to take more measurements than seem necessary at the time since information not collected is impossible to retrieve **at the later date**.
公理としては，「疑わしい場合，その時点で必要と思われるよりも多くの測定を行うことを選びなさい．なぜなら，収集されなかった情報は，後に取り戻すことができないからである」。

3. Seasonal variability was tested by monthly sampling year-round at two streams, and showed that between-month comparisons should not be done, but upstream-downstream sampling **on the same date** is valid year-round.
季節変動性は，2つの水流での通年サンプリング(毎月)により検定されて，月ごとの比較は，すべきでないが，同一日の上流地点-下流地点サンプリングが一年中有効であることを示した。

● **To date** ：今日までに

4. **To date**, a few states have developed comprehensive and sophisticated biological criteria programs that play a critical role in protecting water resource quality.
今日までのところ，水資源の質保護において重要な役割を演じるような総合的で精巧な生物クライテリアプログラムを策定した州は，ごく少数である。

5. **To date**, qualitative methods, although used for many applications where the usefulness of the results is more important than the scientific rigor of the techniques used, have not been widely accepted.
今日までのところ，質的方法は，技法の科学的厳密性よりも結果の有用性が重要な多くの適用例で用いられているが，まだ広く容認されてはいない。

6. **To date** no formal comparison has been made between data collected by volunteers and data collected by biologists.
現在までのところ，ボランティアによる収集データと，生物学者による収集データについて，公式な比較はなされていない。

7. The focus of this discussion is primarily on stream systems, because that is the waterbody type where most of the developmental work has been done, **to date**.
本論の焦点は，主に水流系に置かれている。なぜなら，それは，今日までの開発作業の大半が対象としてきた水域タイプだからである。

Day：日

1. Lobsters were not fed during their captivity and were kept in the aquarium for **at least 10 days before** being used in the experiment.
   ロブスターは飼育中はえさを与えず，少なくとも実験の前10日間は水槽の中で飼われた。
2. There was a significant correlation between the calculated free zinc concentration and the number of cells present **after 5 days of** growth.
   算出した遊離亜鉛の濃度と培養5日後の細胞数との間に有意な相関関係があった。
3. **In day-to-day activities** of a regulatory agency it is important to point out to client(e.g., dischargers) that short stretches of modified stream do not preclude application of stringent water quality rules.
   管理機関の日常的活動において，(排出者など)クライアントに対し，改変水流の短区間は，厳しい水質ルールの適用を阻まないことを指摘することが重要である。

   ● **during the course of the day** ：一日の間に

4. The concentrations of metals in raw sewage may vary considerably **during the course of the day**.
   生下水中の金属濃度は一日の間にかなり変化するかもしれない。

Deal：取り扱う；量；程度

1. Much has been written about the effects of petroleum on marine life, but few experiments have **deal**.
   原油が海洋生物に与える影響については多く書かれてきたが，それに対する実験はほとんどされていない。

   ● **a great deal of** ..... ：強い；多くの；かなりの

2. This language was not trivial and caused **a great deal of** concern regarding how to define "integrity" (especially biological integrity) and the measurements to be applied.
   この文言は，些細なものでなく，「健全性」(特に生物学的健全性)の定義づけと適用すべき測定値について強い関心を生じさせた。
3. Although **a great deal of** effort has been devoted to studying the problem of pollution, most work has concentrated on freshwater and terrestrial

systems.
多くの努力が汚染問題の研究に捧げられてきたが，ほとんどの研究は真水と陸生系に専念されてきた。

4. Examination of the literature on sorption reveals **a great deal of** confusion regarding the time to reach the sorption equilibrium.
吸着についての文献調査によれば，吸着均衡に達するまでの時間に関してかなりの混同がある。

5. The increasing threat of contamination and the potential hazard have focused **a great deal of** attention on the fate of mercury in the environment.
汚染による増加する脅威と潜在的危険性が，環境における水銀の衰退に多くの注意を注いできた。

- **deal with ……**：… を扱う

6. The latter **can be dealt with** on a case-by-case or site-specific basis if necessary.
後者は，必要なら個別もしくは地点特定むけの基準で対処されうる。

7. The previous volume **dealt** primarily **with** the use of statistical techniques.
前書では主に統計学的手法の使用について述べられた。

8. While we **have dealt with** most of these issues in Texas, these and other issues will arise elsewhere, thus regionally consistency in achieving a resolution of these issues will be needed.
筆者は，テキサス州におけるこれらの問題の大半を扱ってきたが，これらおよび他の問題は，他地域でも生じることで，これらの問題の解決における地域的整合性が必要だろう。

9. Classification forms the foundation of science and management because neither **can deal with** all objects and events as individuals.
分類は，科学と管理の基礎をなす。なぜなら，すべての対象や事象を個別に扱うことはできないからである。

10. The difficulty of **dealing with** a large mass of information is that of the many interactions occurring within the community, some may be related to water quality while others may not.
大量の情報を取り扱う場合の困難は，生物群集において多くの相互作用が生じることであり，それらは，水質に関係したり，関係しなかったりする。

Decade：10年

1. **Two decades of** Federal controls have sharply reduced the vast outflows of sewage and industrial chemicals into America's rivers and streams, yet the

life they contain may be in deeper trouble than ever.
20年間にわたる連邦政府による防止策は，米国の河川・水流への下水および産業化学物質の大量流出を激減させてきたが，河川・水流に生息する生物は，以前よりも深いトラブルを抱え込んでいる。

- **the past decade**：これまでの 10 年間

2. **The past decade** has witnessed a dramatic increase in the development, testing, and application of mathematical modeling for analysis of water resources problems.
これまでの 10 年間に，水資源問題の解析のための数学的モデリングの構築，試験，そしてその適用の急激な増加を目撃することになった。

- **in the decade of the 1900's**：1900 年代の 10 年間
- **in the past couple of decades**：ここ数十年
- **in the first decade or two**：この 10 年あるいは 20 年初期に

3. **In the decade of the 1970's** water quality modeling has come of age.
1970 年代の 10 年間に水質モデリングが成熟してきた。
4. A variety models have been developed **in the past decade**.
これまでの 10 年間，多種多様なモデルが構築されてきた。
5. Much progress and advancement has been made **in the past decade**.
多くの進歩と前進がこれまでの 10 年間になされてきた。
6. The idea that conditions were pristine in North America prior to European settlement has been convincingly challenged **in the past couple of decades**.
欧州人の入植前の北米では自然（無傷）の状態だったという見解にたいして，ここ数十年，説得力ある異議が申し立てられてきた。
7. Calcite leaching and gypsum precipitation occurred **in the first decade or two**.
この 10 年あるいは 20 年初期に，Calcite の浸出と gypsum の析出が生じた。

- **for decades**：数十年間
- **for the rest of this decade and beyond**：
この残りの 10 年間，そして更に

8. It has been known **for decades** that certain groups of macroinvertebrates are, in general, more tolerant of pollution than are others.
大型無脊椎動物の一部集団が一般的に他集団に比べ汚濁耐性が強いことは，数十年前から知られている。
9. Many petrochemical companies will have to operate **for the rest of this**

decade and beyond.
この10年間の残り，そして更にもっと長いあいだ，数多くの石油化学会社が操業し続けなければならないであろう。

- **within several decades**：数十年以内に
- **by the end of the decade**：この10年代末までに
- **during the last decade and a half**：過去15年間に

10. At this rate, low-pH conditions will reach the bottom of the tailings **within several decades**.
この速度では数十年以内に，低pH状態が選鉱廃棄堆積物の底まで届くであろう。
11. It is envisioned that **by the end of the decade**, nearly all of the state water resource agencies will have some form of biological monitoring.
この10年代末までに，ほぼすべての州の水資源担当機関が何らかの形の生物モニタリングを有するようになるものと予想される。

- **over the past few decades**：ここ数十年にわたって
- **over the last decade**：過去10年間
- **over a decade ago**：10年以上前に

12. **Over the past few decades**, a considerable number of studies have been conducted on the effects of temperature on the growth of nitrifying bacteria.
ここ数十年にわたって，温度が硝化菌の増殖に及ぼす影響に関する研究がかなり行なわれてきた。
13. Work **over the last decade** by investigators, such as Bencala (1993), have emphasized the importance of riparian areas for maintaining the quality and function of the hyporheic zones of streams.
Bencala (1993) などの調査者による過去10年間の研究は，水流のhyporheic地帯の質・機能の維持にとっての水辺区域の重要性を強調した。

Decision：決定

1. The likelihood of a false negative **decision** is higher due to uncertainty regarding the vertical extent of soil contamination and the potential for groundwater contamination.
縦方向の土壌汚染と地下水汚染の可能性に関して確信がないために，誤った否定の決定になる可能性がより高い。
2. The uncertainty associated with the definition of probable conditions and the consequences of a false positive or false negative error has been considered as part of the **decision** selection process.

ありそうな条件の定義と虚偽の肯定的であるか，あるいは虚偽の否定的なエラーの結果と結び付けられた不確実性は決定選択過程の一部であると考えられました。

- A ..... decision is to ..... ：…の決定は…することである

3. A second **decision is to** define the set of sites to which a criterion applies.
   第二の決定は、クライテリアが適用される地点セットを定めることである。

   - a decision was made to ..... ：…することが決定された
   - decisions on (about) ..... is made ：…について決定をする
   - decisions on ..... is made ：…について決定をする
   - decisions on ..... are not made on ..... ：
     …に関する決定は…では下されない

4. **A decision was made to** record this information even though the eventual importance of its use was not immediately apparent.
   たとえその利用の最終的重要性が直ちに明らかでなくても，この情報を記録することが決定された。

5. In most other cases, aquatic life use **decisions are made** with biosurvey and habitat data.
   他の大半の事例では，水生生物使用の決定が生物調査および生息場所データでもって下される。

6. By calibrating over large areas, **decisions about** level of protection **can be made** with a clear understanding of the differences among areas over which differing criteria might be established.
   広い地域での較正による保護レベルに関する決定は，異なるクライテリアが設定されうる各地域の差異を明確に理解したうえで下されるようになる。

7. However, final **decisions on** management actions are not made on the single, aggregated number alone.
   しかしながら，管理行動に関する最終決定は，単一の集計数値のみでは下されない。

   - make decisions on (about) ..... ：…について決定をする
   - in making decisions on (about) ..... ：…について決定を下す際

8. The protocol provided the means for obtaining such an answer in a defensible way, and the results have been used to **make** regulation **decisions**.
   このプロトコルは，そうした回答を弁護可能な形で導くための手段となった。そして，結果を用いて規制決定が下された。

9. It is important to note that numerous sites along a stream are examined before **making** a use **decision**.
   使用決定を下す前に水流沿いの多数地点が検討されることに留意すべきである。
10. Modeling is important in **making decisions about** water resources.
    モデルの構築は，水資源についての決定を下す際には重要である。

> **Decline** (See also **Decrease, Increase**) ：
> 減少；減退；減少する；減退する（➡ Decrease, Increase）

1. The **decline** in dieldrin from the reservoir outlet is due to the decreased aldrin application rates.
   貯水池出口でのディルドリンの減少はアルドリンの散布量の減少による。
2. Since many of the species were migratory, it cannot be definitely concluded that this contamination caused **the decline**.
   多くの種が移動性であるため，この汚染が種の減退をもたらしたとは確定的には結論づけられない。

★ Decline の動詞としての使用例文を下に示します。

3. The annual catch of these two species has increased as the populations of lobster **have declined**.
   ロブスターの個体数が減少するとともに，これら2種の年度収穫が増加してきた。

> **Decrease** (See also **Increase**) ：
> 減少；短縮；減少する；短縮する（➡ Increase）

1. Sample processing time **decreases** as the processor gains taxonomic expertise.
   サンプル処理時間は，処理担当者が分類群知識を得るにつれて短縮される。
2. If the contamination were to be assumed to reach the screen depth, the concentration **would decrease** due to dilution up to an order of magnitude.
   もし汚染が井戸のスクリーンの深さに達すると仮定されるなら，その濃度は希釈のために一桁減少するであろう。
3. The most accurate way **to decrease** sample variability is to collect from only one type of habitat within a reach and to composite many samples within that habitat.
   サンプルのばらつきを減じるための最も正確な方法は，ある河区において1種類だけの生息場所から採集し，その生息場所における多数のサンプルを複合することである。

4. The addition of AAA led to **decreased** BBB intermediate concentrations and increased CCC intermediate concentrations.
   AAAの添加は，BBB中間体の濃度減少とCCC中間体の濃度上昇を導いた。

Define (Defining)：定義づけする（定義している）

1. Pilot studies or small-scale research may be needed **to define**, evaluate, and calibrate the model structure.
   モデルの構成を確定し，評価し，較正するには，予備的研究または小規模研究が必要であろう。
2. A second decision is **to define** the set of sites to which a chemical criterion applies.
   第二の決定は，化学クライテリアが適用される一連の地点を定めることである。
3. Data are evaluated within the ecological context (waterbody type and size, season, geographic location, and other elements) that **defines** what is expected for similar waterbodies.
   データは，生態学的状況（水域の種類・規模，季節，地理的位置，および他の要素）において評価され，これによって類似の水域について何が期待しうるかが決まる。
4. Approaches that **are poorly defined** or based on a flawed conceptual basis provide erroneous judgments with the potential for erroneous management decisions.
   きちんと定められていない手法や，欠陥ある概念の基準に基づいた手法は，誤った判断を生じさせ，管理上の判断ミスを招く可能性がある。

    - **defining**：定義している
    - **for defining**：決定するため
    - **in defining.....**：…の確定に；…の定義づけに

5. If the coefficients **defining** this process are much greater than the other kinetic coefficients, the assumption of instantaneous equilibrium is appropriate.
   もしこのプロセスを定義している係数が他の速度係数よりかなり大きければ，瞬時均衡の仮定は適切である。
6. A recent article describes some approaches **for defining** how well various percentiles can be estimated and notes pitfalls in the use of extreme percentiles for these purposes.
   最近の論文では，様々なパーセンタイルがどれほど十分に定められるか決定する手法を述べており，これらの目的のための極値パーセンタイルの利用における落

とし穴を記している。

7. Another reason for using single-purpose frameworks, as mentioned in the preceding section, stemmed from the belief that a scientifically rigorous method **for defining** ecological regions must address the processes that cause ecosystem components to differ from one place to another and from one scale to another.
前章で述べたとおり，単一目的枠組みを用いる別の理由は，エコリージョンを定義づけるための科学的に厳密な方法が場所ごと，大きさごとに異なる生態系構成要素を生じさせる過程に取り組むべきだという考えにあった。

8. The problem has been **in defining** the regions.
問題点は，地域の定義づけであった。

9. The general principles used **in defining** metrics seem consistent over wide geographic areas.
メトリック確定に用いられる一般原理は，広範な地区に共通なようである。

10. In most areas, the use of watersheds is an obvious necessity **in defining** and understanding special patterns of aquatic ecosystem quality and addressing ecological risk.
大半の区域において，流域の利用は，水生生態系の質での特別のパターンの確定・理解，ならびに生態学的リスクへの対処において明らかに必要不可欠である。

Definition：定義

1. Most agree with a general **definition that** .....
… という一般的な定義にはほとんどの人が同意した。
2. This holistic **definition** appears to be gaining acceptance.
この全体論的な定義は，容認されつつあるようだ。
3. **Definition of** these regions must consider the regional tolerance, resilience, and attainable quality of ecosystems.
これらの地域の定義は，生態系の地域的な耐性，弾性，到達可能な質を考慮せねばならない。
4. **The definition of** degradation responses a priori is justified if clear patterns emerge from specific metrics.
特定のメトリックから明確なパターンが現れたならば，劣化応答という定義は先験的に正当化される。

- **under the definition of** ..... : … の定義下
- **in accordance with the definition** : 定義に従い
- **by definition, ～ is** ...... : 定義上，～は … である

5. These potential release sites do not fall **under the definition of** Solid Waste Management Units ; however, they are areas of environmental concern.
   これら潜在的流出地点は廃棄物管理ユニットの定義に該当しない。しかしながら，それらは環境への懸念のある地点である。
6. The numeric water quality criteria for the designated use vary by region **in accordance with the narrative definition** for each site type.
   指定使用についての数値的水質クライテリアは，各地点タイプでの記述的定義に従い，地域ごとに異なる。
7. The differentiation on the basis of experimental results may not be as straight forward as implied **by these definitions**.
   実験結果を基にしての区別は，これらの定義によって暗示されるほど単純ではないかもしれない。

Degree：程度

1. The model is less successful in predicting **the degree of** lateral dispersion.
   このモデルは，横断方向の拡散の度合いを予測する点でよい結果をださない。
2. The condition of the immediate riparian zone was correlated with **the degree of** impairment.
   隣接水辺地帯の状態は，損傷度に相関していた。
3. This distinction is made necessary by **the widespread degree** to which macrohabitats have been altered among the headwater and wadable streams in the HELP ecoregion.
   この区別が必要になった理由は，HELP生態地域（エコリージョン）における源流・水中を歩ける河川で，大型生息地が大幅に改変されたことである。

　　● **to a greater or lesser degree**：多かれ少なかれ

4. These chemicals are associated, **to a greater or lesser degree**, with suspended and sedimented particles.
   これらの化学物質は，多かれ少なかれ，浮遊粒子および堆積粒子と関係している。
5. Engel (1997) demonstrated that in some case, fuel utilizing bacteria concentrated metal ions **to a lesser degree** when killed, than when metabolically active.
   Engel (1997) は，あるケースで，燃料（ガソリン）を利用しているバクテリアが活発に新陳代謝しているときより，滅菌されているときの方が，より少ない量の金属イオンを濃縮させていたことを明示した。

　　● **to some degree**：ある程度まで

6. These observations raised several questions which we attempted to answer **to some degree** in the experiments described below.
これらの観察はいくつかの質問を提起した。その質問に，我々は下に記述した実験である程度答えを出そうと試みた。

- **to varying degree**：様々に

7. These processes have been incorporated **to varying degree** in three models.
これらのプロセスは様々に3つのモデルに取り入れられてきた。

8. While it is recognized that individual waterbodies differ **to varying degree**, the basis for having regional reference sites is the similarity of watersheds within defined geographical regions.
個々の水域が様々に異なることは，認識されるものの，地域参照地点を設けることの理由は，所定の地理的地域内での流域の類似性にある。

- **..... with a degree of certainty**(See **Certainty**)：
  …の確実度で(➡ **Certainty**)

Demonstrate (Demonstrating)：明示する；実証する(実証すること)

1. These data clearly **demonstrate** the presence of two different fish communities on the same waterbody.
これらのデータは，明らかに同一水域に2種類の魚類群集が存在することを実証している。

2. The results **demonstrate** the utility of using the theoretical range of the IBI to differentiate between and interpret different types of impacts.
結果は，異なる影響種類を差別化するためにIBIの値域を利用することの有用性を実証している。

3. This analysis **demonstrates** the need to access and interpret community information beyond the ICI and individual metrics.
この分析は，ICIおよび各メトリックを超えて生物群集情報にアクセスし解釈する必要性があることを示している。

- **～ demonstrate that .....**：～は…を明示する

4. The data summarized in Fig. 3 **demonstrate that** .....
図3に要約されているデータは…を明示している。

5. Egawa(1997) **demonstrated that** fuel utilizing bacteria concentrated metal ions to a lesser degree when killed, than when metabolically active.

Egawa(1997)は，燃料（ガソリン）を利用しているバクテリアが活発に新陳代謝しているときより，滅菌されているときのほうが，より少ない量の金属イオンを濃縮させていたことを明示した。

6. Our experience **demonstrates that** the implementation and enforcement of NPDES permits is enhanced by the site-specific information provided by biosurveys.
筆者の経験によれば，NPDES 許可の実施・施行は，生物調査から提供された地点特定情報によって強化される。

- **demonstrate how** ..... : どれほど…か実証する

7. Research must be conducted to **demonstrate how** the two approaches are complementary.
これら2つの手法がどれほど補完的か実証するための研究を行わねばならない。

8. Figure 7 also **demonstrates how** different types of impacts can be layered together in a segment.
図-7は，また，ある区間においてどれほど異なる種類の影響が積み重ねられうるかも示している。

- **as demonstrated above** : 上に明示されるように
- **as ～ have demonstrated** : ～が明示したように
- **What has been demonstrated in ～ is** ..... :
  ～で証明されていることは…である

9. **As demonstrated above**, the selection of the appropriate criterion depends on the content of the database.
上に明示されるように，適切なクライテリアの選定は，データベースの内容に左右される。

10. **As** Nishikawa and Koishi(2004) **have demonstrated**, the ideal number of reference sites must be balanced against budget realities.
NishikawaとKoishi(2004)が明示したように，参照地点の理想的な数は，現実的予算に照らして決定せねばならない。

- **demonstrating** ..... : 実証すること

11. The Biological Response Signatures can be particularly useful in **demonstrating** that the observed degradation is likely related to specific discharges, especially those involving Complex Toxicity impact type.
生物応答サインは，特に観察された劣化が複雑毒性影響に関わる具体的排出に関係する見込みが強いことを実証するうえで有用である。

Denote：示す；意味する；印である

1. The circle **denotes** the mean value obtained from a number of experiments.
   丸は多くの実験から得た平均値を示す。
2. The grey shaded area **denotes** the ± 1 standard deviation spread, whereas the thick line represents the overall mean measured soil concentration.
   灰色に塗られた範囲は±1標準偏差を示し，一方濃く塗られた線は土壌の測定濃度の総合平均値を示す。

Depend：依存する；左右される

- ~ **depend on** ..... ：~は … に依存する；~は… に左右される
- ~ **depend to a larger extent on** ..... ：~は … に大きく左右される
- ~ **depend in large part on** ..... ：~は … にかなり依存する

1. The selection of the appropriate criterion **depends on** the content of the database.
   適切なクライテリアの選定は，データベースの内容に左右される。
2. The boundaries between classes of wastes **depend** both **on** the isotope's half-life and the specific activity.
   廃棄物類の分類は，その同位体の半減期と特定活性の両方によって決まる。
3. Whether the models we create are good models or poor models **depends on** the extent to which they aid us in developing the understanding which we seek.
   我々が作るモデルが良いモデルであるか，あるいは貧弱なモデルであるかどうかは，それらのモデルが我々が求める理解力をつける手助けする程度による。
4. The type of data required **depends to a larger extent on** the chemical being tested.
   必要とされるデータのタイプは，試験される化学物質に大きく左右される。
5. The electric output of the plant **depends in large part on** the amount of heat which can be transferred in the steam generator and the condensers.
   発電所の電気出力は，蒸気発生・濃縮装置においての移送可能な熱量にかなり依存する。

Dependent (See also Independent)：
依存する；左右される（→ Independent）

1. At concentration less than 2 mg/L, the rate is concentration-**dependent**.

濃度が 2 mg/L 以下では，その反応速度は濃度に依存する。
2. The reaeration rate constant is a temperature-**dependent** parameter.
再曝気係数は温度に左右されるパラメータである。

- ~ **is dependent of** ..... ：～は … に左右される
- ~ **is dependent on**(**upon**) ..... ：～は … に左右される

3. The rate of TCE disappearance **is dependent of** pH.
TCE の消失速度は pH に左右される。
4. Faunal similarity between Midwestern river systems **is dependent on** river mile distance, along with other factors.
中西部河川系における動物相の類似は，河川距離(リバーマイル)および他の因子に左右される。

- **depending on**(**upon**) ..... ：… に応じ

5. The range of inhibition concentrations tested was chosen arbitrarily, **depending upon** the effect of each.
それぞれの効果によって，この試験濃度内での抑制濃度範囲が恣意的に選択された。
6. It might be more reasonable to argue for differing percentiles for different regions **depending upon** the nature of regional-scale degradation.
地域規模劣化の性質に応じ，異なる地域についてパーセンタイルを違えることを主張した方が合理的であろう。

Depict：描写する

1. The situation **is depicted** in Figure 3.
その状態が図 3 に描写されている。
2. Chemical interactions of the model **are depicted** in Figure 1.
このモデルの化学的な相互作用が図 1 に描写されている。
3. The problem with using this type of framework for geographic assessment and targeting is that it does not **depict** areas that correspond to regions of similar ecosystems or even regions of similarity in the quality and quantity of water resources.
この種の枠組みを地理アセスメントおよびターゲット設定に用いるうえでの問題点は，類似生態系の地域や，水資源の質・量において類似の均一地域に符合する区域を描かないことである。

Depth (See also **Dimension**) ：深さ (→ **Dimension**)

1. By accounting for additional water, **a depth** is calculated.
   追加の水を考慮することによって，深さが計算される。
2. Jamaica Bay occupies approximately 52 km^2. **Average depth** is about 4 m at mean low water.
   ジャマイカ湾の面積ははおよそ 52 km^2 である。平均水深は平均の低水位においておよそ 4 m である。
3. The soil core analyses correspond to values averaged over 10 cm intervals from the surface to 60 cm **depth**.
   この土壌コア分析結果は，地表から深さ 60 cm まで 10 cm 間隔の平均値である。
4. If the contamination were to be assumed to reach the screen **depth**, the concentration would decrease due to dilution up to an order of magnitude.
   もし汚染が井戸のスクリーンの深さに達すると仮定されるなら，その濃度は希釈のために一桁減少するであろう。

   - **at the same depth** ：同じ深さで
   - **at a depth of** ..... ：…の深さで

5. The design of the tubes allowed them to be placed into the sonicator horn **at the same depth**.
   このチューブの設計によって，それらをソニケータ（超音波細胞破砕機）—ホーンの中に同じ深さで設置することができた。
6. A background soil sample collected **at a depth of** 40 to 41 ft had a TPH concentration below the 3 mg/kg detection limit.
   40 から 41 フィートの深さから収集されたバックグランド土壌サンプルでは，TPH の濃度が検出限界の 3 mg/kg 以下であった。

   - **with a mean depth of** ..... ：…の平均水深が
   - **with a maximum depth of** ..... ：…の最大水深が
   - **with depth** ：水深と共に
   - **with soil depth** ：土壌の深さによって
   - **~ has a mean depth of** ..... ：～は平均水深が…である

7. The western basin is the shallowest, **with a mean depth of** 11 m.
   西水域は最も浅く，平均湖深が 11 m である。
8. The pH dramatically increases **with depth** with a corresponding increase in calcium and magnesium concentrations.
   pH は，カルシウムとマグネシウム濃度の増加に対応し，水深が増すと共に劇的

に増加する。
9. Another likely explanation for the biphasic degradation is that the differences in degradation rates **with soil depth**.
二相分解についてのもう1つのありえる説明としては，土壌の深さによって分解速度に相違があるということである。
10. The central basin is the largest in terms of area (16, 317 km^2) and has a mean depth of 25 m, while the eastern basin is the deepest, **with a maximum depth of** 64 m.
中央水域は面積では最大 (16,317 km^2) で，平均湖深が 25m である。一方，東水域は最も深く，最大湖深が 64 m である。

Derive：導出する

1. Tanaka and Okubo (2002) **have derived** the following solution.
TanakaとOkubo ((2002)は次の解を得た。
2. A relationship **was derived** on the basis of energy considerations.
ある関係が，エネルギーを考慮することによって得られた。
3. Narrative and numerical criteria **are derived** to ensure protection for some of these uses.
これらの利用の一部について保護を保証するため，記述クライテリアおよび数値クライテリアが導出される。

　　● ～ **is derived from** ‥‥‥：～は…に由来する

4. The total pollutant load to a waterbody **is derived from** point, nonpoint, and background sources.
水域への全汚濁負荷量は点源と面源，それにバックグランド源からなる。
5. While model applications may differ greatly in scope and purpose, it is hoped that the representative data **derived from** pilot studies will be useful in this process.
モデルのアプリケーションが応用範囲と目的で大いに異なるかもしれない一方，予備試験的研究から得られる代表的データがこのプロセスに有用あろうことが期待される。
6. Usually, expectations have been developed on a watershed or regional basis, and **have been derived from** recent field data.
通常，期待値は，流域または地域基準で設定されてきており，最近の現地データから導出されてきた。

　　● **in deriving** ‥‥‥：…を導出する際

7. The selection of reference sites from which attainable biological performance can be defined is a key component **in deriving** numerical biological criteria.
達成可能な生物学的実績が定義されうるような参照地点の選定は，数量的生物クライテリアを導出する際の重大な構成要素である。

Describe ：記述する；述べる；説明する

1. The exponential term **describes** the settling while C' (x, y) describes the dispersion.
C' (x, y)が拡散を説明する一方，指数の項は沈降を説明する。
2. This paper **describes** the construction and application of a numerical model representing mercury transformations in the aquatic environment.
この論文には，水生環境での水銀の変化を説明する数値モデルの構築とアプリケーションが記述されている。
3. A recent article **describes** some approaches for defining how well various percentiles can be estimated and notes pitfalls in the use of extreme percentiles for these purposes.
最近の論文では，様々なパーセンタイルがどれほど十分に定められるか決定する手法を述べており，これらの目的のための極値パーセンタイルの利用における落とし穴を記している。

- **briefly describe** ..... ：…を簡潔に説明する
- **poorly describe** ..... ：…は十分に説明していない

4. The purpose of this article is to **briefly describe** some critical research and information needs in surface water quality.
この論文の目的は，表流水の水質における重要な研究のいくつかと情報の必要性を簡潔に説明することである。

- ～ **is described** ：～が記述されている

5. All areas storing containerized wastes **are described**.
コンテナに入れられ貯蔵されている廃棄物のすべてのエリアが記述されている。
6. The installation methods **are described** in more detail by Ito (1999).
設置方法は，Ito (1999) によってもっと綿密に記述されている。
7. The chlorophenol transformation kinetics **is described** in Eqn 3.
塩化フェノールのこの変化速度は式3で記述される。
8. The UVUS reactor **has been described** elsewhere.
UVUS反応器は，ほかのいくつもの論文に記述されてきた。

Design：計画；設計

1. The **design** of the tubes allowed them to be placed into the sonicator horn at the same depth.
   このチューブの設計によって，それらをソニケーター（超音波細胞破砕機）ホーンの中に同じ深さで設置することができた。
2. As with all sampling **designs**, the ideal number of reference sites must be balanced against budget realities.
   あらゆるサンプリング計画と同様に，参照地点の理想的な数は，現実的予算に照らして決定せねばならない。
3. Accurate representation of peak flow may be critical to **a design study** for a flood control structure.
   正確なピーク流量の表記は洪水防止構造物の設計研究に欠かせないかもしれない。
4. The guidelines are followed step-by-step to demonstrate how the guidelines can aid in **design** of a field study.
   いかにこの指針が現地研究の計画の助けになることができるかを明示するのに，この指針が一歩一歩追髄された。

- **～ be designed to .....：**
  ～は…のために構成されている；～は…のために意図されている

5. This integrated assessment **was designed to** address the following questions: 1..., 2..., etc.
   この統合化されたアセスメントは次の質問を扱うよう意図された。1…，2…，などである。
6. EXAMS **is** specifically **designed to** simulate the fate and persistence of organic chemicals in aqueous ecosystems.
   EXAMSは，水性生態系での有機化学物質の衰退と持続性をシミュレートするために特に構成されている。
7. This system **is designed to** promote more efficient use of ambient monitoring resources and to ensure timely results.
   この手法の目的は，環境モニタリング資源の効率的利用を促すこと，タイムリーな結果を確保すること，である。
8. The traditional use of performance standards is inadequate because they **are** not **designed to** address broad goals such as ecological integrity.
   実績標準の伝統的な利用は，生態系健全性などの広範な目標に取り組んでいないためまだ不十分である。
9. This project **was designed to** involve the county's citizens in ongoing

waterways education and restoration and to foster the relationship between citizens and government to ensure that both are more responsive to the needs of the environment.
このプロジェクトの意図は，郡民が進行中の水路教育・再生活動に関与するよう促すこと，市民と政府の関係を育んで，両者が環境ニーズにより対応的であるようにすることであった．

Designate (Designation)：指定；指定する(指定)

1. The eight lysimeters **were designated by** the letters J-Q.
   八個のライシメターはJからQまでのアルファベットで指定された．
2. Substrate particle size is obtained by making observational estimates **to designate** percentage of each of the seven EPA size categories, as listed in Weber (1993).
   基質の粒子サイズは，Weber (1973) により列挙された7つのUSEPAサイズ区分それぞれのパーセンテージを指定するための観測的推定で得られる．
3. The numeric water quality criteria for the **designated use** vary by region in accordance with the narrative definition for each site type.
   指定使用についての数値的水質クライテリアは，各地点タイプでの記述の定義に従い，地域ごとに異なる．

   - **designation**：指定

4. Two of the proposed **use designation** changes were challenged in court because upgrades in designated use could possibly result in more stringent permit limitations.
   指定変更案のうち2つは，法廷で異議申立てされた．使用指定の昇格は，より厳しい許可限度をもたらす可能性があったからである．

Despite：にもかかわらず

1. **Despite** the impaired condition of the receiving water, the impact of the sewage discharge resulted in criteria exceedance in biotic index.
   受入水の悪化した状態にもかかわらず，下水排出の影響は，生物指数におけるクライテリア超過を招いた．
2. **Despite** the aforementioned impoundments and flow alterations, overall habitat conditions in the Green River are good to excellent.
   前述の人工湖と流量改変にもかかわらず，Green川の全般的生息場所状態は，良ないし優である．

3. Compared to the Scioto River, the Ottawa River exhibits evidence that, **despite** some improvements, makes it one of the most severely impaired rivers in the state.
Scioto 川に比べ，Ottawa 川は，若干の改善にもかかわらず，同州で最もはなはだしく悪化した河川の一つという証拠を示している。
4. **Despite** our growing knowledge of CP anaerobic dechlorination pathways and bacteria, few studies have rigorously examined the dechlorination kinetics of Cps.
CP の嫌気性脱塩素化反応経路およびバクテリアに対する知識が増えているにもかかわらず，CPs の脱塩素化反応速度論が徹底的に研究がされてきていない。

- **despite the fact that ……(See** Fact）：
  … という事実にもかかわらず（➡ **Fact**）

Detail（Detailed）：詳細；細部（詳細な）

1. **Details** concerning the experimental procedure are given elsewhere and are described only briefly here.
実験手順に関する詳細は，ほかの様々なところで示されているので，ここではただ手短かに記述する。
2. **The details of** the studies and results have been presented elsewhere (12–14) and are only summarized here.
その研究と結果の詳細は，ほかの様々なところに提出されている (12–14)。それなので，ここではただ要約するだけにする。
3. **A more detailed** explanation of Bailey's approach is given later in this chapter.
Bailey のアプローチのいっそう詳細な説明がこの章の後に示される。
4. **A more detailed** explanation is found in USEPA (1993) and Rankin (1994).
より詳細な説明は，USEPA (1993) および Rankin (1994) に記されている。
5. **Detailed** examination of Figure 5 shows that MEK elimination was over estimated.
図 5 の詳細な調査が，MEK 除去が過剰に見積もられていたということ示している。
6. Such **detailed** understanding of the interactions is impossible from steady-state models.
相互作用のこのような詳細な理解は，定常状態モデルからは不可能です。
7. More extensive and **detailed** investigations are necessary to gain a further understanding of these processes.
いっそう大規模かつ詳細な調査が，これらのプロセスをいっそう理解するために

必要である。
8. **A detailed** discussion of these results is given in Section 2.3.
これらの結果について，後ほど2.3節で詳述する。
9. **More detailed** discussions of biological criteria development for the States of Maine, New York, and Texas can be found elsewhere in this volume.
メイン州，ニューヨーク州，テキサス州での生物クライテリア設定に関する詳細な考察は，本書のほかのところにも記されている。

- **in detail**：詳しく；詳細に

10. Factors influencing the partition of heavy metals are considered **in detail**.
重金属の分配に影響を与えている要因が詳細に考慮される。
11. The ecological damages caused by the river modification needs to be examined **in detail**.
河川改修工事での生態系への被害について詳しく調査する必要がある。
12. The development and implementation of rapid assessment programs is discusses **in detail** by Resh and Norris (1994).
迅速アセスメントプログラムの策定・実施については，Resh と Norris (1994) が詳述している。
13. The choice of groups to be identified and analyzed is related to the measures that will be used, as was discussed **in detail** above.
識別・分析されるべき集団の選択は，前に詳細に述べたように，用いられる指標に関係している。

- **in more detail**：より詳しく；より詳細に
- **in greater detail**：より詳しく；より詳細に

14. Each of these three tasks is considered **in more detail** below.
これらの3つの仕事のそれぞれが下にもっと詳細に考慮されている。
15. The installation methods are described **in more detail** by Ito (1999).
設置方法は Ito (1999) によってもっと詳細に記述されている。
16. The risk and benefits of this proactive strategy are discussed **in more detail** below.
この先取的戦略の持つ危険と便益については，後ほど詳述する。
17. The process can be divided into three major tasks 1)...., 2)......, and 3)...... Each of these three tasks is considered **in more detail** below.
このプロセスは3つの主要な作業に分けられる。それらは1)…，2)…，3)…である。これらの作業の一つ一つは下でより詳細に考慮することにする。
18. Each of these decay pathways will be discussed **in greater detail** below.
これらの分解過程のそれぞれが下にもっと詳細に論じられるであろう。

- **in some detail**：若干詳細に
- **in any detail**：若干詳細に

19. Sonoluminescence from aqueous solutions has been studied **in some detail**.
    水溶液からの Sonoluminescence は若干詳細に研究されてきた。
20. Residence times for individual PCBs in the atmosphere have been investigated **in some detail**.
    個々の PCB の大気中での滞留時間が若干詳細に調査されてきた。
21. This assumption has not been examined **in any detail**.
    この仮定は，まだ詳しく検討されていない。

- ～ **is detailed**：～が詳述される

22. The available field data **were not sufficiently detailed** to allow an assessment of spatial variations.
    入手可能なフィールドデータは，空間的変動のアセスメントをするには十分に詳述されていなかった。

Detect：検出する

1. Arsenic **was** not **detected** in the reference wells.
   ヒ素は対照井戸で検出されなかった。
2. TCE **was detected** at concentrations in excess of the drinking water criteria.
   TCE が飲料水基準を超過した濃度で検出された。
3. Cadmium **was detected** above the recommended drinking water criteria.
   カドミウムが推奨された飲料水クライテリアを超過した濃度で検出された。
4. Chromium **was** not **detected** in excess of recommended criteria in any wells.
   クロムは，推奨されたクライテリアを超過してはどの井戸からも検出されなかった。
5. Cyanide **was** not **detected** in levels exceeding recommended drinking water criteria in any wells.
   シアン化物は，飲料水推奨されたクライテリアを超過するレベルではどの井戸からも検出されなかった。
6. Mercury **was detected** at levels close to the recommended drinking water criteria.
   水銀が飲料水推奨されたクライテリアに近いレベルにおいて検出された。

> **Determine** (Determining) (Determination)：
> 決定する；判断する（決定すること）（決定；判断）

1. One objective of the initial site investigation should be **to determine** the mass of contaminants adsorbed to the soil.
   最初の現地調査の1つの目的は土壌に吸着した汚染物質量を決定すべきことである。
2. The condition of the pavement in the area needs to be evaluated **to determine** the potential for continued infiltration and leaching.
   その区域の舗装面の状態を評価し，継続的な浸入・浸出の可能性が判断されるべきである。
3. The plots for each parameter were examined **to determine** if any visual relationship with drainage area existed.
   各パラメータについての図が検討され，流域面積との目測関係が存在するか否か決定された。

   ● **it is determined that .....**：… が決定される

4. In ancillary experiments, **it was determined that** .....
   補助的な実験で … が決定された。
5. **It has not yet been determined** which level would be most appropriate for the water quality standard.
   どのレベルが水質基準に最も適しているかは，まだ未決定である。

   ● **determining .....**：… を決定することに

6. The environmental research laboratory contributes to this information through research programs aimed at **determining** the effects of toxic organic pollutants.
   環境調査研究所は，毒性有機汚染物の影響を決定することを目的とした研究プログラムを通してこの情報に貢献する。

   ● **in determining .....**：… を決定するうえで

7. It may be crucial **in determining** the validity of soil samples.
   土壌サンプルの妥当性を決定することにおいて，それは欠くことができないかもしれない。
8. **In determining** the effect of pH on the arsenic removal in river water, ...
   河川水でのヒ素の除去に及ぼす pH の影響を決定するうえで，…
9. This could be important **in determining** how well real impairment can be reasonably expected to be detected.

これは，現実の悪化がどれほど合理的に検出されると思われるかを決定するうえで重要であろう。

10. Laboratory data, although not typically viewed as useful **in determining** reference conditions, can be valuable.
    実験データは，通常の場合，参照状態設定に有用だとみなされないが，有益にもなりうる。
11. One major factor **in determining** the cost and efficiency of the remediation of hydrocarbons is the mass of the organic compounds present in the subsurface.
    炭化水素の浄化のコストと効率を決定するうえでの一つの重要な要因は，地表下に存在している有機化合物の総量である。
12. It is increasingly recognized that photochemical processes may be important **in determining** the fate of organic pollutants in aqueous environments.
    水環境において有機汚濁物質の衰退を決定するのに，光化学プロセスが重要であるかもしれないことがますます認められている。
13. Although this was not part of the original experimental design, the response of the process to an original load is critical **in determining** its utility for this application.
    これは本来の実験計画法の一部ではなかったけれども，本来の負荷量に対するプロセスの反応は，このアプリケーションではその利用を決定することにおいてきわめて重要である。

    ● **for determining** ..... ：…を決定するため

14. Regional reference sites can fulfill a dual role as the arbiter of regionally attainable biological performance and as an upstream reference **for determining** the significance of any longitudinal changes.
    地点の選定は，数量的生物クライテリアを導出する際の重大な構成要素である。地域参照地点は，地域で達成可能な生物学的実績の規範として，また，縦断的変化の有意性を決定するための上流の参照として，2つの役割を果たしうる。
15. Without a sufficient theoretical basis it would be very difficult, if not impossible, to develop meaningful measures and criteria **for determining** the condition of aquatic communities.
    十分な理論的基礎がなければ，水生生物群集の状態を決定するための有意義な指標および基準（クライテリア）を設定することは，仮に不可能でなくても，非常に困難だろう。

    ● **by determining** ..... ：…を決定することによって

16. To a large extent, the status of biological criteria programs across the nation

can be described **by determining** the presence of these activities in each of the states.
おおむね全米の生物クライテリアプログラムの現状は，各州におけるこれらの活動の存在を決定することによって示されうる．

- **determination**：決定

17. Additionally, a t-test was added to provide statistical strength to **a determination of** significant biological impairment.
さらに，t 検定が追加され，有意の生物学的な損傷の決定に統計的効力を与えた．
18. The t-test does not form the basis of **the determination of** impairment, but tightens the conditions for a genuine exceedance.
t 検定は，悪化決定の根拠にならないが，純粋な超過についての状態を厳しくする．

- **~ make the determination**：~が決定する
- **~ permit a determination**：~が決定する

19. Persons involved in possible field tests must **make this determination** on a case-by-case basis.
可能な現地テストに関係している人々が，状況によってこの決定をしなくてはならない．
20. The data do not **permit a determination** as to whether the toxicity was due to a slug of something toxic in the water or was continuously present.
この毒性が，水に短時的に存在する毒性物質によるのか，または継続して存在する毒性物質によるのか，このデータでは決められない．

- **determination is made**：判断が下される

21. **Determination** as to whether a release has occurred **is** generally **made** at the time a waste line is decommissioned.
排出が起こったかどうかについての判断は，一般的に廃棄ラインが使用済みになる時に下される．

- **in the determination of** .....：… を決定するのに
- **for the determination of** .....：… の決定について

22. Definite patterns in biological community data exist and can be used **in the determination of** whether or not a waterbody is attaining its designated use and in identifying the predominant associated causes of impairment.
生物群集データにおける明確なパターンは，現に存在し，また，ある水域が指定使用を達成しているか否か決定するため，ならびに優勢な関連の悪化原因を確認

するため利用できる。
23. Biological data has always played a central role in the Arizona water quality standards, particularly **for the determination of** appropriate and attainable aquatic life use designations.
生物学的データは，常に，適切かつ達成可能な水生生物使用指定の決定について，特に，アリゾナ州の水質基準において中心的な役割を演じてきた。

> **Develop**（Development）：
> 構築する；進展する；発展する；定める；設ける；編み出す（設定；構築；進展；発展）

1. Yoshida (1999) **developed** a relationship between the IBI and the QHEI sufficient to forecast nonattainment due to habitat degradation with a reasonable degree of certainty.
Yoshida (1999) は，生息場所劣化による未達成を妥当な確実度で予測しうる IBI と QHEI との関係を定めた。
2. In order to maximize the meaningfulness of extrapolations from these studies and the use of data collected from national or international surveys, we must **develop** a clear understanding of ecosystem regionalities.
しかし，これらの研究からの外挿や，国内調査・国際調査での収集データ利用の意義を最大化するため，我々，生態系地域性へのより明確な理解を進展させるべきである。
3. There is a need **to develop** models which can be used to predict water quality changes within distribution systems.
給水システム内での水質変化を予測するために使うことができるモデル構築の必要性がある。
4. Assuming that the substance under consideration is nonconservative and decay or increases in accordance with a first-order reaction, the following equation **is developed**.
考慮中の物質は反応性があり，一次反応に従って分解または生成するとの仮定で，次の式が構築される。
5. Procedures **should be developed** that allow for both regional- and subregional-scale deviations from broadly established water quality criteria.
広範に定められた水質クライテリアから，地域規模および小地域規模の偏差を示すための手順を設けるべきである。
6. The ecoregions concept **has been developed** through the effort of many people, too many to acknowledge individually here.
生態地域（エコリージョン）という概念は，大勢の方々の努力によって構築されて

きたので，ここで個々にあげて感謝することができない。

- **in developing**：設定において
- **when developing**：策定時に

7. **In developing** biological criteria for water quality programs, states have undertaken a wide range of efforts to improve bioassessment methods.
水質プログラムのための生物クライテリア設定において，諸州は，生物アセスメント方法を改良する広範な努力を行ってきた。
8. The time spent by states **in developing** biological criteria includes both the bioassessments conducted and the activities related to the implementation of biological criteria within water quality standards.
諸州が生物クライテリア設定に費やした時間は，生物アセスメントの実行と，水質基準における生物クライテリア実施に関わる活動を含んでいる。
9. The efforts that states invest **in developing** biological criteria begin with biological surveys conducted to characterize the condition of the state's water resources.
諸州が生物クライテリア設定に行う努力は，まず，その州の水資源の状態の特性把握をするために行われる生物調査でもって開始される。
10. States may choose to concentrate on one or more of these uses of biological criteria **when developing** their programs.
諸州は，プログラム策定時に，生物クライテリアの上記用途のうち1つないし複数に専念することを選ぶ場合もある。

- **development**：設定
- **during the development of .....**：…の設定中

11. **Development** and implementation of water quality criteria consists of four primary steps:
水質クライテリアの設定・実施は，以下の4つの主要ステップからなる。
12. There are technical challenges for industry **during the development of** state-specific biocriteria.
州別の生物クライテリアの設定中，産業界にとって技術的課題が存在する。

- **～ have been on the development**：～は設定に置かれてきた
- **～ is under development**：～は設定中である；～は案出中である

13. The emphasis of both state and federal efforts **has been on the development** and implementation of bioassessment methods for streams and small rivers, as both the scientific and experimental databases are greatest for these waterbodies.

州および連邦による努力の主眼は，水流・小河川についての生物アセスメント方法の設定・実施に置かれており，また，科学データベースと経験データベースのいずれも，これらの水域について最も大切である。

14. Metrics for the periphyton assemblage **are** currently **under development**.
付着生物群集についてのメトリックは，目下設定中である。

15. Approaches for assessing the response of stress proteins in organisms when exposed to chemicals **are** currently **under development**, but they must be cost-effective before they are likely to be widely used in biological assessment and criteria program.
化学物質に曝された際の生物体内のストレス蛋白質の応答を評価するための手法を目下案出中である。だが，費用効果的でなければ，これらが生物アセスメントおよび生物クライテリアプログラムにおいて多用される見込みは，低いと思われる。

Deviation：逸脱；偏差；ばらつき

1. One must take into consideration the season, as well as precipitation and temperature **deviations** (both long and short term).
季節，並びに降水量・温度の偏差（長期的／短期的）を考慮しなければならない。

2. The question arose whether these individual **deviations** could be the result of measurable differences between the fishes.
これらの個々のばらつきが魚の間の測定可能な相違の結果で有るのか否かという質問があがった。

- **deviations between A and B**：AとBの間の逸脱
- **with minor deviations**：やや逸脱した

3. For such extreme operating conditions, **deviations between** model prediction **and** experiments can be expected.
このような過酷な操作条件では，モデル予測値と実験値の間のずれが予想されえる。

4. Biosurveys were conducted near Coralville Station from 1988 to 1991 using methods that conformed to Idaho DEQ protocols or **with minor deviations**.
Idaho州EPAプロトコルに準拠した，もしくはやや逸脱した方法を用いて，1989年から1991年までCoralville発電所付近で生物調査が行われた。

- **standard deviation**：標準偏差

5. Box plots have important advantages over the use of means and **standard deviations**.
ボックス図は，平均や標準偏差の利用に優る重要な利点を持つ。

6. The grey shaded area denotes the ± 1 **standard deviation** spread, whereas the thick line represents the overall mean measured soil concentration.
灰色に塗られた範囲は±1標準偏差を示し，一方濃く塗られた線は土壌の測定濃度の総合平均値を示す。

Devote：捧げる

● ～ is devoted to ……：～が…に捧げられている

1. Primary emphasis **is devoted to** groundwater.
地下水が主に重要視されている。
2. Much of the literature concerned with waste treatment applications of nitrification **has been devoted to** municipal sewage.
硝化処理適用に関連する文献の多くが都市下水に捧げられてきた。
3. Although a great deal of effort **has been devoted to** studying the problem of pollution, most work has concentrated on freshwater and terrestrial systems.
多くの努力が汚染問題の研究に捧げられが，ほとんどの研究は真水と陸生系に専念されてきた。

Differ (Differing)：異なる（異なる）

1. Microorganisms **differ** with respect to their aluminum toxicity tolerance.
微生物はアルミニウム毒性許容性において異なっている。
2. This approach assumes that the control and study site **differ** only in the presence of pollution.
この手法は，対照地点と研究地点が汚濁の有無においてのみ異なると仮定している。
3. Population dynamics of bacteria were studied in two reservoirs that **differed** in organic load.
バクテリアの個体群動態が，有機物負荷が異なった2つの貯水池で研究された。
4. While it is recognized that individual waterbodies **differ** to varying degree, the basis for having regional reference sites is the similarity of watersheds within defined geographical regions.
個々の水域が様々に異なることは，認識されるものの，地域対照地点を設けることの理由は，所定の地理的地域内での流域の類似性にある。

● ～ **differ from** ……：～ は … と異なる

- **~ differ essentially from ..... in that ..... :**
  ～は…という点で…と本質的に異なっている

5. Their result **differs** significantly **from** ours.
   彼らの結果は際立って我々のとは違う。
6. These results **differ from** earlier studies.
   これらの結果は，初期の研究(の結果)とは異なっている。
7. This method **differs from** that used by Toguchi et al. (1996).
   この方法はToguchiら(1996)によって用いられた方法とは違う。
8. These new versions had a multimetric structure, but **differed from** the original version in the number.
   これらの新バージョンは，多メトリック構造を有するが，メトリックの数，個性，スコアリングにおいてオリジナルバージョンと違っていた。
9. A scientifically rigorous method for defining ecological regions must address the processes that cause ecosystem components to **differ from** one place to another and from one scale to another.
   生態学的地域(エコリージョン)を定義づけるための科学的に厳密な方法が場所ごと，大きさごとに異なる生態系構成要素を生じさせる過程に取り組むべきである。

- **differ among ..... :** …の間で異なる

10. Fish community composition in streams has been shown to **differ among** ecoregions.
    河川中の魚類群集の構成は，生態地域間で異なることが示されている。
11. Stream classification provides relatively homogeneous classes of streams for which biocriteria may **differ among** the classes.
    河川分類は，比較的同質な河川階級を生じさせるが，河川階級ごとに生物クライテリアが異なることもある。
12. Opinions **differ among** states as to whether formal incorporation of biological criteria into state water quality standards should be the ultimate goal of all biological criteria programs.
    生物クライテリアを州の水質基準へ正式に盛り込むことが生物クライテリアプログラムの究極目標か否かに関して，諸州で意見の相違がある。
13. It might be more reasonable to argue for **differing** percentiles for different regions depending upon the nature of regional-scale degradation.
    地域規模劣化の性質に応じ，異なる地域についてパーセンタイルを違えることを主張した方が合理的だろう。

Difference：差；相違；分化

1. Greater **differences** were observed for other issues.
   他の問題点で，より大きい相違が見られた。
2. One explanation for this **difference** was loss of membrane integrity.
   この相違についての説明の1つは膜の完全性の低下にあった。
3. Whatever the reasons for the **differences**, it is important to keep in mind that........
   相違の理由が何であるとしても，…を念頭におくことが重要である。
4. If conditions are similar among regions, or if the **differences** can be associated with management practices that have a chance of being altered, it seems wise to combine the regions for the purposes of establishing biocriteria.
   仮に，各地域における状態が似ている場合や，その差違が変化の可能性のある管理慣行に関連している場合には，生物クライテリア設定のために諸地域を組み合わせた方が賢明だろう。

   ● **differences among** ..... ：…の間の差

5. **Differences among** sites would be much greater than year-to-year variability at the same site.
   地点間の差の方が同一地点での年度変化よりも大きいかもしれない。
6. By calibrating over large areas, decisions about level of protection can be made with a clear understanding of the **differences among** areas over which differing criteria might be established.
   広い地域での較正による保護水準に関する決定は，異なるクライテリアが設定されうる各地域の差異を明確に理解したうえで下されるようになる。

   ● **differences in** ..... ：…の相違；…の差

7. Some of this disagreement stems from **differences in** individual perceptions of ecosystems, the uses of ecoregions, and where humans fit into the picture.
   意見相違の一因は，生態系に対する各自の認識の差，生態地域（エコリージョン）の利用人間へ適用する場所の差である。
8. This may not have been **a difference in** state practice. It may have reflected differences in ownership, say, between federal and state or federal and private.
   これは，各州慣行の相違ではなかったかもしれない。所有における差，つまり，

連邦所有地と州有地の差，連邦所有地と私有地の差を反映していたのかもしれない。

9. The problem is that little effort is being expended on studying ecosystems holistically and attempting to define **differences in** patterns of ecosystem mosaics.
問題点は，生態系の全体論的研究や，生態系モザイクのパターン差を定義づける試行にあまり力が入れられていないことである。

10. In many arid areas, spatial **differences in** subsurface watershed characteristics have a stronger influence on water quality than the size or characteristics of the surface watershed.
多くの乾燥区域において，地表下流域特性における空間差は，地表流域の規模または特性よりも水質に対し強い影響を及ぼす。

  ● difference between A and B ：AとBの間の相違

11. There is a fundamental **difference between** the two .....
その二つの … には根本的な相違点がある。

12. There appears to be considerable **difference between** the samples A, B, **and** C.
サンプルA，B，Cの間にかなりの差があるように思われる。

13. It is of interest to point out the fundamental **difference between** A and B.
AとBの間の基本的な相違を指摘することは重要である。

14. Loss of membrane integrity was not proven as a cause for the observed **difference in** removal **between** Tests 1 **and** 2.
膜の完全性の低下は，試験1と試験2の間で観察された除去の相違が原因であるとは証明されなかった。

15. Chloride is in the range of 6 to 41 mg/L and exhibits no significant **difference between** the alluvium **and** the shale.
塩化物が6から41 mg/Lの範囲にあり，沖積層と頁岩の間に重要な相違を示さない。

16. The **difference** in TPH concentrations **between** the initial **and** four-week samples was statistically significant at the 0.025 probability level.
最初のサンプルと4週間後のサンプルでのTPH濃度の差は0.025の確率レベルにおいて統計学的に有意であった。

Different (Differently)：異なる(異なって；違って)

1. Index scores can be compared among streams of **different** sizes.
指数スコアは異なる規模の水流において比較されうる。
2. The approach provides valuable consistency **among the many different**

programs.
   この手法は，多くの様々なプログラムにおける貴重な整合性をもたらす。
3. These data clearly demonstrate the presence of two **different** fish communities on the same waterbody.
   これらのデータは，明らかに同一水域に2種類の魚類群集が存在することを実証している。
4. Each state traditionally has conducted biomonitoring in **a slightly different** manner.
   各州は，従来それぞれ若干異なる形で生物モニタリングを行っている。
5. **Different** classes of algae have different proportions of internal structure occupied by vacuoles.
   異なる綱の藻類は，液胞に占められた内部構造の比率が異なっている。
6. The comparison of the **two different time period** showed that the increased concentrations for the second period were highly significant.
   これら2つの時期の比較は，第二期間での濃度増大がきわめて有意なことを示した。

   - ～ **is different**：～は異なっている
   - ～ **is significantly different**：～が有意に異なっている
   - ～ **is somewhat different**：～がいくぶん異なっている
   - ～ **is different from .....**：～は…と異なっている

7. When comparing all of the stations with Station 6, only Station 7 **was not significantly different**.
   観測地点のすべてと観測地点6との比較では，観測地点7だけが有意に異なっていなかった。
8. We hope the reader will bear in mind that the real world **is somewhat different**.
   我々は，実世界がいくぶん異なっているということを読者が心に留めておくであろうことを希望する。
9. All things **are somewhat different**, yet some things are more similar than others, offering a possible solution to the apparent chaos.
   あらゆる事物は，それぞれ若干異なるが，あるものは他のものよりも類似性が強く，見かけの無秩序に対する潜在的解決策となる。

   - **different than .....**：…と異なっている

10. Its goals were somewhat **different than** those of most engineering applications.
    その目的はたいていの工学的適用の目的とはいくらか異なっていた。
11. The estimates are somewhat **different than** previously reported.

この推定値は前に報告されている値よりいくらか異なっている。

- **differently**：違って

12. Stated **differently**, ...
   言い換えれば，…
13. Organic suspension is affected **differently** by dissolved substances.
   有機懸濁物質は，溶存物質によって違った影響をもたらされる。

Differentiate (Differentiation)：区別する；差別化する（分化）

- ～ **differentiate between A and B**：～はAとBを区別する

1. We **differentiate between** homogeneous **and** heterogeneous reactions.
   我々は均一反応と不均一反応を区別する。
2. The results demonstrate the utility of using the theoretical range of the IBI to **differentiate between** and interpret different types of impacts.
   結果は，異なる影響の型式を差別化するために理論的IBIの値域を利用することの有用性を実証している。

- **differentiation**：分化；区別

3. The **differentiation** on the basis of experimental results may not be as straight forward as implied by these definitions.
   実験結果を基にしての区別は，これらの定義によって暗示されるほど単純ではないかもしれない。

Difficult (Difficulty)：難しい；困難な；難問な（難しさ；難問）

1. Remote, **difficult** to reach areas have also been monitored by citizen volunteers when they were too difficult for state personnel to reach on a timely basis.
   カバーしづらい遠隔地も，州職員がタイムリーに到着しかねる場合には，市民ボランティアによってモニターされてきた。

- ～ **become difficult**：～が難しくなる

2. The decision can **become difficult** when ......
   …の場合に決定が難しくなりえる。
3. Toxicity assessment of complex industrial effluents **has become an increasingly difficult task**.

複雑な産業排水の毒性アセスメントは，ますます難しい仕事になってきた。

- **~ is difficult**：~は難しい

4. Since many taxa **are difficult to** distinguish even with a microscope in a laboratory, the levels of taxonomic identification in the field are limited.
実験室で顕微鏡を使っても多くの分類群が区別しづらいので，現地での分類識別は，限られている。

5. Courts have noted that the implementation of water quality criteria that are not based on chemical-specific numeric criteria **can be difficult** as a matter of practice.
それでも法廷は，化学特定数値クライテリアに基づいていない水質クライテリアは，実際問題として難しいと指摘してきた。

- **It is difficult to .....**：…することは難しい；…することは困難である
- **It is difficult to apply A to B**：
  AをBに当てはめることは困難である
- **It would be difficult to .....**：…は難しいであろう

6. **It is difficult to** apply methods of statistical analysis to their study.
統計学的分析法を彼らの研究に当てはめることは困難である。

7. **It is difficult to** ensure, a priori, that implementing nonpoint source controls will achieve expected load reductions.
点源制御の実施が予想される負荷量の削減を達成するであろうことを先験的に保証することは難しい。

8. Without accounting for natural geographic variability **it would be difficult to** establish numerical indices that were comparable from one part of State to another.
自然地理的変動性を説明できなければ，ある州の一地域と別の地域を比較可能な数値指数を定めることは難しいであろう。

9. Without a sufficient theoretical basis **it would be very difficult**, if not impossible, to develop meaningful measures and criteria for determining the condition of aquatic communities.
十分な理論的基礎がなければ，水生生物群集の状態を決定するための有意義な測度および基準（クライテリア）を設定することは，仮に不可能でなくても，非常に困難だろう。

- **difficulty**：難しさ

10. The procedure was abandoned since it did not appear that these **difficulties** could be easily overcome.

これらの困難が容易に克服されるとは思われなかったので，この手順は断念された。
11. Other problems associated with chlorophyll a are high temporal variability and **difficulty** in interpreting trends.
クロロフィルaに関わる他の問題は，時間的変動性の強さと傾向解釈の難しさである。
12. One common source of error arises from the **difficulty of** obtaining a representative soil sample and the lack of reproducibility of organics analysis of soils.
代表的な土壌サンプルを得ることの難しさ，および土壌有機分析の再現性の欠如が一般に生じるエラーの一因となっている。
13. **The difficulty of** dealing with a large mass of information is that of the many interactions occurring within the community, some may be related to water quality while others may not.
大量の情報を取り扱う場合の困難は，生物群集において多くの相互作用が生じることであり，それらは，水質に関係したり，関係しなかったりする。

- **the difficulty comes from .....** ：…から難しさが生じる

14. The **difficulty comes from** trying to model a biological process chemically.
生物プロセスを化学的にモデル化しようとすることから難しさが生じる。

- **difficulty arises**：困難が生じる

15. The writers want to show herein other **difficulties** which **can arise** in toxicity measurements.
著者は，毒性測定の際に発生可能な他の難問題をここに示したい。

Dimension (Dimensional) ：次元（次元の）

1. The **dimensions of** each basin are 125 feet long, 67 feet wide, and 21 feet deep.
各ため池の寸法は長さ125フィート，幅67フィートと深さ21フィートである。
2. Two of the cells have inside **dimensions of** 2 m wide by 5 m high by 2 m deep.
これらのセルの2つは，内側寸法が幅2メートル，高さ5メートル，深さ2メートルである。
3. The numeric indices used to help define the narrative classification system were comprised of **single-dimension** measures.
記述的分類システムを定義するのに用いられた数量的指数は単一次元の指標からなっていた。

- **dimensional**：次元の
4. We also examined all nine of the impact types in a **two-dimensional** framework.
筆者は，また，9種類の影響すべてを二次元の枠組みでも検討した。
5. Copper toxicity profile is shown in Figure 1 by a **three-dimensional** plot of the specific growth rates as a function of the two variables, total ammonia concentration and total copper concentration.
銅毒性特徴を図1に示す。それは全アンモニア濃度と全銅濃度の二変数を関数として比増殖速度を表す3次元のプロットである。

Direct：向ける

1. In Section 304(a)(8), Congress **directed** USEPA to "develop and publish information on methods for establishing and measuring water quality criteria for toxic pollutants on other basis than pollutant-by-pollutant criteria, including biological monitoring and assessment methods."
304条(a)(8)において，連邦議会は，USEPAが「生物モニタリング方法および生物アセスメント方法など，汚濁物質別クライテリア以外の，毒性汚濁物質についての水質クライテリアを設定・測定するための方法に関する情報を設定し発表する」よう指示した。
2. Despite the large amount of effort that **has been directed towards** IBI development, much remains to be done, both in terms of generating new versions for different regions and habitat types, and in terms of validating existing versions.
IBI案出へ向けて多大な労力が払われてきたにもかかわらず，様々な地域や生息場所種類について新バージョンを生じること，ならびに既存バージョンを確証することの両方においてまだ多くの課題が残されている。
- **~ is directed toward ___ing**：~が___することに向けられる
3. Site studies **will be directed toward answering** that question.
その質問に答えるために，現地研究が行われるであろう。
4. The major effort to control eutrophication **has been directed towards reducing** the input of......
富栄養化をコントロールする主要な努力が…の投入量を減らすことに向けられてきた。
5. Relatively little work **has been directed towards modifying** the IBI for use on rivers too large to sample by wading.

水の中を歩いてサンプリングするには大きすぎる河川で用いるためのIBI修正に関する研究は，比較的少なかった。

6. Therefore, it has been argued, insufficient attention **is directed toward assessing** cumulative impact of these policies in terms of costs and outcomes.
   したがって，費用と成果の面でこれらの政策の累積影響評価にあまり注意が払われない，という主張がなされてきた。

   - **much attention has been directed at** ..... (See Attention)：
     …に多くの注意が向けられてきた(➡ Attention)
   - **insufficient attention is directed toward** ..... (See Attention)：
     …にあまり注意が払われていない(➡ Attention)
   - **particular attention should be directed to** ..... (See Attention)：
     特定の注意が…に向けられるべきである(➡ Attention)

Direction：指導；方向

1. **The direction of** groundwater flow beneath the site is toward the northeast.
   この地点で地下水が流れる方向は，北東の方角である。
2. The groundwater flow **direction** is generally to the northwest.
   地下水が流れる方向は，一般に北西の方角である。
3. A significant observation of the contaminant plume is that it attenuates rapidly **in a riverward direction**.
   汚染物プルームの観察から重要な点は，それが川の方角に向かって急激に弱くなるということである。

   - **under the direction of** ..... ：…の指導のもとで

4. Management decisions based on biological criteria must be made **under the direction of** an aquatic biologist expert with the specific methods, organism group, indices, and criteria being used.
   生物クライテリアに基づく管理決定は，水生生物学専門家の指導のもと，具体的な方法，生物集団，指数およびクライテリアを用いて下さなければならない。

★ direction（方向）に関連する表現例文を下に示します。

5. This creek drains from the northwestern slope of the Black Hills.
   この小流はBlack Hills（ブラックヒルズ）の北西部の斜面から流れ出る。
6. The Columbia River flows through the northern part of the Hanford Site,

and turning south, it forms part of the site's eastern boundary.
コロンビア川は Hanford 地域の北の部分を通過し，そして南に曲がり，地域の東境界線の一部を形成する。

7. The Yakima River runs along part of the southern boundary and joins the Columbia River south of the city of Richland.
ヤキマ川は南境界線の一部に沿って流れて，そして Richland 市の南でコロンビア川と合流する。

8. North of the line in Wyoming, patterns of logging activity were apparent, whereas south of the line they were not.
ワイオミング州境の北では伐採活動パターンが顕著だったのに対し，州境の南ではパターンが認められなかったのである。

9. The Hocking River headwaters are in glacial deposits of Fairfield County southeast of Columbus, Ohio, and it flows southeasterly through unglaciated, rugged topography to the Ohio River.
Hocking 川の源流は，オハイオ州 Columbus 市南東部の Fairfield 郡の氷河沈積物にあり，氷河作用を受けていない高低ある地形を抜けて南東へ流れ，Ohio 川へ達している。

Discover(Discovery)：発見する(発見)

1. Twice as many impaired waters **have been discovered** by using biological criteria and chemistry assessments together than were discovered using chemistry assessments alone.
化学アセスメントの単独利用に比べて，生物クライテリアと化学アセスメントの併用によって悪化した水域が2倍も多く発見された。

   ● **it was discovered that ……**：…が発見された

2. In examining the measured data further, **it was discovered that** ……
観測されたデータをさらに調べると，…が発見された。

   ● **discovery**：発見

3. The **discovery** has considerable environmental significance.
この発見は，かなり環境上の重要性をもつ。

4. For a period of approximately 100 years, from the original **discovery** of gold at Deadwood Gulch in 1875 until the late 1970's, huge volumes of mining and milling wastes were discharged into Whitewood Creek.
1875年に Deadwood Gulch での最初の金の発見から1970年代後期までおよそ100年の間，大量の鉱山採掘廃棄物と粉砕廃棄物が Whitewood Creek に放出された。

Discrepancy：矛盾；相違；不一致

1. Differences in conditions can account for the apparent **discrepancies**.
   条件の相違によって，この明白な矛盾が説明されうる。
2. Any **discrepancies** should be resolved before actually drawing a sample to prevent as much confusion as possible.
   いかなる不一致も，混乱をできる限り防ぐため，実際のサンプリング前に解決されるべきである。

　● another discrepancy is that ……：
　　もう 1 つの矛盾は… ということである

3. **Another discrepancy is that** no gas evolution was observed, nor were any other products detected by GC in either the gas or liquid phase.
   もう 1 つの矛盾は，ガスの発生が観察されなかったし，また，他のいかなる生成物も GC によっては気体相または液体相で検出されなかったということである。

　● discrepancy between A and B：A と B の間の矛盾

4. The obvious question concerns the **discrepancies between** this **and** earlier reports.
   この当然な質問は，この報告書と前の報告書の間の矛盾に関係している。
5. This **discrepancy between** the samples collected two years apart suggests that significant variations in the magnitude of soil contamination can occur either over very short horizontal distances or over a relatively short time period.
   2 年間隔てて収集されたサンプルの間でのこの矛盾は，土壌汚染度の有意な相違が非常に短い水平距離で，あるいは比較的短期間にわたって起こりえることを示唆する。

Discuss：論じる

1. McCarthy (1990) briefly **discussed** several studies that have shown biomarker responses correlate with predicted levels of contamination and with site rankings based on community level measures of ecological integrity.
   McCarthy (1990) は，予測された汚染レベルや，生物学的健全性の生物群集レベル測度に基づく地点順位と相関するバイオマーヤー応答を明らかにした諸研究について簡潔に論じた。

Discuss

2. The mechanisms constituting ABCD processes **have been discussed extensively** elsewhere in the literature (12–18).
ABCDプロセスを構成している仕組みは，文献(12–18)の至る所で広範囲に論じられてきた。
3. The development and implementation of rapid assessment programs **is discusses in detail** by Noguchi and Toda (1994).
迅速アセスメントプログラムの策定・実施については，NoguchiとToda(1994)が詳述している。

   - **~ is discussed below**：~は後ほど詳述する；~は下に論じられている

4. Possible reasons for these divergences **are discussed below**.
これらの相違の可能な理由は以下で論じていく。
5. Water quality data collected by citizen monitors **are discussed below** by the traditional categories of physical, chemical, and biological conditions.
市民モニターが収集した水質データについて，物理的状態，化学的状態，生物的状態という従来のカテゴリーによって以下で論じていく。
6. Each of these loss pathways **will be discussed in greater detail below**.
これらの消失経路それぞれがもっと詳細に下に論じられるであろう。
7. The risk and benefits of this proactive strategy **are discussed in more detail below**.
この先取的戦略の持つ危険性と便益については，後ほど詳述する。

   - **as discussed in the previous section**：前節で述べたとおり
   - **as was discussed in detail above**：前に詳細に述べたように
   - **as discussed earlier**：前に論じたように
   - **as discussed below**：下に，論じられるように
   - **as discussed previously**：前に，論じられたように
   - **as will be discussed later**：後に，論じられるように
   - **as will be discussed in subsequent section**：次のセクションで論じられるが，

8. **As discussed in the previous section**, the status of biological criteria programs across the United States can be thought of as the sum of all the states in various stages of developing and implementing biological criteria.
前節で述べたとおり，全米での生物クライテリアプログラムの現状は，生物クライテリア設定・実施の様々な段階における諸州すべての総和として考えうる。
9. The choice of groups to be identified and analyzed is related to the measures that will be used, **as was discussed in detail above**.
識別・分析されるべき集団の選択は，前に詳細に述べたように，用いられる指標

に関係している。
10. The sedimentation term is calculated from the **above discussed** transport capacity.
堆積の項は，上で論じた輸送容量から算出される。

Discussion：考察；論議

1. **The following discussion presents** examples of methods presently used in evaluating the periphyton assemblages.
以下の考察は，付着生物群集を評価する際に現在用いられている方法の諸例である。
2. **The preceding discussion** of the analysis of the capabilities of biological survey data to discriminate different types of impacts is contrary to several of the assertions of Suzuki (1999) in a critique of community indices.
生物学的調査データが様々な影響の種類を識別できる能力への分析に関する以前の考察は，生物群種指数を批判した Suzuki (1999) の主張の一部に反している。
3. Three states have not yet begun to formulate their bioassessment approach, but **have initiated discussion** within their own agencies and with the USEPA.
3州は，まだ生物アセスメント手法の作成を開始していないが，州機関内での討論や USEPA との論議に着手した。

- **discussion is provided**：論議が記述されている
- **discussion can be found in** .....：論議が … に見出される
- **discussion is limited to** .....：論議は … に制限される

4. Further **discussion is provided** elsewhere.
さらなる論議が至る所で記述されている。
5. A complete **discussion of** the data **can be found in** the HLA reports.
このデータの完全な論議が HLA 報告書に見出される。
6. The following **discussion is limited to** the data obtained up to 25 January 1998.
以下の論議は 1998 年 1 月 25 日までに得られたデータに制限される。

- ..... **is open to discussion**：… は議論の余地がある

Distance：距離

1. Aquatic communities gradually recover **with** increased **distance** downstream from impacts, although the pattern may vary according to the

type of impact and discharge.
水生生物群集は，影響からの下流距離が増すにつれて徐々に回復する。ただし，そのパターンは，影響および放流の種類に応じ異なる。

2. The improvement is even more remarkable when conditions since the turn of the century, when the river lacked any fish life for **a distance of** nearly 40 mile, are considered.
この川で約40mileにわたり魚類が生息していなかった今世紀初頭移行の状態を考慮すれば，この改善は，一層明らかである。

3. This discrepancy between the samples collected two years apart suggests that significant variations in the magnitude of soil contamination can occur either over very short **horizontal distances** or over a relatively short time period.
2年間隔てて収集されたサンプルの間でのこの矛盾は，土壌汚染度の有意な相違が非常に短い水平距離で，あるいは比較的短期間にわたって起こりえることを示唆する。

4. Faunal similarity between Midwestern river systems is dependent on **river mile distance**, along with other factors.
中西部河川系における動物相の類似は，河川距離（リバーマイル）および他の因子に左右される。

★ Distance（距離）に関連する例文をいくつか下に示します。

5. As a result many thousands of miles of United States streams have been and continued to be degraded each year.
その結果，米国の河川は，年に数千mileも劣化してきたし，今も劣化しつづけている。

6. The long-term fixed station is located on the mainstream approximately 1 mi (1.6 km) downstream from the confluence of the East and West Branches and upstream of the Elyria wastewater treatment plant.
East小流とWest小流の合流点から約1mile(1.6km)下流の本流と，Elyria廃水処理場の上流に長期的な固定観測場所がある。

7. The lake has a total area of 25,300 km^2, total volume of 470 km^3, **length of 386 km**, and mean width of 17 km.
この湖は，全面積が25,300 km^2，全容積が470 km^3，湖長が386 km，平均湖幅が17 kmある。

Distinct (Distinction) ：別の；個別の（弁別；区別）

1. The lake has a total area of 25,300 km^2, and is divided into three major,

distinct subbasins.
この湖は，全面積が 25,300 km² あり，三つの主要な水域に明確に分けられる。

- ～ **is distinct from** ……：～と … とは別のものである

2. Our experimental conditions **are distinct from** theirs.
   我々の実験条件と彼らの条件とは別のものである。
3. These results **are distinct from** earlier studies.
   これらの結果は，初期の研究（の結果）とは別のものである。

- **distinction**：弁別；区別

4. This **distinction** is made necessary by the widespread degree to which macrohabitats have been altered among the headwater and wadable streams in the HELP ecoregion.
   この区別が必要になった理由は，HELP生態地域（エコリージョン）における源流並びに水中を歩ける河川大型生息地が大幅に改変されたことである。

Distinguish（Distinguishing）：区別する（際立った）

1. ～ do not **distinguish** the different interpretation …..
   ～は，異なった意味を区別してはいない。
2. Since many taxa are difficult to **distinguish** even with a microscope in a laboratory, the levels of taxonomic identification in the field are limited.
   実験室で顕微鏡を使っても多くの分類群が区別しづらいので，現地での分類識別は，限られている。
3. This paper **distinguishes** itself **from** …..
   この論文は，… と異なっている。
4. Regardless, such within-ecoregion differences in land use and land cover must be **distinguished from** ecoregional characteristics.
   ともあれ，土地利用と土地被覆におけるこのような生態地域（エコリージョン）内での差異を生態学的特性と区別しなければならない。

- **distinguishing**：際立った

5. A key **distinguishing** feature of our model is …..
   我々のモデルの重要で際立った特徴は … である。

Divide：分ける

- ～ **is divided into** ……：～は … に分けられる

D

1. The lake has a total area of 25,300 km^2, and **is divided into** three major, distinct subbasins.
   この湖は，全面積が25,300 km^{2}あり，三つの主要な水域に明確に分けられる。
2. For modeling purposes, the study area **was divided into** nine pervious land segments.
   モデル構築のために，研究地域は9つの浸透性区域に分けられた。
3. Data quality **can be** roughly **divided into** three categories based on program objectives:
   データの質は，大体プログラム目標に基づいて以下の3つのカテゴリーに分けられる。
4. Biological processes **can be divided into** several categories for consideration as potential metrics.
   生物学的プロセスは，潜在的メトリックとして考慮するに当たり，いくつかの範疇に分けることができる。
5. The process **can be divided into** three major tasks 1)...., 2)......, and 3)...... Each of these three tasks is considered in more detail below.
   このプロセスは3つの主要な作業に分けられる。それらは1)…，2)…，3)…である。これらの作業の一つ一つは下でより詳細に考慮することにする。

Document (Documentation)：
資料；文献；文書；文書化する（文書化）

1. The primary emphasis of this **document** is on the transport in the vadose zone.
   この文献では，通気層における物質輸送が主に強調されている。
2. This **document** has been prepared at the ISU Environmental Research Laboratory in Pocatello, Idaho, through contract #24−C5−0117, with Kurokawa Environmental Technology, Inc.
   本章は，Idaho州Pocatello市にあるISU環境研究所において，Kurokawa Environmental Technology, Inc.との契約 # 24−C5−0077によって作成された。
3. We are indebted to the state representatives who provided information for this **document**, as well as to Shigeru Saito, two anonymous reviewers, and the editors of this book who commented on this chapter.
   我々は，本章のために情報提供してくれた各州代表者に恩義を感じている。また，Shigeru Saito，匿名査読者2名，そして本章にコメントをくださった本書の監修者らにも感謝したい。

★Documentの動詞としての使用例文を下に示します。

4. Widespread contamination of the stream system **has been documented**.
   この河川系の広範囲にわたる汚染が文書化された。
5. Obviously, these choices must be made with considerable care and **documented** so that they do not include fundamentally different communities.
   明らかに，こうした選択をする際，根本的に異なる群集が含まれないようにするため，かなり入念に行って文書化しなければならない。
6. Data **are** tabulated in the field (fish and habitat) and laboratory (macroinvertebrates), and **documented** via chain-of-custody procedures.
   データは，現地（魚類と生息地）および実験室（大型無脊椎動物）において表にされ，過程管理（chain-of-custody）手順を通じて文書化される。
7. In 1970–1971, a series of studies by the U.S. EPA (1973) were undertaken **to document** and characterize the discharge of tailings to Whitewood Creek.
   1970年から1971年までに，Whitewood Creekへの選鉱廃物の排出を文書化し，特徴づけするための一連の研究がU.S. EPA (1973) によって着手された。
8. Bioassessments can be used **to document** instream improvements that result from wastewater facility upgrades and the implementation of best management practices.
   生物アセスメントは，廃水施設の向上および最善管理慣行の実施に起因した流入改善の文書化に利用できる。

- **～ is well documented**：〜は文書化されている

9. The health hazards of environmental mercury **are well documented** in the literature.
   環境での水銀の健康障害は，文献に明確に文書化されている。
10. The major industrial sources of metals in wastewaters **are well documented**.
    廃水中に存在する金属の主要な産業源は，明確に文書化されている。
11. Sedimentation is widely held as responsible for degradation of fish communities in warmwater and coldwater streams, and many of the mechanisms of this degradation **have been well documented**.
    堆積は，温水河川ならびに冷水河川において魚類群集劣化の原因と広く認められており，劣化メカニズムの多くが十分に文書化されている。

- **documentation**：文書化

12. Emphasis is on explanation and intellectual stimulation, rather than on comprehensive **documentation**.
    包括的な文書化よりもむしろ，解説と知的な刺激を強調している。

13. This overview is developed from available **documentation on** state bioassessment and monitoring programs and an informal survey of state staff biologists.
ここで示す全体像は，州の生物アセスメント，生物モニタリングプログラム，ならびに州所属生物学者による非公式調査に関する入手可能な文書から導いたものである。

Doubt：疑い

- **There is no doubt that.....**：…することに疑いはない
- **There is no doubt about .....**：…については疑いはない
- **There is little doubt that.....**：…することにほとんど疑いはない
- **When in doubt**：疑わしい場合

1. **There is no doubt that** gypsum is ubiquitous in the tailings deposits along Whitewood Creek.
Whitewood Creek に沿って存在する選鉱堆積物に石膏が遍在することには疑いがない。
2. **There is little doubt that** monitoring enhances the ability to detect degradation.
モニタリングが解析検出能力を向上させることには，疑問の余地がほとんどない。
3. **There is little doubt that** much of the gypsum has formed in-situ as a result of this overall process of pyrite oxidation.
石膏の多くが，黄鉄鉱の総合酸化プロセスの結果として，現地で生成したことにほとんど疑いがない。
4. The axiom follows "**When in doubt** choose to take more measurements than seem necessary at the time since information not collected is impossible to retrieve at the later date."
公理としては，「疑わしい場合，その時点で必要と思われるよりも多くの測定を行うことを選びなさい。なぜなら，収集されなかった情報は，後に取り戻すことができないからである」。

Drawback：欠点

1. A significant **drawback to** the mineralogic analysis of waste material is that ......
廃棄物の鉱物学的分析への重要な欠点は，…
2. Although UV light has very good germicidal capabilities, its main

drawbacks are its lack of residual and its inability to oxidize organic compounds.
紫外線は非常に良い殺菌能力があるが，紫外線の主な欠点は殺菌力残余の欠如と有機化合物を酸化させる能力のなさである。

Drive：誘導する；左右する

1. Assignment of an appropriate aquatic life use to waterbody **is** primarily **driven by** an assessment of biotic integrity.
   水域への適切な水生生物利用指定は，主に生物の健全性アセスメントに左右される。
2. Consequently, Arkansas determined that the water quality standard **driving** this process needed revision.
   その結果，アーカンソー州は，この過程を導く水質基準の改正が必要だと決定したのである。

Due to ： に起因する；に原因がある

1. **Due to** a lack of extensive knowledge of pesticide toxicity, ...
   殺虫剤毒性に関する広い知識が欠如しているために…
2. **Due to** variability in microbial densities between samples, these changes were not statistically significant.
   微生物の密度はサンプル間でばらつきがあるため，これらの変動は統計的に有意ではない。
3. Long-term ecological effects **due to** shoreline oiling were highly unlikely.
   海岸線での油田事業による長期の生態的影響は，ほとんどありそうもない。
4. The risk of error **due to** inappropriate classification across heterogeneous regions should be avoided.
   異質な各地域にわたる不適切な分類に由来した誤差リスクを避けるべきである。
5. This is especially true in such cases where a waterbody may be perceived as being at risk **due to** new dischargers.
   ある水域が，新しく生じた排出のために危険な状態であるものと見なされるかもしれない場合，これは特に事実である。
6. The likelihood of a false negative decision is higher, **due to** uncertainty regarding the vertical extent of soil contamination and the potential for groundwater contamination.
   縦方向の土壌汚染と地下水汚染の可能性に関して確信がないために，誤った否定の決定になる可能性がより高い。

7. The influence of water vapor on the reaction rate derived from the low adsorption of ethylene **due to** its low adsorption affinity relative to water.
反応速度に対する水蒸気の影響は，水と比較して低い吸着性に起因するエチレンの低い吸着性から生じた。

- ～ **is due to** ……：～は … に起因する；～は … に原因がある
- ～ **is primarily due to** …..：～は主として … による
- ～ **would be most likely due to** …..： … に起因する見込みが高い
- ～ **are largely due to** …..：～の大部分は … に起因する

8. It is not clear from their results whether the effect of ～ **is due to**…..
～の効果が … によるものかどうかは，彼らの結果からは明らかではない。
9. It is believed that the decline in dieldrin from the reservoir outlet **is due to** the decreased aldrin application rates.
貯水池出口でのディルドリンの減少は，アルドリンの使用度によると思われる。
10. Since external impacts are minimal, measured variability **would be most likely due to** the sampling inconsistencies.
外部影響が最小であるから，測定された可変性は，サンプリングの不整合性に起因する見込みが高い。
11. Significant improvement in the biological indices has been observed during the past twenty years, which **has been due** in large part to significant reductions in point source loadings.
生物指標における大幅な改善が過去20年間に観察されており，その主因は，点源負荷の大幅な削減であった。
12. The data do not permit a determination as to whether the toxicity **was due to** a slug of something toxic in the water or was continuously present.
この毒性が，水に短時的に存在する毒性物質によるのか，または継続して存在する毒物によるのか，このデータでは決められない。

Each：それぞれ

1. **Each** phase is taking into account the interaction with the other.
   それぞれの段階が他との相互作用を考慮に入れている。
2. **Each** remedial action will be screened to determine its effectiveness in achieving the specified objectives.
   指定目的を達成するにあたってその効果を決めるために，それぞれの環境修復活動が検査されるであろう。
3. There are advantages and disadvantages **with each**.
   それぞれに利点と欠点がある。
4. The range of inhibition concentrations tested was chosen arbitrarily, depending upon the effect **of each**.
   それぞれの効果によって，この試験濃度内での抑制濃度範囲が恣意的に選択された。
5. Discrimination analysis provides a weight for **each** variable in the form of a coefficient.
   判別分析は，係数の形で各変数に加重値を与える。
6. **For each** index there has been a significant positive change over time at most sites.
   各指標について，大半の地点で，経年的に有意なプラス変化が見られた。
7. As a result many thousands of miles of United States streams have been and continued to be degraded **each year**.
   その結果，米国の水流は，年に数千 mile も劣化してきたし，今も劣化しつづけている。

   ● **each of .....** ：… のそれぞれ

8. **Each of** these loss pathways will be discussed in greater detail below.
   これらの消失経路それぞれがもっと詳細に下に論じられるであろう。
9. Samples from **each of** these sites have been analyzed for mercury.
   これらの各地点からのサンプルで，水銀分析がおこなわれてきた。
10. Between 16 and 52 samples from **each of** these sites have been analyzed for total arsenic.

これらの各地点から採取された16から52のサンプルで全ヒ素の分析がされてきた。

- at each step (See Step)：それぞれのステップにおいて（➡ Step）

Easy：容易な

- **It is relatively easy to .....**：…することは比較的容易である

1. **It is relatively easy to** read off the proportion of sites that are above or below a selected score.
   選ばれた値を上回る地点もしくは下回る地点の比率を読みとることは，比較的容易である。
2. The advantage of taking this additional step is that **it is now relatively easy to** see the proportion of sites that do not meet their respective biocriterion scores.
   この追加措置による利点は，それぞれの生物クライテリアスコアを満たさない地点の比率が比較的見やすくなることである。

Effect：影響；効果

1. The basic process and **the cause and effect relations** are still poorly understood.
   その基本的なプロセスとその因果関係は，まだ良く理解されていない。
2. There is no evidence that any such **effects** translate into impacts on fish populations.
   そのような効果が，魚類集団に対する影響につながるという証拠はない。
3. There have been many attempts to explain, describe and quantify these **effects**.
   これらの影響を説明し，記述し，そして数量化する多くの試みがされてきた。

   - **the effects of ～**：～の影響
   - **the effects of ～ on .....**：～が…に与える影響

4. Research has been undertaken to increase the level of understanding of **the effects of** heavy metals.
   重金属の影響に関する理解力レベルを向上させるために研究がなされてきた。
5. The dispersion coefficient lump together **all the effects of** averaging.
   分散係数はすべての影響を平均し，それをひとまとめにしたものである。
6. The environmental research laboratory contributes to this information

through research programs aimed at determining **the effects of** toxic organic pollutants.
環境調査研究所は，有毒な有機汚染物の影響を決定することを目的とした研究プログラムを通してこの情報に貢献する。

7. This research provides background information which would be of use in interpreting **the effects of** heavy metals in wastewater treatment systems.
この研究は，廃水処理システムにおいて重金属の影響を解釈するのに役立つであろうバックグラウンド情報を提供する。

8. A considerable amount of information is available on **the effect of** inhibitors on culture of *Nitrosomonas sp.*
Nitrosomonas 種の菌株に対する抑制剤の影響に関するかなりの量の情報が利用可能である。

9. Much has been written about **the effects of** petroleum on marine life, but few experiments have dealt.
原油が海洋生物に与える影響については多く書かれてきたが，それに対する実験はほとんどされていない。

- **identify the effect of ..... on.....**：…に対する…の影響を判別する
- **explain the effects of ..... on .....**：…に対する…の影響を説明する
- **~ have some effect on .....**：～は…に何らかの影響を及ぼしている
- **~ have a significant effect on .....**：
  ～は…に重大な影響を与えている
- **~ have a profound effect on .....**：～は…に深い影響を与えている
- **~ have a stronger effect on .....**：…は…に強い影響を与えている
- **~ have little effect on .....**：～は…にほとんど影響を与えない

10. This study is an attempt to **identify the effect of** selected inorganic ions **on** a fixed-bed catalyst.
この研究は固定床の触媒に対する選択された無機イオンの影響を判別する試みである。

11. The primary purpose of this research work has been to attempt to **explain the effects of** copper toxicity on aerobic biological systems.
この研究の主要な目的は，好気性生物システムに対する銅の毒性影響を説明しようと試みることであった。

12. Nonpoint sources probably **have some effect on** biological performance at many ecoregion reference sites.
面源は，おそらく，多くの生態地域参照地点での生物学的性能に何らかの影響を及ぼしてきた。

Effective：効果的

1. Further research into sediment transport is needed to develop more **effective** tools to better address these issues.
   これらの問題をより良く扱うためのより効果的なツールを開発するには，堆積物輸送に対するいっそうの研究が必要である。
2. In recent years there has been an increasing awareness that **effective** research, inventory, and management of environmental resources must be undertaken with an ecosystem perspective.
   近年，環境資源の効果的な研究・調査一覧作成・管理は，生態系を視野に入れて行うべきだ，との認識が強まってきた。

Effort：努力

1. **Additional effort was required** at sites where only one collection method was used.
   1つの採集方法しか使えない地点では，一層の労力が必要とされた。
2. **Data gathering efforts** for the Kiso River yielded adequate stream flow.
   Kiso川でのデータ収集の努力において，適切な流量が得られた。
3. **There have been efforts** to define ecologically critical flow thresholds using a toxicological rationale.
   毒物学的根拠を用いて生態臨界流量の閾値を定める努力がなされきた。
4. **Further efforts are needed to** enable use of IBI's over broad geographic areas in other Great Lake and coastal estuaries.
   他の五大湖および海岸河口域の広い地域でのIBI利用を可能にするには，さらなる努力が必要とされる。
5. Insofar as past state biological criteria **efforts** grew out of biological monitoring efforts, **future efforts will likely build on** current monitoring programs.
   生物モニタリング努力から発した過去の州の生物クライテリア努力に関する限り，将来の努力は，現行のモニタリングプログラムに立脚するであろう。
6. The ecoregions concept has been developed through **the effort of many people**, too many to acknowledge individually here.
   生態地域という概念は，大勢の方々の努力によって編み出されてきたので，ここで個々にあげて感謝することができない。

- **efforts focus on ……**：…に主眼を置く；…に集中する
- **effort centered on ……**：…に集中する

7. **The overall effort should focus on** building improved "line of evidence".
   努力全体は，改善された「連の証拠」の構築に主眼を置くべきである。
8. Current monitoring **efforts focus on** only a few of the hundreds of chemicals known to be present in drinking water supplies.
   現在のモニタリングの取り組みが，飲料給水への存在が知られている数百という化学物質だけに集中している。
9. **The focus of this effort is** to analyze exhaustively the solutions.
   この努力の焦点は，徹底的に解決策を分析することである。
10. Although **the effort centered on** surface water quality, limited sampling was conducted on tailings deposits.
    努力が地表水の水質に集中したけれども，サンプリングが選鉱屑堆積層でわずかではあるが実施された。

    - efforts are ongoing：努力が進行中である
    - efforts to ..... are carried out：…する努力が行われる
    - efforts are underway：努力が払われている
    - effort has been devoted to.....：努力が…に捧げられてきた

11. Laboratory work addressing some aspects of these problems and related modeling **efforts are ongoing** at the Idaho State University.
    これらの問題のある局面を扱う実験研究と関連したモデリングの努力がアイダホ州立大学で進行中である。
12. **Efforts to** assess, research, and manage the ecosystems **are** normally **carried out** via extrapolation from data gathered from single-medium/single-purpose research.
    生態系を評価・研究・管理する努力は，通常，単一媒体／単一目的の研究で収集されたデータからの推定によって行われる。
13. Although **substantial** national **efforts are underway** to provide consistency in the implementation of biological criteria, the actual development and implementation of biological criteria remains an activity of state water quality programs.
    生物クライテリアの実施における整合性をもたらすために相当の全米的な努力が払われているものの，生物クライテリアの実際の設定・実施は，いまだ州の水質プログラム活動にとどまっている。
14. Although **a great deal of effort has been devoted to** studying the problem of pollution, most work has concentrated on freshwater and terrestrial systems.
    多くの努力が汚染問題の研究に捧げられてきたが，ほとんどの研究は淡水と陸上のシステムに専念されてきた。

- **No effort was made to .....** ：…する試みがなされなかった
- **little effort is being expended on .....** ：
  …にあまり力が入れられていない
- **the large amount of effort that has been directed towards .....** ：
  …へ向けて多大な労力が払われてきた．

15. **No effort was made to** partition background variability by using ecoregions since the technology was not ready for use at that time.
当時は，技術がまだ調っていなかったので，生態地域を用いて背景可変性を仕分けする試みはなされなかった。

16. The problem is that **little effort is being expended on** studying ecosystems holistically and attempting to define differences in patterns of ecosystem mosaics.
問題点は，生態系の全体論的研究や，生態系モザイクのパターン差を定義づける試行にあまり力が入れられていないことである。

17. Despite **the large amount of effort that has been directed towards** IBI development, much remains to be done, both in terms of generating new versions for different regions and habitat types, and in terms of validating existing versions.
IBI案出へ向けて多大な労力が払われてきたにもかかわらず，様々な地域や生息場所種類について新バージョンを生じること，ならびに既存バージョンを確証することの両方においてまだ多くの課題が残されている。

- **～ have undertaken a wide range of efforts to .....** ：
  ～は…するため，広範な努力を払ってきた

18. In developing biological criteria for water quality programs, states **have undertaken a wide range of efforts** to improve bioassessment methods.
水質プログラムのための生物クライテリア設定において，諸州は，生物アセスメント方法を改良するため，広範な努力を払ってきた。

- **the long-term effort of ～ on .....** ：…における～の長期にわたる努力

19. It represents a tool to estimate **the long-term effort of** human activities **on** ....
それは…における人間活動の長期にわたる努力を推定する手段を表す。

- **in an effort to .....** ：…するため

20. **In an effort to** protect the site resources, USGS is conducting a study of the White River.

その地域の資源保護のため，USGS は White 川の研究を行なっている。
21. **In an effort to** determine the effect of using this more definitive, less subjective assessment methodology, evaluations using the original narrative protocol (i.e., exceptional, good, fair, or poor) from 431 sites sampled between 1981 and 1987 were compared to ICI-based biocriteria calibrated using regional reference sites.
より明確な，およびより主観的でないアセスメント方法論用いた場合の効果を決定するため，1981 年から 1987 年までの間にサンプリングされた 431 地点からのオリジナルの記述プロトコル(優，良，可，不可)を用いた評価が地域参照地点を用いて較正された ICI ベースの生物クライテリアと比較された。

Either：いずれか

1. Historical data may present a problem because the site selection generally is not conductive to **either of** these processes.
   史的データは，問題を生じうる。地点選定は，通常これらの過程のいずれかを資するわけではないからである。
2. We need to examine the use of art in science, rather than assuming an **either/or** scenario.
   我々は，二者択一シナリオを仮定する代わりに，科学における芸術の利用を検討する必要があろう。

    ● **either A or B** ：A か B かどちらか

3. Currently, most dredged material disposal is accomplished **either** by discharge into open water **or** by land disposal.
   現在，ほとんどの浚渫土砂の処分は開放水域への放出あるいは埋立によって達成されてる。
4. The surveying of ambient biota takes the form of **either** coordinated monitoring networks **or** a series of special studies.
   環境生物相に対する調査は，調整された監視網もしくは一連の特別研究の形をとる。
5. No gas evolution was observed, nor were any other products detected by GC in **either** the gas **or** liquid phase.
   ガスの発生が観察されなかったし，また，他のいかなる生成物も GC によっては気体相または液体相で検出されなかった。
6. An additional 23 states appear to be developing biological criteria for use **either** in water quality standards **or** in water resource management.
   さらに 23 州は，水質基準または水資源管理のいずれかでの利用のために生物クライテリアを設定中とみなされる。

7. It should be noted that in major portions of the country, topographic watersheds **either** cannot be defined **or** their approximation has little meaning.
米国の大部分において，地形流域は，画定されえないか，その概算がほとんど無意味であることに留意すべきである。
8. Key information about the environmental setting, characteristics of the receiving waterbody, and insights into the chemical/physical dynamics of the discharge are **either** directly **or** indirectly reflected by the biota.
環境背景に関する主要情報，受入水域の特性，そして排出の化学的／物理的動態に関する洞察は，生物相によって直接的または間接的に反映される。

Elsewhere：ほかのところにも

1. Further discussion is provided **elsewhere**.
さらに進んだ考察がほかのところにも記されている。
2. The BBB reactor has been described **elsewhere** (3–6).
この BBB 反応器はほかのところにも記述されている (3–6)。

★ここでの (3–6) は参照文献の 3 から 6 までを示しています。

3. Procedures for these calculations are also presented **elsewhere** (2, 3).
これらの計算のための手順はほかのところにも提出されている (2, 3)。
4. The mechanisms constituting ABCD processes have been discussed extensively **elsewhere** in the literature (12–18).
ABCD プロセスを構成している仕組みは，文献 (12–18) の至る所で広範囲に論じられてきた。
5. More detailed discussions of biological criteria development for the States of Maine, New York, and Texas can be found **elsewhere** in this volume.
メイン州，ニューヨーク州，テキサス州での生物クライテリア設定に関する詳細な論考は，本書のほかのところにも記されている。
6. The details of the studies and results have been presented **elsewhere** (4–9) and are only summarized here.
この研究とその結果の詳細は，ほかのところにも (4–9) 提示されている。ここではそれらの要約が提示されるだけである。
7. This is similar to observations that we have made in the Scioto River downstream from Columbus, Ohio, and **elsewhere**.
この観察結果は，オハイオ州 Columbus 市の Scioto 川下流および他の場所における筆者の観察結果と似通っている。
8. While we have dealt with most of these issues in Texas, these and other

issues will arise **elsewhere**, thus regionally consistency in achieving a resolution of these issues will be needed.筆者は，テキサス州におけるこれらの問題の大半を扱ってきたが，これらおよび他の問題は，他地域でも生じることで，これらの問題の解決における地域的整合性が必要だろう。

Emerge (Emerging)：出現する (新規の)

1. Interest **has emerged** recently in the possibility of using sonication to remediate groundwater.
   地下水 (水質) を回復させる目的のために超音波反応を使う可能性について，関心が最近また高まってきた。
2. Beginning about 1973 to 1975, groundwater quality models **have been emerging** in the literature.
   およそ 1973 から 1975 に始まって，地下水水質モデルが文献に出現していた。
3. Chemical applications of ultrasound are just beginning **to emerge**.
   超音波の化学的アプリケーションがちょうど出始めてきた。
4. The definition of degradation responses a priori is justified if clear patterns **emerge from** specific metrics.
   具体的メトリックから明確なパターンが現れたならば，劣化応答という定義は先験的に正当化される。
5. The following section describes current state efforts in biological criteria development and illustrates both the various differences and the many similarities of existing and **emerging** biological criteria programs.
   次に生物クライテリアにおける現行の諸州での努力を述べ，既存・新規の生物クライテリアプログラムにおける様々な相違点と多くの類似点を示すことにする。

Emphasis：強調；主眼

- ~ **is the emphasis of** ..... ：~が … の主眼である

1. Safe drinking water **has been** the primary **emphasis of** these programs.
   安全な飲料水がこれらのプログラムの主眼であった。

- ~ **place emphasis on** ..... ：~ は … に力点を置いている
- ~ **provide emphasis on** ..... ：~ は … を重視する

2. One model development **places particular emphasis on** the kinetics of chemical weathering in the watershed as the primary mechanisms.
   1つのモデル構築では，主要な機構として，流域における化学的風化速度に力点

が置かれている。
3. We intend to **place the emphasis on** understanding real systems, formulating problems and interpreting the results of the analysis.
我々は，実際のシステムを理解し，問題を定式化し，そして分析結果を解釈することに力点を置く傾向にある。
4. Aquatic life standards were proposed to improve impact assessment and **provide greater emphasis on** the management of ecological resources.
影響評価を改善し，生態資源管理をより重視するため，水生生物学的基準が提言された。

- **emphasis is on .....**：…が強調されている

5. **Emphasis is on** explanation and intellectual stimulation, rather than on comprehensive documentation.
包括的文書化より，解説と知的刺激を強調している。
6. **The primary emphasis of** this document **is on** the transport in the vadose zone.
この記録書では，不飽和透水層における(物質)輸送が主に強調されている。
7. For the first ten years, **the primary emphasis was on** groundwater flow modeling.
最初の10年間，地下水流のモデリングに重点がおかれた。
8. **The emphasis of** both state and federal efforts **has been on** the development and implementation of bioassessment methods for streams and small rivers.
州および連邦による努力の主眼は，水流・小河川についての生物アセスメント方法の設定・実施に置かれてきた。

- **emphasis is placed on .....**：…に主眼がおかれている

9. **Emphasis is placed on** the experimental conditions.
実験条件に主眼がおかれている。
10. **Emphasis is placed on** simple treatment and disposal schemes.
単純処理・処分計画に主眼がおかれている。

- **emphasis is given to .....**：…が強調される
- **emphasis is devoted to .....**：…が強調される

11. **Special emphasis is given to** highly hydrophobic chemicals.
疎水性の化学物質を特に重視している。
12. **Particular emphasis will be given** in this review **to** natural water samples.
このレビューでは，自然水サンプルが特に重視されている。

13. **Primary emphasis is devoted to** the water quality models of groundwater.
 地下水の水質モデルに主眼がおかれている。

    ● ..... with emphasis on..... ：… に力点を置いて

14. This will require new approaches to management **with an emphasis on** the assessment of a wide expression of ecological impacts.
 これには，生態影響の広範な発現への評価に力点を置いて新たな管理手法を要するだろう。

15. Environ Company sponsored a laboratory study of the geochemical behavior of tailings from the Marsh Creek area, **with emphasis on** the effects of oxidation and the controls on arsenic behavior.
 Environ 社は，Marsh Creek 域からの選鉱滓がおよぼす地質化学的挙動に対する室内研究に出資した。その研究は，ヒ素の挙動に対する酸化反応の影響と抑制に力点を置いていた。

Emphasize：強調する；重視する

1. The model **emphasizes** the water and sediment transport in aqueous systems.
 このモデルは水系での水輸送と堆積物輸送を強調している。

2. The discussion **will emphasize** assessments of stream environments, but aspects of monitoring ponds and lakes also will be mentioned.
 本章の力点は，水流環境アセスメントに置かれているが，湖沼モニタリングについても言及していく。

3. Work over the last decade by investigators, such as Murata (1993), **have emphasized** the importance of riparian areas for maintaining the quality and function of the hyporheic zones of streams.
 Murata (1993) などの調査者による過去 10 年間の研究は，河川の河床間隔層の質・機能の維持にとっての水辺区域の重要性を強調した。

4. In September 1987, USEPA published a management study entitled, Surface Water Monitoring: A Framework for Change, that **strongly emphasized** the need to accelerate the development and application of promising biological monitoring techniques in state and USEPA monitoring programs.
 1987 年 9 月に USEPA は，『地表水モニタリング：変化のための枠組み』と題する管理研究論文を発表した。同書は，州および USEPA のモニタリングプログラムにおける有望な生物モニタリング技法の案出・適用を促す必要性があることを力説していた。

- **It is worth emphasizing that ......** : … を強調する価値がある

5. **It is worth emphasizing that** technology based performance standards tend to deemphasize receiving system condition and even human health.
技術ベースのパフォーマンス基準が，受系状態ならびに人間の健康状態さえにも重点を置かない傾向があるということを強調する価値がある。

- **~ cannot be overemphasized** ：
~は，どんなに強調してもしすぎることはない

6. The importance of clean water **cannot be overemphasized**.
清水の重要性は，どんなに強調してもしすぎることはない。

Encompass：カバーする；包含する

1. The values fell within the range **encompassed by** sampling variation.
この値はサンプリング変動による範囲内で低下した。
2. A total of 27 stations were selected **to encompass** a wide range in sedimentary and trace metal gradients.
堆積性で微量な金属濃度の勾配にて広範な範囲をカバーするように，合計27の地点が選ばれた。
3. It was felt that most of the impacted reference results **should be encompassed by** the baseline WWH use designation for Montana's inland rivers and streams.
影響をうけた対照結果の大半は，モンタナ州の内地河川・水流についてのベースラインWWH使用指定によって成し遂げられるべきだと思われた。

Encounter：生じる；遭遇する

1. A similar problem **is encountered** in estimating pesticide model parameters.
殺虫剤の環境モデルのパラメータを推定する過程にて，類似の問題が生じた。
2. The following is a description of the problematic areas **encountered** during this survey.
次に，この調査の間に生じる問題点の多いエリアが記述されている。
3. This sampling did not **encounter** evidence of excessive concentrations of arsenic.
このサンプリングでは，ヒ素の濃度が過剰であるという証拠はあがらなかった。
4. Health problems related to the intake of contaminated water **have been**

encountered in some regions of Taiwan.
台湾のある地域では，汚染された水の摂取に関連した健康問題に直面してきた。

5. The majority of fish species **encountered** in warmwater rivers and streams are essentially sedentary during the summer and fall months.
温水の河川・水流で見られる魚種の大半は，基本的に夏から秋にかけて固着性である。

6. To predict whether a behavioral response to a chemical pollutant will occur, one must ask whether the organism can detect the pollutant at concentrations likely **to be encountered** in field situations.
化学汚染物質に対する挙動反応が起こるかどうか予測するには，現場の状況で起こりえる濃度で生物が汚染物を検知できるかどうか尋ねなくてはならない。

7. The question of chemical speciation poses one of the most difficult problems to be resolved by the chemist, especially in the context of synergistic effects **as encountered** in natural water.
化学的種分化への問いは，特に自然水域で見られる相乗効果という文脈において，化学者によって解決されるべき最も難しい問題の1つを提出している。

Ensure：保証する；確保する

1. It is difficult **to ensure**, a priori, that implementing nonpoint source controls will achieve expected load reductions.
点源制御の実施が予想される負荷量の削減を達成するであろうことを先験的に保証することは難しい。

2. Narrative and numerical criteria are derived **to ensure** protection for some of these uses.
これらの利用の一部について保護を保証するため，記述クライテリアおよび数値クライテリアが導出される。

3. This system is designed to promote more efficient use of ambient monitoring resources and **to ensure** timely results.
この手法の目的は，環境モニタリング資源の効率的利用を促すこと，タイムリーな結果を確保すること，である。

4. This project was designed to involve the county's citizens in ongoing waterways education and restoration and to foster the relationship between citizens and government **to ensure that** both are more responsive to the needs of the environment.
このプロジェクトの意図は，郡民が進行中の水路啓発・再生活動に関与するよう促すこと，市民と政府の関係を育んで，両者が環境ニーズにより対応的であるようにすることであった。

5. It is especially useful to select panelists and reviewers with opposing biases and different professional backgrounds **to ensure** different points of view, and to increase the credibility of the product in the eyes of the public.
様々な視点を確保し，判断結果を一般市民が眺めた際の信頼度を高めるには，反対意見や異なる専門的背景を持つ委員および査読者を選ぶことが特に重要であろう．

6. It is important to note that most models represent an idealization of actual field conditions and must be used with caution **to ensure that** the underlying model assumptions hold for the site-specific situation being modeled.
ほとんどのモデルは，実際の現地状態を理想化していること，そしてモデルの対象となっている特定な状況に対してモデルの基礎仮定が有効であることを保証するため注意して使われなくてはならないということ，を指摘することは重要である．

Entitled：題する

1. In September 1987, USEPA published a management study **entitled**, Surface Water Monitoring: A Framework for Change, that strongly emphasized the need to accelerate the development and application of promising biological monitoring techniques in state and USEPA monitoring programs.
1987年9月にUSEPAは，『Surface Water Monitoring: A Framework for Change（地表水モニタリング：変化のための枠組み）』と題する管理研究論文を発表した．同書は，州およびUSEPAのモニタリングプログラムにおける有望な生物モニタリング技法の案出・適用を促す必要性があることを力説していた．

Envision：予想する；構想する

1. The QCTV **is envisioned** as having application when a quick turnaround is needed to problem assessment or when a screening-level, less definitive technique is desired in lieu of the more complex ICI process.
QCTVは，問題あるアセスメントに速やかな対処が必要な場合，もしくは，より複雑なICI過程の代わりに選別レベルの決定的でない技法が用いられた場合に，適用されるものとして構想されている．

● **it is envisioned that .....** ：… と予想される

2. **It is envisioned that** by the end of the decade, nearly all of the state water resource agencies will have some form of biological monitoring.

この 10 年代末までに，ほぼすべての州の水資源担当機関が何らかの形の生物モニタリングを有するようになるものと予想される。

Equal (Equally)：等しい（同じく）

1. Setting this term **equals to** z yields ..... .
   この項を z に等しいとおくと … となる。
2. This three-tiered scheme works well for groups like stoneflies in which all species **are** nearly **equal** in tolerance to organic enrichment pollution.
   この三段層方式は，カワゲラのようにすべての生物種が有機物濃縮汚濁への耐性度がほぼ等しい集団では，うまく機能する。

   - **equally important is** ..... (See **Important**)：
     同じく重要なのは…である（→ **Important**）

Equation (See also Solution)：式（→ Solution）

1. **The equation for** a loading to a well-mixed lake is ....
   よく混合している湖への汚濁負荷は … の式で示される。
2. The material balance **equation** can be written as .....
   物質収支の式は … と書かける。
3. If this assumption holds true, the following **equation** can be used.
   もしこの仮定が真であるなら，次の式を使うことができる。
4. Algebraically, this **equation** can be rearranged to yield ......
   代数的に，この式は再度変形すれば … となるであろう。
5. Integrating this **equation** for the interval 0 to t yields ......
   この式を 0 から t まで間で積分すると， … となる。

   ★式を番号で呼ぶ場合，equation 2, Equation 2, Eqn 2, Eq. 2 など，出版協会，出版社によってさまざまです。著者のための手引き書(Author's Guide)を出版協会等から入手し参照することを勧めます。

6. **Equation** 1 becomes .....
   式 1 は … となる。
7. **Eqn** 2 can be reformulated as .....
   式 2 は … と再式化される。
8. **Eq.** 3 can be reexpressed as .....
   式 3 は … と再表現される。
9. **Eqn** 2 can be rewritten as.....

式2は … として書き直される。
10. **Eqs**. 1 **and** 2 can be substituted into **Eq**. 3 and solved for.....
    式1と2を式3に代入し，....に対して解いた。
11. The observed value of the K is definitely higher than calculated from **Eq**. 12.
    観察によるKの値は式12による計算値より確かに高い。
12. Inspection of **Equation** 8 indicates that the suspended sediment distribution is dependent on four transport coefficients: .....
    式8は浮遊堆積物の分布が4つの輸送係数に依存していることを示している。
13. Therefore, by rearranging **Equation** 6, the following equation may be obtained.
    したがって，式6を変形することによって，次の式が得られるであろう。
14. **Eqn**. 2 can be integrated between limits to yield .....
    式2を極限で積分すると，… を得る。
15. The CP transformation kinetics is described in **Equation** 3.
    CP変化速度は式3で説明される。
16. **Equation** 3 is the general solution describing dispersion and settling in a river.
    式3は，川の拡散と沈降を説明する一般的な解である。

★ Equationに関連する表現例文を下に示します。

17. Taking its natural logarithm and dividing by Y gives .....
    自然対数をとり，Yで割ると … になる。
18. The rate of movement of mass can be represented as.....
    物質移動速度は … と記される。
19. Dividing by V and rearranging gives .....
    Vで割って，変形すれば … が得られる。

Equilibrium：均衡

1. Examination of the literature on sorption reveals a great deal of confusion regarding the time to reach the sorption **equilibrium**.
   吸着についての文献調査によれば，吸着均衡に達っするまでの時間に関してかなりの混同がある。
2. The system had approached almost complete **equilibrium** for mercury, but two weeks were required for complete **equilibrium** to be achieved.
   水銀では，このシステムはほとんど完全な均衡状態に近づいていたが，完全均衡に達するには2週間は必要とされた。
3. The cupric ion concentration was measured and a comparison was made

with cupric ion concentrations predicted by a chemical **equilibrium** model.
銅イオン濃度を測り，そして化学均衡モデルによって予測した銅イオン濃度と比較した。

Error (Erroneous)：エラー；欠陥（欠陥ある）

1. It allows for **a** very small **margin of error**.
   それは非常に小さい範囲の誤差を許している。
2. The failure of the model to predict it **may be attributed to errors** in the model input data.
   それを予測するモデルの失敗は，モデル入力データのエラーに帰されるかもしれない。
3. Perfect charge balance would not be expected because of **analytical errors** in measuring the many parameters used.
   多数の使用されたパラメータを測るにいたっての分析エラーのために，完全なチャージバランスは期待されないであろう。

   ● **source of error arises from** ……：…からエラーが生じる

4. One common **source of error arises from** the difficulty of obtaining a representative soil sample and the lack of reproducibility of organics analysis of soils.
   エラーを生じさせる一般的原因1つは，代表的な土壌サンプル採集の困難さ，それに土壌の有機分析の再現性の欠如である。

   ● **erroneous**：欠陥ある

5. Metrics that are poorly defined or based on a flawed conceptual basis provide erroneous judgments with the potential for **erroneous** management decisions.
   きちんと定められていないメトリックや，欠陥ある概念ベースに基づいたメトリックは，誤った判断を生じさせ，管理上の判断ミスを招く可能性がある。

Especially：特に

1. This reaction is **especially** important from a global standpoint.
   この反応は世界的な見地から特に重要である。
2. Studies to provide these data may be prohibitively expensive, **especially** in a large watershed.
   これらのデータを提供する研究は，大きな流域の場合には，特にひどく高価にな

3. This work should be **especially** helpful in addressing impacts from sedimentation.
   この作業は，特に堆積による影響への対処に有益なはずである。
4. This is **especially** true in such cases where a waterbody may be perceived as being at risk due to new dischargers.
   ある水域が，新しく生じた排出のために危険な状態であるものと見なされるかもしれない場合，これは特に事実である。
5. It is hoped that the medium can be used for toxicity screening tests for various compounds, **especially** for heavy metals.
   希望的に，この媒質が種々の化合物，特に重金属の毒性スクリーニングテストに使うことができる，と期待される。
6. Numerous pollutant studies, **especially** with respect to heavy metals, have already been performed on the < 63– μ m fractions, allowing better comparison of results.
   多くの研究が，特に重金属においては，<63– μ m 部分でなされてきたので，他の結果との比較をより可能にしている。
7. Monitored groundwater concentrations were not found at levels exceeding the detection limits of BTEX constituents, **especially** benzene, at groundwater monitoring wells.
   監視下の地下水での濃度では，BTEX構成要素，特にベンゼン，の検出限度を超えるレベルでは地下水モニタリング井戸においては，みられなかった。

Essential（Essentially）：不可欠な（基本的に）

1. Perpetual evaluation of metrics and indices is an **essential** feature of the use of biocriteria.
   メトリックおよび指標の永続的評価は，生物クライテリアの利用に不可欠な特色である。
2. **Essential to this** is the development of ecological regions and indices of ecosystem integrity.
   このために不可欠なのは，生態地域ならびに生態系健全性指標の設定である。
3. Use of standardized methods **is essential**.
   標準化された方法の利用が不可欠である。
4. Comparison to reference conditions **is essential** to evaluate the extent to which study sites are influenced by human actions.
   研究対象地点が人間の活動の影響を受ける度合を評価するには，参照状態との比較が不可欠である。

- it is essential to ..... : …することが不可欠である
- it is essential that ..... : …の必要がある

5. **It is essential to** recognize that the proposed model depends on its intended purpose.
   提案されたモデルが，それが意図した目的に依存することを認識することが不可欠である。
6. Owing to the great extent of the fish industry in the Illinois River, **it is essential that** the condition of the river be kept as good as possible.
   Illinois川では漁業が盛んなため，この川の状態をできるだけ良く保つ必要がある。

- **essentially** : 基本的に

7. In contrast to many other human impacts, habitat loss can be **essentially** irretrievable over a human time frame.
   他の多くの人為的影響と異なり，生息場所消失は，基本的に人間の時間枠では取り返しがつかない。
8. Some minor revisions were made in 1985, but the approach remained **essentially** the same through 1987.
   1985年にいくつか細かな改正が施されたが，この手法は，基本的に1987年まで存続した。
9. The majority of fish species encountered in warmwater rivers and streams are **essentially** sedentary during the summer and fall months.
   温水の河川・水流で見られる魚種の大半は，基本的に夏から秋にかけて固着性である。
10. The purpose of the narrative classification system was **essentially** twofold: (1) ....., and (2) .....
    記述的分類システムの目的は，基本的に2つある。(1)…(2)…
11. The use of each attribute is **essentially** based on a hypothesis about the relationship between instream condition and human influence.
    各属性の利用は，基本的に流入水状態と人為的影響の関係に関する仮説に基づいている。

Establish (Establishing) : 確証する (設定すること；確定すること)

1. This information **will establish** the pollutant status of these organisms and facilitate interpretation of the other data.
   この情報によって，これらの生物の汚染状態が確証され，そして他のデータの解釈が容易にされる。

Establish

2. Idaho DEQ **has established** a rule that attainment of biocriteria should be granted disproportionate weight in demonstrating overall use attainment, due to the fact that chemical-specific and WET criteria are only surrogate measures of biological integrity.
Idaho州DEQは，化学クライテリアおよびWETクライテリアが生物完全性の唯一の代用測度であるという事実により，全般的な用途達成の実証において生物クライテリア達成が不相応な加重を与えられるべきだとのルールを確定した．

- ～ **is established** ：～が確立されている
- ～ **is well established** ：～は，十分に確立されている
- **it is a well-established fact that ……** ：
 …は十分に確立された事実である
- **it has been well established that ……** ：
 …ということは，しっかり確立されてきている

3. The distribution of pesticides in reservoirs **is established** by application of the principle of continuity or mass balance.
貯水池における殺虫剤の分布は，連続性の原則あるいは物質収支の適用によって確立される．

4. The following conclusions **were established** from the process emissions survey and bench-scale testing: 1)..., 2),.....
プロセス排気テストの調査，そしてベンチスケール実験から次の様な結論が確立された．1)…，2)…

5. The need for a cost-effective "rapid" biological assessment method **was established**.
費用効果的で「迅速な」生物アセスメント方法の必要性が確立された．

6. The framework within which water quality criteria **were established** and used to evaluate Connecticut rivers and streams includes the following major steps:
コネチカット州の河川・水流を評価するために水質クライテリアが設定・利用された枠組みには，以下の主要ステップが含まれる．

7. The best and most representative sites for each stream class are selected and represent the best set of reference sites from which the reference condition **is established**.
各水流等級について最善かつ最も代表的な地点が選ばれて，参照地点セットにされ，それに基づき参照状態が定められる．

8. The standard procedure outlined above **was established** so that the problems which often obscure interpretation of results are minimized.
結果の解釈をしばしば不明瞭にする問題点が最小になるように，上に概説された

標準的な手順は確立された。

9. By calibrating over large areas, decisions about level of protection can be made with a clear understanding of the differences among areas over which differing criteria **might be established**.
広い域での較正によって，保護レベルに関する決定は，異なるクライテリアが設定されうる各地域の差異を明確に理解したうえで下されるようになる。

10. The use of macroinvertebrates **is well established** in state programs and the advantages are well known.
大型無脊椎動物の利用は，州プログラムにおいて十分に確立されており，その利点がよく知られている。

11. **It is a well-established fact that** the odor in the untreated air are caused to a large extent by reactive chemical species.
浄化されていない空気の臭気は主に反応性のある化学種によって引き起こされる，ということは十分に確立された事実である。

12. **It has been well established** in the research literature that …..
… ということは，研究文献でしっかり確立されてきている。

- to establish ….. : … を定めること

13. Without accounting for natural geographic variability it would be difficult **to establish** numerical indices that were comparable from one part of State to another.
自然の地理的変動性を説明できなければ，ある州の一地域と別の地域を比較可能な数値指標を定めることは難しいだろう。

14. The advantage of this approach is that it would allow land-use planners and managers **to establish** criteria for the conservation of high-quality watersheds and the restoration of degraded areas.
この手法の利点は，土地利用の企画者・管理者が高質流域の保全ならびに劣化区域の回復についてクライテリアを定められるようにすることである。

- establishing : 設定すること；確定すること

15. For the purposes of **establishing** numerical biocriteria, the two most important uses are Warmwater Habitat (WWH) and Exceptional Warmwater Habitat (EWH).
数量的生物クライテリア確定のために最重要な2つの用途指定は，温水生息地 (WWH)，例外的温水生息地 (EWH) である。

16. If conditions are similar among regions, or if the differences can be associated with management practices that have a chance of being altered, it seems wise to combine the regions for the purposes of **establishing**

biocriteria.
仮に，各地域における状態が似ている場合や，その差違が変化可能性ある管理慣行に関連している場合には，生物クライテリア設定のために諸地域を組み合わせた方が賢明だろう。

Establishment：確立

1. Nebraska hopes that this **will lead to the establishment of** numeric biological criteria in the future.
   ネブラスカ州は，これが将来，数値生物クライテリアの確定に通じることを期待している。
2. What we are aiming toward is **establishment of** the validity of the concept of controlled catalytic biomass.
   我々が狙いを定めているものは，制御触媒性のある生物量概念の正当性を立証することである。
3. Need for **establishment of** modified biocriteria due to limitations in habitat quality will become evident in applications of habitat assessment routines.
   生息地質における限界故に，改善された生物クライテリアを定める必要性は，型どおりの生息地アセスメントルーティンの適用において顕著となろう。

Estimate (Overestimate) (Underestimate)：
推定；概算；見積る；推定する；概算する（過大評価する；過大に見積る）（過小評価する）

1. The **estimates** are somewhat different than previously reported.
   この見積もりは前に報告されてい値るよりいくぶん異なっている。
2. Quantifiable **estimates** of land use can be made with a known degree of certainty.
   土地利用の計量可能な推定は，既知の確実度でもって得られる。
3. The model **estimates** do not take into account the spatial and temporal variability of the real Rock Creek.
   このモデル予測値は Rock Creek の空間と時間の可変性を考慮に入れていない。
4. No attempt is made here to give a quantitative **estimate** of the error.
   ここでは，エラーの量的に推定をする試みはなにもされなかった。
5. The BIOMODEL **estimates** of biotic DDTR concentration correspond fairly closely to measured concentrations.
   BIOMODEL によって推定された生体内の DDTR の濃度は，測定された濃度にかなり近い。

6. The assumption will result in very conservative **estimates** for nutrient concentrations.
   この仮定は栄養塩濃度に対しかなり余裕ある推定値をもたらすであろう。
7. Substrate particle size is obtained by making observational **estimates** to designate percentage of each of the seven EPA size categories, as listed in Weber (1993).
   基質の粒子サイズは，Weber (1973) により列挙された7つのUSEPAサイズ区分それぞれのパーセンテージを指定するための観測的推定で得られる。

   ● **estimated** ..... ：推定の

8. The **estimated** rate constants are listed in Table 2.
   推定速度定数を表2にリストする。
9. The higher the value, the better the **estimated** water quality.
   この値が高いほど推定水質が優れていることになる。
10. Using all the **estimated** rate constants, the calculated accumulation of CO2 is plotted in Figure 2.
    すべての推定速度定数を使って計算されたCO2の蓄積量を図2に図示する。
11. On the basis of **estimated** rate constants, the oxidation rate of formic acid is faster than oxalic acid over a pH range of 4–8.
    推定された速度定数によれば，ギ酸の酸化速度はpH4–8の範囲でシュウ酸より速い。

    ● **〜 is estimated** ：〜が推定される
    ● **It is estimated that** ..... ：… と推定されている

12. In 1968, the total world production of mercury **was estimated to be** 8000 tons.
    1968年に，世界の水銀の全生産量は8000トンであると推定された。
13. Several site-specific hydrologic input data required were not collected and **had to be estimated** on the basis of model calibration using historical climatological data.
    必要とされる地点特定の水文入力データのいくつかが集められなく，歴史的気候学データを使いモデルのキャリブレーションを基に推測されなければならなかった。
14. A recent article describes some approaches for defining how well various percentiles **can be estimated** and notes pitfalls in the use of extreme percentiles for these purposes.
    最近の論文では，様々なパーセンタイルがどれほど十分に定められるか決定する手法を述べており，これらの目的のための極値パーセンタイルの利用における落とし穴を記している。

15. **It is estimated that** some 800 tons of DDTR currently exist in the sediments.
およそ800トンのDDTRが，現在，堆積物に存在すると推定されている。

- **to estimate** ：推定するため

16. There have been some attempts **to estimate** copper toxicity in seawater using bioassay.
バイオアッセイによって，海水での銅の毒性を推測する試みが若干されてきた。

17. **To estimate** the average velocity, use the plot and identify the velocity that corresponds to the average annual discharge.
平均速度を推定するために，プロットを使い，そして平均年度排水量に対応する速度を確認してください。

- **overestimate** ：過大評価する；過大に見積もる

18. Detailed examination of Figure 15 shows that MEK elimination **was overestimated**.
図15の詳細な検査によって，MEKの除去が過剰に見積もられていたということが示される。

19. The 100％ absorption assumed for diesel fuel above is clearly an **overestimated**, and is likely to have **overestimated** risks about 15 times.
上記のディーゼル燃料で想定された100％吸着は明らかに過大見積もりであり，リスクをおよそ15倍過大に見積もった可能性が高い。

20. The exposure periods **are** most likely **overestimated** and therefore the corresponding soil cleanup guidance should be considered conservative.
暴露期間が過大に見積もられた見込みが最も高く，そのため，対応する土壌浄化指導は保守的であると考えられるべきである。

- **underestimate** ：過小評価する；過小に見積もる

21. In some cases, biological impairment criteria may detect water quality problems that **are** undetected or **underestimated** by other methods. In these cases, biological impairment criteria may be used as the sole basis for regulatory action.
一部の場合には，生物損傷クライテリアが他の方法で探知されなかった水質問題，もしくは過小評価された水質問題を探知できる。これらの場合，生物損傷クライテリアは，規制措置の唯一の根拠となろう。

Evaluate (Evaluating) ：評価する(評価すること)

1. The second study reviewed here is that of *Kuraki et al.* (2003), who **evaluated** the PRID model.
   ここでレビュゥーされた2番目の研究はKurakiら(2003)によるもので，彼らはPRIDモデルを評価した。

   ● ～ is evaluated ：～が評価される

2. The data **will be evaluated** to determine whether the contaminant concentrations in the soils have the potential to cause unacceptable concentrations in the groundwater.
   これらのデーターは土壌での汚染物の濃度が地下水を許容外の(規準をこえる)濃度にする可能性があるかどうか決定するために評価されるであろう。

3. The condition of the pavement in the area needs to **be evaluated** to determine the potential for continued infiltration and leaching.
   その区域の舗装面の状態を評価し，継続的な浸入・浸出の可能性が判断されるべきである。

4. Periphyton has received recent attention related to monitoring program and its use in bioassessment **is being evaluated by** three states.
   付着生物は，最近，モニタリングプログラムに関連して注目を浴びており，生物アセスメントにおける付着生物の利用が3州によって評価中である。

5. However, the use of larger-scale (covering smaller areas) materials is also expensive and they **must be carefully evaluated for** representativeness.
   しかし，大縮尺(狭い面積をカバー)資料の利用は，高価でもあり，代表性に入念に配慮しなければならない。

   ● to evaluate ：評価するには

6. Clearly, much work needs to be done **to evaluate** and polish the proposed technique.
   明らかにかなりの研究が，提案された技巧を評価して，そしてそれに磨きをかけるためには必要である。

7. Comparison to reference conditions is essential **to evaluate** the extent to which study sites are influenced by human actions.
   研究対象地点が人間の活動の影響を受ける度合を評価するには，参照状態との比較が不可欠である。

8. Pilot studies or small-scale research may be needed **to** define, **evaluate**, and calibrate metrics.

メトリックを確定し，評価し，較正するには，予備的研究または小規模研究が必要だろう．

9. The framework within which water quality criteria were established and used **to evaluate** Blue rivers and streams includes the following major steps:
Blue 州の河川・水流を評価するために水質クライテリアが設定・利用された枠組みには，以下の主要ステップが含まれる．

- **evaluating**：評価すること

10. A qualitative/narrative system of **evaluating** biological data in the 1970s and early 1980s shifted to more quantitative/numerical framework in the mid−1980s.
1970 年代から 1980 年代前期までの質的/記述的な生物学的データ評価方式は，1980 年代中期により量的/数量的な枠組みへ移った．

11. Past efforts to evaluate the utility of individual metrics illustrate procedural approaches for **evaluating** the validity of a set of metrics.
個々のメトリックの有用性を評価してきた従来の努力は，一連のメトリックの妥当性を評価するための手順手法を示している．

12. Because of the nature of ecoregions, the ideal way of **evaluating** them would be through use of an ecological index of integrity.
生態地域の性質故に，その理想的な評価方法は，生態系の完全性指標を用いることだろう．

- **In evaluating .....**：… を評価する場合；評価する際に

13. **In evaluating** the performance of a model relative to some field measurement or another model, qualitative rather than quantitative statements are often made.
あるフィールド測定，またはもう 1 つのモデルとの比較からモデルの性能を評価する場合，量より質的な報告がしばしばなされる．

14. **In evaluating** the accuracy of the estimated parameter, more weight should be given to the ability of the model to fit the data points near segment D than to its ability to fit the points near segment G.
推定パラメータの正確さを評価する場合，モデルが川の切片 G の近くのデータポイントで合う能力により，切片 D の近くでデータポイントに合うことができる能力に，もっとウエートが置かれるべきである．

15. The following discussion presents examples of methods presently used **in evaluating** the periphyton assemblages.
以下の論考は，付着生物群を評価する際に現在用いられている方法の諸例である．

Evaluation (Reevaluation) : 評価 (再評価)

1. **Evaluation** and optimization of specific benthic invertebrate indicators **is** currently **underway**.
   具体的な底生無脊椎動物指標の評価と最適化が，目下進行中である。
2. Perpetual **evaluation of** metrics and indices is an essential feature of the use of biocriteria.
   メトリックおよび指標の永続的評価は，生物クライテリアの利用に不可欠な特色である。
3. Stream habitat modification was identified as the third leading cause of aquatic life impairment in the 1992 Indiana Water Resource Inventory, although the database was oriented primarily to **the evaluation of** point sources.
   水流生息地改変は，1992年のインディアナ州水資源調査一覧において，水生生物損傷の第三の主因と確認された。ただし，このデータベースは，主に点源評価を志向している。
4. Such information can influence decisions to control certain substances or processes that might have been overlooked or underrated in **an evaluation** based on only one group.
   そのような情報は，1集団のみに基づく評価において看過または過小評価されてきた物質や過程を制御する決定に影響を及ぼしうる。
5. This designation of impairment was supported by toxicity testing results, and subsequently resulted in **a reevaluation of** the treatment process of this facility.
   この損傷指定は，毒性試験結果によって裏づけられ，後日，この施設の処理過程への再評価が行われた。

Event : イベント

1. We hypothesize that a specific sequence of **events** will occur in response to chronic nitrogen amendments.
   長期にわたる窒素の土壌改良に応じて，複数のイベントが特定の順序で起こるであろう，と我々は仮定する。
2. Examination of local gaseous and liquid concentrations permitted an explanation of the complex **events** occurring under transient state operating conditions.
   局地的なガスと液体の濃度の検討から，瞬間状態操業条件下で生じる複雑なイベ

3. For most part, ICI scores were consistent and different by no more than 6 points over the sampling intervals which spanned 7 to 8 years and included 3 to 6 sampling **events**.
たいていの場合，ICI スコアは，7〜8 年間で 3〜6 回のサンプリングが行われるサンプリング間隔で，かなり一貫しており，6 点以上の差がなかった。

- **In any event**：とにかく
- **In the event of** .....：…の場合
- **During** ..... **events**：…の間に

4. **In any event**, it was very clear that .....
とにかく，…のことは非常に明確であった。
5. **In the event of** a spill emergency, the gate valves at the B-1 bypass have the capacity of diverting the creek flows to Pond B-1.
漏出緊急時の場合，B-1 バイパスにおいての水門弁は，小川の流れを B-1 池にそらす能力を持っている。
6. **During** major precipitation **events**, storm water may be discharged directory to the river.
主要な降水時には，嵐水は川に直接に放流されるかもしれない。

Every：毎回；あらゆる

1. Soil samples at the field sites were collected **every** month over a 12-month period.
土壌サンプルが，野外地点においての 12 カ月間にわたって毎月収集された。
2. Additional issues concerning level-of-detail are critical to **every** step of the simulation process.
細部レベルに関する追加問題はシミュレーション過程のすべてのステップに重要である。
3. Although **every** program provides some form of educational benefits, the extent of educational opportunities varies among the programs.
どのプログラムも何らかの啓発効果を有するものの，啓発機会の度合は，プログラムごとに異なっている。
4. The obvious need for **every** organism group to have a dispersal mechanism is critical if that species is to be sustained.
その生物種が持続されるべきならば，あらゆる生物集団が分散メカニズムを持つことの明らかな必要性がきわめて大事である。

- **in every case**：すべてにおいて
- **in nearly every case**：ほとんどの場合；ほぼ決まって
- **not every ～ is .....**：すべての～が…というわけではない

5. **In every case** where impairment was indicated, we would not have been able to assign the source without bracketing the discharge with upstream-downstream sites.
損傷が示された事例すべてにおいて，我々は，上流地点-下流地点での排出を一括せずには損傷源を指定できなかっただろう。
6. Response to the question **in nearly every case** has been that ecoregions are the desired regional framework.
この質問への返答は，ほとんどの場合，生態地域が望ましい地域枠組みだというものだった。
7. Some of the original models have been changed **in nearly every** subsequent version, whereas others have been largely retained.
オリジナルモデルのいくつかは，その後のバージョンほぼすべてで変更されたが，そのまま保たれたモデルも多い。

Evidence：証拠

1. It should not be adopted without experimental **evidence**.
それは実験的な証拠なしで採用されるべきではない。
2. The study areas were selected on the basis of **evidence** suggesting that they are areas of relatively intense geochemical activity
これらの調査地域は，比較的激しい地質化学的活動の地域であることが示唆されている証拠をベースに選択された。
3. Many other natural restrictions to achievement can be identified, and care must be taken that actual degraded conditions are not included as **evidence** for regional scale biocriteria modification.
達成への多くの他の自然的制限が確認される場合もあり，実際の劣化状態が地域的な生物クライテリア改善の証拠として含まれないよう，注意を払うべきである。

- **evidence is growing that.....**：…という証拠が増えてきている
- **evidence accumulated that.....**：…という証拠が蓄積した
- **evidence has been gathered**：証拠が集められてきた
- **as evidence continues to mount that .....**：
…の証拠が増え続けるにつれて
- **Evidence mounts that .....**：…という証拠が増えている

4. **Evidence is growing that** the present high levels of acid deposition originate predominantly from human activities.
   現状の高濃度の酸性降下物は主に人間活動によるという証拠が増えてきている。
5. **Evidence has been gathered**, and solutions have been proposed.
   証拠が集められ，そして解決策が提案されてきた。
6. Abundant **evidence** eventually **accumulated that** this approach was unsatisfactory.
   この手法が満足のいかないものであったという数多くの証拠が結局は蓄積した。
7. It is regrettable that no substantive body of **evidence** confirming ..... **has been generated**.
   … を確証する実質的な一連の証拠をつくれなかったことは残念である。
8. **Evidence mounts that** the Earth`s climate is undergoing significant change.
   地球の気候が大きく変わってきているという証拠が増えている。

- **evidence indicates that**.....：証拠が … のことを示唆している
- **Several lines of evidence indicates that** ..... ：
  数連の証拠は… であることを示唆している

9. A large amount of circumstantial **evidence indicates that** .....
   大くの付随的な証拠が … のことを指摘する。
10. The fact is that substantial **evidence indicates that** metabolites are toxic to animal systems.
    事実は，かなりの証拠が代謝産物が動物のシステムにとって有毒であることを示唆している，ということである。
11. **Several lines of evidence**, including analyses of denitrification products (Ukita, 2001), nitrogen isotope investigations (Imai, 2002), and tracer studies (Okamoto, 2004) **indicates that** aquifer nitrogen loss may be a result of denitrification.
    脱窒反応生成物の分析 (Ukita, 2001)， 窒素同位体の調査 (Imai, 2002)，及びトレーサーによる研究 (Okamoto, 2004) を含む数連の証拠は，帯水層における窒素の減少が脱窒作用の結果であるかもしれないことを示唆している。

- **There is evidence of** ..... ：…の証拠がある
- **There is strong evidence that** ..... ：… という強い証拠がある
- **There is enough evidence to show that** ..... ：
  … を示す十分な証拠がある
- **There is enough evidence against** ..... ：
  … に対して十分な証拠がある

- **There is no evidence to support .....**：… を支持する根拠はない
- **....., but there is as yet no clear evidence of ......**：
  …，しかしまだ … の明確な証拠がない
- **There appears to be sufficient evidence that ......**：
  … を示すに足る証拠だと思われる
- **There is increasing evidence that .....**：… の証拠が増えてきている

12. **There is evidence of** forest damage in New York's Adirondack Mountains.
    ニューヨーク州のアジロンダック山脈に森林破壊があるという証拠がある。
13. **There is strong evidence that** these substances reach the atmosphere from emissions.
    これらの物質は排出されると大気中に達するという強い証拠がある。
14. **There is enough evidence against** their opinion.
    彼らの意見に反する十分な証拠がある。
15. **There is some empirical evidence that** instream biological response and instream ambient toxicity are closely associated.
    経験的証拠によれば，水流内生物応答と水流内環境毒性は，密接に関連している。
16. The likelihood of a false positive error at ST57 is relatively low since **there is visual evidence of** contamination.
    現場 ST57 において汚染が無いのに有ると間違うがう可能性は比較的低い。なぜなら目に見える汚染の証拠があるからである。
17. **There appears to be sufficient evidence that** IBI metrics will differ for the upper and middle Snake River due to zoogeographic factors.
    これらは，IBI メトリックが動物地理因子故に Snake 川の上流と中流で異なることを示すに足る証拠だと思われる。
18. On the other hand, **there is increasing evidence that** hydroxyl radical is the primary species to oxidize the VOC present in aqueous solutions.
    一方，ヒドロキシル遊離基が水溶液中の VOC を酸化する主要な化学種である証拠が増えてきている。

★ evidence は一般に数えられない名詞とされ，there is evidence であり，冠詞は付きません。

- **〜 provide evidence for .....**：〜は … について証拠をあげる
- **〜 provide evidence to .....**：〜は … をするための証拠を提供する
- **〜 provide further evidence of .....**：
  〜は … にさらなる証拠を提供する
- **〜 exhibits evidence**：〜は証拠を示している

- ~ **found no evidence of** .....：~では … の証拠は見つけられなかった
19. These data **provide evidence that** .....
    これらのデータは … という証拠をもたらす。
20. The weathering characteristics of the tailings deposits **provide visual evidence of** effects of pyrite oxidation in the tailings.
    選鉱滓堆積層の風化特性が，その堆積層における黄鉄鉱の酸化反応影響を明らかに証拠づけている。
21. Examination of Table 2 **provides** little **evidence** of any correlation between influent metal concentrations and percentage removal.
    表2からは，流入水中の金属濃度と除去率の間に相互関係がある，という証拠はほとんど得られなかった。
22. Compared to the Scioto River, the Ottawa River **exhibits evidence that**, despite some improvements, makes it one of the most severely impaired rivers in the state.
    Scioto川に比べ，Ottawa川は，若干の改善にもかかわらず，同州で最もはなはだしく影響された河川の一つという証拠を示している。
23. This study **found no evidence of** pollution influencing potable wells.
    この研究では，飲料水用井戸に影響を与える汚染の証拠は見つけられなかった。

Evident：明白である；顕著である

- ~ **is evident**：~は顕著である

1. The toxicity **was most evident** at cadmium concentration of 0.4 mg/L.
   毒性はカドミウムの濃度が0.4 mg/Lで最も明白であった。
2. Although Cu–NH$_3$ complexes occurred at small activities, their toxic effect on N. europaea **was evident**.
   Cu–NH$_3$ 錯体の活性度は低かったけれど，N. europaeaに対するそれらの毒性影響は明白であった。
3. The magnitude and significance of chemical contamination of aquatic environments **are increasingly evident**.
   水生環境の化学汚染の規模と重要性がますます明かになっている。
4. The Scioto River has never had any significant problem with toxics, which **is certainly evident** in the biological responses.
   Scioto川は，重大な毒物問題を抱えたことがなく，それが生物応答において顕著である。

- ~ **become evident**：~が明白になる；顕著となる

- **it has become increasingly evident that** ..... ：
  … は，ますます明白になってきた

5. For many situations this will only **become evident** through an iterative process.
   多くの事態について，これは，反復過程でもって初めて顕著になるだろう。
6. When the program expands to all or a major part of a watershed, the effects on one area of adding pollution in another **become more evident**.
   そのプログラムが流域の全部または主要部分へと拡大された場合，他での汚濁追加が一区域に及ぼす効果は，一層顕著になる。
7. The magnitude and significance of the problem has **become increasingly evident**.
   この問題の規模と重要性は，ますます明白になってきた。
8. Need for establishment of modified biocriteria due to limitations in habitat quality **will become evident** in application of habitat assessment routines.
   生息地質における限界故に，改善された生物クライテリアを定める必要性は，生息地アセスメントの手順の適用において顕著となろう。
9. **It has become increasingly evident that** the environmental impact of a particular metal species may be more important than the total metal concentration.
   特定の金属種の環境に対する影響が，金属すべての濃度の影響よりいっそう重要であるかもしれないということがますます明白になってきた。

- **It is evident that** ..... ： … は明白である
- **It is evident from** ..... ： … から明白である

10. **It is evident that** only metals present in an insoluble form will be removed.
    不溶性の金属だけが除去されるであろうことは明白である。
11. **It is evident from** core-sample descriptions that ....
    … はコアサンプルの記述から明白である。

Examination：試験

1. The validity of these predictions could be determined by **an examination of** the biotic condition of the stream.
   これらの予測の正当性は，水流の生物状態の検定によって決定されうる。

- **An examination of ~ revealed that** ..... ：
  ～の検討から… が明らかになった

Examination

2. **An examination of** this process **revealed that** the reduction process was sequential.
このプロセスの検討から，この還元プロセスが連続的であったことが明らかになった。

3. **An examination** by *Toda et al.* (2002) of oysters and other shellfish exposed to an oil spill **revealed** the presence of large quantities of petroleum compounds in the body tissues.
カキと他の貝に流出した石油をさらしたToda ら(2002)の研究によれば，貝の体組織に大量の石油化合物が存在することが明らかになった。

4. **Examination of** the literature on sorption **reveals** a great deal of confusion regarding the time to reach the sorption equilibrium.
吸着についての文献調査によれば，吸着均衡に達っするまでの時間に関してかなりの混同がある。

- **examination of** ～ **shows that** ‥‥‥ ：～検査によって … が示された
- **examination of** ～ **provides** ‥‥‥ ：～の考察から … が出される
- **examination of** ～ **permitted** ‥‥‥ ：～の検討から … ができる

5. **Preliminary examinations** and comparisons **show** good agreement between the two at most common stations.
予備的な検討・比較によれば，最も一般的な観察所では，これら両者がかなり合致する。

6. **Detailed examination** of Figure 15 **shows that** MEK elimination was overestimated.
図15の詳細な検査によって，MEKの除去が過剰に見積もられていたということが示される。

7. **Examination of** Table 2 **provides** little evidence of any correlation between influent metal concentrations and percentage removal.
表2の考察から，流入水の金属濃度とパーセンテージ除去率との間の相互関係に関する証拠はほとんどなにも出てこなかった。

8. **Examination of** local gaseous and liquid concentrations **permitted** an explanation of the complex events occurring under transient state operating conditions.
局地的なガスと液体の濃度の検討から，瞬間状態操業条件下で生じる複雑なイベントの説明ができる。

- **examination is made of** ‥‥‥ ：… について検討される
- **examination of** ～ **is performed** ：～の検討が行われる

9. **An examination was made of** the level of B(a)P in mussels taken at and

near two locations that appeared to be point sources of PAH pollution.
PAH汚染の点源であると思われた地点とその付近の2地点から採集されたムラサキ貝体内のB(a)Pのレベルについて検討がなされた。
10. The longitudinal **examination of** biological sampling results **is** also **performed** in an attempt to interpret and describe the magnitude and severity of departure from the numerical biological criteria.
生物サンプリング結果の縦断的検討も，数量的生物クライテリアからの逸脱の規模・程度を解釈し記述するために行われる。

Examine（Examining）：調べる；検討する（調べる；検討すること）

1. Hattori(1977) **examined** the effects of domestic and industrial wastewaters on nitrifying organisms.
   Hattori(1977)は，硝化(微)生物に対する生活・産業廃水の影響を調べた。
2. The modeling framework WASTOX permits the user **to examine** the transport of a toxic chemical.
   モデリング枠組みWASTOXを使うことによって，ユーザーは毒性化学物質の輸送を検討できる。
3. Another way to visualize these trends is **to examine** changes in the cumulative frequency distribution(CFD) of biological index scores between each time period.
   これらの傾向を視覚化するための別方法は，各時期間の生物学的指標スコアの累積度数分布(CFD)における変化を検討することである。

    - **This paper examines** ..... ：この論文では … を検討する
    - **The investigation examined** ..... ：この調査で … が検査された

4. **This paper examines** the ecological damages caused by the river modification.
   この論文では河川修正工事が生態におよぼした損傷を検討する。
5. Previous **investigation have examined** the ecological damages causedy the river dredging.
   以前の調査で河川浚渫が生態におよぼした損傷が検査されてきた。

    - **We examined** ..... ：筆者は … を検討した．
    - **In this paper we examine** ..... ：この論文では，我々は … を検討する
    - **We need to examine** ..... ： … を検討する必要がある

6. **We** also **examined** all nine of the impact types in a two-dimensional framework.

筆者は，また，9種類の影響すべてを二次元枠組みでも検討した。

7. **In this paper we examine** physical transport and chemical transformation processes
   この論文では，我々は物理的輸送と化学変化プロセスを検討している。
8. **We need to examine** the use of art in science, rather than assuming an either / or scenario.
   我々は，二者択一シナリオを仮定する代わりに，科学における芸術の利用を検討する必要があろう。

- ~ **is examined** ... ：~が調べられる
- ~ **needs to be examined** ..... ：~を検査する必要がある
- ~ **must be examined for** ..... ：
  ~は…について検査されなければならない
- ~ **has not been examined** ：~はまだ検討されていない

9. Chemical transport and transformation processes **are examined more closely.**
   化学物質の輸送・変化プロセスがいっそう綿密に検討される。
10. This **will be examined further** in the next chapter (in the following chapter).
    このことは次の章でさらに検討される。
11. The ecological damages caused by the river modification **needs to be examined in detail.**
    河川修正工事が生態におよぼした損傷について詳しく調査する必要がある。
12. All fish **must be examined for** external deformities, skeletal anomalies, eroded fins, tumors, and lesion
    すべての魚は，外部奇形，骨格異常，ひれ腐食，腫瘍，病変について検査されなければならない。
13. This assumption **has not been examined in any detail.**
    この仮定は，まだ詳しく検討されていない。

- **examining** ：検討すること；調べること

14. **In examining** the measured data further, it was discovered that ......
    観測データをさらに調べると，…が発見された。
15. **Examining** the variance structure **can give insight into** the extent over which particular biocriteria might be applicable.
    変動構造を検討すれば，ある生物クライテリアが適用可能な度合に関する洞察が得られよう。
16. An advantage of **examining** the trophic status of component populations is

that it also provides information on functional aspects of the community as well.
構成要素たる個体群の栄養状況を検討することの利点は，その生物群集の機能面に関する情報も提供することである。

Example：例

1. There are many **examples** to support this theory.
   この理論を支持する例が多くある。
2. **Examples of** such sources are backyard trash burning, copper smelting, and dioxins in sediments.
   このような出所の例としては，裏庭ゴミの焼却，銅の製錬，堆積物のダイオキシンである。
3. The following discussion presents **examples of** methods presently used in evaluating the periphyton assemblages.
   以下の論考は，付着生物群を評価する際に現在用いられている方法の諸例である。

    - **as an example**：一例として
    - **another example of ～ is .....**：～のもう1つの例は … である
    - **a noteworthy example is .....**：注目すべき例は … である
    - **to take an example**：例をあげると

4. **As an example**, consider a study which calls for an evaluation of the effects of tertiary treatment of domestic wastewater on the quality of downstream receiving waters.
   一例として，下水の三次処理が，下流において受水流域の水質に及ぼす影響を評価する研究を考慮してみてください。
5. The derivation of the impairment for species richness is shown **as an example** in Table 1.
   生物種豊富度についての損傷(クライテリア)の導出は，表-1に例示されている。
6. **Another example of** advanced oxidation processes **is** phtocatalysis.
   促進酸化プロセスのもう1つの例は，光触媒反応である。

    - **for example**：例えば

7. **For example**, seasonality is a well-understood concept, therefore, it is not necessary to sample in multiple seasons for the sake of data redundancy.
   例えば，季節性は，十分に理解されている概念であるから，データ冗長性のために複数季節でサンプリングする必要がない。

8. A primary release is one from the primary contaminant source ; a secondary release is one that occurs, **for example**, from the contaminated soil to the groundwater.
主要な流出は第一汚染源から生ずるもので，第二の流出は例えば，汚染土壌に生じ地下水まで浸透するものである。
9. The reaction of gold with hemoglobin is of interest, **for example**, because studies have shown that gold accumulates in the red blood cells of proteins.
例えば，ヘモグロビンと金の反応は興味がある。なぜなら，研究が金がタンパク質の赤血球に蓄積することを示したからである。
10. Subregions of the Southern Mountains Ecoregion are, **for example**, characterized by different combinations of vegetation, elevation, land use, and climate characteristics.
例えば，Southern Rocky 山脈生態での小地域は，異なる植生，高度，土地利用，および気候特性の組合せによって特性把握される。

Exceed（Exceeding）：超える；過度に（越える）

1. Iron **exceeded** recommended concentrations in two wells at the Rockland sites.
鉄分が Rockland 地域では 2 つの井戸で勧告濃度を超えていた。
2. Benzene **exceeded** the 5 μg/L MCL in well B–8 by a factor of 13 in 1998 to 2002.
ベンゼンは，1998 から 2002 の間，井戸 B–8 では MCL（最大許容濃度）5 μg/L の 13 倍以上だった。
3. TCE **exceeded** the 0.005 mg/L MCL by over 2 orders of magnitude.
TCE が 0.005 mg/L MCL を二桁以上超えていた。
4. Surface soil TPH concentrations **exceed** 100 mg/kg around most of the tank farm perimeter.
表面土壌の TPH 濃度がタンク場周囲の大部分で 100 mg/kg を超えていた。
5. Several wells **exceeded** the Primary Drinking Water Standards proposed by the EPA in arsenic and sulfate.
いくつかの井戸で，EPA によって提案されている主要な飲料水基準が，ヒ素と硫酸塩で越えていた。
6. Although all indices showed a worsening trend, none **exceeded** the criteria, and no significant impairment was indicated.
すべての指標が悪化傾向を示したものの，いずれもクライテリアを超えておらず，有意の損傷が認められなかった。
7. For example, what are the antidegradation ramifications where water quality

exceeds levels necessary to protect existing uses, but uses do not exist, or vice versa?
例えば，現行使用を保護するのに必要なレベルを水質が上回るけれど使用が存在しない場合，もしくは逆の場合，劣化防止の結果は，何なのか？

- 〜 **is exceeded** ……：〜が…を超える

8. Both the risk-based and regulatory-based criteria for benzene **were exceeded** at groundwater probe FW18.
リスクベースと規制ベースクライテリアの両方が，ベンゼンにおいて，地下水プローブFW18で超えていた。

- **exceeding**：超える

9. Cyanide was not detected in levels **exceeding** recommended drinking water criteria in any wells.
シアン化物は，望まれる飲料水クライテリアを超えるレベルではどの井戸でも検出されなかった。

10. Monitored groundwater concentrations were not found at levels **exceeding** the detection limits of BTEX constituents, especially benzene, at groundwater monitoring wells.
監視下の地下水での濃度では，BTEX構成要素，特にベンゼンの検出限度を超えるレベルでは，地下水モニタリング井戸においては見られなかった。

Exceedance：超過

1. The t-test does not form the basis of the determination of impairment, but tightens the conditions for a genuine **exceedance**.
t検定は，損傷決定の根拠にならないが，純粋な超過について状態を強化する。

2. Despite the impaired condition of the receiving water, the impact of the sewage discharge resulted in criteria **exceedance** in EPT, biotic index, and model affinity.
受入水の損傷状態にもかかわらず，下水排出の影響は，EPT，生物指数，モデル類縁度におけるクライテリア超過を招いた。

Excessive (Excessively)：過度；極度（はなはだしく）

1. This sampling did not encounter evidence of **excessive** concentrations of arsenic.
このサンプリングでは，ヒ素の濃度が過剰であるという証拠はあがらなかった。

2. Regions characterized by karst topography, extensive sandy soils, or **excessive** aridity are examples of areas where watersheds are less important.
カルスト地形，広大な砂質土壌，極度の乾燥を特色とする地域は，流域があまり重要でない区域の例である。

- **excessively**：はなはだしく

3. Stream and most riverine fish species are not **excessively** mobile to the point where they are unusable as indicators.
水流および大半の河川の魚種は，指標に適さないほど移動性が高くない。
4. The concentrations of chromium and sulfate are not **excessively** high compared to values expected in soils, but the elevated cadmium concentration is more dramatic.
クロムと硫酸塩の濃度は，土壌で予測される値と比較してはなはだしく高くはない。しかし，カドミウムの高い濃度は劇的である。

Except：除いて；以外は

- **except for** .....：… を除いて；… 以外は
- **except that** .....：ただし，…

1. Table 6 indicates that, **except for** one sample, all samples contain between 1 and 20 weight percent of gypsum.
表6は，1つのサンプルを除いて，すべてのサンプルが1と20の間の重量パーセントの石膏を含んでいることを示している。
2. All values were obtained for a time period of 1900 to 1960 **except for** the Detroit River flow which was averaged over the period from 1939 to 1960.
すべての値は，1939から1960までの期間にわたって平均されたデトロイト川の流量以外は，1900から1960までの期間に収集された。
3. The criteria were also applied to multiplate samples, **except that** percent model affinity was not used, because variability among percent contribution of major groups in multiplate samples from nonimpacted sites was found to be too great to establish a model community.
このクライテリアは，多プレートサンプルにも適用された。ただし，パーセントモデル類縁度は適用されず，その理由は，無影響地点での多プレートサンプルにおける主要集団のパーセンテージの変移が高すぎて，モデル生物群集を定められないと考えられたからである。

Exception：例外

1. Two **notable exceptions** are calcium and sulfates.
   2つの顕著な例外はカルシウムと硫酸塩である。
2. One **notable exception** is cadmium which was detected at higher than commonly expected values throughout the study area.
   1つの顕著な例外はカドミウムで，一般的予想値より高い値が調査地域全体で検出された。
3. Although there are **a few exceptions**, all metals are soluble to some extent.
   少数の例外はあるけれども，すべての金属はある程度溶解する。
4. This chemical reaction is **no exception**.
   この化学反応も例外ではない。

   ● **exception to .....**：… に対する例外

5. There is an **exception to** this adsorption rule.
   この吸着の法則には一つの例外がある。
6. One important **exception to** ..... is .....
   … に対する一つの重要な例外は … である。
7. One **exception to** this is nuclear reaction.
   これに対する一つの例外は原子核反応である。

   ● **exception with .....**：… との例外

8. Two **exceptions with** cadmium and sodium are seen in the data.
   データに見られる2つの例外は，カドミウムとナトリウムである。

   ● **with the exception of .....**：… を除いて
   ● **with the exception that .....**：… を例外として
   ● **without exception**：例外なく

9. The formulation is based upon a similar concept **with the exception that** .....
   … ということは例外として，この公式化は類似の概念に基づいている。
10. **With the exception of** a few experiments made with KI, all experiments suggest that the reactor generates hydroxyl radicals.
    KIを使用した少数の実験を除いて，すべての実験はこの反応器がヒドロキシル遊離基を発生させることを示唆する。
11. **With the exception of** facilities on specific waterbodies, all regulated dischargers in Georgia must consider their compliance with numeric

biocriteria during the NPDES permit renewal period.
特定水域の施設を除き，ジョージア州の被規制排出者は，みな NPDES 許可更新期間中に数量的生物クライテリアの遵守を各自考慮しなければならない。

12. The concentration ratios of the two PAH compounds in mussels from the area are very similar, **with two notable exceptions** (Stations E and F).
その地域から採集されたムラサキ貝では2つの PAH 化合物の濃度比率が，（地点 E と F で）2つの顕著な例外があるが，非常に類似している。

13. **Without exception**, all living organisms require water.
例外なく，すべての生き物は水を必要とする。

Exchange：交換

1. It is the purpose of this research to determine the rate of **ion exchange**.
イオン交換速度を決定することがこの研究の目的である。
2. The model formulation incorporates chemical decay and transport mechanisms of particulate and **diffusive exchange** between water column and sediment.
モデル式は化学的分解，微粒子の輸送メカニズム，水コラムと堆積物の間の拡散的交換が盛り込まれている。

Exclude (See also **Include**)：除外する(→ **Include**)

1. One area that **was** specifically **excluded from** this study was human health effects.
この研究から特に除外された1つの領域は，人間の健康への影響であった。
2. Special care must be taken **to exclude** anomalous sites.
異常地点を排するよう特に配慮すべきである。

Exhibit：示す

1. The linear discriminant function is expected **to exhibit** a multivariate normal distribution.
線形判別関数は，多変量正規分布を示すと予想される。
2. Chloride is in the range of 6 to 41 mg/L and **exhibits** no significant difference between the alluvium and the shale.
塩化物が6から41 mg/Lの範囲にあり，沖積層と頁岩の間に重要な相違を示さない。
3. Vertical profiles also **exhibited** order of magnitude differences between the

upper and lower layer average Lindane concentrations.
縦断プロフィールにおいて，Lindane 濃度の上層の平均と下層の平均で桁違いの相違を示した。

4. These regions generally **exhibit** similarities in the mosaic of environmental resources, ecosystems, and effects of humans and can therefore be termed ecological regions or ecoregions.
これらの地域は，一般的に環境資源，生態系および人為的影響のモザイクにおいて類似性を示すので，生態学的地域，もしくは生態地域と呼ばれる。

5. As described by *Kuraki et al.* (1989) and *Imura et al.* (1989), the range of pollution sensitivity **exhibited** by each metric differs among metrics.
Kuraki ら (1986) および Imura ら (1989) が述べたとおり，各メトリックによって示された汚濁感度値域は，それぞれ異なる。

6. Compared to the Scioto River, the Ottawa River **exhibits** evidence that, despite some improvements, makes it one of the most severely impaired rivers in the state.
Scioto 川に比べ，Ottawa 川は，若干の改善にもかかわらず，同州で最もはなはだしく影響された河川の一つという証拠を示している。

7. Core matrics provide useful information in discriminating among sites **exhibiting** either good or poor quality ecological conditions.
コアメトリックは，良質または悪質な生態学的状態のいずれかを示す地点を区別するうえで，有益な情報をもたらす。

Exist (Existing) (Existence) : 存在する (既存の) (存在)

1. Data **exist** from several years of studies that support these conclusions.
数年間の研究からなるデータが存在し，この結論を支持している。

2. Eleven dams **exist** along the Columbia River in the United States.
11 のダムが合衆国のコロンビア川に沿って存在する。

3. **There exist** three possible levels at which the biological impairment criteria could be instituted in New York State.
生物損傷クライテリアがニューヨーク州で制度化されえる3つの潜在的レベルが存在する。

4. The preceding results demonstrate that discernable patterns in biological community information **do exist**.
前述の結果は，生物群集情報における区別可能なパターンが存在することを実証している。

5. The plots for each metric were examined to determine if any visual relationship with drainage area **existed**.

各メトリックについての図が検討され，流域面積との目測関係が存在するか否か決定された。

- **existing**：既存の

6. **Existing** herbicides transport models can be classified into physical models.
   既存の除草剤輸送モデルは物理的モデルに分類される。
7. **Existing** knowledge is too limited to provide a sound basis for planning to prevent accelerated eutrofication and its adverse effects.
   既存の知識は，加速する富栄養化とその悪影響の防止計画のための効果ある基礎を呈するにはあまりにも限定されている。
8. This could be accomplished via a redirection of **existing** state agency resources.
   これは，州機関の既存資源の方向転換によって達成されうる。
9. Short term studies should be undertaken to evaluate **existing** source water monitoring programs.
   短期の研究が，既存の源水モニタリングプログラムを評価するために着手されるべきである。
10. Based on the **existing** data and the conceptual model, the following data needs have been identified: 1) ....., 2) ......, and 3) ....
    既存のデータと概念モデルに基づいて，データの必要性が次のように識別された。1)…, 2)…, 3)…である。

- **existence**：存在

11. Resampling of these wells by EPA confirmed the **existence of** arsenic concentrations elevated above background concentration.
    これらの井戸でのEPAによる再度のサンプリングによって，ヒ素の濃度がバックグランド濃度以上であることが確認された。

Expect：期待する

1. Soil metal concentrations do not appear to be elevated above **expected** natural background levels.
   土壌での金属濃度は，自然に存在するとされる背景レベル以上に上昇しそうも無い。
2. These measurements show generally good agreement with **expected** concentration values.
   これらの測定値は，予期していた濃度値と一般に良い一致を示している。

3. One notable exception is cadmium which was detected at higher than commonly **expected** values throughout the study area.
   1つの顕著な例外はカドミウムで，一般的予想値より高い値が調査地域全体で検出された。

   ● ～ **is expected**：～が期待される

4. For such extreme operating conditions, deviations between model prediction and experiments **can be expected**.
   このような極端な操作条件では，モデル予測値と実験値の間の逸脱が予想されうる。
5. Perfect charge balance **would not be expected** because of analytical errors in measuring the many parameters used.
   多数の使用されたパラメータを測るにいたっての分析エラーのために，完全なチャージバランスは期待されないであろう。
6. Generally, less variability **is expected** among surface waters within the same region than between different regions.
   一般に，異なる地域の地表水域よりも，同じ地域の地表水域の方が可変性が低いと予想される。
7. This could be important in determining how well real impairment **can be** reasonably **expected to** be detected.
   これは，現実の損傷がどれほど合理的に探知されると思われるかを決定するうえで重要だろう。
8. Regional variation in metric details **is expected** but the general principles used in defining metrics seem consistent over wide geographic areas.
   メトリックの細部における地域的変化が予想されるが，メトリック確定に用いられる一般原理は，広範な地区に共通なようである。
9. Data are evaluated within the ecological context that defines **what is expected** for similar waterbodies.
   データは，生態学的関係において評価され，これによって類似の水域について何が期待しうるかが決まる。

   ● ～ **is not unexpected**：～は予想外ではない

10. Actually, from a biological perspective, the differences in index score parameters **are not unexpected**.
    実際，生物学上の視点から，指標評点パラメータにおける差異は，予想外ではない。

    ● **it is expected that** ......：…が期待される
    ● **it is reasonable to expect** ......：…であると予想するのが合理的である
    ● **as would be expected**：予想どおり

11. By the turn of this century, **it is expected that** .....
    この新世紀の到来時までに，… が期待される。
12. **It is expected that** the sulfide represented in the total sulfide analyses is arsenopyrite, pyrite and pyrrhotite, all of which will be referred to henceforth in this paper as pyrite.
    全硫化物分析で表される硫化物は硫ヒ鉄鉱，黄鉄鉱，磁硫鉄鉱と思われ，この論文ではこれからはそのすべてが黄鉄鉱として述べられるであろう。
13. **It is reasonable to expect** some inhibition due to the elevated levels of BOD5 and COD.
    高いレベルのBOD5とCODによれば，若干の抑制があると予想できる。
14. **It is reasonable to expect that** the process of pyrite oxidation and gypsum formation will continue to occur for a long time period.
    黄鉄鉱の酸化と石膏生成プロセスが長期間起こり続けるであろうと思われる。
15. **As would be expected**, the fish community found at each site reflects the predominant habitat features.
    予想どおり，各地点で見られた魚類群集は，優勢な生息地特色を反映している。

Expectation：期待

1. Insufficient knowledge about regional **expectations** can result in misinterpretations about the severity of impacts in streams.
   地域期待に関する知見不足は，水流中の影響度に関する誤った解釈を招きうる。
2. Regional modifications using this data set may raise this MPS **expectation** by approximately one third of one point for the spring dominated streams in the central part of the state.
   このデータセットを用いた地域修正は，MPS期待値を同州中部の春季優勢水流について3分の1点ほど上昇させるだろう。
3. Industry personnel should become aware of the process that states are using to derive biocriteria and communicate their concerns when technical flaws may result in unrealistic, or invalid, biological **expectations** outside of reference sites.
   業界人は，諸州が生物クライテリアを導出する過程を認識して，技術上の欠陥が非現実的もしくは無効な参照地点外の生物学的期待値を生じそうな場合には，懸念を伝えるべきである。

Experience：経験

1. Perhaps most important to the enhancement of biological criteria efforts,

however, are the lessons being learned from state **experiences**.
しかしながら，生物クライテリア努力の増強にとっておそらく最も重要なのは，諸州の経験から教訓を学ぶことである。

2. Understanding the underlying concepts and theoretical assumptions of biocriteria may be a challenging to environmental managers that have little or no **experience** with biological assessments.
生物クライテリアの根本的概念と理論的仮定を理解することは，生物アセスメントの経験に乏しい，もしくは経験のない環境管理者にとって課題になろう。

3. Therefore, permitting **experiences** for AEP facilities are unique and cannot be considered representative of Virginia's regulated industry as a whole.
したがって，AEP施設の許可経験は，独自のものであり，バージニア州の被規制業界全体を代表しているとみなすことができない。

- **experience has shown that** ..... ：経験が … を示してきた
- **experience from** ..... **provides** ..... ：
  … によって得られた経験は … を提供する
- **our experience demonstrates that** ..... ：
  我々の経験によれば，… である

4. **Experience has shown that** several criteria must be considered.
経験が，いくつかのクライテリアが考慮されなくてはならないことを示してきた。

5. This wealth of data and **experience from** the HSPF application on Iowa River provide major benefits for the Study.
アイオワ川におけるHSPFの適用によって得られたこの豊富なデータと経験は，この研究に主要な利益を提供する。

6. However, **our experience demonstrates that** the implementation and enforcement of NPDES permits is enhanced by the site-specific information provided by biosurveys within a biocriteria framework.
しかし，筆者の経験によれば，NPDES許可の実施・施行は，ある生物クライテリア枠組みにおける生物調査から提供された地点特定情報によって強化される。

7. **In our experience** the following are the situations where conflicts have arisen in New Mexico.
我々の経験によれば，ニューメキシコ州で抵触が生じた事態は以下のとおりである。

Experiment：実験

1. Insight into the factors controlling the chemistry of arsenic in the tailings is provided by the results of the **laboratory experiments**.
選鉱廃物中のヒ素の化学的性質をコントロールしている要因が，室内実験の結果

によっての洞察される。

- ● 〜 **carried out experiments**：〜は実験を実施した
- ● 〜 **run experiments**：〜は実験を実施した
- ● 〜 **conducted experiments**：〜は実験を実施した
- ● 〜 **performed experiments**：〜は実験を実施した

2. *Cheng et al.* (1998) **carried out experiments** to compare the metal uptake in active and non-active sludge.
   Cheng ら (1998) は，活性汚泥と非活性汚泥で金属取り入れを比較する実験を実施した。
3. To help us understand the earlier experiments, we **are** now **running** more controlled **experiments**.
   今我々は，以前の実験を理解するため，さらに制御された実験を行なっている。

   - ● **experiments were performed**：実験が実施された
   - ● **experiments were conducted**：実験が実施された
   - ● **experiments were run**：実験が実施された
   - ● **experimental work has been done on** ……：
     … の研究が実験がなされてきた

4. A series of laboratory **experiments were performed to** assess factors that …..
   … の要因を評価するために一連の室内実験が行なわれた。
5. The **experiments were conducted** in glass bottles at room temperature.
   この実験は，室温でガラスのビンの中で行なわれた。
6. **Experiments were run** changing residence time in the UV chamber in an attempt to determine if certain chemicals in the system are more completely destroyed as the residence time is increased.
   滞留時間が増えるにつれて，システム内のある特定の化学物質がいっそう完全に分解されるかどうかを決定する試みで，UV 反応器の滞留時間を変えて実験が実施された。
7. **Very little experimental work has been done on** the vertical dispersion coefficient.
   縦断拡散係数の研究がほとんどされていない。

   - ● **experiments were initiated**：実験が始められた
   - ● **experiments concentrated on** …..：実験は … に的をしぼった

8. A series of **experiments were initiated** in an attempt to obtain results similar to run F5.

実験 F5 と類似の結果を得る試みで一連の実験が始められた。
9. The **experiments** described in this article **concentrated on** the leaching characteristics of selected arsenic-bearing wastes.
この論文での実験は，選択されたヒ素含有廃棄物のヒ素浸出性に的をしぼった。

- **in the experiment**：この実験で
- **in some experiments**：いくつかの実験で．
- **in other experiments**：他の実験で
- **in all experiments**：すべての実験で
- **in ancillary experiments**：補助的な実験で

10. These samples were stored at room temperature in these bags prior to use **in the experiments**.
これらのサンプルは，実験使用前に室温でこれらの袋に保管された。
11. Lobsters were not fed during their captivity and were kept in the aquarium for at least 10 days before being used **in the experiment**.
ロブスターは飼育中はえさを与えず，少なくとも実験の前 10 日間は水槽の中で飼われた。
12. **In all experiments** the following were monitored.
次（のパラメータ）がすべての実験でモニターされた。
13. **In ancillary experiments**, it was determined that .....
補助的な実験では… が決定された。

Explain：説明する

1. There seems to be no established theory **to explain** this phenomenon.
この現象を説明する定説はないとおもわれる。
2. At each step in the application process, **we will first explain** what need to be done.
アプリケーション過程のそれぞれのステップにおいて，何をしなければならないかを最初に説明する。
3. Nonetheless, to guard against potential legal attacks on water quality criteria based on chemistry or toxicity, agencies should be prepared to **explain fully** the limitations as well as the capabilities of bacteria.
それでも，化学的特性または毒性に基づく水質クライテリアへの法的攻撃に対し防護するため，諸機関は，細菌の力の限界と能力を十分に説明できるよう備えるべきである。

- **~ can be explained by .....**：～は … によって説明することができる

- ~ **could be explained**：~が説明されうるかもしれない
- ~ **attempt to explain** .....：~は…を説明しようと試みる
- ~ **has long attempted to explain** .....：
  ~は…を説明しようと試みてきた

4. This **can be explained by** the fact that .....
   これは…という事実によって説明できる。
5. This observation **could be explained** if multiple dechlorinating microorganisms were present in the filter.
   もし多種の脱塩素微生物がこのフィルターに存在していたなら、この観察は説明できるかもしれない。
6. The kinetic model **attempts to explain** the mechanics of sorption.
   この運動のモデルは、脱吸着機構の説明を試みる。
7. There have been many **attempts to explain**, describe and quantify these effects.
   これらの影響を説明し、記述し、そして数量化する多くの試みがされてきた。

Explanation：説明；解説

1. **There is no definitive explanation for** the fact that .....
   …という事実についての決定的な説明がない。
2. Our finding should be considered **one of the most realistic explanations**.
   我々の発見は、最も現実的な説明の一つと考えられるべきである。
3. Emphasis is on **explanation** and intellectual stimulation, rather than on comprehensive documentation.
   包括的文書化より、解説と知的刺激を強調している。
4. Examination of local gaseous and liquid concentrations permitted an **explanation of** the complex events occurring under transient state operating conditions.
   局地的なガスと液体の濃度の検討から、瞬間状態操業条件下で生じる複雑なイベントの説明ができる。

- **One explanation for** ~ **is** .....：~に対する一つの説明は…である
- **A more detailed explanation is** .....：いっそう詳細な説明は…である
- **The best known explanation among** ~ **is** .....：
  ~の中で最もよく知られている説明は…である
- **The most popular explanation in** ~ **is** .....：
  ~の最も一般的な説明は…である

5. **One explanation for** this difference **was** loss of membrane integrity.
   この相違についての1つの説明は，薄膜の完全性の損失であった。
6. **A more detailed explanation** of Bailey's approach **is** given later in this chapter
   Baileyのアプローチのいっそう詳細な説明がこの章の後に示される。
7. **A more detailed explanation is** found in USEPA and Maruyama (1989).
   より詳細な説明は，USEPA (1988) およびMaruyama (1989) に記されている。
8. **The best known explanation among** those explanations proposed by 〜 **is** ……
   〜によって提示された説明の中で最もよく知られているのは … である。

   - **Another likely explanation for** ……：
     … についての，もう1つのありえる説明は
   - **A logical explanation is that** ……：
     論理的な説明としては… ということだ
   - **A more plausible explanation is that** ……：
     いっそう妥当な説明は … ということである
   - **One plausible explanation for** 〜 **is that** ……：
     〜についての妥当な説明の1つは … ということである

9. **Another likely explanation for** the biphasic degradation **is that** the differences in degradation rates with soil depth.
   二相分解についてのもう1つのありえる説明としては，土壌の深さによって分解速度に相違があるということである。
10. **A logical explanation is that** ・OH oxidized Ph-Cl and PCB mixture without dehalogenation.
    論理的な説明としては，・OHがPh-ClとPCBの混合物を脱塩化無しで酸化させたということことである。
11. **Possible explanation for** the strong PMPA response in the presence of the pesticides include: 1) ……, 2) ……
    殺虫剤が存在するときのPMPAの強い反応については，次のような説明が可能である。

    - **Whatever the explanation**：説明が何であるとしても

12. **Whatever the explanation**, the weak and inconsistent detection of naphthalene did not allow estimation of a threshold concentration by the method used here and elsewhere.
    説明が何であるとしても，微弱，かつ不規則なナフタリンの検出は，ここで使われた方法，そしてほかのところでも使われた方法による閾値濃度の推測を難しくした。

Explore：探査する；探究する

★ Explore は Examine「審査する；考察する」, investigate 「調査する」などと共に, 研究・調査を示す場合の表現です。

1. In this study **we explored** .....
   この研究では，我々は … を探査した。
2. The project **explored** three important aspects of the subjects.
   このプロジェクトでは，3つの重要な局面を探究した。
3. The effect of ozone and p-nitrophenol on CCl_4 degradation is also **explored**.
   オゾンと p-ニトロフェノールが CCl_4 の分解におよぼす影響も探究される。
4. Potential uses of biological criteria in the total maximum daily loads (TMDL) process **are** also **being explored** at both the state and federal levels.
   日最大負荷量(TMDL)過程における生物クライテリアの潜在的利用も，州レベルおよび連邦レベルで追求されている。
5. We **are** currently **exploring** the use of ultrasound in destroying chlorinated hydrocarbons in dilute aqueous solutions.
   現在，我々は，希釈水溶液中での塩化炭化水素を分解するのに，超音波の使用を探究している。

Express：表現する

1. The results **can be expressed** in a fashion analogous to an adsorption isotherm.
   結果は吸着等温線に類似した方法で表現されうる。
2. Some concern **was expressed** by the author of the report over the accuracy of the majority of the data.
   大多数のデータの正確さに関して，若干の懸念の色が報告書の著者によって表された。
3. Concern **is** frequently **expressed** about the potential for biological criteria to be underprotective.
   生物クライテリアが保護過少な可能性について，懸念がしばしば表明される。

   ● ～ **is expressed as** ….. ：～は … と表現される
   ● ～ **can be expressed as** ….. ：～は … として表現できる

4. This **is expressed** mathematically as .....
   これは数学上 … として表現される。

5. The mass balance **can be expressed as** .....
   物質収支は … として表現できる。
6. Eq. 2 **can be expressed as** .....
   式 2 は … と表現される。

Expression：表現

1. The kinetic **expression for** growth is modified in order to account for environmental effects.
   成長の速度表現は，環境の効果を説明するために修正される。
2. This challenge resulted in a search for numerical **expressions** in a form simpler to understand than long species lists and well-thought but lengthy technical **expressions of** the data.
   この課題により，生物種の長大なリストや，綿密だが長々しいデータの技術的説明よりも理解しやすい形での数値表現が追求されるようになった。
3. This will require new approaches to management with an emphasis on the assessment of **a wide expression of** ecological impacts.
   これには，生態影響の広範な表現への評価に力点を置いて新たな管理手法を要するだろう。

Extend：進展する；伸びる

1. This analysis **has been extended** in their study.
   この分析は，かれらの研究で進展してきた。
2. A 5-ft-thick concrete wall **extends** 9 ft toward the ceiling.
   厚さ 5 フィート，高さ 9 フィートのコンクリートの壁が天井に向かって立っている。
3. The soil benzene plume **extended** northeastward from the intersection of J drive and K Street, and bifurcated to the north and east around the power plant.
   ベンゼンのプルームが J 通りと k 通り交差点から北東方向にのび，発電所のあたりで北と東に分かれていた。
4. At this level a line of some twenty to thirty elongated depressions, each measuring about 100 m in length, **extends** from north to south.。
   20 から 30 くらいの細長いくぼみからなる線が北から南にのびている。窪みの長さはそれぞれ 100 m くらいである。

Extent：程度；度合；範囲

1. The **extent of** the contamination is estimated to be approximately 200 feet by 100 feet.
    汚染地帯の広さは，およそ 200 フィートかける 100 フィートであると推定される。
2. However, the **extent** and form of biological survey data vary widely among states.
    しかし，生物調査データの程度と形式は，州ごとで大幅に異なる。
3. The use of nonstatistical surveys to describe populations always involves the question of the existence and **extent of** bias in the population inferences.
    母集団を記述するための非統計的調査の利用は，常に母集団推測における偏向の存在および程度に関する疑問を伴う。
4. Although every program provides some form of educational benefits, the **extent of** educational opportunities varies among the programs.
    どのプログラムも何らかの啓発効果を有するものの，啓発機会の度合は，プログラムごとに異なっている。
5. The likelihood of a false negative decision is higher, due to uncertainty regarding the **vertical extent of** soil contamination and the potential for groundwater contamination.
    縦方向の土壌汚染と地下水汚染の可能性に関して不確実であるために，誤った否定の決定になる可能性がより高い。

　　● extent over which ..... : … である度合（範囲）

6. It is important to define the regional **extent over which** a particular biocriterion is applicable.
    ある特定の生物クライテリアが適用可能な地域的度合を定めることが重要である。
7. Examining the variance structure can give insight into the **extent over which** particular biocriteria might be applicable.
    変動構造を検討すれば，ある生物クライテリアが適用可能な度合に関する洞察が得られよう。
8. A critical issue is to determine the regional **extent over which** a particular biological attribute is applicable.
    重大問題は，ある特定の生物学的属性が適用可能な地域範囲を決定することである。

　　● extent to which ..... : … する度合（程度）

9. **The extent to which** this subpopulation reflects the target population would elicit discussion.

この部分母集団がターゲット母集団を反映する度合は，論議を呼ぶだろう。
10. Comparison to reference conditions is essential to evaluate the **extent to which** study sites are influenced by human actions.
研究対象地点が人間の活動の影響を受ける度合を評価するには，参照状態との比較が不可欠である。
11. Therefore, the **extent to which** the information can be used to characterize the resources statewide is open to question.
したがって，全州の資源を特性把握するため情報が用いられうる度合が疑問視されるのである。
12. Whether the models we create are good models or poor models depends on the **extent to which** they aid us in developing the understanding which we seek.
我々が作るモデルが良いモデルであるか，あるいは貧弱なモデルであるかどうかは，それらのモデルが我々が求める理解力をつける手助けする程度による。

   ● **to what extent** ..... ：どの程度まで … か？

13. **To what extent** is mercury in the form of mercuric sulfide available for biological methylation?
硫化水銀としての水銀は，どの程度まで生物によってメチル化されるのか？
14. The question concerning drinking water are addressed in this article : 1) **to what extent**, if any, does acid precipitation cause a drinking water problem ; 2) if a problem is present, what can be done about it?
この論文では，飲料水に関する問題を扱います。1) もしあるとするなら，どの程度まで酸性雨が飲料水問題になるのか; 2) もし問題があるなら，何がそれについて何ができるか？

   ● **to some extent** ：ある程度まで
   ● **to a significant extent** ：かなり
   ● **to a large extent** ：大部分は;おおむね;大いに
   ● **to a great extent** ：大部分は;多くは
   ● **to lesser extent** ：より少なく
   ● **to a limited extent** ：限度内で

15. Although there are a few exceptions, all metals are soluble **to some extent**.
少数の例外はあるけれども，すべての金属はある程度溶解する。
16. It is important to recognize that the proposed model depends **to a significant extent** on its intended purpose.
提案されたモデルが，そのモデルの意図的な目的にかなり左右されることを認識することが重要である。

17. The odor in the untreated air is caused **to a large extent** by reactive chemical species.
   浄化されていない空気の臭気は，主に反応性のある化学種によって引き起こされる。
18. The type of data required depends **to a larger extent** on the chemical being tested.
   必要とされるデータのタイプは，被試験化学物質に大きく左右される。
19. **To a large extent**, the status of biological criteria programs across the nation can be described by determining the presence of these activities in each of the states.
   おおむね全米の生物クライテリアプログラムの現状は，各州におけるこれらの活動の存在を決定することによって示されうる。

Extrapolate（Extrapolation）：推定する（推定）

● extrapolate ～ from A to B ：～をAからBまで推定する

1. The user should be aware that such effects may complicate attempts to **extrapolate** data for photolysis rates **from** one aquatic medium **to** a very different medium.
   利用者は，このような影響が光反応速度データを1つの水生媒質からかなり異質の媒質まで外挿法によって推定する試みを複雑にするかもしれないということを認識すべきである。

● via extrapolation from ..... ：…からの推定によって

2. Efforts to assess, research, and manage the ecosystems are normally carried out **via extrapolation from** data gathered from single-medium/single-purpose research.
   生態系を評価・研究・管理する努力は，通常，単一媒体／単一目的の研究で収集されたデータからの推定によって行われる。

● make meaningful extrapolations from ..... ：
   …からの推定を有意義にする

3. We must develop an understanding of ecosystem regionalities at all scales, in order to **make meaningful extrapolations from** site-specific data collected from case studies.
   我々は，ケーススタディで収集された地点特定データからの推定を有意義にするために，すべての縮尺での生態系地域性への理解を発展させなければならない。

F

Facilitate：容易にする

1. This information will establish the pollutant status of these organisms and **facilitate** interpretation of the other data.
   この情報によって，これらの生物の汚染状態が確証され，そして他のデータの解釈が容易にされる。

Fact：事実

- **The fact is that .....**：その事実は … である
- **It is a well-established fact that .....**：
  … は十分に確立した事実である
- **The fact provides clearly that .....**：
  この事実は … のことを明らかに提示している

1. **The fact is that** substantial evidence indicates that metabolites are toxic to animal systems.
   事実は，新陳代謝廃物が動物システムにとって有毒であることを示す実質的な証拠があるということである。
2. **It is a well-established fact that** the odor in the untreated air are caused to a large extent by reactive chemical species.
   浄化されていない空気の臭気は主に反応性のある化学種によって引き起こされる，ということは十分に確立された事実である。

- **for the fact that .....**：… という事実について
- **with the fact that .....**：… という事実と
- **lies in the fact that .....**：… という事実に基づいている

3. There is no definitive explanation **for the fact that** .....
   … という事実についての決定的な説明がない。
4. When selecting these sites one must account **for the fact that** minimally disturbed conditions often vary considerably from one subregion to another.
   これらの地点を選ぶ際には，最小攪乱状態が小地域ごとに大幅に異なりうるとい

5. These results are consistent **with the fact that** ......
   これらの結果は … という事実と一貫している。
6. The answer to those questions **lies in the fact that** …….
   それらの質問への答えは … という事実に基づいている。

   ● **due to the fact that** ..... ： … であるという事実により
   ● **despite the fact that** ..... ： … という事実にもかかわらず

7. USEPA has established a rule that attainment of biocriteria should be granted disproportionate weight in demonstrating overall use attainment, **due to the fact that** chemical-specific and WET criteria are only surrogate measures of biological integrity.
   USEPAは，化学クライテリアおよびWETクライテリアが生物完全性の唯一の代用測度であるという事実により，全般的な用途達成の実証において生物クライテリア達成が不相応な加重を与えられるべきだとのルールを確定した。
8. **Despite the fact that** the large amount of work has completed for wadable warmwater streams, much remains to be done.
   徒渉可能温水水流について多くの研究が終わったにもかかわらず，残された課題は多い。

   ● **in fact** ： 実際；要するに；事実上；実のところ

9. **In fact**, in finding that dams were not required to have NPDES permits, one court rule that the fact that dams may cause "pollution" does not necessitate an NPDES permit where there is no addition of "pollutants."
   実のところ，ダムがNPDES許可を必要としないという所見において，ある裁判所は，「汚濁物質」の追加がない場合，ダムが「汚濁」を生じうるという事実は，NPDES許可を必要とせしめない，と裁定した。
10. The U.S. Supreme Court and lower courts have indicated repeatedly that water quality standards must be enforced through NPDES permits, and **in fact**, that NPDES permits are the primary means of translating water quality standards into enforceable requirements.
    米国最高裁判所およびその下位裁判所は，これまで再三，水質基準がNPDES許可を通じて実施されねばならないこと，そして実際，NPDES許可が水質基準を実施可能な要求へと移し換える主要手段であることを主張してきた。

Factor ： 要因

1. **Factors** influencing the partition of heavy metals are considered in detail.

重金属の分配に影響を与える要因が詳細に考慮される。
2. A model has been structured to account for the relevant **factors**.
モデルが関連要因を説明するのに構築されてきた。
3. **A number of factors** contribute to the differences in measured values of DDTR concentrations in biota.
多くの要因が，生物体内 DDTR の測定濃度の相違に起因している。
4. Insight into the **factors** controlling the chemistry of arsenic in the tailings is provided by the results of the laboratory experiments.
選鉱廃物中のヒ素の化学的性質をコントロールしている要因が，室内実験の結果によっての洞察される。
5. Knowledge of the **factors** affecting the movement and transformations of these substances are of obvious importance in understanding and controlling the hazard.
これらの物質の移動・変化に影響を与えている要因に対する知識は，危険を理解し，またコントロールするうえで明らかに重要である。
6. Faunal similarity between midwestern river systems is dependent on river mile distance, along with other **factors**.
中西部河川系における動物相の類似は，河川距離および他の因子に左右される。
7. Although **factors** such as geology and soils are also important, the other **factors** appear to be the most important in this ecoregion.
地質や土壌などの因子も重要だが，この生態地域では，他の因子が最も重要に思われる。

- **by a factor of** ..... ： … の倍率で

8. Benzene exceeded the 5 μg/L MCL in well B-8 **by a factor** of 13 in 1998 to 2002.
ベンゼンは，1998 から 2002 の間，井戸 B-8 では MCL（最大許容濃度）5 μg/L の 13 倍以上だった。

Failure：失敗

1. The **failure of** the model to predict it may be attributed to errors in the model input data.
それを予測するモデルの失敗は，モデル入力データのエラーに帰されるかもしれない。
2. Although studies have been made on the cause of nitrification **failure**, there is little agreement on how it was failed.
硝化作用が生じなかった原因についての研究はおこなわれてきたが，その原因に

ついての意見はほとんど一致していない。
3. Coralvill Reservoir does not thermally stratify to any great extent, so the **failure** to include a hypolimnion compartment is not viewed as a serious problem.
Coralvill 貯水池は大きくは温度層化しない，それゆえ（このモデルに）深層部が含まれていないということが重大な問題だとは見なされない。

Fall ： …にある

1. These potential release sites do not **fall under** the definition of Solid Waste Management Units ; however, they are areas of environmental concern.
これら潜在的流出地点は廃棄物管理ユニットの定義下ではない。しかしながら，それらは環境にとって関心のある地点である。

- **fall into** ….． ： … に分類される

2. These organic chemicals **fall into** nine groups.
これらの有機化学物質は9つのグループに分類される。
3. His article certainly **fell into** that category.
彼の論説は確かにその部門に分類された。
4. These processes **fall into** three general categories: 1)………., 2)………., 3)………..
これらの過程は3つの一般的な部門に分類される。それらは，1)…, 2)…, および3)…, である。
5. The approaches used in most numerical models **fall into** three broad classes: Eulerian methods, Eulerian-Lagrangian hybrid method, and Lagrangian(particle) method.
ほとんどの数値解析モデルで使われる手法は3つの広範囲なクラスに分類される。それらは，Eulerian 方法，Eulerian-Lagrangian ハイブリッド方法および Lagrangian 方法，である。
6. In the view of the Task Force, the research questions of priority importance **falls into** two general categories: 1)….., 2)….., and 3)…..
特別委員会の観点では，研究質問の優先的重要性は2つの一般的な部門に分類される。それらは，1)…, 2)…, および3)…, である。

Far ： より；はるかに；ずっと；ほど遠い

- **far less** ：より少なく
- **far more** ：より多く

- ~ is far from ..... : ~は…にはほど遠い
- thus far : 今までのところ
- as far as ..... is concerned : …に関する限り

1. Results for lower depths showed **far less** variation, owing to the influence of the constant water table elevation.
   地下水面の高さが一定であるため，その影響で深いところの結果はより変化が少ない。
2. During the first two decades of the CWA, regulatory agencies paid **far more** attention to chemical than to biological integrity of the nation.
   CWAの当初20年間に規制機関は，米国の水域の生物完全度よりも化学完全度に注意を払った。
3. It is **not too far from** the truth
   それは真実からそれほど遠くはない。
4. The acid rain problem in this region **is far from** a solution.
   この地域の酸性雨問題は解決にはほど遠い。
5. This project, **thus far**, has concentrated on two- and three-dimensional analyses of IBI, Miwb, and ICI metrics and selected subcomponents.
   このプロジェクトでは，今までのところIBI, Miwb, ICIメトリックおよびいくつかの部分構成要素の二次元分析または三次元分析が重視されてきた。

★ Farに関連した例文を下に示します。

6. **Inso far as** past state biological criteria efforts grew out of biological monitoring efforts, future efforts will likely build on current monitoring programs.
   生物モニタリング努力から発した過去の州の生物クライテリア努力に関する限り，将来の努力は，現行のモニタリングプログラムに立脚するであろう。

Fashion (See also **Manner**) : 様式；型；方法 (➡ also **Manner**)

- In a fashion..... : …の様式で

1. The results can be expressed **in a fashion** analogous to an adsorption isotherm.
   結果は吸着等温線に類似した方法で表現されうる。
2. It appears that there is nothing gained by plotting the data **in this fashion**.
   データをこの様式で作図することによっては，何も得られないように思われる。
3. Each element of material flows downstream **in a** unique discrete **fashion**.
   物質の要素それぞれが，独特の断続的な形で下流に流れる。

Feasible：可能な；実行できる

1. Regional scale modifications in management practices **may be feasible**, allowing for the significant recovery of impaired aquatic resources.
   管理慣行における地域スケール修正も実行可能であり，これは傷ついた水生資源をかなり回復させる。
2. Results of such additional studies could determine if **it would be feasible** for small systems to treat entire volumes of water with a flow-through reactor.
   このような追加研究の結果によって，水の全量処理が流動反応器で可能かどうか決定することが小型のシステムで出来るかもしれない。

Feature：特徴

1. A comparison of the two groups showed two **prominent features**.
   この2つのグループの比較によって，2つの顕著な特徴が示された。
2. **Novel features of** the EXAMS model include the introduction of "canonical" environment.
   EXAMSモデルの新奇な特徴は，「規準的な」環境の導入である。
3. Some **known features of** the behavior of stable iodine in the environment are not included in the model, because they are not relevant to the problem of interest.
   自然環境下ではヨードは安定しているが，その挙動に関してすでに知られている特徴の一部モデルに含められていない。なぜならそれらはここで取り上げている問題点とは直接に関係が無いからである。

Feel：感じる；思う

- it was felt that ..... ：…だと感じられた
- we feel that ..... ：我々は…だと思う
- they felt ..... ：…と考えた；…だと思った

1. Due to the above considerations, **it is felt that** .....
   上記の考慮から，…と思われる。
2. **It was felt that** this data would be particularly meaningful with respect to the .....
   このデータが…に関して特に重要な意味を持つであろうと感じられた。
3. **It was felt that** most of the impacted reference results should be

encompassed by the baseline WWH use designation for Wisconsin's inland rivers and streams.
最小影響参照結果の大半は，ウィスコンシン州の内地河川・水流についてのベースライン WWH 使用指定によって成し遂げられるべきだと感じられた。
4. **We feel that** ..... is incomplete.
我々は … は不完全だと思う。
5. Many other states did not undertake substantial bioassessment programs because **they felt** the cost of bioassessment outweighed the benefits.
他の多くの州は，生物アセスメントの費用が便益をはるかに上回ると考えたせいで，実質的な生物アセスメントプログラムを行わなかった。

Few：少数の

1. However, **few** comparative studies have been done to investigate the distribution of PCB in marine organisms.
しかしながら，海洋生物体内での PCB の分配を調査するための比較研究がほとんど実施されなかった。
2. At present, **few** states can characterize even a small fraction of their water resources with biological survey data.
現在，生物調査データで各自の水資源のごく一部でも特性把握できる州は，きわめて少数である。

- **only a few**：ほんの少数の

3. **Only a few** species of fish can survive in this river.
ほんの数種類の魚類だけがこの川では生息できる。
4. The most heavily polluted zone was characterized by **only a few** highly tolerant species.
最もひどく汚染された地域はただ少数の強いに抵抗力のある種だけによって特徴づけられた。
5. **Only a few** rocks need to be sampled, because each has a periphyton assemblage with hundreds of thousands of individuals.
サンプリングに必要なのは，数個の岩石だけである。なぜなら，1個の岩石に個体数が数百ないし数千もの付着生物群があるからである。

- **few studies**（See Study）：ごく少数の研究（→ Study）

Figure (See also **Table**,)：図（→ **Table**）

- **Figure N shows** ..... ：図 N は … を示している

1. **Figure 1 shows** the result of the nitrification test for the media.
   この培養液での硝化反応テストの結果を図1に示す。
2. **Figure 2 shows** a plot of changing nitrification rates versus time.
   図2に，硝化速度の変化・対・時間のプロットを示す。
3. **Inspection of Figure 3 shows that** .....
   図3の調査から … を示していることがわかる。
4. **Figure 5 shows** the variation in parathion concentration in ppm with the irradiation time at various pHs.
   種々の pH 値での照射時間に対するパラチオンの濃度変化を ppm で図5に示す。
5. **Figure 6 shows** good agreement between the predicted and observed suspended solids concentrations.
   図6は，浮遊固体濃度の予測値と観察値の間でかなりの一致を示している。

- **This figure shows clearly that** ..... ：
  この図は … を明らかに示している
- **examination of Figure N shows that** ..... ：
  図 N の検査によって … が示される

6. **This figure shows clearly that** A, B, and C give comparable results.
   この図はA，B，Cが類似の結果を出すことを明らかに示している。
7. Detailed **examination of Figure 7 shows that** MEK elimination was overestimated.
   図7の詳細な検査によって，MEK の除去が過剰に見積もられていたということが示される。

- ～ **is shown in Figure N** ：～を図 N に示す

8. Soil concentrations for the coring periods **are shown in Figures 4**.
   コアリング調査期間での土壌における濃度を図4に表わす。
9. The results of this experimental run **are shown graphically in Fig. 5**.
   この実験の結果をグラフで図5に示す。
10. The computational results along with the observed data **are shown in Fig. 4**.
    計算結果を観察データとともに図4に示す。
11. Plots of the glucose concentrations against the sorbose concentrations **are shown in Figure 6**.

グルコースの濃度対ソルボースの濃度のプロットを図6に示す。
12. Copper toxicity profile **is shown in Figure 6** by a three-dimensional plot of the specific growth rates as a function of the two variables, total ammonia concentration and total copper concentration.
銅毒性プロフィールを図6に示す。それは全アンモニア濃度と全銅濃度の二変数を関数として成長速度定数を表す3次元のプロットである。

- **Figure N depicts** ..... ：図Nは … を描写している
- **Figure N illustrates** ...... ：図Nは … を例証する
- **Figure N portrays** ..... ：図Nは … を表したものである
- **Figure N outlines** ..... ： … を図Nに概説する
- **Figure N presents** ..... ：図Nに … を提示する
- **Figure N demonstrates** ..... ：図Nは … を明示している

13. **Fig 4 depicts** the situation.
図4はその状態を描写している。
14. **Figure 4 illustrates** representative results.
図4に代表的な結果を例証する。
15. **Figure 9 portrays** the biosurvey results from the Scioto River downstream from Columbus, Ohio。
図-9は，オハイオ州Columbus市のScioto川下流での生物調査結果を表したものである。
16. **Figure 2 outlines** the investigation and remedial process which consists of six basic components.
6つの基本的なコンポーネントから成り立つ調査と改善プロセスを図2に概説する。
17. **Figure 2 presents** the relationship between A and B.
図2にAとBの関係を提示する。
18. **Fig. 3 presents** a typical reaction profile observed during a test run.
試運転の間に観察された典型的な反応プロフィールを図3に提示する。
19. **Figure 7 also demonstrates** how different types of impacts can be layered together in a segment. The darker shading represents the impact from a point source discharge, the lighter from a nonpoint source.
図-7は，また，ある区間においてどれほど異なる種類の影響が積み重ねられうるかも示している。より暗い影は，点源汚濁影響を示し，より明るい影は，面源汚濁影響を表す。

- **～ is illustrated in Fig. N** ：～が図Nに例証されている
- **～ is depicted in Figure N** ：～が図Nに描写されている
- **～ is plotted in Figure N** ：～は図Nに図示（プロット）されている

- ~ **is given in Figure N**：～を図Nに示されている
- ~ **is presented in Figure N**：～を図Nに示す

20. Representative results **are illustrated in Fig. 4**.
代表的な結果が図4に例証されている。
21. The situation **is depicted in Figure 3**.
その状態が図3に描写されている。
22. Chemical interactions of the model **are depicted in Figure 1**.
このモデルの化学的な相互作用が図1に描写されている。
23. Using all the estimated rate constants, the calculated accumulation of CO_2 **is plotted in Figure 2**.
すべての推定速度定数を使って計算されたCO_2の蓄積量を図2に図示する。
24. Quantitatively, the results **plotted in Figure 6** may be seen to be in general agreement with the theory.
数量的に，図6にプロットされた結果はその理論と一般的な一致にあるように見られるかもしれない。
25. The model results and experimental values **are given in Figure 5**.
モデルの結果と実験値を図5に示す。
26. The following conclusions can be drawn from the data **given in Fig. 5**.
図5に示されたデータから次の結論が導出されうる。
27. A comparison of predicted and measured Alachlor residues in the soil **is given in Figure 3**.
土壌におけるアラクロール残余の予測値と測定値の比較を図3に示す。
28. The data **presented in Figure 3** are .....
図3に示されたデータは … である。
29. An abstraction of the urban drainage system **is presented in Fig. 1**.
都市排水設備システムの抽象概念を図1に示す。
30. Limited data on pesticide residues **are presented in Figure 9**.
数少ない殺虫剤残余のデータを図9に示した。
31. Typical plots of the nitrate concentration versus time **are presented in** Fig. 4.
硝酸濃度対時間の典型的なプロットが図4に提示されている。

- **Figure N is** ..... ：図Nは … である

32. **Figure 6 is** a display of the data plotted along with this hypothetical line.
図6は，この仮説線と一緒に作図したデータの図示である。
33. **Fig. 2 is** a plot of five growth curves that shows both the experimental and computed results.

図2は5つの(バクテリア)成長カーブのプロットであり，実験結果と計算結果の両方を示している．

- **Figure N summarizes ......** ：… を図Nに要約する
- **～ are summarized in Figure N** ：～は図Nに要約されている

34. **Fig. 3 summarizes** some of the available data that illustrate the relationship between the rate of oxidation of ammonia ion and the pH of the ambient water.
アンモニアイオンの酸化速度と周囲の水のpHとの関係を例証する入手可能なデータのいくつかを図3に要約する．
35. The data **summarized in Fig. 3** demonstrate that ......
図3に要約されているデータは … を実証している．

- **in this figure** ：この図に
- **in Figure N** ：図Nに

36. **In this figure** the A concentration is plotted versus the B concentration.
この図には，Aの濃度対Bの濃度が図示されている．
37. **In Figure 5**, conversion of mercury compounds in an aquatic ecosystem is illustrated in the form of a flow-diagram where the arrows illustrate important transformation reactions.
図5には，水生生態系での水銀化合物の変換がフローチャートのかたちで例証され，矢印は重要な変換反応を例証している．
38. Ordinate values **in Figure 2** are plotted logarithmically to show linear correspondence to logarithmic increase of bacterial concentration.
図2の縦座標には，バクテリア濃度の増加が対数直線関係にあることを示すために対数値がプロットされている．

- **from Figure N** ：図Nから
- **as can be seen from the figure** ：この図から見られるように

39. Furthermore, **from Fig. 2**, it can be seen that the magnitude of the effect is erratic.
さらに，図2から，その効果の規模が不規則であるといえる．
40. **As can be seen from the figure**, the slopes of the lines are parallel and constant.
この図から見られるように，線の勾配は平行ならびに一定である．
41. Model predictions show reasonable agreement **as can be seen from both Figures 2 and 3**.
図2と図3の両方に見られるように，モデルによる予測が妥当な合意を示している．

- **Figure N reveals that .....** ：図Nが…を明らかにしている

42. **Fig. 4 and Fig. 5** strikingly **reveal that** the toxicity of the A is also influenced by the substrate concentration.
図4と図5が，Aの毒性が同じく基質の濃度に影響されることを極めて明らかにしている。

- **As shown in Figure N** ：図Nに見られるように
- **As Figure N demonstrates** ：図Nが実証するように

43. **As shown in Figure 4**, there is no significant difference between A and B.
図4に見せられるように，AとBの間に，有意な相違がない。
44. **As Figure 5 demonstrates**, there is no significant difference between A and B.
図5が実証するように，AとBの間に，有意な相違がない。

Filter ：ろ過する；フィルター

1. Water sample was processed with hydrochloric acid, then neutralized and **filtered** as previously described.
水のサンプルは塩素で処理された後，前述したように中和され，そしてろ過された。
2. This observation could be explained if multiple dechlorinating microorganisms were present **in the filter**.
もし多種の脱塩素微生物がこのフィルターに存在していたなら，この観察は説明できるかもしれない。
3. Most **filter** research work completed to date has been based on the measurement of suspended solids (SS).
今日まで完了したほとんどの濾過の研究は浮遊物質 (SS) の測定に基づいていた。

Finally ：最後に

1. **Finally**, one of the major concerns with regional reference sites is their acceptable level of disturbance.
最後に，地域参照地点に関わる重要事の一つは，擾乱の許容可能レベルである。
2. **Finally**, and perhaps most importantly, many existing versions have as yet not been properly validated with independent data.
最後に，そしておそらく最も重要なことに，多くの既存バージョンは，また，独立データで適切に確証されていない。

Find (Finding)：判明する；わかる (所見；調査結果)

- **It was found that** ..... : … ということがわかった

1. **It was found** from the result that .....
   この結果から … ということがわかった。
2. **It was found that** the observed processes are described satisfactory by this model.
   観察されたプロセスがこのモデルによって十分に説明されることがわかった。

   - ~ **was found that** ..... : ~は … であることがわかった
   - ~ **was found to** ..... : ~は … することが判明した

3. These rivers are generally of moderate to high quality and few sites **were found that** were rated either fair or poor.
   これらの河川は，おおむね質が中度(良)ないし高度(優)なものであり，可または不可の地点は，ごく少数であることがわかった。
4. These constant values **were found to** have a satisfactory fit to all the experimental data.
   これらの定性係数値は，満足のいく程度で，実験データすべてに一致することが判明した。
5. Students' samples **were found to** be very similar to those collected and analyzed at the same sites by professionals at the Alberta Environment, Water Quality Branch.
   高校生によるサンプルは，その地点でアルバータ州環境・水質部の専門家が採集・分析したものとよく似ていた。
6. Variability among percent contribution of major groups in multiplate samples from nonimpacted sites **was found to** be too great to establish a model community.
   無影響地点での多プレートサンプルにおける主要集団のパーセンテージの変異が高すぎて，モデル生物群集を定められないと考えられた。

   - **findings** : 所見
   - **findings to date** : 今日までの調査結果

7. The inclusion of Oligochaeta as a toxic tolerant group contrasts with the original findings of Yamamoto (3), but agree with **the findings of** Sugita et al. (4) and others (5, 6, 7).
   貧毛綱を毒物耐性集団として含めることは，Yamamoto (3)のオリジナル所見に反しているが，Sugita ら(4)および他,(5, 6, 7)の所見に合致している。

8. **Findings to date** indicate no relationship between the original source of DDT and PCBs.
今日までの調査結果は，DDT と PCBs の最初の汚染源の関係を示していない。

   ● **our finding on** ..... ：… に関する我々の発見

9. **Our finding on** ~ can be thought of as .....
~に関する我々の発見は，… とみなすことができる。
10. **Our finding on** ~ can be viewed as .....
~に関する我々の発見は，… とみなすことができる。
11. **Our finding on** ~ should be considered ......
~に関する我々の発見は，… と考えられるべきである。
12. **Our finding** should be considered very significant.
我々の発見は，非常に意義深いものと考えられるべきである。
13. **Our finding** should be considered one of the most realistic explanations.
我々の発見は，最も現実的な説明の一つと考えられるべきである。

First ： 第一；最初

1. **The first part of** this paper provides .....
この論文の最初の部分は … を示している。
2. The **first** investigator to pay attention to the reaction was Yamanaka (1930).
その反応に最初に注意を払った調査者は Yamanaka (1930) であった。
3. Figure 3 shows good agreement between the predicted and observed suspended solids concentrations for the **first** 150 meters downstream.
図3が，最初の150メートル下流で浮遊固体濃度の予測値と観察値の間で良い合意を示している。
4. At each step in the application process, we will **first** explain what need to be done.
アプリケーション過程のそれぞれのステップにおいて，何をしなければならないかを最初に説明する。

   ● **First**, ... **Second**, ... ：第一に …，第二に …

5. Three main conclusions can be drawn from these profiles. **First**, ... **Second**, ... **Third**, ...
これらのプロフィールから3つの主な結論が導出されうる。第一に …，第二に …，第三に …
6. From an analysis of these plots two general conclusions may be drawn. **First**, ... **Second**, ...

これらのプロットの解析から2つの一般的な結論が引き出されるかもしれない。第一に…，第二に

7. The advantages of these increased efforts are threefold. **First**, ... **Second**, ... **Third**, ...
   こうした努力増強の利点は，3つある。第一に … 第二に … 第三に …

   - **First and foremost**：まず第一に

8. **First and foremost**, we thank Dr. Takashi Yamamoto, USEPA for the lengthy discussions and his insights on the biocriteria process and appropriate bioassessment approaches.
   まず第一に，我々が感謝を捧げる相手は，USEPAのDr. Takashi Yamamotoである。彼は，長時間にわたる論議や，生物クライテリア過程ならびに適切な生物アセスメント手法に関する洞察を提供してくれた。

   - **A first step**：第一歩
   - **first of all**：まず第一に
   - **in the first stage**：最初の段階では

9. This report is **a first step** toward meeting this goal.
   この報告は，この目的達成に向かっての第一歩である。

Fit：一致

1. The kinetic data and the model **fits** of the acetylene transformation experiment are shown in Figure 2.
   アセチレン変化実験での速度データとモデルの一致を図2に示す。

2. In evaluating the accuracy of the estimated parameter, more weight should be given to the ability of the model to **fit** the data points near segment D than to its ability to fit the points near segment G.
   推定パラメータの正確さを評価することにおいて，モデルが川の切片Gの近くのデータポイントで合う能力により，切片Dの近くでデータポイントに合うことができる能力に，もっとウエートが置かれるべきである。

   - **～ is fitted by** .....：～は…によってフィットされる

3. The data **were fitted by** linear regression and generally the correlation coefficients were greater than 0.90.
   このデータは線形回帰によってフィットされ，相関係数は一般に0.90以上であった。

   - **fit to** .....：…に一致する

- **fit into .....** ：…に組み込む

4. These constant values were found to **have a satisfactory fit to** all the experimental data.
   これらの定性係数値は，十分に実験データすべてに一致することが判明した。
5. This correlation coefficient was 0.99 indicating a surprising good **fit to** the data points employed.
   この相関係数は 0.99 で，使用したデータポイントに驚くほど一致することを示唆している。
6. This relationship has been identified by several other workers, and **fitted to** Langmuir and Freundlich isotherms.
   この関係は数人の他の研究者によって確認されて，そして Langmuir と Freundlich 等温線に照合させられた。
7. A newly proposed toxicity test **will fit into** the overall hazard management system.
   新たに提案された毒性テストが総合的な危険管理システムに組み込まれるであろう。
8. Some of this disagreement stems from differences in individual perceptions of ecosystems, the uses of ecoregions, and where humans **fit into** the picture.
   意見相違の一因は，生態系に対する各自の認識の差，生態地域の利用における人間の位置などである。

- **fit between ..... and .....** ：…と…の間に合致に達する

9. The goal is to achieve a suitable **fit between** the planned modeling effort **and** the data, time, and money available to perform the study.
   目標は，計画されたモデリングの努力とデータ，時間，それに研究を実行するために利用可能な資金の間で，適当な折り合いをつけることである。
10. **The fit between** the model **and** data is certainly good ; the least square line presented by *Watanabe et al.* (1997) is of slightly steeper slope.
    モデルとデータは確かに一致している。Watanabe ら (1997) によって示された最少二乗法による線は，少し勾配が大きい。

- **a line of best fit** ：ベストフィットのライン（線）

11. If a relationship was observed, **a 95 % line of best fit** was determined and the area beneath trisected following the method recommended by *Fausch et al.* (1984).
    関係が見られたならば，ベストフィットの 95 ％ラインが決定され，その下方の

面積は，Fauschら(1984)の推奨方法に従って三等分された。

Focus（See also **Center**）：
焦点を合わせる；集中する；主題とする；焦点；主眼（➡ **Center**）

● **The focus is on** ..... ：焦点は … にある

1. **The focus is on** contemporary industrial chemicals.
   焦点は工業用化学物質にある。
2. **The focus is on** industrial chemicals and the mechanisms that affect their distribution in the aquatic environment.
   焦点は工業用化学物質と水生の環境でそれらの分配に影響を与える機構（メカニズム）にある。
3. In some cases, **the focus** of the surveys **is on** the chemical condition of lakes or streams；in the others, **the focus is** biological.
   一部の場合，調査の主眼は，湖沼・河川の化学的状態に置かれている。また，生物的状態に主眼が置かれている場合もある。
4. **The focus of** this discussion **is primarily on** stream systems, because that is the waterbody type where most of the developmental work has been done, to date.
   本論の焦点は，主に河川系に置かれている。なぜなら，それは，今日までの開発作業の大半が対象としてきた水域タイプだからである。

● **The focus of** 〜 **is to** ..... ：〜の焦点は … することである

5. The **focus** of this effort **is to** analyze exhaustively the solutions.
   この取り組みの焦点は，徹底的にその溶液を分析することにある。

● 〜 **was focused on** ..... ：〜は … に集中した；〜は … に焦点が置かれた

6. The field investigations **were focused on** the local study area.
   この野外調査は地方の研究地域に集中された。
7. In particular, the investigation **was focused on** evaluating the influence of physical hydrology on .....
   特に，この調査は … に対する物理水文学的影響を評価することに焦点を当てた。

● 〜 **is the focus** ：〜が焦点である

8. Although freshwater rivers and streams **are the focus of** this book, the same framework is being applied for other resource types.
   淡水河川・水流が本書の主題であるが，この枠組みは，他の資源類にも適用され

ている。

- **We focus on .....** :
  我々は … に焦点を当てている；我々は … に主眼を置いている

9. **We focused on** data from 56 sites on streams.
   我々は，水流の 56 地点でのデータに主眼を置いた。
10. **We focus on** various aspects of crop production and produces annual summaries available to the public.
    我々は，穀物生産の諸面に主眼を置き，一般市民が利用可能な年ごとの概要を作成している。
11. **We** also use a survey approach and **focus on** characteristics related to timber production.
    我々もサンプル調査手法を用いており，材木生産に関わる特性に主眼を置いている。

- **〜 focus on .....** : 〜は … に焦点を合わせる；〜は … に集中する

12. The study **focused on** sampling surface water.
    この研究は，地表水の採取に焦点が置かれた。
13. The investigation **focused on** an 18-mile segment of the creek.
    この調査は，小川の 18 マイル区間に焦点を合わせた。
14. The debate **focused on** the stream standards.
    論議の的は，水流基準であった。
15. Some volunteer programs **focus on** environmental advocacy or on enforcement of environmental permits and pollution deterrence.
    一部のボランティアプログラムは，環境擁護または環境許可および汚濁阻止の施行に主眼をおいている。
16. Considerable environmental concern **has focused on** these nonbiodegradable and toxic heavy metals.
    かなりの環境の関心がこれらの生物的分解が不可能な，しかも有毒な重金属に集中した。
17. Current monitoring efforts **focus on** only a few of the hundreds of chemicals known to be present in drinking water supplies.
    現在のモニタリングの努力が，飲料給水への存在が知られている数百という化学物質だけに集中している。
18. The increasing threat of contamination and the potential hazard **have focused** a great deal of attention **on** the fate of mercury in the environment.
    汚染と可能な危険性に対する増加する脅威は，環境での水銀の衰退に多くの関心を集めた。
19. This program **will focus** its survey and assessment activities **on** a total of 60

study units, each study unit a targeted watershed.
同プログラムは，調査・評価活動の焦点を合計 60 の研究単位に絞ることになっており，各研究単位がターゲット流域に当たる。

- **focusing on .....**：… に主眼をおいた

20. Over a 2-year period, the criteria were tested, **focusing on** the following questions: (1) …, (2) ….
    2 年間にわたってクライテリアの検定が行われ，力点は，以下の諸問題に置かれていた。(1) …, (2) ….
21. Biological assessment, **focusing on** population and community level response, addresses impact rather than only discharger performance.
    個体数および生物群集レベル応答に主眼をおいた生物アセスメントは，排出者実績ばかりでなく影響にも取り組んでいる。

Fold：倍の

1. Rate enhancements of more than **tenfold** are common.
   反応速度の増速は，通常 10 倍以上である。
2. The purpose of the narrative classification system was essentially **twofold**: (1)....., and (2).....
   記述的分類システムの目的は，基本的に 2 つある。(1) … (2) …
3. High-intensity ultrasound dramatically increases the rates of interaction by **as much as 200-fold**.
   強度の超音波は，反応速度の相互作用を劇的に増やし，それはおよそ 200 倍にも達する。
4. The focus at this step is **twofold**: to locate and reject disturbed areas, and to seek and retain minimally disturbed areas.
   このステップでの主眼は 2 つある。それは擾乱区域を位置特定し排除すること，それに，最小擾乱区域を探して確保することである。
5. The advantages of these increased efforts are **threefold**. First, ... Second, ... Third, ...
   こうした努力増強の利点は，3 つある。第一に，… 第二に，… 第三に，…
6. It has been concluded from the chemical form of the heavy metals that atmospheric contributions play the most important role and that a **3.5-fold** increase of the lead concentration and a **2.5-fold** increase of the cadmium concentration are from this source.
   大気中からの寄与が重要な役割を果たしていること，そして，大気中からの寄与が鉛の 3.5 倍上昇やカドミウムの 2.5 倍上昇の原因であることが重金属の化学形

体から結論づけられている。

Follow：続く；追う

- ～ are followed ：～は追ってされる
- ～ that follow ：次に続く～で
- It follows that ..... ：…ということになる

1. The guidelines **are followed** step-by-step to demonstrate how the guidelines can aid in design of a field study.
   この指針書は，これがどのように現場研究の計画に役立つのかを明示するために一歩一歩段階を追って説明している。
2. The method for interpret ting the sign will be illustrated in the discussion and example problems **that follow**.
   このサインを解読する方法が考察と次に続く例題で例証されるであろう。
   我々は，本章で，数量的生物クライテリアが州ごとに設定・実施されるための枠組みを示してきた。だが，今後の研究を要する重要な領域がいくつか残っており，例えば，以下のとおりである。

- ～ are as follows: ..... ：～は次のとおりである

3. Although we have presented here a framework from which numerical biocriteria can be developed and implemented by states, there remain important areas for future development and research. Some of these **follow**：
4. Assumptions associated with the assessment exercise **are as follows**: 1)....., 2)....., 3).....
   アセスメントの実施と関連する仮定は次のとおりである。1)…，2)…，3)…
5. The original goals of this effort **were as follows**: 1)....., 2)....., 3).....
   この活動の当初の目標は，以下のとおりである。1)…，2)…，3)…
6. Assumptions associated with the assessment exercise **are as follows**：
   アセスメントの実施と関連する仮定は，次のとおりである。
7. The objectives of the limited sampling program at ST6 are **as follows**：.........
   ST6地点での，限定されたサンプリングプログラムの目的は次のとおりである。
8. This systematic process involves discrete steps, which **are** described **as follows**：
   この体系的過程は，非連関的なステップを伴うものであり，それについて以下に述べる。

- **As a follow-up note**：追記すれば

9. **As a follow-up note**, this sewage treatment plant subsequently underwent a major upgrade, and later macroinvertebrate sampling revealed an improved community downstream.
追記すれば，この下水処理場は，その後大幅な改善がなされ，後の大型無脊椎動物サンプリングで下流生物群集における向上が見られた。

Following：次の

1. The **following** measurement equipment was used: .....
   次の測定装置を使用した。…
2. The **following** experimental parameters were investigated: .....
   次の実験パラメータを調査した。…
3. The **following** discussion is limited to the data obtained up to 25 January 1998.
   次の論議は，1998年1月25日までに得られたデータに制限される。
4. The **following** discussion presents examples of methods presently used in evaluating the periphyton assemblages.
   以下の論考は，付着生物群がりを評価する際に現在用いられている方法の諸例である。
5. It will concentrate on the **following** issues: a) ....., b) ....., and c) .....
   次の事に的をしぼるであろう，すなわち a)…，b)…，それに c)…
6. The procedure included the **following** step: 1) ....., 2) ....., 3) .....
   手順は次のステップを含んでいた。1)…，2)…，3)…。
7. This integrated assessment was designed to address the **following** questions: 1..., 2..., etc.
   この統合化されたアセスメントは次の質問を扱うよう意図された。1…，2…，などである。
8. Also, I thank the **following** individuals for contributing useful comments on a draft version: Akio Watanabe, Kousuke Iwata, Ichko Ohkura .
   また，わが草稿について有意義な意見を述べてくださった次の諸氏にも感謝したい。Akio Watanabe, Kousuke Iwata, Ichko Ohkura。

   ● **the following conclusions**：次の結論

9. **The following conclusions** can be drawn from the data given in Fig. 5.
   次の結論は図5で与えられたデータから導出されうる。
10. From the results obtained in these investigations, **the following conclusions** were reached.
    これらの調査で得られた結果から，次の結論に達した。

Form

11. Extensive laboratory investigations and data analysis provided the **following conclusions**.
    大規模な研究室での調査とデータ分析が次の結論を提示した。

    ● **the following ～ is .....** : 次の～は … である

12. **The following is** a list of his criticisms and a response to a potentially limited viewpoint of the IBI.
    以下に，彼の批判と，IBIの潜在的に限られた観点への反応を列挙していく。

13. In an analysis of resources expended during FFY 1987 and 1988 **the following were** revealed.
    1987～88連邦会計年度に行われた資源分析において，以下が明らかになった。

14. In our experience **the following are** the situations where conflicts have arisen in New Mexico.
    筆者の経験によれば，ニューメキシコ州で抵触が生じた事態は以下のとおりである。

15. If this assumption holds true, the **following** equation **can be** used.
    もしこの仮定が正しければ，次の式が使われる。

16. Assuming that the substance under consideration increases in accordance with a first-order reaction, the **following** equation **is** developed.
    考慮中の物質が一次反応にしたがって増加すると想定して，次の式をつくった。

17. Although the mechanisms of acoustic foam disintegration are not entirely understood, the **following** acoustic effects **are** believed to be most important in foam breakage.
    音波によって生じる泡の崩壊機構は完全に理解されていないが，次に示す音波効果は気泡崩壊過程上最も重要であると思われる。

    ● **～ did the following** : ～は以下をした
    ● **～ are the following** : ～は次のことである

18. The analysis **revealed the following**: 1) ....., 2) ....., 3) .....
    この分析は，以下を明らかにした。1)…, 2)…, 3)…,

19. The criteria for selection of sites **included the following**: 1) ....., 2) ....., 3) .....
    地点選択のための基準は，次のことを含んでいた。1)…, 2)…, 3)…。

20. The basic steps of the risk assessment process **are the following**: 1) ....., 2) ....., 3) .....
    リスクアセスメントの基本的なステップは，次のことである。1)…, 2)…, 3)…。

Form ： 形；型；方式；形成する；なす

1. Classification **forms** the foundation of science and management because

neither can deal with all objects and events as individuals.
分類は，科学と管理の基礎をなす。なぜなら，どちらもすべての対象や事象を個別に扱うことはできないからである。

2. These classifications **formed** the initial basis of the early narrative biological criteria developed in 1980.
これらの分類は，1980年に設定された初期の記述的生物クライテリアの第一の基礎となった。

3. The Columbia River flows through the northern part of the Hanford Site, and turning south, it **forms** part of the site's eastern boundary.
コロンビア川はHanford地域の北の部分を通過し，そして南に曲がり，地域の東境界線の一部を形成する。

- **in a form .....** ：…の形で

4. This challenge resulted in a search for numerical expressions **in a form** simpler to understand than long species lists and well-thought but lengthy technical expressions of the data.
この課題により，生物種の長大なリストや，綿密だが長々しいデータの技術的説明よりも理解しやすい形での数値表現が追求されるようになった。

5. Public participation **in the form** of public comment and public hearing is also an important part of Sction 404 process, which addresses wetlands use and protection.
般市民による論評や公聴会などの形での公衆参加も404条の重要な一部であり，同条は，湿地利用や湿地保護を扱っている。

- **some form of .....** ：何らかの形の…

6. It is envisioned that by the end of the decade, nearly all of the state water resource agencies will have **some form of** biological monitoring.
この10年間のおわりまでに，ほぼすべての州の水資源担当機関が何らかの形の生物モニタリングを有するようになるものと予想される。

- **～ take the form of .....** ：～は…の形をとる

7. The surveying of ambient biota **takes the form of** either coordinated monitoring networks or a series of special studies.
環境生物相に対する調査は，調整された監視網もしくは一連の特別研究の形をとる。

- **～ form the foundation for .....** (See **Foundation**)：
  ～は…の基礎となる(➡ **Foundation**)：
- **～ form the foundation of .....** (See **Foundation**)：

〜は … の基礎をなす（➡ **Foundation**）

Former (See also Latter)：前者（➡ **Latter**）

1. **The former** question acknowledges that .....
   前の質問は … を認めている。
2. **The former** can be used just as much as the latter as a means of experimentation.
   前者は後者とまったく同じぐらいよく実験手段として使用できる。
3. The mixture of chemicals had an EC10 of 3.7 % for **the former** and 5.2 % for the latter test.
   その混合化学物質では，EC10 が，前のテストでは 3.7 %，後のテストでは 5.2 %であった。
4. We differentiate between homogeneous and heterogeneous reactions, and in this chapter focus on **the former**.
   我々は，均一系反応と不均一系反応を区別し，そしてこの章では前者に焦点を置いている。

Formulate (Formulation)：調合する；作成する；定式化する（式）

1. Eqn. 3 **can be formulated** as .....
   式 3 は … と式化される。
2. There are two known limitations of the synthetic leachate as currently **formulated**.
   現在調合されている合成浸出液には 2 つの周知の限界がある。
3. Three states have not yet begun **to formulate** their bioassessment approach, but have initiated discussion within their own agencies and with the USEPA.
   3 州は，まだ生物アセスメント手法の作成を開始していないが，州機関内での討論や USEPA との論議に着手した。
4. We intend to place the emphasis on understanding real systems, **formulating** problems and interpreting the results of the analysis.
   我々は，実際のシステムを理解し，問題を定式化し，そして分析結果を解釈することに力点を置くつもりである。

   ● **formulation**：式

5. The **formulation** is based upon a proven concept.
   この公式化はすでに証明されている概念に基づいている。

6. The **formulation** is based upon a similar concept with the exception that .....
 … ということは例外として，この公式化は類似の概念に基づいている。
7. The model **formulation** incorporates chemical decay and transport mechanisms of particulate and diffusive exchange between water column and sediment.
 このモデル公式化は化学的分解，微粒子の輸送メカニズム，水コラムと堆積物の間の拡散的交換が盛り込まれている。

Foundation：基礎

- ～ **laid the foundation for** .....：～は … を支えてくれた
- ～ **has a good foundation** .....：～は確かな基礎を持つ

1. This work is dedicated to the memory of Genki Nakata whose years of service **laid the foundation** for our efforts.
 本稿を我々の努力を長年支えてくれた Genki Nakata の思い出に捧げる。
2. Thus the model **has a good foundation** in the observation of actual environmental conditions and associated biological community responses.
 故に，このモデルは，実際の環境状態と関連の生物群種応答の観察において確かな基礎を持つ。

- ～ **form the foundation for** .....：～は … の基礎となる
- ～ **form the foundation of** .....：～は … の基礎をなす

3. Together, they **form the foundation for** a sound, integrated analysis of the biotic condition.
 これらは，共に生物状態の確かな統合的分析の基礎となる。
4. Classification **forms the foundation of** science and management because neither can deal with all objects and events as individuals.
 分類は，科学と管理の基礎をなす。なぜなら，どちらもすべての対象や事象を個別に扱うことはできないからである。

Fraction：ごく一部；部分；分数

1. At present, few states can characterize even **a small fraction of** their water resources with biological survey data.
 現在，生物調査データで各自の水資源のごく一部でも特性把握できる州は，きわめて少数である．
2. Numerous studies have already been performed on the $< 63-\mu m$ **fractions**,

allowing better comparison of results.
多くの研究が<63-μm 部分でなされてきたので，他の結果との比較をより可能にしている。

★ Fraction（分数）に関連する例文を下に示します。

3. Currently, about **one-third** of the world's population lives in countries that face moderate to severe water shortages.
今日，世界の人口の約3分の1が農村に住んでおり，中度または重度の水不足に直面している。
4. Regional modifications using this data set may raise this MPS expectation by approximately one third of one point for the spring dominated streams in the central part of the state and **one sixth of** one point for the eastern ecoregions.
このデータセットを用いた地域修正は，MPS 期待値を同州中部の春季優勢水流について3分の1点ほど上昇させ，州東部生態地域について6分の1点ほど上昇させるだろう。

Full (Fully)：十分；完全（十分に；完全に）

1. A modeling strategy which makes **full** use of available data must be devised.
利用可能なデータを十分に活用するモデリング戦略が考案されなくてはならない。
2. It can be argued that **full** protection of aquatic life requires the use of biological criteria.
水生生物の完全保護には，生物クライテリアが必要だとの主張もありえる。

● **fully**：十分に；完全に

3. The literature on the nitrification process was **fully** reviewed by .....
硝化プロセスに関する文献が …によって十分にレビューされた。
4. Because of a lack of extensive knowledge on pesticide toxicity, the methodology was not **fully** verified.
殺虫剤毒性に関する広い知識が欠如していたために，この方法論は十分に立証されなかった。
5. Naphthalene produced this inhibition no matter whether the melanin was **fully** aggregated or more less dispersed at the time of initial exposure.
最初にさらされた時にメラニンが完全に凝結していたか，あるいはある程度分散していたかの状態にはかかわらず，ナフタリンはこの抑制効果を呈した。
6. In order to **fully** understand the potential impact of local resource management decisions, data should be available on the status of all the

waterbodies within a watershed.
地方の資源管理決定による潜在的影響を十分に理解するため，データは，ある流域における水域すべての現状について利用可能とされるべきである。

7. Despite these proven uses the role of biological criteria in surface water resource management and policy has yet to be **fully** implemented by most states and the USEPA.
これらの立証済み用途にもかかわらず，地表水資源の管理・政策における生物クライテリアの役割は，大半の州や米国環境保護庁によって十分には実現されていない。

Function ： 相関；関数

1. The relationship between cellular copper content and cupric ion activity shows good agreement with the hyperbolic **function**.
   細胞の銅含有量と銅イオン活性の間の関係は，双曲線関数で示すと良く一致する。

   ● ～ **is a function of** ..... ：～は … と相関している

2. The coefficient **is a function of** the size, shape, and density of the suspended particles.
   この係数は浮遊粒子の大きさ，形と密度に相関している。
3. The observed heavy metal content of waterways **is a function of** many variables.
   水路にて観察された重金属含有量は，多くの変数と相関関係にある。
4. The actual number of regional reference sites needed **is a function of** regional variability and size, the desired level of detectable change, resources, and study objectives.
   実際に必要な地域参照地点の数は，地域の変化性や規模，望ましい水準の探知可能変化，資源，研究目標に相関している。

   ● **as a function of** ..... ：… の関数として

5. Copper toxicity profile is shown in Figure 1 by a three-dimensional plot of the specific growth rates **as a function of** the two variables, total ammonia concentration and total copper concentration.
   銅毒性プロフィールを図1に示す。それは全アンモニア濃度と全銅濃度の二変数を関数として成長速度定数を表す3次元のプロットである。
6. Patterns in human activities must be considered as well, and these often vary **as a function of** ownership or political unit, as well as ecoregion.
   人間の活動におけるパターンも考慮すべきであり，それは，しばしば所有単位や

政治単位，そして生態地域に相関して異なる。

Fund：資金；資金供給される

1. Only limited **funds are available** to support research on the development of inexpensive analytical methods.
   ほんの限られた資金か，低価な分析方法の開発研究の援助のために利用できる。
2. These projects **are funded,** in part, through the AAS program.
   これらのプロジェクト資金の一部は，AASプログラムを通じて供給されている。
3. The research described in this report **has been funded** by the USEPA.
   この報告書で述べてきた研究は，USEPAからの資金供給によるものである。

Fundamental：基本

1. The **fundamental** equation can be found elsewhere.
   その基本式はほかでも見いだされる。
2. **Fundamental** studies are still needed to assess the possibilities of this method.
   この方法の可能性を評価するために基本的な研究がまだ必要とされる。
3. It is of interest to point out the **fundamental** difference between A and B.
   AとBの間の基本的な相違を指摘することは重要である。

Further：さらに

1. **Further** discussion is provided elsewhere.
   さらに進んだ考察がほかのところにも記されている。
2. No **further** attempts were made to calibrate the model on the study site.
   この研究地域におけるモデル校正のために，それ以上の試みはされなかった。
3. There is a **further** complication to the analysis of flow in the unsaturated zone.
   不飽和帯での地下水流の解析にはそれ以上の複雑さがある。
4. **Further** analyses of the data are planned that will employ additional existing geographical classification.
   他の既存地理分類を用いるいっそうのデータ分析が予定されている。
5. In examining the measured data **further**, it was discovered that .....
   測定データをさらに調べると，…が発見された。
6. More extensive and detailed investigations are necessary to gain a **further** understanding of these processes.

いっそう大規模かつ詳細な調査が，これらのプロセスをいっそう理解するために必要である。

- **further research**(See Research)：いっそうの研究(➡ Research)

Furthermore：さらに；しかも

1. **Furthermore**, from Fig. 2, it can be seen that the magnitude of the effect is erratic.
さらに，図2から，その効果の規模が不規則であるといえる。
2. The smallmouth bass (Micropterus dolomieui) is one of the most important game species in midwestern rivers and streams. **Furthermore**, this is a species that requires little or no external support in the way of supplemental stocking.
コクチバス(Micropterus dolomieui)は，中西部の河川水流における最重要な漁獲対象種の1つである。さらに，補足的放流という外部支援をほとんどまたは全く必要としない魚種である。

Future：将来の

1. **Future studies** will focus on .....
将来の研究は … に集中するであろう。
2. We recommend that the highest priority for **future experimental work** is to validate the existing models.
我々は，将来の実験的研究の最も高い優先順位が既存モデルの実証におくことを勧める。
3. **Future work is needed** to better utilize the ADV as part of resource value assessment such as NRDAs and in assessing other types of environmental damage claims.
NRDAsといった資源価値アセスメントの一部として，また，他の種類の環境損害請求の評価においてADVをより活用するには，今後の作業が必要である。
4. Insofar as past state biological criteria efforts grew out of biological monitoring efforts, **future efforts** will likely build on current monitoring programs.
生物モニタリング努力から発した過去の州の生物クライテリア努力に関する限り，将来の努力は，現行のモニタリングプログラムに立脚するであろう。
5. If a simple relationship could be established, it would be of great in planning the **future direction** of any development work on a new chemical.

もし単純な関係が確証されえるなら，いかなる新化学物質開発研究の将来方針を計画するにあたって，それは偉大なことであろう。

6. Among all the techniques of waste management, waste reduction is the common sense solution to the prevention of **future** hazardous waste problems.
廃棄物マネージメントのすべてのテクニックのなかで，廃棄物削減は将来の危険廃棄物問題の防止の常識的な解決策である。

7. Although we have presented here a framework from within which numerical biocriteria can be developed and implemented by states, there remain important areas for **future development and research**. Some of these follow:
我々は，本章で，数量的生物クライテリアが州ごとに設定・実施されるための枠組みを示してきた。だが，今後の研究を要する重要な領域がいくつか残っており，例えば，以下のとおりである。

● **in the future** ：将来には

8. Nebraska hopes that this will lead to the establishment of numeric biological criteria **in the future**.
ネブラスカ州は，これが将来，数値生物クライテリアの確定に通じることを期待している。

9. Biological information can be expected to have a greater role in water resource decisions **in the future** for several reasons.
生物学的情報は，将来，いくつかの理由で水資源決定においてより重大な役割を演じるものと予想される。

10. The arsenic concentrations will not change very much **in the near future** regardless of any channel shifting.
水路が移動したとしても，ヒ素の濃度は近い将来ではそれほど変化しないであろう。

● **well into the future** ：未来に向けて；ずっと将来まで

11. There is reason to believe that it will continue to be so **well into the future**.
それが未来に向けて良くあり続けるであろうと確信する理由がある。

12. This is a critical juncture in the process since these initial decisions will determine the overall effectiveness of the effort **well into the future**.
これは，過程における重大時である。これらの初期決定が労力の効果性をずっと将来まで左右するからである。

● **prediction of future trends** ..... ：将来の傾向

13. A need exists for **prediction of future trends** and control in contaminated

areas.

汚染された地域に対し、これからの傾向の予測と管理の一つの必要性が存在する。

14. Understanding the reactions of PCBs with OHC is crucial to **predicting the future persistence** of PCBs and the hazards associated with their presence.

PCBの将来の持続性およびPCBの存在による危険性を予測するのに、PCBとOHCの反応を理解することが不可欠である。

- **key to future success**（See **Key**）：将来の成功を握る鍵（➔ **Key**）

Gain：得る

1. Sample processing time decreases as the processor **gains** taxonomic expertise.
サンプル処理時間は，処理担当者が分類群知識を得るにつれて短縮される。
2. Additional information **is gained from** population, bacteriological, and productivity studies.
追加情報は，個体群研究，細菌学研究，生産度研究から得られる。
3. The key objective of this analysis is to determine whether or not the feedback **gained from** the biological community can communicate about and characterize these differences.
この分析の主要目標は，生物群集から得られたフィードバックがこれらの差を伝達し特性把握できるか否か決定することである。

   - **there is nothing gained by .....**：… によっては何も得られない
4. It appears that **there is nothing gained by** plotting the data in this fashion.
データをこの様式で作図することによっては，何も得られないように思われる。

   - **insight is gained on .....** (See **Insight**)：
   … についての洞察が得られる (➡ **Insight**)
   - **to gain an understanding of .....** (See **Understand**)：
   … を理解するために (➡ **Understan**)
   - **gaining**：得られる
5. This holistic definition appears **to be gaining** acceptance.
この全体論的な定義は，容認されつつあるようだ。
6. The method used here may be a new approach not only for **gaining** better results, but also for acquiring new knowledge and insights about toxic behavior.
ここで使われた方法は，良い結果が得られるだけではなく，毒性挙動について新しい知識と洞察が得られることで，新しいアプローチであるかもしれない。

Gather (Gathering)：収集する（収集）

- ～ are gathered from ..... :… から～が収集される
- ～ are gathered on ..... :… の～が収集される

1. The objectives of the research were **to gather** all possible information available to define potential contaminant source area.
 この研究の目的は，汚染源の可能性がある場所を明らかにするために可能な情報すべてを収集することであった。
2. Data **were gathered from** a number of experiments using a pilot-scale reactor.
 試験規模の反応器を使った多数の実験から，データが収集された。
3. Data **were gathered on** fish, sediment, water, and aquatic plants in the area.
 その区域で魚類，堆積物，水，それに水生植物のデータが収集された。
4. Evidence **has been gathered**, and solutions have been proposed.
 証拠が集められ，そして解決策が提案されてきた。
5. Useful information on farming practices **had been gathered** for the Blackfoot River study.
 農業に関する有用な情報が Blackfoot 川の研究のために集められていた。

- data gathered from ..... :… から集められたデータ

6. **Data gathered from** the piezometers indicate that .....
 ピエゾメーターから集められたデータが … のことを示している。
7. Efforts to assess, research, and manage the ecosystems are normally carried out via extrapolation from **data gathered from** single-medium/single-purpose research.
 生態系を評価・研究・管理する努力は，通常，単一媒体／単一目的の研究で収集されたデータからの推定によって行われる。
8. Based on the information **gathered** and the publication in preparation, a first approximation of reference ecoregions and subregions is compiled.
 収集情報に基づき，また，準備中の刊行物の概説に基づき，改良された生態地域および小地域の初の概要が作成される。

- data gathering ..... :データ収集 …

9. **Data gathering efforts** for the Boise River yielded adequate stream flow.
 Boise 川でのデータ収集の努力が，適切な流量をもたらした。
10. During the past three weeks, staffs have participated in detailed **data gathering sessions**.

過去3週間に，スタッフが詳細なデータ収集セッションに参加した。
11. At the onset of each project, and at the **data gathering meetings**, the question of whether ecoregions or special purpose regions are desired is always asked.
各プロジェクトの発端で，また，データ収集会合で，常に生態地域と特殊目的地域のいずれが望ましいか，という質問が発せられる。

Gear：装置；道具；適合させる；目的とする

1. For bedrock samples, **gear** usually includes a 3- to 4-in. diameter PVC pipe with a rubber seal that can be held in place where the rock is brushed clean and the sample removed by suction.
床岩サンプルの場合，サンプリング装置には，通常，直径3～4inのラバーシールドPVC(ポリ塩化ビニル)管が含まれており，これを設置すれば，岩石を磨いてサンプルを吸引除去できる。

- ～ **is geared toward** .....：～は…を目的としている

2. Most of the early legislation **was geared toward** the protection of waters for human use(beneficial uses).
初期の法律の大半は，人間による利用(有益用途)のための水域保護を目的としていた。

General(Generally)：一般的な(一般的に)

1. The idea being brought forward in this research is to measure several of these **general** parameters.
この研究にて前進的考え方は，これらの一般的なパラメータのいくつかを測定するということである。
2. These processes fall into three **general** categories: 1) ....., 2) ....., 3) .....
これらの過程は3つの一般的な部門に分類される。それらは，1)…，2)…，および3)…，である。
3. The first term on the right-hand side of the equal sign is the **general** solution and the second term is the particular solution.
等号の右側の最初の項は一般的な解で，そして2番目の項は特定の解である。

- **in general**：全般；一般的に

4. It has been known for decades that certain groups of macroinvertebrates are, **in general**, more tolerant of pollution than are others.

大型無脊椎動物の一部集団が一般的に他集団に比べ汚濁耐性が強いことは，数十年前から知られている。

5. This need was part of a larger concern for a framework to structure the management of aquatic resources **in general**.
この必要性は，水生資源全般の管理を構成する枠組みへの大きな関心の一部である。

6. Dan Ozaki and Jim Yoshida contributed extensively to the early development and review of important concepts of biological integrity, ecoregions, reference sites, and biological monitoring **in general**.
Dan Ozaki と Jim Yoshida は，生物完全性，生態地域，参照地点，生物モニタリング全般という重要な概念の初期の設定・再検討に大いに貢献してくれた。

- **generally**：一般に

8. These measurements show **generally** good agreement with expected concentration values.
これらの測定値は，予期していた濃度値と一般に良い一致を示している。

9. **Generally**, less variability is expected among surface waters within the same region than between different regions.
一般に，異なる地域の地表水域よりも，同じ地域の地表水域の方が可変性が低いと予想される。

10. These rivers are **generally** of moderate to high quality and few sites were found that were rated either fair or poor.
これらの河川は，おおむね質が中度（良）ないし高度（優）なものであり，可または不可の地点は，ごく少数である。

11. Historical data may present a problem because the site selection **generally** is not conductive to either of these processes.
史的データは，問題を生じうる。なぜなら地点選定は，通常これらの過程のいずれかを資するわけではないからである。

Given：に鑑みて；を考えれば

1. **Given** tolerance and identification considerations, biotic indices may not be appropriate to use in developing countries at this time.
耐性および識別への考慮に鑑みて，生物指数は，現時点では開発途上諸国での利用に適さないかもしれない。

2. Simply referring to a waterbody as a high-quality or significant resource is inadequate **given** the penchant for these characterizations to have unique attributes according to the individual making the pronouncements.

意見表明者によれば，これらの特性が独自な属性を帯びがちな傾向に鑑みて，ある水域を単に高質または有意資源と評するだけでは不十分である。
3. For macroinvertebrates the issue of identifying midges to the genus/species level (as opposed to the family level) proved likewise to be a farsighted decision **given** the value of this group in diagnosing impairments.
大型無脊椎動物について，小虫を(科レベルでなく)属／種レベルで識別する問題は，同様に，損傷判定におけるこの集団の価値に鑑みて，先見的決定であることが判明した。

- **given that .....** ：… であることを考えれば

4. This overlap is not surprising **given that** many of the CSO Toxic impacted segments were in the same streams and rivers as some of the Complex Toxic impacted segments.
この重複は，CSO 毒性影響区間の多くが複雑毒性影響区間の一部と同じ水流・河川にあったことを考えれば，驚くに当たらない。

Goal (See also **Aim**, **Objective**, **Purpose**) ：
目標；ゴール；目的（→ **Aim, Objective, Purpose**）

1. The protection of aquatic life is **a goal**.
水生生物保護を目標とする。
2. This report is a first step toward meeting this **goal**.
この報告書は，この目標達成に向かう第一歩である。

- **address the goal** ：目標に取り組む

3. These data will be analyzed to assess the status of biological condition and to **address the goals** as stated above.
これらのデータが分析されて，生物状態の現状を評価し，上述の目標に取り組むことになろう。
4. The traditional use of performance standards is inadequate because they are not designed **to address broad goals** such as ecological integrity.
実績標準の伝統的な利用は，生態完全度などの広範な目標に取り組んでいないためまだ不十分である。

- **The goal is to .....** ：目的は … することである
- **The goal can be accomplished** ：目標は達成されうる
- **It is not the goal of this research to .....** ：
… は，この研究の目標ではない

- **Our goal in this paper is to** ..... :
  この論文における我々の目的は…である

5. **The goal is to** examine .....
   目的は…を検討することである。
6. The **goal of** biological criteria **is to** provide additional support for the state's water quality standards.
   生物クライテリアの目標は，州の水質基準に追加サポートを提供することである。
7. The original **goals of** this effort **were** as follows: 1)....., 2)....., 3)......
   この活動の当初の目標は，以下のとおりである。1)…, 2)…, 3)…
8. **The main goal of** any risk assessment **is to** determine those conditions likely to produce harm.
   どんなリスクアセスメントでも主要な目標は，害を作り出す可能性があるそれらの条件を決定することである。
9. The **goal is to** achieve a suitable fit between the planned modeling effort and the data, time, and money available to perform the study.
   目標は，計画されたモデリングの努力とデータ，時間，それに研究を実行するためにアクセス可能な資金の間で，適切な合致に達することである。
10. These **goals** eventually **can be** accomplished.
    これらの目標は，最終的に達成されえる。
11. **It was not the goal of this research to** investigate the interactions between A and B.
    AとBの間の相互作用を調査することは，この研究の目標ではなかった。

- **in defining goals for** ..... : …のための目標を定義するにあたって
- **With clearly defined goals for** ..... : …のための明確な目標によって

12. **In defining a clear set of goals for** the Blue River Study, the following factors were significant: 1)....., 2).....
    Blue川研究のために一連の明確な目標を定義するにあたって，次の要因は重要であった。1)…, 2)…
13. **With clearly defined goals for** the maintenance of different levels of integrity through impact standards, the overall success of water resource management program can be evaluated.
    影響標準を通じて様々な水準の完全度を保つという明確な目標によって，水資源管理プログラムの全般的成功が評価されうる。

Governing：左右する；左右している

1. Knowledge of the rate constants **governing** the uptake and clearance of chemicals in fish is required.
   魚体内で，化学物質の入取・排出を左右している速度定数の知識が必要とされる。
2. With the increased need for predicting the movement and distribution of chemicals in an aquatic environment, knowledge of the rate constants **governing** the uptake and clearance of chemical in fish is required.
   水生環境での化学物質の移動・分配を予測する必要性が増加するとともに，魚体内での化学物質の摂取・除去を左右している速度定数の知識が必要とされる。

Great (Greatly)：大きい；偉大な（大幅に）

1. The bacterial luminescence bioassay shows **great** promise as it responds to a wide range of compounds.
   広範囲の化合物に対して反応するため，バクテリアの蛍光バイオアッセイは有望である。
2. While the quality of some regions has improved during recent years, there is still **great** concern regarding the marine environment.
   近年の間に若干の地域環境が良くなってきた一方，海洋環境に関してまだ大きな懸念がある。
3. If a simple relationship could be established, **it would be of great** in planning the future direction of any development work on a new chemical.
   もし単純な関係が確証されえるなら，いかなる新化学物質開発研究の将来方針を計画するにあたって，それは偉大なことであろう。

   - **〜 is as great as .....**：〜 は … ほど大幅である
   - **too great to .....**：〜 は … には高すぎる；〜 は … には大きすぎる
   - **so great that .....**：〜 があまりに広いため … となる；〜 があまりに大きいため…となる

4. Macroinvertebrate changes **were not as great as** in previous years.
   大型無脊椎動物の変化は，それまでの数年間ほど大幅ではなかった。
5. The criteria were also applied to multiplate samples. This is partly because variability among percent contribution of major groups in multiplate samples from nonimpacted sites was found **to be too great to** establish a model community.
   このクライテリアは，多プレートサンプルにも適用された。その理由は，一つに

は，無影響地点での多プレートサンプルにおける主要集団のパーセンテージの変移が高すぎて，モデル生物群集を定められないと考えられたからである。

6. From a biological standpoint, the range of any given class **is so great that** a criterion established for an expected biological community would be so broad as to be ineffectual.
生物学的な見地から，ある等級の値域があまりに広いため，予想生物群集について定められたクライテリアも広範すぎて無効となる。

- **to the great extent** ：大きく；盛んな

7. Coralvill Reservoir does not thermally stratify **to any great extent**, so the failure to include a hypolimnion compartment is not viewed as a serious problem.
Coralvill 貯水池は大きくは温度層化しない，それゆえ（このモデルに）深層部が含まれていないということが重大な問題だとは見なされない。

8. Owing **to the great extent of** the fish industry in the Illinois River, it is essential that the condition of the river be kept as good as possible.
Illinois 川では漁業が盛んなため，この川の状態をできるだけ良く保つ必要がある。

- **a great deal of** ......(See **Deal**)：かなりの…；多くの…（➡ Deal）
- **greatly** ：大幅に

9. By offering a more robust framework, the chance for deriving an inappropriate criterion is **greatly** reduced.
より確かな枠組みを提供することによって不適切なクライテリアが導出される確率が大幅に低下する。

10. We appreciated critical review comments from three anonymous reviewers, which **greatly** improved an earlier draft of this manuscript.
筆者は，3人の匿名査読者からの批評的論評に感謝する。それは，本章の草稿を大幅に改善してくれた。

11. While model applications may differ **greatly** in scope and purpose, it is hoped that the representative data derived from pilot studies will be useful in this process.
モデルのアプリケーションが応用範囲と目的で大いに異なるかもしれない一方，予備試験的研究から得られる代表的データがこのプロセスに有用あろうことが期待される。

Greater（See also Less）： より大きい（→ Less）

1. **Greater** differences were observed for other issues.
   より大きい相違が他の問題で観察された。
2. **Greater** attempts are now being made to relate metal toxicity to speciation.
   金属毒性を化学種に関連づけるために，より大きい試みが今なされている。
   - **the higher A, the greater B will be** ：
     A が高ければ高いほど，B がより大きくなるであろう
   - **to a greater or lesser degree** ：多かれ少なかれ
   - **in greater detail** ：もっと詳細に
3. **The higher** the coefficient of performance, **the greater** the effect for a given work input **will be**.
   性能係数が高ければ高いほど，特定研究のインプットのための影響がより大きくなるであろう。
4. These chemicals are associated, **to a greater or lesser degree**, with suspended and sedimented particles.
   これらの化学物質は，おおかれ少なかれ，浮遊粒子および堆積粒子と関係している。
5. Each of these decay pathways will be discussed **in greater detail** below.
   これらの分解過程路のそれぞれがもっと詳細に，下に論じられるであろう。
   - **～ is greater than ……** ：～は… より大きい
   - **～ is much greater than ……** ：～は… よりかなり大きい
6. The value for the correlation coefficient for each lines **was greater than** 0.96.
   線それぞれの相関係数の値は 0.96 より大きかった。
7. Under steady-state conditions, CTEX removal rates **were greater than** 98％ in both reactors.
   定常状態の条件下では，CTEX の除去速度は両方の反応器で 98％以上であった。
8. The data were fitted by linear regression and generally the correlation coefficients **were greater than** 0.90.
   このデータは線形回帰によってフィットされ，相関係数は一般に 0.90 以上であった。
9. The effluent nickel concentration averaged **greater than** 95 percent of the influent nickel concentration.
   廃水中のニッケルの濃度は平均して流入水中のニッケル濃度の 95 パーセント以上であった。
10. Differences among sites **would be much greater than** year-to-year

variability at the same site.
地点間の差の方が同一地点での年度変化よりもかなり大きいかもしれない。
11. If the coefficients defining this process **are much greater than** the other kinetic coefficients, the assumption of instantaneous equilibrium is appropriate.
もしこのプロセスを定義している係数が他の速度係数よりかなり大きければ，瞬時均衡の仮定は適切である。

- ~ **is order of magnitude greater than** ..... (See **Magnitude**)：
  ~は … と比べて桁外れに大きい（➡ **Magnitude**）
- ~ **have a greater role in** ..... (See **Role**)：
  ~は … においてより重大な役割を演じる（➡ **Role**）
- **a greater understanding** (See **Understanding**)：
  より良い理解（➡ **Understanding**）

Group：グループ；分類する

- **A group of** ..... ： … の1グループ
- ~ **fall into** ..... **groups** ：~は … に分類される
- ~ **belong to the group of** ..... ：~は … の部類に属する

1. **A group of** compounds that produced large peaks in the total ion chromatograms were identified.
   全イオンクロマトグラムに大ピークを生じさせた構成物の一グループが識別された。
2. These organic chemicals **fall into** nine **groups**.
   これらの有機化学物質は9つに分類される。
3. Naphthalene **belongs to the group of** PAH.
   ナフタリンはPAHの部類に属する。

- ~ **is grouped into** ..... ：~は … に分類される
- ~ **are grouped by** ..... ：~は … ごとに分類される
- ~ **are grouped as** ..... ：~は … としてまとめられる

4. PCE, TCE, and DCE **are grouped into** the same VOC group.
   PCE，TCE，そしてDCEは同じVOCの部類に分類される。
5. Fish assemblage metrics **are** generally **grouped into** three classes.
   魚類群がりメトリックは，一般に3つの等級に類別される。
6. Existing concepts **can be grouped** basically **into** two different approaches: .....
   基本的に，既存の概念は2つの異なったアプローチに類別される。それらは …

7. Descriptions of solids waste management units at the facility **are grouped by** technical areas.
施設場での固体廃棄物処理設備の説明が専門的な分野ごとにまとめられている。
8. Samples from compost piles **were grouped by** "age of compost."
コンポーストからのサンプルは，「コンポーストの年齢」によってグループ分けされた。
9. All components of the treatment system **are grouped as** one unit.
処理システムの構成要素すべては，1つのユニットとしてまとめられる。

- ～ **grouped together** ..... ：～は … をグループ分けした
- ～ **grouped with** ..... ：～は … とともにグループ分けした

10. The agency **grouped together** these minor waters under the same preliminary aquatic life designation.
TNRCCは，これらの小水域を同じ予備的な水生生物使用指定のもとでグループ分けした。
11. Sites from a northern ecoregion **grouped with** sites located within the northern areas of neighboring coregions.
州北部の生態地域の地点は，近隣生態地域の北部区域に所在する地点とともにグループ分けされた。

Grow (Growing) ：成長する；発する；増える（成長している）

1. Insofar as past state biological criteria efforts **grew out** of biological monitoring efforts, future efforts will likely build on current monitoring programs.
生物モニタリング努力から発した過去の州の生物クライテリア努力に関する限り，将来の努力は，現行のモニタリングプログラムに立脚するであろう。

- **growing** ：成長している；増えている

2. Little as yet known about the mechanism of carbon dioxide fixation by **growing** cells.
増殖している細胞によって二酸化炭素が固定されるメカニズムについて，ほとんどがまだ知られていない。
3. The environmental concern with endocrine disruptors **has been growing** for the last several years.
ここ数年の間，内分泌かく乱物質について環境的関心が高まってきている。
4. **There is a growing consensus** among researchers that excess nutrients are the primary cause of the reduced oxygen levels observed in the lake.

栄養塩類の過剰がこの湖で観察される酸素レベル減少の主要な原因であるというのが研究者の間で増えている一致した意見である。

- evidence is growing that ……(See Evidence)：
  …という証拠が増えてきている(➡ Evidence)
- ～ are matter of growing concern(See Concern)：
  ～は，これからますます懸念される事態である(➡ Concern)

Growth：成長

1. The kinetic expression for **growth** is modified in order to account for environmental effects.
   増数速度の表現は，環境の効果を説明するために修正される。
2. This model is capable of correlating **growth** data from many situations in a satisfactory manner.
   このモデルは，多様の条件下からの成長データを満足のいく方法で関連づけることができる。
3. The role of trace metals in regulating **the growth rates** of marine phytoplankton is not well understood.
   海洋藻類の成長速度制御における微量金属の役割はよく理解されていない。
4. Scales are collected so that **growth rates** as well as general size can be calculated and compared.
   鱗を採集して生長度や一般的休寸が計算・比較されうる。
5. Over the past few decades, a considerable number of studies have been conducted on the effects of temperature on the **growth** of nitrifying bacteria.
   ここ数十年にわたり，温度が硝化菌の成長に及ぼす影響に関する研究がかなり行なわれてきた。

H

Half：半数

1. Of these 45 states, **about half** have documentation (mostly in draft form) supporting the methods and analyses and providing program rationale.
   これら45州のうち，ほぼ半数は，方法・分析を裏づけ，プログラムの根拠を提示する文書（主に草稿）を備えている。
2. For example, **over half of** the 400 lakes that are monitored in Illinois and included in the state's 305(b) reports are sampled by citizen volunteers rather than state water quality specialists.
   例えば，イリノイ州では，400のモニタリング対象湖沼のうち，半数以上が州の水質専門家でなく，市民ボランティアの手でモニターされ，同州の305(b)報告に収められている。
3. Atmospheric **half-lives** for PCBs may be on the order of weeks.
   大気中でのPCBの半減期は数週間であろう。
4. The boundary between classes of wastes depends both on the isotope's **half-life** and the specific activity.
   廃棄物類の分別は，その同位体の半減期と特定活性の両方によって決まる。

Hand：手

- **by hand** ：手で
- **on the right-hand side of.....** ：…の右側の
- **hand-in-hand with.....** ：…と一緒に

1. A 10-mL sample of calibration standards was contacted with 1 mL of hexane in conical bottom test tubes followed by shaking for 2-5 minutes **by hand**.
   コニカルボトム試験管で，10 mLの標準校正サンプルに1 mLのヘキセンを加えてから，2-5分間手で振った。
2. The first term **on the right-hand side of** the equal sign is the general solution and the second term is the particular solution.
   等号の右側の最初の項は一般的な解で，そして2番目の項は特定の解である。
3. Bank erosion problems often occur **hand-in-hand with** riparian

vegetation disturbance.
河岸浸食(bankerosion)問題は，しばしば水辺植生攪乱と一緒に発生する。

- **on the other hand** ..... ：その一方で；他方
- **on one hand A, on the other hand B** ：一方ではA，他方ではB

4. **On the other hand**, the model developed in this paper indicates that .....
   他方，この論文にて構築したモデルは … を示す。
5. **On the other hand**, it is imperative to management decisions that data be available on each watershed.
   他方，流域それぞれに関するデータを利用可能にするという決定が不可避である。
6. **On the other hand**, their low stiffness and tendency to creep are disadvantages that must be carefully weighed against their advantages.
   他方，それらの剛度の低さおよびクリープの傾向は不利点であり，それらの利点と慎重に比較考察されなくてはならない。
7. These processes are affected by the concentration, size, and valence of metal ions **on one hand** and the redox potential, ionic strength, and pH of the solution **on the other**.
   これらのプロセスは，金属イオンの濃度，サイズ，結合価の影響を受ける一方，他方では溶液の酸化還元電位，イオン強度，pHの影響を受ける。

Help (Helpful)：助ける；役立つ（有益な）

- **to help understand** ..... ：… の理解を助けるために
- **to help define** ..... ：… を定義するのを助けるために
- **to help determine** ..... ：… を決定するのを手伝うため
- **to help organize** ..... ：… を組織化するのを手伝うために

1. We use mathematics **to help understand** the behavior of some physical process or system.
   我々は，物理的プロセス挙動あるいは物理的システム挙動の理解を助けるために数学を使う。
2. Predictive models can be used **to help define** the relationships between water pollutants and their sources.
   予測モデルは，水汚染物とそれらの流入源との関係を定義するのを助力するために使うことができる。
3. The numeric indices used **to help define** the narrative classification system were comprised of single-dimension measures.
   記述的分類システムを定義するのに用いられた数量的指数は，単一次元の測度か

らなっていた。

4. **To help organize** the investigator's conceptualization of the waste site, the analyst may want to establish a set of questions.
廃棄物地域に対する調査者の概念化を組織化するのを手伝うために，アナリストは一連の質問を確立することを望むかもしれない。

5. The results of this assessment are used **to help determine** the need for remedial action, to select the remedial action, and to determine risk-based cleanup levels for the remedial action.
このアセスメントの結果は，修復活動の必要性を決定するのを手伝うため，修復活動を選択するため，そして修復活動のリスクベースの浄化レベルを決定するため，に使われる。

6. A robust statistical treatment of these data often **can help determine** whether or not the biological response is attributable to the measured pollutant.
これらのデータへの確かな統計処理は，生物応答が測定された汚濁物質に帰せられるか否か決定するのに役立つことが多い。

7. In some cases the site-specific habitat attributes are used **to help separate** where the transition from one ecoregion to the other takes place.
一部の場合，地点特定生息地属性を用いて，ある生態地域から他の生態地域への推移発生場所が分けられる。

8. Volunteer monitoring **can help provide** that data.
ボランティアモニタリングは，そのようなデータを提供するのに役立ちうる。

● **help us understand** ..... ： … を理解するのに役立つ

9. **To help us understand** the earlier experiments, we are now running more controlled experiments.
以前の実験を理解するために，我々は今いっそう制御された実験を行なっている。

10. All our attempts at classification are human constructs that **help us understand** a spatially and temporally varying landscape.
分類における我々の試みは，すべて人為的構成物であり，空間的・時間的に異なる景観を理解するのに役立つ。

● **to help ～ determine** ..... ： ～が … を決定するのを手伝うのに

11. To help organize the investigator's conceptualization of the waste site and **to help** the investigator **determine** the most appropriate transport pathways and exposure routes, the analyst may want to establish a set of questions.
廃棄物地域に対する調査者の概念化を組織化するのを手伝うために，そして調査

者が最も適切な輸送経路と露出ルートを決定するのを手伝うために，アナリストは一連の質問を確立することを望むかもしれない。

- ～ is helpful ：～は助けになる；～は役立つ；～は有益である

12. This work should be especially **helpful** in addressing impacts from sedimentation.
この作業は，特に堆積による影響への対処に有益なはずである。
13. The outcome of this effort **has been helpful** to design engineers and treatment plant operators in solving a treatment problem resulting from heavy metal ion toxicity.
重金属イオン毒性に起因している処理問題を解決するにおいて，これらの努力の結果は，デザインエンジニアと処理場オペレーターに役立った。

Herein：ここに

1. The writers want to show **herein** other difficulties which can arise in toxicity measurements.
著者は，毒性測定の際に発生可能な他の難問題をここに示したい。
2. The methodology described **herein** makes it possible to gain insight into water quality changes to be expected over the life of a groundwater recharge project.
ここに記述された方法論は，地下水涵養プロジェクト期間中に生じるであろうと思われる水質変化の洞察を可能にする。

Higher：より高い

- higher than ...... ：…より高い

1. The observed value of the K is definitely **higher than** calculated from Equation 2.
観察によるKの値は式2による計算値より確かに高い。
2. Volunteers generally rated water quality **higher** at most sites **than** did state biologists.
一般的に，ボランティアは，州の生物学者よりもほとんどの地点で水質を高めに位置付けた
3. Chromium was detected at **higher than** commonly expected values throughout the study area.
クロミウムが一般的予想値より高い値で調査地域全体で検出された。

4. The predicted CO₂ concentration is **higher than** the observed concentration at later time.
   予測された CO_2 濃度は，後に観察された濃度より高い。

Highlight：重視する；強調する；目立たせる

1. The model results **highlight** the importance and utility of a modeling framework.
   モデル結果は，モデリングの枠組みの重要性と実用性を重視する。
2. The model results underscore the significance of chemical partitioning on chemical fate and **highlight** the importance.
   モデルの結果は，化学物質の衰退におけるその物質分配性の意義を強調し，そしてその重要性を重視する。
3. Some **highlights of** this analysis of the IBI metrics and fish community indices are as follows:
   IBIメトリックおよび魚類群集指数に関するこの分析の重要点は，以下のとおりである。
4. The Pollution Tolerance Index, developed by Lange-Bertalot(1979), **highlights** organisms sensitive to toxic substances.
   Lange-Bertalot(1979)が設定した汚濁耐性指数は，毒性物質に対し敏感な生物を目立たせる。
5. This chapter introduces the elements of a sound survey design and illustrates the techniques for surface waters (lakes and streams) by describing several examples **highlighting** various components of sample surveys.
   本章では，サンプル調査の様々な構成要素を強調した事例を述べることによって健全な調査設計の要素を紹介し，地表水域(湖沼・水流)についての技法を示すことにする。

History (Historically)：歴史(歴史的に)

1. **Case histories** are presented below showing the outcome of impairment criteria testing in a variety of situations.
   様々な状況での損傷クライテリア検定の結果を示した事例史が下に述べられている。
2. Considerable quantities of these constituents were discharged in the tailings during much of the mining **history**.
   採鉱史上，長年の間に，これらの成分のかなりの量が選鉱滓に排出された。

- ～ have a long history：～は長い歴史を持っている

3. Arsenic **has a long history of** use for its toxic and medicinal properties.
ヒ素は，それがもつ毒性および薬効性のため，その利用は長い歴史を持っている。
4. The surveying of ambient biota **has a long history** in many states and usually takes the form of either coordinated monitoring networks or a series of special studies.
環境生物相に対する調査は，多くの州で長年行われており，通常，調整された監視網もしくは一連の特別研究の形をとる。

- **historically**：歴史的に

5. **Historically**, after water was used for societal needs, it was labeled as sewage or wastewater and treated for discharge into receiving water or for land disposal.
歴史的にいって，水は，社会的需要として使用された後，下水または廃水と呼ばれ，処理されてから受水域へ排水されるか土地処分される。

Hope：希望する

- We hope …..：我々は…を希望する

1. **We hope** the reader will bear in mind that the real world is somewhat different.
我々は，読者が実世界がいくぶん異なっているということを心に留めておくであろうことを希望する。
2. Kansas **hopes** that this will lead to the establishment of numeric biological criteria in the future.
カンサス州は，これが将来，数値生物クライテリアの確定に通じることを期待している。

- **it is hoped that** …..：…と希望する；…であろうことが希望される

3. **It is hoped that** the medium can be used for toxicity screening tests for various compounds, especially for heavy metals.
希望的に，この媒質が種々の化合物，特に重金属の毒性スクリーニングテストに使うことができる，とおもわれる。
4. While model applications may differ greatly in scope and purpose, **it is hoped that** the representative data derived from pilot studies will be useful in this process.
モデルの適用が利用範囲と目的で大いに異なるが，予備的研究から得られたこの

代表的データがこのプロセスで有用であろうことが希望される。

Hold：適用される；考えられる；認められる

1. One early hypothesis **held** that .....
   初期の仮説の一つは … と考えられていた。
2. For bedrock samples, gear usually includes a 3- to 4-in. diameter PVC pipe with a rubber seal that **can be held** in place where the rock is brushed clean and the sample removed by suction.
   床岩サンプルの場合，サンプリング装置には，通常，直径3〜4inのラバーシールドPVC(ポリ塩化ビニル)管が含まれており，これを設置すれば，岩石を磨いてサンプルを吸引除去できる。

   - 〜 **holds for**..... ：〜は，… について言える
   - 〜 **hold in all cases** ：〜は，すべての場合に適用される

3. The converse **holds** for the case 2.
   逆のことは，ケース2について言える。
4. These two principles **hold in all cases** under any conditions.
   これらの2つの原則は，どんな状況であってもすべての場合に適用される。

   - 〜 **is widely held as** ..... ：〜 は … と広く認められている

5. Sedimentation **is widely held as** responsible for degradation of fish communities in warmwater and coldwater streams, and many of the mechanisms of this degradation have been well documented.
   堆積は，温水水流ならびに冷水水流において魚類群集劣化の原因と広く認められており，劣化メカニズムの多くが十分に文書化されている。

   - **assumption holds true**(See **Assumption**)：
     仮定が正しい；仮定が有効である(➡ **Assumption**)
   - **assumption holds for** .....(See **Assumption**)：
     … に対して仮定が有効である(➡ **Assumption**)

Hours：時間

1. Chlorobenzene and all intermediate products disappeared within the first **2 hr** of the reaction.
   クロロベンゼンとすべての中間生産物は分解反応し，最初2時間以内に消失した。
2. Dissolved oxygen, pH, temperature, and specific conductance were

measured in the field hourly **over a 24-h period** with either a Hydrofield (R) Surveyor III or Data Sonde.
溶存酸素，pH，水温，および導電率が Hydrofield (R) Surveyor III または Data Sonde のいずれかで，24 時間にわたり，1 時間ごとに現地で測定された。

3. Keiko Tomoyasu provided **many hours of** support in the development of the basic computer programs.
Keiko Tomoyasu は，長時間を割き基礎的なコンピュータプログラムの作成を支援してくれた。

How：どのように；どれほど

- how ～ is .....：～がどれほど…であるか
- how ～ does .....：～はどのように…するのか

1. Figure 3 shows **how** water **is** used on a global scale.
地球規模で水がどのように使用されているかを図3に示す。
2. In this figure, the dashed line represents **how** V **varies** with N.
この図では，点線はNに対してVがどのようにで変化するかを示す。
3. They do not have to be designed with an idea of **how** a lake **works**.
湖がどのような働きをしているかという考えは，計画上無用である。
4. One must take into consideration the season, as well as precipitation and temperature deviations (both long and short term) and **how** they **may have affected** vegetation and land cover patterns.
季節や降水量・温度の偏差(長期的／短期的)を考慮に入れたうえで，それらが植生および土地被度パターンにどれほど影響してきたかを検討すべきである。

- how to .....：どのように…する

5. The problem of **how to** maintain optimal nitrification rates is not only a question of the presence of aerobic conditions.
最適な硝化反応速度をどのように維持するかという問題は単に好気性状態であるか，ないかだけではない。
6. This language was not trivial and caused a great deal of concern regarding **how to** define "integrity" (especially biological integrity) and the measurements to be applied.
この文言は，些細なものでなく，「完全性」(特に生物完全性)の定義づけと適用すべき測定値について強い関心を生じさせた。

- **of how** .....：どれほど…するかについて；様相の

7. Since that time, a much greater understanding has developed **of how** each level of treatment manifests itself in the aquatic environment.
   それ以降，各水準の処理が水生環境でどれほど効果を発揮するかについて，理解がかなり深まった。
8. USEPA's User's Manual (USEPA, 1989) provides a good overview **of how** reference site data were translated into regional biocriteria.
   USEPAの『Users Manual（利用者用マニュアル）』』(USEPA, 1989)は，参照地点データが地域的生物クライテリアへ変換される様相に関する適切な概観を提供している。

   - **how much ......** ：どれほど
   - **how well** ：どれほどよく

9. A measure of **how much** the quality can be improved can be derived through changing management practices in selected watersheds where associations were determined.
   水質がどれほど改善されうるかを示す測度は，関連性が決定された流域における管理慣行の変更を通じて得られるだろう。
10. This could be important in determining **how well** real impairment can be reasonably expected to be detected.
    これは，現実の損傷がどれほど合理的に探知されると思われるかを決定するうえで重要であろう。
11. It is likewise unclear **how well** limestone reactors will neutralize more highly concentrated solutions.
    石灰岩反応器が，より濃厚な溶液をどれほどよく中和するかは同じく不明確である。
12. Studies are needed to identify current prevention effort and to assess **how well** they are working.
    現在の防止努力を識別し，それらがどれほど良く機能しているかを評かするために，研究が必要とされる。
13. A recent article describes some approaches for defining **how well** various percentiles can be estimated and notes pitfalls in the use of extreme percentiles for these purposes.
    最近の論文では，様々なパーセンタイルがどれほど十分に定められるか決定する手法を述べており，これらの目的のための極値パーセンタイルの利用における陥穽を記している。

    - **as to how 〜** ：〜が，いかに…という
    - **as to how much ......** ：いくら…だろうかという

14. The question arises **as to how** we can believe that ......
    … ということを我々がいかに信じることができるかという質問が起こる。
15. The question was raised **as to how much** the scores would change if the HPI were changed.
    もしHPIが変えられたなら，スコアがいくらを変わるだろうかという質問が提起された。

    - **demonstrate how** ..... (See **Demonstrate**)：
      どれほど … か実証する（→ **Demonstrate**）

Hundreds：数百

1. Current monitoring efforts focus on only **a few of the hundreds of** chemicals known to be present in drinking water supplies.
   現在のモニタリングの努力が，飲料給水への存在が知られている数百という化学物質だけに集中している。

    - **hundreds to thousands of** .....：数百から数千の …
    - **hundreds of thousands of** .....：数百ないし数千もの

2. Sulfate occurs in the **hundreds to thousands of** milligram per liter range.
   硫酸塩が数百 mg/L から数千 mg/L といった範囲で生じる。
3. Only a few rocks need to be sampled, because each has a periphyton assemblage with **hundreds of thousands of** individuals.
   サンプリングに必要なのは，数個の岩石だけである。なぜなら，1個の岩石に個体数が数百ないし数千もの付着生物群がりがあるからである．

Hypothesis：仮説

1. One early **hypothesis** held that .....
   初期の仮説の一つは .....と考えていた。
2. Thus, a weight-of-evidence approach can become increasingly obvious when **hypothesis testing** demonstrates the unnecessary usage of independent application
   それ故，証拠加重値手法は，仮説検定が独立適用の不要さを実証した場合には，ますます明瞭になりうる。

    - **This hypothesis is supported by** .....：
      この仮説は … によって支持される
    - **This hypothesis is tested in** .....：この仮説は … でテストされる

Hypothesis

- **This hypothesis has not been validated**：
  この仮説は実証されていない

3. This **hypothesis is supported** by studies of pure cultures.
   この仮説は純粋な培養細菌の研究によって支持される。
4. This **hypothesis is tested** in highly chelated seawater cultures.
   この仮説は高度にキーレートした海水で培養した細菌でテストされる。
5. This **hypothesis has not been validated**, and this phenomenon remains an open question.
   この仮説は実証されていないし，それにこの現象は未解決の問題のままである。

- **~ is based on a hypothesis about .....**：
  ～は…に関する仮説に基づいている

6. The use of each attribute **is** essentially **based on a hypothesis about** the relationship between instream condition and human influence.
   各属性の利用は，基本的に流入水状態と人為的影響の関係に関する仮説に基づいている。

- **We hypothesize that .....**：我々は…であろうと仮定する

7. **We hypothesize that** a specific sequence of events will occur in response to chronic nitrogen amendments.
   長期にわたる窒素の土壌改良に応じて，複数のイベントが特定の順序で起こるであろう，と我々は仮定する。
8. **We hypothesize that** the gill tissue of mussels has a micellar layer which absorbs the hydrocarbons.
   我々は，ムラサキ貝のエラの組織が炭化水素を吸収する micellar 層を持っていると仮定する。

- **~ test the hypothesis that .....**：～は…という仮説をテストする
- **~ test the hypothesis of .....**：～は…という仮説をテストする

9. A series of batch tests was carried out to **test the hypothesis that** the same enzymes are used to attack chlorophenols with the same position-specific dechlorination reactions, resulting in competition between these compounds.
   位置特定の同じ脱塩反応において，同じ酵素が塩化フェノールを攻撃するために使われ，これらの化合物の間に競合反応をもたらす，という仮説をテストするために一連のバッチ試験が行われた。
10. Washington EPA, as a minimum, **should have tested the hypothesis of**

no differences in biological performance for reference sites on impounded and free-flowing sections.
ワシントン州EPAは，最低限でも貯水部および流水部の参照地点について生物能力に差がないとする仮説をテストすべきだった。

- ～ **confirm the hypothesis that** ..... : ～は…という仮説を確証する

11. The results **confirm the hypothesis that** copper uptake by algae is related to the free cupric ion activity and is independent of the total copper concentration.
この結果は，藻類による銅の摂取は自由な銅イオン活性と関係があり全銅濃度とは関係がない，という仮説を確証する。

12. Further investigations are needed to **confirm this hypothesis**.
この仮説を確証するために，よりいっそうの調査が必要とされる。

- ～ **support the hypothesis that** ..... : ～は…という仮説を支持する

13. These results directly **support the hypothesis that** natural levels of copper in seawater can be toxic to plankton.
これらの結果は，海水中に自然に存在する銅のレベルがプランクトンにとって有毒であり得るという仮説を直接に支持している。

Idea：考え

- **The idea that .....** : … という思考
- **The idea being brought forward is .....** : 前進的考え方は … である

1. **The idea that** conditions were pristine in North America prior to European settlement has been convincingly challenged in the past couple of decades.
   欧州人の入植前の北米では無傷状態だったという思考にたいして，ここ数十年，説得力ある異議が申し立てられてきた。
2. **The idea being brought forward** in this research is to measure several of these general parameters.
   この研究にて前進的考え方は，これらの一般的なパラメータのいくつかを測定するということである。

- **an idea of how .....** : どのような … をしているかという考え
- **an idea of what .....** : 何が … するのかといった考え

3. They do not have to be designed with **an idea of how** a lake works.
   湖がどのような働きをしているかという考えは，計画上無用である。
4. This report should serve to give the reader **an idea of what** is involved in testing a model.
   この報告書は，モデルのテストには何が関与するのかといった考えを，読者に与えるのに役立つはずである。

Ideal (Ideally) (Idealization)：理想的な (理想的に) (理想化)

1. Because of the nature of ecoregions, the **ideal** way of evaluating them would be through use of an ecological index of integrity.
   生態地域の性質故に，その理想的な評価方法は，生態系の完全性指数を用いることだろう。
2. As with all sampling designs, the **ideal** number of reference sites must be balanced against budget realities.

あらゆるサンプリング計画と同様に，参照地点の理想的な数は，現実的予算に照らして決定せねばならない。
3. **Ideally**, reference sites should be as undisturbed as possible.
理想的には，参照地点は，できる限り無攪乱であるべきである。
4. It is important to note that most models represent **an idealization of** actual field conditions and must be used with caution to ensure that the underlying model assumptions hold for the site-specific situation being modeled.
ほとんどのモデルは，実際の現地状態を理想化していること，そしてモデルの対象となっている特定な状況に対してモデルの基礎仮定が有効であることを保証するため注意して使われなくてはならないということ，を指摘することは重要である。

Identical：同一の；一致する

1. This is the **identical** model which we used in the previous study.
これは我々が以前の研究で使用したものと同一のモデルである。
2. The solution of these equations is accomplished using finite difference approximations in **identical** fashion to the Water Quality Analysis Simulation Program (WASP).
これらの方程式は水質解析シミュレーションプログラム(WASP)と同一方法，有限差分近似を使って解かれている。

- **〜 is identical with .....**：〜は…と一致している

3. The results from AMODEL **were identical with** those from BMODEL.
AMODELの結果は，BMODELの結果と一致していた。

Identify（Identification）：識別する；確認する（識別）

1. This relationship **has been identified by** several other workers, and fitted to Langmuir and Freundlich isotherms.
この関係は，数人の他の研究者によって確認され，LangmuirとFreundlich isothermsに当てはめられた。
2. The choice of groups to **be identified** and analyzed is related to the measures that will be used, as was discussed in detail above.
識別・分析されるべき集団の選択は，前述のとおり用いられる測度に関係している。
3. These data provide additional information on previously **identified** Solid Waste Management Units.
これらのデータは，前に識別された廃棄物処理ユニットについての追加情報を供給する。

4. Studies are needed **to identify** current prevention effort and to assess how well they are working.
現在の防止努力を確認し，そしてそれらがどれほどよく働いているかを評価する研究が必要である。
5. It is better **to identify** large areas over which calibrations will be performed rather than small areas.
較正を行うに当たり，狭い地域よりも広い地域を確認した方が良い。
6. Rapid assessment approaches are intended **to identify** water quality problems and to classify aquatic habitats according to a variety of water resource criteria.
迅速アセスメント手法の意図は，水質上の問題点を確認すること，様々な水資源クライテリアに従って水生生息地を分類することである。

- ～ **was identified as** ..... : ～ は … と確認された
- ～ **was identified to** ..... : ～ は … へ識別された

7. Stream habitat modification **was identified as** the third leading cause of aquatic life impairment in the 1992 Tennessee Water Resource Inventory.
水流生息地改変は，1992年のテネシー州水資源調査一覧において，水生生物損傷の第三の主因と確認された。
8. Larger fish **were identified to** species, enumerated in the field, and usually released.
より大型の魚は，種が識別されて現地で数えられ，通常は放流された。

- **identification** : 識別

9. **Identification** of organisms to species allows the use of indices.
生物種を識別できれば，指数の利用が可能になる。

If：もし … ならば

1. **If** separate habitats are sampled, it is important to keep them separate for subsequent analysis.
隔離生息場所でサンプリングするならば，その後の分析でも隔離することが肝要である。
2. **If** a simple relationship could be established, it would be of great in planning the future direction of any development work on a new chemical.
もし単純な関係が確証されえるなら，いかなる新化学物質開発研究の将来方針を計画するにあたって，それは偉大なことであろう。
3. **If** conditions are similar among regions, or if the differences can be

associated with management practices that have a chance of being altered, it seems wise to combine the regions for the purposes of establishing biocriteria.
仮に，各地域における状態が似ている場合や，その差違が変化可能性ある管理慣行に関連している場合には，生物クライテリア設定のために諸地域を組み合わせた方が賢明だろう。

4. **If** may be necessary to select reference sites from another class of stream, lake, or wetland if all sites in a particular class are highly disturbed.
ある等級における地点すべてがきわめて擾乱されていれば，別の等級の水流，湖沼，湿地から参照地点を選ぶ必要もあろう。

5. This observation could be explained **if** multiple dechlorinating microorganisms were present in the filter.
もし多種の脱塩素微生物がこのフィルターに存在していたなら，この観察は説明できるかもしれない。

- **determine if .....**：…するか否か決定する

6. The plots for each metric were examined to **determine if** any visual relationship with drainage area existed.
各メトリックについての図が検討され，流域面積との目測関係が存在するか否か決定された。

7. Results of such additional studies could **determine if** it would be feasible for small systems to treat entire volumes of water with a flow-through reactor.
このような追加研究の結果によって，水の全量処理が流動反応器で可能かどうか決定することが小型のシステムで出来るかもしれない。

8. Experiments were run changing residence time in the UV chamber in an attempt to **determine if** certain chemicals in the system are more completely destroyed as the residence time is increased.
滞留時間が増えるにつれて，システム内のある特定の化学物質がいっそう完全に分解されるかどうかを決定する試みで，UV反応器の滞留時間を変えて実験が実施された。

- **if any**：もしあるとするなら
- **if not impossible**：仮に不可能でなくても

9. The question concerning drinking water are addressed in this article: 1) to what extent, **if any**, does acid precipitation cause a drinking water problem；2) if a problem is present, what can be done about it?
この論文では，飲料水に関する問題を扱います。1) もしあるとするなら，どの程

度まで酸性雨が飲料水問題になるのか；2) もし問題があるなら，何がそれについて何ができるか。
10. Without a sufficient theoretical basis it would be very difficult, **if not impossible**, to develop meaningful measures and criteria for determining the condition of aquatic communities.
十分な理論的基礎がなければ，水生生物群集の状態を決定するための有意義な測度および基準（クライテリア）を設定することは，仮に不可能でなくても，非常に困難だろう。

Illustrate (See also **Figure**)：示す；例証する（➡ **Figure**）

1. The Snake River project **illustrates** many of the decisions, procedures, and results involved in using HSPF.
このスネーク川プロジェクトは，HSPF の使用に関与する決定，手順，結果の多くを例証している。
2. The following section **illustrates** both the various differences and the many similarities of existing and emerging biological criteria programs.
次の節では，既存・新規の生物クライテリアプログラムにおける様々な相違点と多くの類似点を示すことにする。
3. This chapter introduces the elements of a sound survey design and **illustrates** the techniques for surface waters (lakes and streams) by describing several examples highlighting various components of sample surveys.
本章では，サンプル調査の様々な構成要素を強調した事例を述べることによって健全な調査設計の要素を紹介し，地表水域（湖沼・水流）についての技法を示すことにする。
4. The purpose of this section is to present a description of some of these modifications and to **illustrate** the different directions many programs have taken.
本項の目的は，これらの修正の一部を述べることと，多くのプログラムがとった異なる方向性を示すことである。
5. The method for interpreting the sign **will be illustrated** in the discussion and example problems that follow.
このサインを解読する方法が考察と次に続く例題で例証されるであろう。

● **as illustrated by** ..... ： … によって示されるように

6. We also discuss uses of biological data collected by volunteer monitors **as illustrated by** a case study of the Maryland Save Our Stream (SOS)

biological monitoring program.
ボランティアモニターにより収集された生物データの利用についても論じ，メリーランド州「Save Our Streams (SOS)」生物モニタリングプログラムをケーススタディとして示す。

- **Figure N illustrates .....** (See Figure)：
  図 N に … を例証する (➡ **Figure**)
- **..... is illustrated in Figure N** (See Figure)：
  … を図 N に例証する (➡ **Figure**)

Impact：影響

1. The environmental **impacts of** DDT have become evident in Indian Creek in north central Alabama.
   DDT の環境影響は，アラバマ州の北中央にある Indian Creek において明白になった。
2. In contrast to many other human **impacts**, habitat loss can be essentially irretrievable over a human time frame.
   他の多くの人為的影響と異なり，生息場所消失は，基本的に人間の時間枠では取り返しがつかない。
3. Since external **impacts are minimal**, measured variability would be most likely due to the sampling inconsistencies.
   外部影響が最小であるから，測定された可変性は，サンプリングの不整合性に起因する見込みが高い。
4. Despite the impaired condition of the receiving water, **the impact of** the sewage discharge resulted in criteria exceedance in EPT, biotic index, and model affinity.
   受入水の損傷状態にもかかわらず，下水排出の影響は，EPT，生物指数，モデル類縁度におけるクライテリア超過を招いた。
5. Insufficient knowledge about regional expectations can result in misinterpretations about the **severity of impacts** in streams.
   地域期待に関する知見不足は，水流中の影響度に関する誤った解釈を招きうる。

- **impact on .....**：… に強い影響を与える

6. It has a powerful **impact on** .....
   それは.....に強力な影響を与える。
7. There is no evidence that any such effects translate into **impacts on** fish populations.

そのような効果が，魚類集団に対する影響につながるという証拠はない。

8. Its potential **impact on** the scientific community is large and still developing.
   科学者社会への潜在的影響は大きく，そしてまだ進展している。
9. Their activities may relate to thermal changes, flow changes, sedimentation, and other **impacts on** the aquatic environment.
   それらの活動は熱変化，流量変化，堆積，それに水生環境への影響と関連づけられるかもしれない。
10. This finding led to concern over **potential impacts** of PCB contamination **on** recreational fisheries in the Wood River.
    この調査結果は，Wood川のレクリエーション用漁場に及ぼすPCB汚染の潜在的影響が懸念される原因となった。

    ● **impacts from** ..... : …による影響

11. This work should be especially helpful in addressing **impacts from** sedimentation.
    この作業は，特に堆積による影響への対処に有益なはずである。

    ● ~ **recover from impacts** : ~ は影響から回復する

12. Aquatic communities gradually **recover** with increased distance downstream **from impacts**, although the pattern may vary according to the type of impact and discharge.
    水生生物群集は，影響からの下流距離が増すにつれて徐々に回復する。ただし，そのパターンは，影響および排出の種類に応じ異なる。

Imply : 暗示する

1. The results **imply** that complexes of copper with natural organic ligands are not toxic to the algae.
   この結果は，自然に存在する有機物リガンドとの銅の錯体は藻にとって毒ではないことを暗示する。

   ● **as implied by** ..... : …によって暗示されるほど

2. The differentiation on the basis of experimental results may not be as straight forward **as implied by** these definitions.
   実験結果を基にしての分化は，これらの定義によって暗示されるほど単純ではないかもしれない。

Importance：重要性

1. Work over the last decade by investigators, such as Ishii (1993), have emphasized the **importance of** riparian areas for maintaining the quality and function of the hyporheic zones of streams.
   Ishii (1993) などの調査者による過去10年間の研究は，水流のhyporheic地帯の質・機能の維持にとっての水辺区域の重要性を強調した。
2. Advocates seek to persuade others of **the importance of** maintaining water resource quality.
   環境擁護者 (advocacy) たちは，水資源質の維持の重要性を他者に納得させようと努めている。

   - ～ **is of importance** ……：～は重要である
   - ～ **is of central importance** ……：～は主要な重要性をもつ
   - ～ **is of particular importance** ……：～は特に重要である
   - ～ **is of enormous importance** ……：～は非常に重要である
   - ～ **is of considerable importance** ……：～はかなり重要である
   - ～ **is of obvious importance** ……：～は当然重要である

3. The origin and flow of groundwater in the alluvium **is of central importance** in this study.
   この研究では，沖積層においての地下水の水源と流れは主要な重要性をもつ。
4. Information on the toxicity of ammonia to marine crustacea **is of particular importance**.
   海に住む甲殻類に対するアンモニアの毒性情報は，特に重要である。
5. Catalytic reactions **are of enormous importance** in both laboratory and industrial applications.
   触媒反応は，研究的そして産業的な適用の両方で非常に重要である。
6. An equation which **is of considerable importance** in such application is …..
   このようなアプリケーションでかなり重要である式は … である。
7. Knowledge of the factors affecting the movement and transformations of these substances **are of obvious importance** in understanding and controlling the hazard.
   これらの物質の移動と変化に影響を与える要因の知識は，その物質の危険性を理解し制御していくために当然重要である。

   - **It is of importance to** ……：… することは重要である

8. **It is of critical importance** in biological monitoring to collect a consistent and reproducible sample.
整合的かつ再現可能なサンプルを採集することは，生物モニタリングにおいてきわめて重要である。

- **Of particular importance is** ..... ：特に重要なのは…である

9. **Of particular importance** for the photodegradation of water pollutant **is** the highly reactive hydroxyl radical.
水の汚濁物質の光分解において特に重要なのは，反応性が高いヒドロキシル遊離基である。

Important (Importantly) ：重要な（重要なことには）

- **important ～ is** ..... ：重要な～は…である
- **important to ～ is** ..... ：～に重要なのは…である

1. The most **important** aspect **is** the choosing of proper initial concentrations for .....
最も重要な側面は…のために適切な最初の濃度を選択することである。
2. **Important to** waste minimization efforts **is** knowledge of the quantity and characteristics of waste stream.
廃棄減少化の努力に重要なのは，廃物フロー過程での量と特性の知識である。
3. Perhaps most **important to** the enhancement of biological criteria efforts **are** the lessons being learned from state experiences.
生物クライテリア努力の増強にとっておそらく最も重要なのは，諸州の経験から教訓を学ぶことである。

- **equally important is** ..... ：同じく重要なのは…である

4. **Equally important are** examples of minimization.
等しく重要なのは，最小化の例である。
5. Watershed studies are a necessity, but **equally important is** the development of an understanding of the spatial nature of ecosystems, their components, and the stress we humans put upon them.
流域の研究は必要であるが，同じく重要なのは，生態系の空間的性質，その構成要素，そして我々人間が生態系に与えるストレスへの理解を育むことである。
6. The site classification and metric calibration steps are important. **Equally important is** the designation of aquatic life uses for the waterbodies under investigation.

地点分類ステップおよびメトリック較正ステップは，重要である。同じく重要なのは，調査対象水域について水生生物利用を指定することである。

7. Although every program provides some form of educational benefits, the extent of educational opportunities varies among the programs, some of which exist almost entirely to provide education, some for which education is secondary to action or data collection, and others where the objectives of education and data collection **are equally important**.
どのプログラムも何らかの啓発効果を有するものの，啓発機会の度合は，プログラムごとに異なっている。例えば，現存プログラムの一部は，もっぱら啓発を提供していたり，啓発が行動やデータ収集の従属的なものだったり，啓発目標とデータ収集目標が同等に重要だったりする。

- **~ is important** ：~は重要である
- **~ is most important** ：~は最も重要である
- **~ is less important** ：~は，あまり重要でない

8. This **could be important** in determining how well real impairment can be reasonably expected to be detected.
これは，現実の損傷がどれほど合理的に探知されると思われるかを決定するうえで重要だろう。

9. Modeling **is important** in making decisions about water resources.
モデルは，水資源についての決断をくだすために重要です。

10. Although the mechanisms of acoustic foam disintegration are not entirely understood, the following acoustic effects are believed to **be most important** in foam breakage.
音波によって生じる泡の崩壊機構は完全に理解されていないが，次に示す音波効果は気泡崩壊過程上最も重要であると思われる。

11. Regions characterized by karts topography, extensive sandy soils, lack of relief, or excessive aridity are examples of areas where watersheds **are less important**.
カルスト地形，広大な砂質土壌，起伏の欠如，極度の乾燥を特色とする地域は，流域があまり重要でない区域の例である。

- **it is important to .....** ：…することが重要である

12. **It is important to** define the regional extent over which a particular biocriterion is applicable.
ある特定の生物クライテリアが適用可能な地域的度合を定めることが重要である。

13. **It is important to** recognize that the proposed model depends on its intended purpose.

提案されたモデルがそのモデルの意図的な目的に依存する，ということを認識することが重要である．

14. **It is important to** note that numerous sites along a stream are examined before making a use decision.
使用決定を下す前に水流沿いの多数地点が検討されることに留意すべきである．
15. If separate habitats are sampled, **it is important to** keep them separate for subsequent analysis.
隔離生息場所でサンプリングするならば，その後の分析でも隔離することが肝要である．
16. Whatever the reasons for the differences, **it is important to** keep in mind that ....
相違の理由が何であるとしても，… を念頭におくことは重要である．

   - **What is important is to** ..... ：重要なことは… である

17. **What is important is to** examine the effects of endocrine disruptors on the environment.
重要なことは，内分泌撹乱物質が環境におよぼす影響を調査することである．

   - **one of the most important** 〜 **is** ..... ：
   最も重要な〜の一つは … である
   - **〜 is one of the most important** ..... ：
   〜 は最も重要な … の一つである

18. **One of the most important** attributes **is** molecular weight.
最も重要な特質の1つは分子量である．
19. The smallmouth bass (Micropterus dolomieui) **is one of the most important** game species in midwestern rivers and streams.
コクチバス (Micropterusdolomieui) は，中西部の河川水流における最重要な漁獲対象種の1つである．

   - **most importantly**：最も重要なことに

20. Finally, and perhaps **most importantly**, many existing versions have as yet not been properly validated with independent data.
最後に，そしておそらく最も重要なことに，多くの既存バージョンは，また，独立データで適切に確証されていない．

Impossible (See also **Possible**) ：不可能である (➡ **Possible**)

1. Such detailed understanding of the interactions **is impossible** from steady-

state models.
このような相互作用の詳細な理解は定常モデルからは不可能である。
2. The axiom follows "When in doubt choose to take more measurements than seem necessary at the time since information not collected **is impossible** to retrieve at the later date.
公理としては,「疑わしい場合,その時点で必要と思われるよりも多くの測定を行うことを選びなさい。なぜなら,収集されなかった情報は,後に取り戻すことができなからである」。
3. Without a sufficient theoretical basis it would be very difficult, **if not impossible**, to develop meaningful measures and criteria for determining the condition of aquatic communities.
十分な理論的基礎がなければ,水生生物群集の状態を決定するための有意義な測度および基準(クライテリア)を設定することは,仮に不可能でなくても,非常に困難だろう。

Improve：改善する

1. The method described in this paper **can improve** the interpretation of the results.
この論文に記述された方法によって,この結果の解釈を改善することができる。
2. While the quality of some regions **has improved** during recent years, there is still great concern regarding the marine environment.
近年の間に若干の地域環境が良くなってきた一方,海洋環境に関してまだ大きい懸念がある。
3. This concern has arisen lately as analytical techniques for detecting water contaminants **have improved** to the point where it is possible to detect contaminants at the part per-trillion level.
水汚染物質を検出するための分析技術が汚染物質をpptレベルで検出可能なまでに向上したため,最近この関心が高まった。
4. We appreciated critical review comments from three anonymous reviewers, which greatly **improved** an earlier draft of this manuscript.
筆者は,3人の匿名査読者からの批評的論評に感謝する。それは,本章の草稿を大幅に改善してくれた。
5. A measure of how much the quality **can be improved** can be derived through changing management practices in selected watersheds where associations were determined.
水質がどれほど改善されうるかを示す測度は,関連性が決定された流域における管理慣行の変更を通じて得られるだろう。

- **to improve** ...... ： … を改良するため

6. States have undertaken a wide range of efforts **to improve** bioassessment methods.
   諸州は，生物アセスメント方法を改良するため広範な努力を払ってきた。
7. Aquatic life standards were proposed **to improve** impact assessment and provide greater emphasis on the management of ecological resources.
   影響評価を改善し，生態資源管理をより重視するため，水生生物学的基準が提言された。

- **improved** ...... ：改善された …

8. The overall effort should focus on building **improved** "line of evidence".
   努力全体は，改善された「証拠ライン」の構築に主眼を置くべきである。
9. As a follow-up note, this sewage treatment plant subsequently underwent a major upgrade, and later macroinvertebrate sampling revealed an **improved** community downstream.
   追記すれば，この下水処理場は，その後大幅な改善がなされ，後の大型無脊椎動物サンプリングで下流生物群集における向上が見られた。
10. **Improved** detection techniques have now shifted concern to the threat posed by toxic chemicals.
    改善された検出技術は，今は関心を毒性化学物質による脅威に移行させた。

Improvement ： 改善

1. Significant **improvement** in the biological indices has been observed during the past twenty years, which has been due in large part to significant reductions in point source loadings.
   生物指数における大幅な改善が過去20年間に観察されており，その主因は，点源負荷の大幅な削減であった。
2. The **improvement** is even more remarkable when conditions since the turn of the century, when the river lacked any fish life for a distance of nearly 40 mile, are considered.
   この川で約40mileにわたり魚類が生息していなかった今世紀初頭移行の状態を考慮すれば，この改善は一層明らかである。
3. Compared to the Scioto River, the Ottawa River exhibits evidence that, despite some **improvements**, makes it one of the most severely impaired rivers in the state.
   Scioto川に比べ，Ottawa川は，若干の改善にもかかわらず，同州で最もはなは

だしく影響された河川の一つという証拠を示している。
4. Bioassessments can be used to document instream **improvements** that result from wastewater facility upgrades and the implementation of best management practices.
生物アセスメントは，廃水施設向上および最善管理慣行の実施に起因した流入改善の文書化に利用できる。

   ● **Improvement in ..... was made**：…の改善が行われた
5. **Improvements in** the treatment process **were made** in the year before this sampling, and macroinvertebrate changes were not as great as in previous years.
サンプリング前年に処理過程の改善が行われ，大型無脊椎動物の変化は，それまで数年間ほど大幅でなかった。

Include：含む

1. Primary contaminants of concern associated with the fuel leakage **include** BTEX.
燃料漏れと関連する主要な汚染物質としてはBTEXである。
2. The inherent advantages of photocatalysis **include** (a).....; (b).... ; and (c).....
光触媒反応の固有の利点としては，(a)…，(b)…，そして(c)…である。
3. Needs **include** various types of information concerning toxic substances in water supplies.
必要とするものの中には，供給水中の毒性物質に関する種々のタイプの情報が含まれる。

   ● ～ **is included in** .....：～が…に含まれている
   ● ～ **is included below** .....：～が…下に含まれている
4. Some known features of the behavior of iodine **are not included in** the model, because they are not relevant to the problem of interest.
ヨードの挙動に関してすでに知られている特徴の一部は，モデルに含められていない。なぜならそれらはここで取り上げている問題点とは直接に関係が無いからである。
5. New SWMUs identified after this date **will be included in** the next revision to this report.
この日付後に認識された新しいSWMUsは，この報告書の次の修正版に入れられる。
6. A brief summary **is included below** as an aid to understanding the results obtained in this project.

このプロジェクトで得られた結果を理解する手助けとして，短い要約が下に含まれている。

- **to include** ..... : …を含むこと

7. Coralvill Reservoir does not thermally stratify to any great extent, so the failure **to include** a hypolimnion compartment is not viewed as a serious problem.
Coralvill 貯水池は大きくは温度層化しない，それゆえ(このモデルに)深層部が含まれていないということが重大な問題だとは見なされない。

- **including** ..... : …を含む；…といった

8. Several new approaches, **including** multivariate analysis and weighted average metrics, make periphyton analysis a better indicator than it was 5 or 1o years ago.
多変量分析や加重平均メトリックといった新手法のおかげで，付着生物分析は，5〜10年前よりも優れた指標となっている。

Incorporate：盛り込む；取り入れる

1. The model **incorporates** the effects of mass transport, chemical interactions, hydrodynamic dispersion, and radioactive decay.
このモデルには，物質輸送，化学的な相互作用，水力学的拡散，放射性崩壊の影響が盛り込まれている。
2. Existing multimetric approaches are robust in their ability to measure biological condition because they **incorporate** biological and ecological principles that enable an interpretation of exposure/response relations.
既存の多メトリック手法は，生物学的状態の測定能力において確かである。その理由は，曝露／応答関係の解釈を可能にするような生物学的・生態学的原理を盛り込んでいることである。

- 〜 **incorporate A into B** : 〜はAをBに取り入れる

3. It is current USEPA policy that all states **incorporate** biological criteria **into** their water quality standards.
USEPAの現行方針は，諸州すべてが生物クライテリアを各自の水質基準に取り入れることである。
4. States have undertaken a wide range of efforts **to incorporate** biological criteria **into** water quality standards or water resource management programs.

諸州は，生物クライテリアを水質基準または水資源管理プログラムに盛り込むため，広範な努力を払ってきた。

- **～ is incorporated into** ……：～は … に取り入れられている
- **～ is being incorporated into** ……：～は … に盛り込まれつつある

5. Kinetic coefficients from the single-Cp tests **were incorporated into** a Michaelis–Menten competitive inhibition model.
単一 Cp テストからの動力学係数が Michaelis–Menten 競合抑制モデルに取り入れられた。

6. In addition, assessments of the quality of physical habitat structure **are** increasingly **being incorporated into** the biological evaluation of water resource integrity.
さらに，物理的生息地構造の質アセスメントがますます水資源完全性の生物学的評価に盛り込まれつつある。

7. The appropriate techniques **can** easily **be** described and **incorporated into** a sampling protocol.
それらの適切な技法は容易に記述され，サンプリングプロトコルへ盛り込まれる。

Increase (Increased)(See also **Decrease**)：
増加；上昇；増加する；上昇する（増強の）(➡ **Decrease**)

- **an increase of** ……：… の増加

1. Aggregation effects are taken into account as **an increase of** sedimentation velocities of the particles.
集合効果が粒子の堆積速度の増加の原因として考慮されている。

- **increase in** ……：… の増加

2. This has resulted in a steady **increase in** the production of heavy metals during the last fifty years.
これまでの 50 年間，これは重金属の生産の着実な増加をもたらした。

3. The pH dramatically increases with depth with a seemingly corresponding **increase** in calcium and magnesium.
pH は深さとともに劇的に上昇し，うわべはカルシウムとマグネシウムの上昇と対応するかのようである。

- **Increase (decrease) in concentration** (See **Concentration**)：
濃度の増加（減少）(➡ **Concentration**)

- **Increase(decrease) of concentration** (See **Concentration**)：
  濃度の増加(減少)(➡ **Concentration**)
- **increased** ..... ：増強の

4. The advantages of these **increased** efforts are threefold. First, ..… Second, ... Third, ...
   こうした努力増強の利点は，3つある。第一に，…第二に，…第三に，…
5. The addition of AAA led to **decreased** BBB intermediate concentrations and increased CCC intermediate concentrations.
   AAAの添加は，BBB中間体の濃度減少とCCC中間体の濃度上昇に導いた。
6. The problem is receiving **increased** attention throughout the country.
   国中で，この問題に対する注目が増している。

- **with the increased need for** ..... ：…する必要性が増加するとともに
- **with increased distance from** ..... ：…からの距離が増すにつれて

7. **With the increased need for** predicting the movement and distribution of chemicals in an aquatic environment, knowledge of the rate constants governing the uptake and clearance of chemical in fish is required.
   水生環境での化学物質の移動・分配を予測する必要性が増加するとともに，魚体内での化学物質の摂取・除去を左右している速度定数の知識が必要とされる。
8. Aquatic communities gradually recover **with increased distance downstream from** impacts, although the pattern may vary according to the type of impact and discharge.
   水生生物群集は，影響からの下流距離が増すにつれて徐々に回復する。ただし，そのパターンは，影響および排出の種類に応じ異なる。

★ Increase の動詞としての使用例文を下に示します。

9. Water temperature **has increased** by 2 ℃
   水温が2度上昇した。
10. Carbon dioxide concentrations in atmosphere **have increased** nearly 30%.
    大気中での二酸化炭素の濃度がおおよそ30％増加した。
11. Collection of this type of data **will likely increase** over the next few years.
    この種のデータ収集は，今後数年間に増える見込みが強い。
12. High-intensity ultrasound dramatically **increases** the rates of reaction by as much as 200-fold.
    強度の超音波は，反応速度を劇的に増やし，それはおよそ200倍にも達する。
13. The flow did not **increase** significantly in this segment of Blackwood Creek.
    Blackwood Creekのこの区間では，流れはそれほど増加しなかった。

14. Assuming that the substance under consideration **increases** in accordance with a first-order reaction, the following equation is developed.
考慮中の物質が一次反応にしたがって増加すると仮定して，次の式はつくられている。
15. As we **increase** our awareness that a holistic ecosystem approach to environmental resource assessment and management is necessary, we must also develop a clearer understanding of ecosystems and their regional patterns.
我々は，環境資源の評価・管理への全体論的生態系手法が必要だと深く認識するにつれて，生態系とその地域パターンも明確に理解しなければならない。

- **to increase** ..... ：…を向上させるため

16. Research has been undertaken **to increase** the level of understanding of the effects of heavy metals.
重金属の影響に関する理解力レベルを向上させるために研究が着手された。

Increasing (Increasingly)：上昇する；増加する（ますます；だんだん）

1. **Increasing** temperature decreases the plutonium solubility below 10^{-8} mol L^{-1}.
温度の上昇がプルトニウム溶解度を 10^{-8} mol L^{-1} 以下に減少させる。
2. An issue of **increasing** importance in the design and operation of wastewater treatment facilities is the control of endocrine disruptors.
排水処理場の設計と操作においてますます重要になってきている問題は，環境ホルモンの抑制である。

- **an increasing number of** ..... ：ますます増えている…

3. **An increasing number of** multistate and multiagency cooperative efforts are focusing on the development of biological criteria.
生物クライテリア設定に力点を置いた多州・多機関の協力活動の数が増えている。

- **It has become increasing evident that** ..... ：…がますます明白になってきた

4. **It has become increasing evident that** an activity of particular metal species may be more important than total metal concentration.
特定の金属種の活性が全金属濃度よりいっそう重要であるかもしれないことが，ますます明白になってきた。
5. **It is increasingly recognized that** photochemical processes may be

important in determining the fate of organic pollutants in aqueous environments.
水環境において有機汚濁物の衰退を決定するのに，光化学プロセスが重要であるかもしれないことがますます認められている。

- **there has been an increasing 〜 that ..... :**
  …だ，との〜が強まってきた
- **increasingly** : ますます；だんだん
- **It is increasingly recognized that ..... :**
  …がますます認められている
- **It has become increasing important that ..... :**
  …がますます重要になってきた

6. In recent years **there has been an increasing** awareness **that** effective research, inventory, and management of environmental resources must be undertaken with an ecosystem perspective.
近年，環境資源の効果的な研究・調査一覧作成・管理は，生態系を視野に入れて行うべきだ，との認識が強まってきた。

7. The magnitude and significance of chemical contamination of aquatic environments are **increasingly** evident.
水生環境の化学汚染の大きさと重要性は，ますます明白である。

8. It has become **increasingly** desirable to use destruction of hazardous wastes as the site remediation technology.
危険廃棄物の分解をサイト改善技術として使用することが，ますます望ましくなった。

9. **Increasingly**, attempts are being made to integrate all three types (physical, chemical, and biological) of sampling data.
だんだん3種類（物理的，化学的，生物的）のサンプリングデータを統合する試みがなされつつある。

10. Toxicity assessment of single chemical compounds and of complex industrial effluents has become an **increasingly** difficult task.
一つの化合物のそして複雑な産業廃水の毒性査定が，ますます難しい仕事になった。

11. In addition, assessments of the quality of physical habitat structure are **increasingly** being incorporated into the biological evaluation of water resource integrity.
さらに，物理的生息地構造の質アセスメントがますます水資源完全性の生物学的評価に盛り込まれつつある。

Independent (See also Dependent) :
独自の；関係のない；依存しない（→ Dependent）

1. The central concept is that two **independent** processes describe the basic motion of a particle.
   この中心的な概念は，2つの独立したプロセスが粒子の基本運動を説明するということである。
2. Similar conclusions had been reached by state and **independent** researchers even earlier.
   同様の結論は，さらに以前にも州および独立の研究者によって述べられてきた。
3. Finally, and perhaps most importantly, many existing versions have as yet not been properly validated with **independent** data.
   最後に，そしておそらく最も重要なことに，多くの既存バージョンは，また，独立データで適切に確証されていない。
4. Current efforts to develop and implement biological criteria are closely related to **independent** state efforts to apply biological monitoring information to water resource quality assessment.
   生物クライテリアを設定・実施するための現在の努力は，各州が生物モニタリング情報を水資源質アセスメントに適用する努力と密接に関わっている。
5. Thus, a weight-of-evidence approach can become increasingly obvious when hypothesis testing demonstrates the unnecessary usage of **independent** application.
   それ故，証拠加重値手法は，仮説検定が独立適用の不要さを実証した場合には，ますます明瞭になりうる。

- ~ **is independent of** ..... : ～は…とは関係がない

6. The results confirm the hypothesis that copper uptake by algae is related to the free cupric ion activity and **is independent of** the total copper concentration.
   この結果は，藻類による銅の摂取は自由な銅イオン活性と関係があり，全銅濃度とは関係がないという仮説を確証する。

Indicate (Indicating) (Indication) :
示唆する；示す（示している；示唆している）（指標）

1. Findings to date **indicate** no relationship between the original source of DDT and PCBs.

今日までの調査結果は，DDTとPCBsの最初の汚染源の関係を示していない。
2. Measurement of TOC during the treatment **indicated** a good agreement between the experimentally determined TOC values and those calculated from the quantified reaction intermediates.
処理中のTOCの測定が，実験的に決められたTOCの値と定量化した反応中間物から計算されたTOCの値の間で充分な一致を示した。

- ～ **indicate that** ..... ：～が…を示唆する

3. A number of reports **indicate that** .....
多くの報告が.....を示唆する。
4. A number of studies **indicate that** there is a relationship between A and B
多くの研究が，AとBの間に関係があることを示唆している。
5. This occurrence **indicates that** the nickel did not remain within the treatment units.
この発生は，ニッケルが処理工程内に残っていなかったことを示唆する。
6. Substantial evidence **indicates that** metabolites are toxic to animal systems.
新陳代謝廃物が動物システムにとって有毒であることを示す実質的な証拠がある。
7. Table 6 **indicates that**, except for one sample, all samples contain between 1 and 20 weight percent of gypsum.
表6は，1つのサンプルを除いて，すべてのサンプルが1と20の間の重量パーセントの石膏を含んでいることを示している。
8. Inspection of Equation 8 **indicates that** the suspended sediment distribution is dependent on four transport coefficients.
式8は浮遊堆積物の分布が4つの輸送係数に依存していることを示している。
9. The U.S. Supreme Court and lower courts **have indicated** repeatedly **that** water quality standards must be enforced through NPDES permits.
米国最高裁判所およびその下位裁判所は，これまで再三，水質基準がNPDES許可を通じて実施されねばならないことを主張してきた。

- ～ **was indicated** ：～が示された

10. Although all indices showed a worsening trend, none exceeded the criteria, and no significant impairment **was indicated**.
すべての指数が悪化傾向を示したものの，いずれもクライテリアを超えておらず，有意の損傷が認められなかった。
11. In every case where impairment **was indicated**, we would not have been able to assign the source without bracketing the discharge with upstream-downstream sites.
損傷が示された事例すべてにおいて，我々は，上流地点−下流地点での排出を一

括せずには損傷源を指定できなかっただろう。
12. In the same Tennessee results cited above, chemical and biosurvey results agreed 58 % of the time, and impairment **was indicated** by chemical but not biological criteria at only 6 % of the sites.
前記のテネシー州でのアセスメント結果において，化学的調査結果と生物的調査結果は，58％で一致した。また，生物クライテリアでなく化学クライテリアで損傷が示された地点は，わずか6％にすぎなかった。

- **indicating** ..... ：示している；を示唆している

13. This correlation coefficient was 0.99 **indicating** a surprising good fit to the data points employed.
この相関係数は0.99で，使用したデータポイントに驚くほど一致することを示唆している。
14. A linear regression of measurement and predicted values yielded an R2 of 0.75, **indicating** a reasonable agreement.
測定値と予測値の線形回帰が0.75のR^2をもたらし，妥当な一致を示した。
15. This chapter has attempted to describe the status of biological criteria efforts in the United States by **indicating** the diversity of programs and activities in the states as of 1994.
本章では，1994年現在の諸州での計画・活動の多様性を示すことにより，米国における生物クライテリア努力の現状を述べようと試みてきた。

- **an indication of** ..... ：…の指標

16. When used in a monitoring program, these methods provide **an indication of** what specific environmental characteristics are changing over time, and can be used to help diagnose the cause of changes in the aquatic system.
監視プログラムで用いた場合，これらは，どの環境特性が経時的に変化しているか示す指標となり，水生系における変化原因を判定するのに役立つ。

Infer ：推測する；推察する；推定する；暗示する

1. The condition of the stream **can be inferred from** the taxa present and what is known of their requirements and tolerance.
水流状態は，常在分類群や，その要件・耐性に関する知見から推断されうる。
2. Multivariate analyses, such as CCA and WA, are useful because they make use of the information in assemblages of quantitatively **infer** ecological characteristics (e.g., pH and BOD).
CCAおよびWAといった多変量分析は，(pH, BODなど)生態特性を量的に推断

するために群がり情報を活用しているので，有用である。

Influence：影響；影響する；影響を与える

1. The **influence of** water vapor on the reaction rate derived from the low adsorption of ethylene.
   水蒸気の反応速度に対する影響は，エチレンの低吸着性から生じた。

   ● ～ **have influence on** ……：～は…に対し影響を及ぼす

2. Work has shown that highly variable and unpredictable flow regimes can **have strong influences on** fish assemblages.
   研究によれば，可変性が強くて予測不能な流量状況は，魚類群がりに強い影響を及ぼしうる。
3. Competitive adsorption between water vapor and a probe contaminant can **have** a significant **influence on** the oxidation rate of the contaminant.
   水蒸気と調査汚染物質の間の競合性吸着が，汚染物質の酸化速度に重要な影響を与え得る。
4. In many arid areas, spatial differences in subsurface watershed characteristics **have a stronger influence on** water quality than the size or characteristics of the surface watershed.
   多くの乾燥区域において，地表下流域特性における空間差は，地表流域の規模または特性よりも水質に対し強い影響を及ぼす。

   ★ Influence の動詞としの使用例文を下に示します。
5. Such information **can influence** decisions to control certain substances or processes that might have been overlooked or underrated in an evaluation based on only one group.
   そのような情報は，1集団のみに基づく評価において看過または過小評価されてきた物質や過程を制御する決定に影響を及ぼしうる。

Information：情報；インフォメーション

1. This **information** will establish the pollutant status of these organisms and facilitate interpretation of the other data.
   この情報によって，これらの生物の汚染状態が確証され，そして他のデータの解釈が容易にされる。
2. Other **sources of information** include interviews and review of internal correspondence.
   他の情報源はインタビュー，そして内部通信のレビューである。

3. Not only with this result in **a loss of valuable information**, it will likely result in the continued degradation of the aquatic resource.
   これは，貴重な情報の損失を招くばかりか，水資源の継続的な劣化を招く見込みが強い。

   - information for ..... ：… についての情報
   - information on ..... ：… についての情報
   - information about ..... ：… についての情報
   - information concerning ..... ：… に関する情報

4. Ecological **information for** many algae, particularly diatom, has been recorded in the literature for over a century.
   多くの藻類，特に珪藻についての生態学的情報は，1世紀にわたり文献に記録されてきた。

5. **Information on** the toxicity of ammonia to marine crustacea is of particular importance.
   海に住む甲殻類に対するアンモニアの毒性情報は特に重要である。

6. The unit description includes all current, reasonably obtainable **information on** the unit.
   このユニットの記述は，このユニットについての現在の，ある程度入手可能な情報すべてを含む。

7. These results, in conjunction with the previous **information on** population structure, support the conclusion that ......
   これらの結果は，人口構造についての前の情報と共に，.....という結論を支持する。

8. Key **information about** the environmental setting, characteristics of the receiving waterbody, and insights into the chemical/physical dynamics of the discharge(s) are either directly or indirectly reflected by the biota.
   環境背景に関する主要情報，受入水域の特性，そして排出の化学的／物理的動態に関する洞察は，生物相によって直接的または間接的に反映される。

9. Needs include various types of **information concerning** toxic substances in water supplies.
   必要とするものの中には，供給水中の毒性物質に関する種々のタイプの情報が含まれる。

   - **information for** ..... **is available**：
     … のための情報が利用可能である

10. Not all the **information for** refining the target population **was available** before field visits.

Information

現地視察の前には，ターゲット母集団を絞り込むための情報すべてが利用可能ではなかった。

- **information is gained from** ..... ：情報は … から得られる
- **information obtained from** ..... ： … から得られた情報
- **information derived from** ..... ： … から導かれた情報

11. Additional **information is gained from** population, bacteriological, and productivity studies.
 追加情報は，個体群研究，細菌学研究，生産度研究から得られる。
12. The multivariate approach provides a powerful means of maximizing ecological **information obtained from** periphyton assemblages.
 この多変量手法は，付着生物群がりから得られた生態学的情報を最大化するための有効手段となる。
13. **The information derived from** the identification and enumeration of the algal component of the periphyton provides a wealth of ecological information for the cost involved in producing it.
 付着生物の藻類要素の識別・列挙から導かれた情報は，その作成に関与した費用に見合った豊かな生態情報をもたらす。

- **gather information** ：情報を収集する
- **information on ～ had been gathered** ：
 ～に関する情報が集められてきた

14. The objectives of the research were to **gather all possible information** available to define potential contaminant source area.
 この研究の目的は，汚染源の可能性がある場所を明らかにするために可能な情報すべてを収集することであった。
15. Useful **information on** farming practices **had been gathered** for the Missouri River study.
 農業に関する有用な情報が Missouri 川の研究のために集められていた。
16. Based on the **information gathered** and the publication in preparation, a first approximation of reference ecoregions and subregions is compiled.
 収集情報に基づき，また，準備中の刊行物の概説に基づき，改良された生態地域および小地域の初の概要が作成される。

- **～ provide information on** ..... ：～は … に関する情報を提供する

17. An advantage of examining the trophic status of component populations is that it also **provides information on** functional aspects of the community as well.

構成要素たる個体群の栄養状況を検討することの利点は，その生物群集の機能面に関する情報も提供することである。
18. These data **provide additional information on** previously identified Solid Waste Management Units.
これらのデータは，前に確認された廃棄物処理施設についての追加情報を提供する。
19. The field investigation has been conducted to **provide adequate information** to draw a conclusion.
現地調査が，適切な情報を提供し，結論を引き出すために実施されてきた。
20. Core metrics **provide useful information** in discriminating among sites exhibiting either good or poor quality ecological conditions.
コアメトリックは，良質または悪質な生態学的状態のいずれかを示す地点を区別するうえで，有益な情報をもたらす。

- **information contained in** ..... : …に含まれている情報

21. These descriptions are based on **information contained in** the EPA reports (EPA, 2002).
これらの記述は，EPA報告書((EPA,2002)に含まれている情報に基づいている。

★ Informationの量に関連する表現例文を下に示します。

- **a large mass of information** : 大量の情報
- **a considerable amount of information** : かなりの量の情報
- **a wealth of information** : 豊かな情報
- **extensive information** : 広範の情報
- **insufficient information** : 不十分な情報

22. The difficulty of dealing with **a large mass of information** is that of the many interactions occurring within the community, some may be related to water quality while others may not.
大量の情報を取り扱う場合の困難は，生物群集において多くの相互作用が生じることであり，それらは，水質に関係したり，関係しなかったりする。
23. **A considerable amount of information** is available on the effect of inhibitors on culture of *Nitrosomonas* sp.
培養Nitrosomonas spに対する抑制剤の影響に関するかなりの量の情報が利用可能である。
24. **Extensive information on** solid waste management units located at LAN Laboratory has been compiled by EPA.
LAN研究所にある廃棄物管理ユニットについての広範の情報がEPAによってまとめられてきた。

25. At this point there is **insufficient information on** source area ST5 to draw a conclusion on probable risk.
この時点では，じゅう分に可能性のあるリスクについて結論を引き出すには，汚染源ST5についての情報が不十分である。
26. **The amount and quality of available information** are not adequate to complete the studies.
利用可能な情報の量と質は，これらの研究を完成させるのに十分ではない。

Inherent：固有の；本質的；内在する

1. The **inherent** advantages of photocatalysis include (a) .....; (b) .... ; and (c) .....
光触媒反応の固有の利点としては，(a)…，(b)…，そして(c)…がある。
2. The better performance of the DHEI compared to the RBPHQ is not attributed to some **inherent** superiority in the QHEI.
QHEIがRBPHQよりも好成績な原因は，QHEIの本質的優越性ではない。
3. The variability **inherent** to each of the three biological indices used by USEPA has been shown to be quite low and within acceptable limits at relatively undisturbed sites.
USEPAが用いている3つの生物学的指数それぞれに内在する可変性は，比較的無攪乱な地点ではきわめて低く，許容限度内であることが示されている。
4. We believe the latter approach may introduce some unintentional bias into the biological criteria calibration and derivation process because of the **inherent** tendency to select the best sites instead of a more representative, balanced cross section of sites that reflect both typical and exceptional communities.
筆者は，後者の手法が生物クライテリアの較正・導出過程に無意識な偏りを持ち込みかねないと考える。なぜなら，一般的生物群集と例外的生物群集の両方を反映した代表的かつ平衡的な地点見本の代わりに，最小地点を選ぶ傾向が内在するからである。

Insight：洞察

- insight into ～ can be obtained：～についての洞察が得られる
- Insight into ～ is provided by .....：～が…によって洞察される

1. Additional **insight into** the fate of DDE **can be obtained** by considering the magnitudes of the parameters affecting chemical fate.
化学物質の衰退に影響を与えるパラメーターの規模を考慮することによって，DDE

の衰退についての更なる洞察を得ることができる。
2. **Insight into** the factors controlling the chemistry of arsenic in the tailings **is provided by** the results of the laboratory experiments.
選鉱廃物中のヒ素の化学的性質をコントロールしている要因が，室内実験の結果によって洞察される。

- insight is gained on ..... ：…についての洞察が得られる

3. **Insight is gained on** what can be expected as the predominant impact(s) in a particular segment change over time and/or space as a result of decreasing or increasing pollution level.
汚濁水準の低下または上昇の結果として，時間的および／または空間的に特定区間における優勢な影響として予想される事態についての洞察が得られる。

- ～ **give insight into** ..... ：～は…に洞察を加えている
- ～ **provide insight into** ..... ：～は…に洞察を加えている
- ～ **offer insight into** ..... ：～は…に洞察を加えている
- ～ **provide little insight into** ..... ：
  ～は…にほとんど洞察を加えていない

4. Examining the variance structure **can give insight into** the extent over which particular water quality criteria might be applicable.
変動構造を検討すれば，ある水質クライテリアが適用可能な度合に関する洞察が得られよう。

5. However, as with most general solutions, these equations **provide little insight into** the solution.
しかしながら，たいていの一般的解決と同じように，これらの式はその解決にほとんど洞察を加えていない。

6. This often **provides insight about** the factors(s) responsible for degradation and offers a diagnostic capability.
これは，しばしば劣化の要因に関する洞察をもたらし，判別能力をもたらす。

7. They **provide direct insight into** the relative importance of the various mechanisms.
それらは，種々のメカニズムの相対的重要性について直接に洞察を加えている。

8. The project **offers good insight into** the anaerobic treatment of wastewater.
プロジェクトは，廃水の嫌気性処理に良い洞察を加えている。

- ～ **gain insight into** ..... ：～は…に洞察が得られる
- ～ **acquire insight into** ..... ：～は…に洞察が得られる

- **acquire insights about .....** ：…についての洞察を得る.

9. The methodology described herein makes it possible **to gain insight into** water quality changes to be expected over the life of a groundwater recharge project.
   ここに記述された方法論は，地下水涵養プロジェクト期間中に生じるであろうと思われる水質変化の洞察を可能にする。
10. The method used here may be a new approach not only for gaining better results, but also for **acquiring** new knowledge and **insights about** toxic behavior.
    ここで使われた方法は，もっと良い結果を得ることだけではなく，毒物の挙動について新しい知識と洞察を得するのに，新しいアプローチであるかもしれない。

Inspection：点検；調査

- **Inspection of** ～ **shows that .....**：～の調査が～を表している
- **Inspection of** ～ **indicates that .....**：～の調査が～を示している

1. **Inspection of** Figure 2 **shows that** .....
   図 2 の検証は … を示す
2. **Inspection of** Equation 8 **indicates that** the suspended sediment distribution is dependent on four transport coefficients.
   式 8 の検証は浮遊堆積物の分布が 4 つの輸送係数に依存していることを示している。
3. **Inspection of** the figures **indicates that** no significant mass of Kepone is discharged to Chesapeake Bay after production terminated.
   これらの図によれば，Chesapeake 湾に排出されている Kepone の量は，生産が終止された後は重大ではないことを示唆している。

Instance：事例；場合

- **in most instances**：大半の事例では
- **in these instances**：これら場合には
- **in many instances**：多くの場合

1. **In most instances**, however, this will not be true.
   しかしながら，大半の事例では，これは事実にならないであろう。
2. Secondary and ancillary types were simultaneously assigned **in these instances**.

これら場合には，同時に二次的・付随的な影響種類が割り当てられた。
3. **In many instances**, mathematical models are essential to generalize the results of laboratory growth studies.
多くの場合，数学モデルは研究室内におけるバクテリア成長の研究結果を一般化するために不可欠である。

Instead：代わりに

- **instead of .....** ：…の代わりに

1. **Instead of** simply counting the number of cells per unit area, one can determine cell biovolume and use it to account for the differences in sizes of cells that are enumerated.
単位面積当り細胞数を単に数える代わりに，細胞の生物体積を決定し，それを用いて列挙される細胞サイズの差を説明しうる。
2. We believe the latter approach may introduce some unintentional bias into the biological criteria calibration and derivation process because of the inherent tendency to select the best sites **instead of** a more representative, balanced cross section of sites that reflect both typical and exceptional communities.
筆者は，後者の手法が生物クライテリアの較正・導出過程に無意識な偏りを持ち込みかねないと考える。なぜなら，一般的生物群集と例外的生物群集の両方を反映した代表的かつ平衡的な地点見本の代わりに，最小地点を選ぶ傾向が内在するからである。

Insure：保証する

1. These criteria **will insure** consistency in water quality assessment, licensing and certification, and enforcement of water quality standards.
これらのクライテリアは，水質評価，認可・認定，ならびに水質基準執行における整合性を保証するであろう。

Integrate (Integrating)：
積分する；盛り込む；統合する（積分すること）

1. Eqn 2 **can be integrated** between limits to yield .....
式2を極限で積分すると，…を得る。

- **to integrate .....** ：…を統合するため

2. Attempts are being made **to integrate** all three types (physical, chemical, and biological) of sampling data.
3種類(物理的，化学的，生物的)のサンプリングデータを統合する試みがなされつつある。

- 〜 **is integrated into** ..... ：〜を … に盛り込む
- **integrate A into B** ：A を B に盛り込む

3. Considerations of statistical power should be **integrated into** the risk assessment process.
統計の利用性への配慮をリスクアセスメント(危険評価)過程に盛り込むべきである。
4. Few states **have fully integrated** biological criteria **into** their water quality standards or water resource management activities.
生物クライテリアを水質基準または水資源管理活動に完全に盛り込んだ州は，ごく少ない。

- **integrated** ..... ：統合的… ; 統合した…

5. This **integrated** assessment was designed to address the following questions: 1......, 2......, and 3 ......
この統合したアセスメントは次の質問を扱うよう計画された。1 …, 2 …, と3 … である。
6. The **integrated** approach offers the key to an understanding of the difficult problem.
総合的アプローチが，その複雑な問題を理解するための鍵をあたえてくれる。
7. Together, they form the foundation for a sound, **integrated** analysis of the biotic condition.
これらは，共に生物状態の確かな統合的分析の基礎となる。

- **integrating** ：積分すること

8. **Integrating** this equation for the interval 0 to t yields ......
この式を0から t まで間で積分すると，… となる。

Intend ：意図的 ; 意図する

- 〜 **is intended to** ..... ：〜は…するように意図される

1. This chapter **is not intended to** serve as a comprehensive synopsis of state activities.
本章は，州活動の総合的概要を述べることを意図していない。

2. Biological impairment criteria **are intended to** supplement existing chemical standards and toxicity testing requirements.
生物損傷クライテリアの意図は，既存の化学基準および毒性試験要件を補完することである。
3. Since the model **is intended to** compute long-term trends, average annual values are simulated and variability within the year is ignored.
このモデルは長期傾向を計算するように意図されるので，年平均の値がシミュレートされ，年内の可変性が無視される。
4. Since the study **was intended to** demonstrate a methodology, its goals were somewhat different than those of most engineering applications.
この研究は方法論を実証するように意図されたため，その目的はたいていのエンジニアリング適用の目的とはいくぶん異なっていた。

- **intended** ……：意図的…

5. It is important to recognize that the proposed model depends to a significant extent on its **intended** purpose.
提案されたモデルが，その意図的目的に大きく依存するということを認識することが重要である。

Intensive：集中的な

1. The contemporary **regulatory-intensive approach** relies heavily on a legal style of rule setting.
現代の規制重視手法は，主に法律スタイルの規則策定に依拠している。
2. The National Water Quality Assessment (NAWQA) program is in the first of three years of **intensive** data collection that include measures of fish, invertebrate, and algal communities.
National Water Quality Assessment (全米水質アセスメント；NAWQA) プログラムは，魚類，無脊椎動物，藻類という測度を含む3箇年の集中的データ収集の初年度である。
3. Volunteer-collected data are used for trend analysis, as screening tools prior to **intensive** investigation by water quality professionals, as a basis for making local zoning and land management decisions, and in state's 305b water resource quality reports.
ボランティアの収集データは，傾向分析に用いられたり，水質専門家によるさらに徹底的な調査に先立つ選別ツールとして，局地的区域 (localzoning) の設定および土地管理決断のための基礎として，また，諸州の305(b)水資源報告において用いられる。

- ~ **is intensive** ：〜は集中的である；集約的な

4. Laboratory analysis of species composition **is labor intensive**.
   生物種構成の実験室内での分析は，労働集約的である。
5. Upon further investigation it was learned that one site was downstream from an experimental agricultural conservation tillage demonstration plot where pesticide usage **was intensive**.
   さらなる調査研究で，一地点が殺虫剤使用が集中的な実験農業保全耕耘実証区画から下流にあることが判明した。

Interaction ：相互作用

1. Chemical **interactions of** the model are depicted in Figure 1.
   このモデルの化学的な相互作用が図1に描写されている。
2. The model incorporates the effects of mass transport, chemical **interactions**, hydrodynamic dispersion, and radioactive decay.
   このモデルには，物質輸送，化学的な相互作用，水力学的拡散，放射性崩壊の影響が盛り込まれている。
3. Kimoto et al (2004) have strengthened the concept through their **interactions** and testing.
   Kimotoら(2004)は，対話や試験を通じてこの概念を補強してくれた。

- **interaction of A and B** ：AとBの相互作用
- **interactions between A and B** ：AとBの間の相互作用

4. Presently, there is little understanding of the **interaction of** surfactants **and** hydrophobic compounds.
   現在，界面活性剤と疎水性化合物の相互作用がほとんど理解されていない。
5. It was not the goal of this research to investigate the **interactions between** surfactants **and** hydrophobic compounds.
   界面活性剤と疎水性化合物の間の相互作用を調査することは，この研究の目標ではなかった。

- **interaction with** ……：…との相互作用

6. Each phase is taking into account the **interaction with** the other.
   それぞれの段階が他との相互作用を考慮に入れている。
7. The planning process can be strengthened by **interaction with** other programs, allowing joint utilization of reference database.
   立案過程は，参照データベースの共同活用を可能にし，他のプログラムとの相互

作用で強化されうる。

- **interactions occurring within** ..... ：…において生じる相互作用
- **interactions among** ..... ：…の間での相互作用

8. The difficulty of dealing with a large mass of information is that of the many **interactions occurring within** the community, some may be related to water quality while others may not.
大量の情報を取り扱う場合の困難は，生物群集において多くの相互作用が生じることであり，それらは，水質に関係したり，関係しなかったりする。

9. Whether or not petroleum entering the marine environment will have substantial or minimal impact depends on **interactions among** complex variables that are only now beginning to be understood.
海の環境に流入している石油が本質的な影響，または最小影響を持つであろうかどうかは，複雑な変数の間での相互作用に依存する。

Interest (Interesting)：興味，関心（興味深い）

- **interest centers on** ..... ：関心が…に集中している

1. **Increased interest in** environmental issues in the mid-1960s, spurred by many factors including Rachel Carson's Silent Spring (1963), led to increased involvement of citizens with their government regarding environmental issues.
1960年代中期における環境問題への関心の強まりは，Rachel Carsonの著作『Silent Spring（沈黙の春）』(1963) など多くの要因に促されたものであり，その結果，環境問題に関して政府への市民参加が盛り上がった。

2. **Our interest centers on** the land-based portion of the N cycle.
我々の関心は窒素循環の地上ベースの部分に集中している。

- **interest in** ..... **has increased**：…への興味が高まってきた
- **interest has emerged**：関心が高まってきた

3. **Interest in** environmental monitoring in some developing countries **has increased** in recent years.
一部の開発途上諸国では，環境モニタリングへの興味が近年高まってきた。

4. **Interest has** also **emerged** recently in the possibility of using sonication to remediate groundwater.
地下水修復のために超音波処理を使う可能性に，関心が最近また高まってきた。

- ～ **of interest is** ..... ：関心のある～は…である

- **Of particular interest is .....**：特に関心があることは… である

5. The primary area **of interest is** the Natsui River.
   関心のある主要な地域は Natsui 川である。
6. Typically, the area **of interest is** an entire state, or a particular river basin or region of a state or multiple regions.
   通常，関心区域は，1州全体，もしくは1州の特定の河川流域，または地域や複数州地域である。
7. **Of particular interest is** the estimation of "recovery times" for the areas.
   特に関心があることは，その地域の「回復時間」の推定である。

- **there has been an increasing interest in .....**：
  … に対する関心が増してきた
- **there was great interest in .....**：… に関して多くの関心があった
- **there is considerable interest in .....**：… にかなりの関心がある
- **there has been much interest in .....**：
  (… 以来)… に関してかなりの関心があった

8. In recent years **there has been an increasing interest in** biological nitrification processes.
   近年，生物学的硝化プロセスに対する関心が増してきた。
9. **There was great interest in** evaluating the uptake and disposition of petroleum constituents.
   石油構成要素の入取と処理性質を評価することに，多くの関心があった。
10. **There is considerable interest in** the in-situ oxidative degradation of chlorinated ethenes.
    塩化エテンの現地での酸化分解にかなりの関心がある。
11. Recently **there has been much interest in** the effect of trace metals and trace metal chelation on the growth of phytoplankton in natural waters.
    最近，微量金属と微量金属キレーションが，自然水域において植物性プランクトン増殖に及ぼす影響に対してのかなりの関心があった。

- **〜 is of interest**：〜は興味深い
- **It is of interest to point out .....**：… を指摘することは重要である
- **It is of interest to note that ....**：… を指摘することは重要である

12. The processes **were of** particular **interest**.
    このプロセスには特に関心があった。
13. The reaction of gold with hemoglobin **is of interest**, for example, because studies have shown that gold accumulates in the red blood cells of proteins.

金とヘモグロビンの反応は興味深い，なぜなら，例えば，これまでの研究では，金がタンパク質の赤血球に蓄積することが示されてきたからである。

14. **It is of interest to point out** the fundamental difference between A and B.
AとBの間の基本的な相違を指摘することは重要である。

- **interesting**：興味深い

15. Of course, application of biocriteria in the antidegradation program will raise **interesting** questions.
当然ながら，劣化防止プログラムにおける生物クライテリアの適用は，興味深い問題を提起するだろう。

16. In practice, however, the use of biocriteria and biological assessments to establish new permit requirements raises **interesting** and challenging questions.
実際問題として，新規許可要件を定めるための生物クライテリアおよび生物アセスメントの利用は，非常に興味深い問題を提起する。

- **It is interesting to note that** ……：…に注目すれば，興味深い

17. **It is interesting to note that** the Washington EPA database also includes sites sampled over multiple years in areas of lesser quality and impacted by human activities.
ワシントン州EPAデータベースが質の劣る区域で複数年度にわたりサンプリングされ，人為的活動に影響された地点も含んでいることに注目すれば，興味深い。

Interpret（Interpreting）：解読する；解釈する（解読の；解釈の）

1. The longitudinal examination of biological sampling results is also performed in an attempt **to interpret** and describe the magnitude and severity of departure from the numerical biological criteria.
生物サンプリング結果の縦断的検討も，数量的生物クライテリアからの逸脱の規模・程度を解釈し記述するために行われる。

- **~ can be interpreted as** ……：~は…と解釈することができる

2. The results **can be interpreted as** synergism.
この結果は相乗作用と解釈することができる。

- **interpreting**：解読の；解釈の

3. The method for **interpreting** the sign will be illustrated in the discussion and example problems that follow.

このサインを解読する方法が考察と次に続く例題で例証されるであろう。
4. Other problems associated with chlorophyll a are high temporal variability and difficulty in **interpreting** trends.
クロロフィルaに関わる他の問題は，時間的可変性の強さと傾向解釈の難しさである。
5. Critical to the process of **interpreting** and integrating the source material is the care that must be taken to avoid defining regionalities of particular ecoregion components such as fish or macroinvertebrate characteristics, or patterns in a single, or a set of, chemical parameters.
資料の解釈・統合過程できわめて重要なのは，魚類または大型無脊椎動物の特性や，単一または一連の化学的パラメータといった特定の生態系構成要素で地域性を画定するのを避けるよう配慮すべきことである。

Interpretation：解読；解釈

1. Many thematic maps are the products of **interpretations** that include consideration of seasonal and year-to-year differences.
多くの主題地図は，季節差や年差への考慮といった解釈の産物である。
2. **Interpretation of** sampling data will require considerations of zoogeography, historical abundance and distribution, and historical ranges of variability.
サンプリングデータの解釈には，動物地理，史的な数度・分布，史的な可変値域を考慮する必要があろう。

- ～ **facilitate interpretation of** ……：～は…の解釈を容易にする
- ～ **enable an interpretation of** ……：～は…の解釈を可能にする
- ～ **improve the interpretation of** ……：～は…の解釈を改善する
- ～ **obscure interpretation of** ……：～は…の解釈を不明瞭にする
- ～ **do not distinguish the different interpretation** ……：
  ～は…異なった解釈を区別しない

3. This information will establish the pollutant status of these organisms and **facilitate interpretation of** the other data.
この情報によって，これらの生物の汚染状態が確証され，そして他のデータの解釈が容易にされる。
4. Existing multimetric approaches are robust in their ability to measure biological condition because they incorporate biological and ecological principles that **enable an interpretation of** exposure/response relations.
既存の多メトリック手法は，生物学的状態の測定能力において確かである。その

理由は，曝露／応答関係の解釈を可能にするような生物学的・生態学的原理を盛り込んでいること，である。
5. In conclusion, the method described in this paper can **improve the interpretation of** the results.
結論として，この論文に記述された方法によって，この結果の解釈を改善することができる。
6. The standard procedure outlined above was established so that the problems which often **obscure interpretation of** results are minimized.
結果の解釈をしばしば不明瞭にする問題点が最小になるように，上に概説された標準的な手順は確立された。

- **There are three interpretations for** ..... ：… には 3 つの解釈がある
- **～ can result in misinterpretations about** ..... ：
  ～は … に関する誤った解釈を招きうる
- **One interpretation of ～ is that** ...... ：
  ～に対する一つの解釈は … ということである

7. **There are two interpretations for** the sampling data.
これらのサンプリングデータには 2 つの解釈がある。
8. Insufficient knowledge about regional expectations **can result in misinterpretations about** the severity of impacts in streams.
地域期待値に関する知見不足は，水流中の影響度に関する誤った解釈を招きうる。

Interrelationship：相互関係

- **interrelationship among** ..... ：… の間の相互関係

1. Mathematical models provide a basis for quantifying **the inter-relationships among** the various toxic chemicals.
数学モデルは，種々の有毒化学物質間の相互関係を数量化する基礎を提供する。
2. Most agree with a general definition that ecoregions comprise regions of relative homogeneity with respect to ecological systems involving **interrelationship among** organisms and their environment.
衆目の一致した一般的定義では，生態地域とは，生物およびその環境の相互関係を伴う生態系に関して比較的同質な地域からなる。

Interval：間隔；区間

1. The 95％ **confidence interval** is included to display the uncertainty of the

sample estimates.
サンプル推定の不確実性を反映するため，95％信頼区間が含まれている。

- **at ..... intervals**：… 間隔で
- **at intervals of .....**：… 間隔で

2. Analysis was done **at** 30-min **intervals**.
分析は30分間隔でなされた。
3. Analysis was done **at intervals of** 30 min.
分析は30分間隔でなされた。

- **over ..... intervals**：… 間隔の

4. The soil core analyses correspond to values averaged **over** 10 cm **intervals** from the surface to 60 cm depth.
この土壌コア分析結果は，地表から深さ60cmまで10cm間隔の平均値である。
5. Reported concentrations are the result of composite sampling **over** 30 to 60 cm **intervals**.
報告された濃度は30から60 cm間隔の複合サンプリングの結果である。
6. For most part, ICI scores were consistent and different by no more than 6 points **over the sampling intervals** which spanned 7 to 8 years and included 3 to 6 sampling events.
たいていの場合，ICIスコアは，7～8年間で3回ないし6回のサンプリングが行われるサンプリング間隔で，かなり一貫しており，6点以上の差がなかった。

- **on regular intervals**：定期間隔で

7. Some states are instituting comprehensive biological monitoring networks based on a rotational basin approach, wherein waterbody assessments rotate among watersheds **on regular intervals**.
いくつかの州は，交代流域手法に基づき総合的な生物モニタリングネットワークの構築に取り組んでいる。この場合，水域アセスメントは，定期間隔で対象分水界が交代するのである。

Introduction：序論；導入

1. As stated **in the introduction**, ...
序論で述べられているように,.....
2. Novel features of the EXAMS model include the **introduction of** "canonical" environment.
EXAMSモデルの新奇な特徴は，「模範的な」環境の導入である。

3. There is merit to Sudo's concern that an arbitrary mixing of variables, without any thought given to unintentional **introductions of** bias, compounding, and variance, is to be avoided.
偏向，妥協および分散の意図せざる導入を考慮せずに変数の恣意的な混合が避けられるべきだ，というSudoの懸念にはメリットがある。

Investigate (Investigating) ：調査する（調査すること；研究すること）

1. In this study, **we investigated** the effects of buffers and influence of Na and Ca concentrations of the test solution.
この研究では，我々は緩衝液の効果と試験液のNaとCa濃度の影響を調査した。
2. In the second part of this study, the effects of buffers and influence of Na and Ca concentrations of the test solution **were investigated**.
この研究の2番目の部分で，緩衝液の効果と試験液のNaとCa濃度の影響が調査された。
3. Residence times for individual PCBs in the atmosphere **have been investigated** in some detail.
個々のPCBの大気中での滞留時間が若干詳細に調査されてきた。
4. That aspect of the problem **is being investigated** by the Center for Disease Control Office.
この問題点の様相が，病原抑制センター事務所によって調査されている。
5. The distribution of aromatic hydrocarbons and PCB in marine organisms **was** first systematically **investigated** by .....
海洋生物体内における芳香炭化水素とPCBの分布が…によって初めて体系だった調査がなされた。
6. The following experimental parameters **were investigated**: .....
次の実験パラメータが調査された。.....

● **to investigate** ..... ：…を調査すること；…を研究すること

7. Few comparative studies have been done **to investigate** the distribution of aromatic hydrocarbons and PCB in marine organisms.
海洋生物体内における芳香炭化水素とPCBの分布を調査する比較研究がほとんどされていない。
8. It was the goal of this research **to investigate** the interactions between surfactants and hydrophobic compounds.
界面活性剤と疎水性化合物の間の相互作用を調査することが，この研究の目標であった。

- **investigating**：… を調査すること；… を研究すること

9. Much attention has been directed at **investigating** the photocatalytic degradation of organic pollutants.
有機汚染物の光触媒分解の研究に多くの注意が向けられてきた。
10. The merits of combining two advanced oxidation processes, viz., sonolysis and photocatalysis, have been evaluated by **investigating** the degradation of PCBs using a high-frequency ultrasonic generator and a UV-lamp.
二つの促進酸化法，すなわち超音波反応と光触媒反応，を結合させるメリットは，高周波数超音波発生器とUVランプによるPCBの分解を研究することによって評価されてきた。

Investigation：調査

1. **The investigation** process consists of six basic components: .....
調査プロセスは6つの基本的な構成要素から成る。それらは.....
2. From the results obtained **in these investigations**, the following conclusions were reached.
これらの調査で得られた結果から，次の結論に達した。
3. Previous **investigation have examined** .....
以前の調査で.....が検査された。
4. Extensive laboratory **investigations** and data analysis **provided** the following conclusions.
大規模な実験室での調査とデータ解析によって次の結論が提示された。

- **investigation exhamines** .....：調査で … が検査される
- **investigation provides** .....：調査によって … が提示される
- **One objective of the investigation is to** .....：
調査の1つの目的は … することである
- **The major aim of this investigation is to** .....：
この調査の主要な目的は… することである

5. **One objective of the** initial site **investigation should be to** determine the mass of contaminants adsorbed to the soil.
最初の現地調査の1つの目的は，土壌に吸着した汚染物質量を決定すべきことである。
6. **The major aim of this investigation is to** test the hypothesis that copper toxicity to algae and copper content of algal cells are functionally related to free cupric ion activity.

この調査の主要な目的は，藻類への銅毒性そして藻類細胞内の銅蓄積量は機能上遊離銅イオン活量と関係があるという仮説を検証することである．

- **the investigation focused on** ..... :
 この調査は，… に焦点を合わせた
- **the investigation was focused on** ..... :
 この調査は，… に焦点を合わせた

7. **The investigation focused** on an 18-mile segment of the creek.
 この調査は，小川の18マイル区間に焦点を合わせた．
8. The field **investigations were focused on** the local study area.
 この野外調査は，地域の研究エリアに集中された．
9. In particular, **the investigation was focused on** evaluating the influence of physical hydrology on .....
 特に，この調査は … に対する物理水文的影響を評価することに集中した．

- **investigations are needed to** ..... :
 調査が…するために必要とされる
- **investigations are necessary to** ..... :
 調査が…するために必要である

10. Further **investigations are needed to** confirm this hypothesis.
 この仮説を確証するために，よりいっそうの調査が必要とされる．
11. More extensive and detailed **investigations are necessary to** gain a further understanding of these processes.
 いっそう大規模かつ詳細な調査が，これらのプロセスをさらに深く理解するために必要である．

- **investigations into ~ have revealed that** ..... :
 ～の調査によって … が明らかになった

12. Further **investigations into** the time course of metal uptake in activated sludge **have revealed that** this process occurs in two stages.
 活性汚泥中で時間とともに変わる金属摂取をさらに調査し，このプロセスが2つの段階で起こることを明らかにした．

- **under investigation** :調査対象の

13. A number of reference sites were sampled to ascertain as accurately as possible the background levels of all parameters **under investigation**.
 調査中のパラメータすべてのバックグラウンドレベルを可能な限り正確に把握するために，多くの参照地域でサンプルを採集した．

14. The site classification and metric calibration steps are important. Equally important is the designation of aquatic life uses for the waterbodies **under investigation**.
地点分類ステップおよびメトリック較正ステップは，重要である。同じく重要なのは，調査対象水域について水生生物利用を指定することである。

- **Upon further investigation**：さらなる調査研究で

15. **Upon further investigation** it was learned that one site was downstream from an experimental agricultural conservation tillage demonstration plot where pesticide usage was atypically intensive.
さらなる調査研究で，一地点が殺虫剤飼養の不定型に集中的な実験農業保全耕耘実証区画から下流にあることが判明した。

- **investigation has been conducted to .....**：
調査が…するために実施されてきた
- **~ carried out an investigation to .....**：
〜は…するための調査を実施した

16. The field **investigation has been conducted to** provide adequate information to draw a conclusion.
現地調査が，適切な情報を提供し，結論を引き出すために実施されてきた。

17. Okakura (2002) **carried out an investigation to** identify industrially significant nitrogen.
Okakura (2002) は工業的に重要な窒素を識別するための調査を実施した。

Involve (Involving)：関与する；関係する；伴う（関与する）

1. This systematic process **involves** discrete steps, which are described as follows: .....
この体系的過程は，非連関的なステップを伴うものであり，それについて以下に述べる。

2. TCE degradation was postulated to **involve** the following sequential reactions.
TCE 分解が，次の一連の反応を伴うと仮定された。

3. Much of this recent work **involves** non-aqueous solvents, but no systematic study of chemical effects of ultrasound on these has been made.
最近の研究の多くが非水溶剤を伴っている，しかしこれら（非水溶剤）に対する超音波の化学的影響が体系的に研究されていない。

- **involved in .....**：… に関与する

4. The major assumption **involved in** this estimation technique is that .....
   この概算手法に関与する主要な仮定は.....である。
5. Persons **involved in** possible field tests must make this determination on a case-by-case basis.
   可能な現地テストに関係している人々が，状況によってこの決定をしなくてはならない。
6. This report should serve to give the reader an idea of what is **involved in** testing a model.
   この報告書は，モデル試験に何が関与するのか，その考え方を読者に提示するのに役立つはずである。
7. The Snake River project illustrates many of the decisions, procedures, and results **involved in** using HSPF.
   このスネーク川プロジェクトは，HSPFの使用に関与する決定，手順，結果の多くを例証している。
8. The information derived from the identification and enumeration of the algal component of the periphyton provides a wealth of ecological information for the cost **involved in** producing it.
   付着生物の藻類要素の識別・列挙から導かれた情報は，その作成に関与した費用に見合った豊富な生態情報をもたらす。
9. Motivating other citizens to **become involved in** water quality issues increases the influence the program will have on governmental policy and resource management decisions.
   他の市民が水質問題に関与するよう動機づけることは，そのプログラムが政府政策および資源管理決定に及ぼす影響力を強めることになる。

- **involving .....**：… に関与する

10. Hydroxyl radical is the major reactive species **involving** the oxidation of PCE.
    ヒドロキシル遊離基は，PCEの酸化反応に関与する主要な高反応性化学種である。

    ★ Involveの使い方としては，"involved in …"または"involving …"としてよく使われますが，"involving in …"とはなりませんので注意して下さい。

Irrelevant：無関係

- **〜 is not irrelevant to .....**：〜 は … とは無関係ではない

1. The amount of water used **is not irrelevant to** global warming.
   水の使用量は，地球温暖化とは無関係ではない。
2. The recent death of horses **is not irrelevant to** the presence of selenium in the grass.
   最近の馬の死は，牧草にセレンが存在していることとは無関係ではない。

Issue：問題；課題

1. **A critical issue is** to determine the regional extent over which a particular biological attribute is applicable.
   重大問題は，ある特定の生物学的属性が適用可能な地域範囲を決定することである。
2. **The most critical** cleanup **issue** facing DOE is the need to adopt a more practical policy on the program's objectives.
   DOE が直面している最も重要な浄化課題は，プログラムの目標のために，いっそう実務的な政策を採用する必要性である。
3. **Another issue** the author brings up is the intended closure of contaminated facilities.
   この著者が持ち出すもう1つの問題は，汚染された施設の意図的な閉鎖である。
4. **An issue of** increasing importance in the design and operation of wastewater treatment facilities is the control of fugitive odor emissions.
   下水処理施設の設計と操作において重要になってきている問題は，さまよう臭気の排気制御である。

　　● **address issues**：問題を扱う
　　● **This issue is addressed**：この問題に触れる

5. Industry has not been forced to **address issues** related to numeric criteria on a national scale.
   産業界は，全国規模では数量的クライテリアに関わる問題への取組みを強いられていない。
6. Further research into sediment transport is needed to develop more effective tools to better **address these issues**.
   もっと良くこれらの問題を扱うためのより効果的なツールを開発するには，堆積物輸送に対するいっそうの研究が必要である。
7. **This issue can be addressed** only briefly.
   この問題にはごく手短に触れる。

　　● **deal with the issues**：この問題を扱う

8. While we have **dealt with** most of these **issues** in South Dakota, these and

other issues will arise elsewhere, thus regionally consistency in achieving a resolution of these issues will be needed.
筆者は，サウスダコタ州におけるこれらの問題の大半を扱ってきたが，これらおよび他の問題は，他地域でも生じることで，これらの問題の解決における地域的整合性が必要だろう。

- **raised the issue of .....**：…の問題を提起した

9. It **raises** a number of important **issues**.
   それは多くの重要な問題を提起する。
10. Osada et al. (1992) **raised the issue of** the potential confounding effect of the proximity of a site to larger waterbodies, particularly for smaller streams.
    Osadaら(1992)は，特に小水流について，ある地点と大水域の近接による潜在的混乱影響の問題を提起した。

- **issues concerning .....**：…に関する問題
- **issues related to.....**：…に関わる問題

11. Additional **issues concerning** level-of-detail are critical to every step of the simulation process.
    細部レベルに関する追加問題はシミュレーションプロセスのすべてのステップに重要である。
12. The industry has not been forced to address **issues related to** BPA in groundwater.
    業界は，地下水に含まれているBPA(bisphenol A)に関わる問題への取組みを強いられていない。

- **～ is less of an issue**：～は，さほど問題にならない

13. In extensively disturbed regions and uniquely undisturbed regions, the method of reference site selection **will** likely **be less of an issue** because of relatively homogeneous conditions.
    はなはだしく攪乱された地域や，珍しく攪乱されなかった地域では，比較的同質な状態の故に，参照地点選定方法がさほど問題にならないだろう。

- **concentrate on the issues**：問題点に的をしぼる

14. It will **concentrate on the following issues**: a) ....., b) ....., and c) .....
    次の事に的をしぼるであろう，すなわちa)…，b)…，それにc)…である。

J

Join：合流する；合同する；協力する

1. The Bell Fourche River joins the Cheyenne River approximately 130 miles farther downstream.
   Bell Fourche 川はおよそ 130 マイルさらに下流で Cheyenne 川に合流する。
2. The Yakima River runs along part of the southern boundary and joins the Columbia River south of the city of Richland.
   Yakima 川は (Hanford 地域の) 南の境界線の一部に沿って流れて、そして Richland 市の南で Columbia 川と合流する。
3. During 1988 to 1990, TPWD joined TNRCC to sample an additional 66 streams.
   1988 年から 1990 年にかけて、TPWD は、TNRCC と共同で、さらに 66 水流でサンプリングを行った。

Just：ちょうど；正確に

1. Chemical applications of ultrasound are just beginning to emerge.
   超音波の化学的アプリケーションがちょうど出始めてきている。
2. Just because a model has been found valid for one use does not mean it is appropriate for some other use.
   ただモデルが一度の使用で有効であることを見いだされたからといって，それが他の使用に適切であるとはいえない。
3. The former can be used just as much as the latter as a means of experimentation.
   前者は後者とまったく同じぐらいよく実験手段として使用できる。
4. Artificial substrates can be made of a variety of materials and can be floated just below the surface or anchored to the bottom of the stream.
   人造底質は，様々な素材で製作し，水面のすぐ下に浮かべたり，水流底に留めたりできる。
5. A total of 27 individual samples were compiled for sites just upstream of Conesville Station; 51 samples were taken within the entire WAP ecoregion.
   Conesville 発電所のすぐ上流の 4 地点について合計 27 のサンプルが集められ，WAP

生態地域全体では 51 サンプルが採取された。
6. Well D is located just to the north of the power plant.
   井戸 D は発電所の北に位置している。
7. Source Area SS37 is located approximately 100 m east of Building 43, just east of F Avenue between Q Road and C Street.
   汚染源エリア SS37 は, Q 道と C 通りの間の F 大通りのすぐ東にある 43 号棟の東, およそ 100 m に位置している。

Justify (Justification)：ちょうど；正確に

1. The definition of degradation responses a priori is justified if clear patterns emerge from specific metrics.
   具体的メトリックから明確なパターンが現れたならば, 劣化応答という定義は先験的に正当化される。
2. Much of the concern about method is with the potential for misuse and abuse by attempts to justify increasing loadings of pollutants.
   この方法に関する懸念の大半は, 汚濁物質の付加増大を正当化する試みによって誤用・濫用される可能性をめぐるものである。

● **Justification**：正当化

3. From a regulatory perspective, these data provide good technical justification for a modified biological expectation.
   規制上の視点から, これらのデータは, 生物学的期待値修正のための適切な技術的正当化事由となる。
4. These played a major role in assigning and evaluating use designations, water quality management plans, and advanced treatment justifications.
   これらは, 使用指定の割当・評価, 水質管理プラン, および高度処理の正当化において重要な役割を演じた。

Keep：保つ；続ける

1. Attempts to calibrate the model and **keep** parameters in a reasonable range based on literature data proved unsuccessful.
   モデルを較正して，パラメータを文献データに基づく妥当な範囲にとどめる試みは，不成功であることが判明した。

   - **it is necessary to keep in mind .....** ：
     … を心に留めておくことが必要である
   - **it is important to keep in mind .....** ：
     … を念頭に置くことが重要である

2. Whatever the reasons for the differences, **it is necessary to keep in mind that** .....
   この相違の理由が何であるとしても，… を念頭に置くことが必要である。
3. In developing a TMDL **it is important to keep in mind** certain constraints on the WLA portion.
   TMDLを策定において，WLA部分における特定の制約を念頭に置くことが重要である。

   - **keep ..... separate**：… を隔離する

4. If separate habitats are sampled, it is important **to keep** them **separate** for subsequent analysis.
   隔離生息場所でサンプリングするならば，その後の分析でも隔離することが肝要である。

Key：鍵；キー；重要な；鍵となる

1. **A key aspect of** this approach is its iterative nature.
   このアプローチの重要な側面は，それの持つ反復性である。
2. **Key information** about the environmental setting, characteristics of the receiving waterbody, and insights into the chemical/physical dynamics of the discharge(s) are either directly or indirectly reflected by the biota.
   環境背景に関する主要情報，受入水域の特性，そして流出の化学的／物理的動態

に関する洞察は，生物相によって直接的または間接的に反映される。

- ● ～ is a key …… : ～は重大な … である

3. The selection of reference sites from which attainable biological performance can be defined **is a key component** in deriving numerical biological criteria.
達成可能な生物学的実績が定義されうるような参照地点の選定は，数量的生物クライテリアを導出する際の重大な構成要素である。

4. Certain species or groups of species **are key indicators** of polluted condition, such as the 20 species of diatoms identified by Lange–Bertalot (1979) as the most polluted–tolerance taxa worldwide.
いくつかの生物種および生物種集団は，汚濁状態のキー指標である。世界で最も汚濁耐性が強いとLange–Bertalot (1979) により識別された20種の珪藻がその例となる。

- ● **key to A is B** : Aへの鍵はBである
- ● ～ **is the key to solving** …… : ～は … の解決の鍵である
- ● ～ **offer the key to** …… : ～は … へ鍵を提供する

5. The **key to** the study **is** the structure of the molecule.
この研究への鍵は，この分子の構造である。

6. International collaboration **is the key to solving** all of these problems.
国際協力がこれらのすべての問題の解決の鍵である。

7. The integrated approach **offers the key to** an understanding of the difficult problem.
総合的アプローチが，その複雑な問題を理解するための鍵を提供する。

- ● **key to future success** : 将来の成功を握る鍵

8. Arguments over the merits of narrative vs. numeric biological criteria will likely remain, and the **key to future success** will continue to be the dedication and ingenuity of state biologists.
記述生物クライテリアと数値生物クライテリアのメリットをめぐる議論は，今後も続く見込みが強い。また，将来の成功を握る鍵は，州所属生物学者の献身や才気であろう。

Kinetics (kinetic) : 動力学速度；速度論（速度論的；動力学的）

1. In accordance with principles of **bacterial growth kinetics**, it was assumed that ……
バクテリアの増殖の速度論の原則にしたがって，… と仮定された。

- **the kinetics of .....** ： … の速度

2. The **kinetics of** biological dechlorination is not yet thoroughly understood.
   生物学的脱塩化反応の速度論はまだ完全には理解されていない。
3. Information about the **kinetics of** bioaccumulation is relatively scarce.
   生物体内蓄積の速度論についての情報は，比較的欠乏しい。

- **kinetic .....** ： 速度論的 …

4. **The kinetic expression for** growth is modified in order to account for environmental effects.
   成長の速度論的表現は，環境の効果を説明するために修正される。
5. A greater understanding of **kinetic behavior** should lead to better predictions of the behavior of PCP during anaerobic bioremediation.
   動力学的挙動をより良く理解することは，嫌気性生物による修復中のPCPの挙動をもっと良く予測することにつながる。

Knowledge ： 知識

- **knowledge of .....** ： … の知識
- **detailed knowledge of .....** ： … の詳細な知識

1. **Knowledge of** the rate constants governing the uptake and clearance of chemicals in fish is required.
   魚体内で，化学物質の入収・排出を左右する速度定数の知識が必要とされる。
2. **Knowledge of** the factors affecting the movement and transformations of these substances are of obvious importance in understanding and controlling the hazard.
   これらの物質の移動・変化に影響を与えている要因に対する知識は，危険を理解し，またコントロールするうえで明らかに重要である。
3. Important to waste minimization efforts is **knowledge of** the quantity and characteristics of waste stream.
   廃棄物減少化の努力に重要なのは，廃棄物フロー過程での量と特性の知識である。
4. The study provided **detailed knowledge of** the site history.
   この研究から，この地域の歴史に関する詳細な知識が得られた。

- **a lack of extensive knowledge of** ： 広い … の知識の欠如
- **insufficient knowledge about .....** ： … に関する知見不足

5. Due to **a lack of extensive knowledge of** pesticide toxicity, ...

殺虫剤毒性に関する広い知識が欠如しているために …
6. **Insufficient knowledge about** regional expectations can result in misinterpretations about the severity of impacts in streams.
地域予測に関する知見不足は，水流中の影響度に関する誤った解釈を招きうる。

- **without this knowledge**：この知識なしでは

7. **Without this knowledge**, a meaningful analysis is very difficult.
この知識なしでは，有意義な分析は非常に困難である。

- **for acquiring new knowledge**：新しい知識を得るために

8. The method used here may be a new approach not only for gaining better results, but also **for acquiring new knowledge** and insights about toxic behavior.
ここで使われた方法は，より良い結果を得ることだけではなく，毒物の挙動について新しい知識と洞察を得るための新しいアプローチであるかもしれない。

- **existing knowledge is limited**：既存の知識は限定されている

9. **Existing knowledge is** too **limited** to provide a sound basis for planning to prevent accelerated eutrophication and its adverse effects.
既存の知識は，加速する富栄養化とその悪影響の防止計画のための効果ある基礎を呈するにはあまりにも限定されている。

Known（Unknown）：知られている（知られていない）

- **all known** …… ：すべての周知の …
- **some known** …… ：知られている … の一部
- **three known** …… ：3つの周知の …

1. **All known** underground storage tanks previously or currently used for the storage of wastes have been included.
以前に，あるいは現在，廃棄物の貯蔵のために使われたすべての周知の地下貯蔵タンクが含まれてきた。
2. **Some known** features of the behavior of iodine are not included in the model, because they are not relevant to the problem of interest.
ヨードの挙動に関してすでに知られている特徴の一部はモデルに含められていない。なぜなら，それらは，ここで取り上げている問題点とは直接に関係が無いからである。
3. There are **two known** limitations of the synthetic leachate as currently

formulated.
現在調合されている合成浸出液には，2つの周知の限界がある。

- ～ **is well known for** ..... ：～は … について十分に知られている
- **The best known** ～ **is** ..... ：最もよく知られている～は … である

4. Spatial variability in population abundance **is well known for** periphyton, and is caused in part by substrate, rate of flow, and light intensity.
個体群数度の空間的変動性は，付着生物について十分に知られており，その一因は，底質，流速，光度である。

5. The use of macroinvertebrates is well established in state programs and the advantages **are well known**.
大型無脊椎動物の利用は，州プログラムにおいて十分に確立されており，その利点がよく知られている。

6. **The best known** explanation among those explanations proposed by ～ **is** .....
～によって提示された説明の中で最もよく知られているのは … である。

- **It has long been known that** ..... ：… はかなり以前から知られている
- **it has been known for decades that** ..... ：
 … は数十年前から知られている

7. **It has long been known that** certain groups of macroinvertebrates are more pollution-tolerant than others.
大型無脊椎動物の一部集団が他集団よりも汚濁耐性が強いことは，かなり以前から知られている。

8. **It has been known for decades that** certain groups of macroinvertebrates are, in general, more tolerant of pollution than are others.
大型無脊椎動物の一部集団が一般的に他集団に比べ汚濁耐性が強いことは，数十年前から知られている。

- **little is known concerning** ..... ：
 … については，ほとんど知られていない
- **little is known about** ..... ：
 … については，ほとんどわかっていない
- **little as yet known about** ..... ：
 … について，ほとんどまだ知られていない

9. **Little is known concerning** the fate of chlorinated hydrocarbon in estuarine ecosystems.
河口生態系での塩化炭化水素の衰退過程については，ほとんど知られていない。

10. **Little is known about** the chemical nature of organic matter in groundwater.
    地下水での有機物質の化学的性質については，ほとんど知られていない。
11. Currently, **little is known about** the chemical transformation of .....
    現在，…の化学的変換については，ほとんど知られていない。
12. **Little as yet known about** the mechanism of carbon dioxide fixation by growing cells.
    成長している細胞によって二酸化炭素が固定されるメカニズムについては，ほとんどまだ知られていない。

    ● **not much is known** ..... ：…のことはあまり多くは知られていない
    ● **～ is poorly known** ：～はあまり知られていない

13. **Not much is known** concerning the nature and behavior of natural systems.
    自然系の性質と行動に関しては，あまり多くは知られていない。
14. The speciation and distribution of chromium in river waters **are poorly known**.
    河川水におけるクロムの化学種分化と分布は，あまり知られていない。

    ● **what is known of** ..... ：…に関する知見

15. The condition of the stream can be inferred from the taxa present and **what is known of** their requirements and tolerance.
    水流状態は，常在分類群や，その要件・耐性に関する知見から推断されうる。

    ● **known to be** ..... ：…が知られている

16. Current monitoring efforts focus on only a few of the hundreds of chemicals **known to be** present in drinking water supplies.
    現在のモニタリングの努力が，飲料給水への存在が知られている数百という化学物質だけに集中している。

    ● **with a known degree of certainty** ：既知の確実度で

17. Quantifiable estimates of land use can be made **with a known degree of certainty**.
    土地利用の計量可能な推定は，既知の確実度でもって得られる。

    ● **Unknown is** ..... ：…は，わかっていない

18. **Unknown** is whether there is free product on the water table.
    地下水面上にフリープロダクト（浮上ガソリン）があるかどうかは，わかっていない。

L

Lack：欠如している；欠如

1. The improvement is even more remarkable when conditions since the turn of the century, when the river **lacked** any fish life for a distance of nearly 40 mile, are considered.
   この川で約40mileにわたり魚類が生息していなかった今世紀初頭移行の状態を考慮すれば，この改善は，一層明らかである。

   - **~ is lacking**：~が欠けている
   - **What seems to be lacking is .....**：
     欠けていると思われることは … である

2. Quantitative data on byproducts analysis **is lacking**.
   副産物分析の定量的データが欠けている。

3. **What seems to be lacking is** a consideration of the ecological background rather than water chemistry background.
   欠けていると思われることは，水化学の背景より生態学の背景を考慮することである。

   - **the lack of data**：データの欠如
   - **a lack of knowledge on .....**：… に関わる知識の欠如
   - **the lack of reproducibility**：再現性の欠如
   - **lack of understanding**：理解不足

4. Because of **the lack of data**, the methodology was not fully verified.
   データの欠如のために，この方法論は十分に立証されなかった。

5. Because of **a lack of extensive knowledge on** pesticide toxicity, the methodology was not fully verified.
   殺虫剤毒性に関する広い知識が欠如していたために，この方法論は十分に立証されなかった。

6. One common source of error arises from the difficulty of obtaining a representative soil sample and **the lack of reproducibility** of organics analysis of soils.
   エラーを生じさせる一般的原因1つは，代表的な土壌サンプル採集の困難さ，そ

れに土壌の有機分析の再現性の欠如である。
7. These omissions were not due to **lack of understanding**.
   これらの手落ちは，理解不足によるものではなかった。

Later：後に

1. As will be discussed **later**, ...
   後に，論じられるように，…
2. These are addressed **later**.
   これらは後に説明される。
3. A more detailed explanation of Saito's approach is given **later** in this chapter.
   Saitoの手法のより詳細な説明を本章の後半で述べる。

   - **at later time**：後に；後の時間に
   - **at the later date**：後に；後の日に
   - **Later in 1995 and into 1998**：1995年後半から1998年にかけて
   - **for the later period**：後期の

4. The predicted CO_2 concentration is higher than the observed concentration **at later time**.
   予測されたCO_2濃度は，後に観察された濃度より高い。
5. The axiom follows "When in doubt choose to take more measurements than seem necessary at the time since information not collected is impossible to retrieve **at the later date**."
   公理としては，「疑わしい場合，その時点で必要と思われるよりも多くの測定を行うことを選びなさい。なぜなら，収集されなかった情報は，後に取り戻すことができなからである」。
6. **Later in 1992 and into 1993** discussions were initiated in order to determine the feasibility of and possible way to restore WWH use attainment.
   1992年後半から1993年にかけて，WWH使用達成復活の実行可能性とその方法をめぐる討論が開始された。
7. The comparison of the two different time period showed that the increased index scores **for the later period** were highly significant ($p<0.0001$).
   これら2つの時期の比較は，後期の指数スコア増大がきわめて有意なことを示した ($p < 0.0001$)。

   - **Later work by** .....：後の…による研究

- **later sampling**：後のサンプリング

8. **Later work** by Gammon et al. (1990) suggests that nonpoint sources are now impeding further biological improvements observed in larger rivers that resulted from reduced point source impacts.
後の Gammon ら (1990) の研究によれば，点源影響の減少によって大河川で観察された一層の生物学的な向上を面源が妨げている。

9. As a follow-up note, this sewage treatment plant subsequently underwent a major upgrade, and **later** macroinvertebrate **sampling** revealed an improved community downstream.
追記すれば，この下水処理場は，その後大幅な改善がなされ，後の大型無脊椎動物サンプリングで下流における生物群集の向上が見られた。

Latter (See also Former)：後者 (➡ Former)

1. We believe the **latter** approach may introduce some unintentional bias into the model calibration and derivation process.
我々は，後者の手法がモデルの較正・導出過程に無意識な偏りを持ち込みかねないと考える。

- **In the latter case**：後者の場合

2. **In the latter case**, the condition of the immediate riparian zone was correlated with the degree of impairment.
後者の場合，隣接水辺地帯の状態は，損傷度に相関していた。

Law (See also Court)：法律 (➡ Court)

1. Section 303 of the 1972 act recordified and expanded the water quality standards provisions of **the 1965 law**.
1972年法の303条は，1965年法の水質基準規定を再び成文化して拡大した。

2. As a basic principle of statutory construction, general statements of statutory goals and objectives have no legal force and effect, absent specific operative provisions **in the law**.
法的解釈の基本原理として，法律上の目的および目標の全般的言明は，法的効力を持たず，法律における具体的な運用既定を欠いている。

- **under the law**：法律のもとで

3. Stream channelized **under** the auspices of the Iowa Drainage **Law** are

subject to routine maintenance activities, which include herbicide application, tree removal, sand bar removal, and the snagging and clearing of accumulated woody debris.
アイオワ州排水法のもとで流路制御された水流は，定期整備活動の対象となる。これに含まれるのは，除草剤散布，高木除去，砂洲除去，木質堆積物の除去・清掃などである。

★法律に関連する表現を下に示します。

4. State differently, criteria developed under Section 304(a)(1) could be viewed as criteria to address the effects of pollutants, while criteria under Section (a)(2) address the broader effects of pollution.
言い換えれば，304条(a)(1)のもとで設定されたクライテリアは，汚濁物質の効果を扱うものとみなされ，304条(a)(2)のもとでのクライテリアは，汚濁の影響を扱うものとみなされる。

5. Activities requiring a permit under Section 404 of the CWA must be certified as meeting provisions of the water quality standards by the state water quality agency.
CWA第404条のもとで許可を要する諸活動は，州水質担当機関によって水質基準の規定を満たしていると認定されなければならない。

6. Section 303, which addresses the establishment of water quality standards, includes requirements for public involvement in public hearings on proposed changes.
303条は，水質基準の設定を扱っており，変更案に関する公聴会への一般市民参加を要求している。

7. The objective of the CWA, stated in Section 101(a), is "to restore and maintain the chemical, physical, and biological integrity of the Nation's waters."
101条(a)で述べられているCWAの目標は，「米国の水域の化学的・物理的・生物学的な生物健全さを回復・維持すること」である。

8. In Section 304(a)(8), Congress directed USEPA to "develop and publish information on methods for establishing and measuring water quality criteria for toxic pollutants on other basis than pollutant-by-pollutant criteria, including biological monitoring and assessment methods."
304条(a)(8)において，連邦議会は，USEPAが「生物モニタリング方法および生物アセスメント方法など，汚濁物質別クライテリア以外の，毒性汚濁物質についての水質クライテリアを設定・測定するための方法に関する情報を設定し発表する」よう指示した。

Lead (Mislead) ：導く；もたらす（誤り導く；誤りをもたらす）

● ～ lead to ….. ：～は…をもたらす；～は…を導く

1. The saprobien system and the other similar stream classification systems **lead to** the investigation of the "pollution status".
腐敗システムや他の類似の水流分類システムは，「汚濁状態」調査をもたらした。
2. Nebraska hopes that this **will lead to** the establishment of numeric biological criteria in the future.
ネブラスカ州は，これが将来，数値生物クライテリアの確定に通じることを期待している。
3. Its simplicity and numeric relationship to the original four zones of stream pollution **lead to** the development of a widely used biotic index in the United States.
それは，単純で，汚濁水流の当初4つのゾーンと数値関係を持つので，米国で広く用いられる生物指標の確立をもたらした。
4. Forbes' insight and application of the principle of natural selection **led to** the establishment of a biological station on the shores of the Illinois River in 1894.
1894年にIllinois川の岸に生物監視測定所が設置されたが，Forbesの洞察と自然淘汰原理の応用が，その導因となった。

● ～ lead us to ….. ：～は，我々を…へ導く
● ～ leads itself to ….. ：～は，それ自身を…へ導く

5. The results **lead us to** the conclusion that the arsenic was derived from bacteria action solubilizing arsenic from the arsenopyrite in the waste deposits.
この結果は我々を，このヒ素は廃棄物中の硫ヒ鉄鉱を溶かしだすバクテリアの働きによるという結論へ導く。
6. This approach **leads itself to** a better understanding of the nature of the impairment, including which elements or processes of the community are most affected.
この手法は，生物集団のどの要素や過程が最も影響されるかなど，損傷の性質をより深く理解するのに役立つ。

● ～ lead to better ….. ：～は，もっと良く…することにつながる

7. A greater understanding of kinetic behavior should **lead to better** predictions of the behavior of PCP during anaerobic bioremediation.

動力学的挙動をより良く理解することは，嫌身的生物学的浄化における PCP の挙動をもっと良く予測することにつながる。

8. This approach **leads** itself **to a better** understanding of the nature of the impairment, including which elements or processes of the community are most affected.
この手法は，生物集団のどの要素や過程が最も影響されるかなど，損傷の性質をより深く理解するのに役立つ。

9. Reportedly, part of the NBS mission is the coordination of federal monitoring activities, which should **lead to better** and more efficient use of biological data and assessments.
伝聞によれば，NBS の任務の一部は連邦モニタリング活動の調整であり，それは生物データおよび生物アセスメントのより優良かつ効率的な利用に通じる。

- **mislead**：誤りに導く；誤りをもたらす

10. The assumption of neglecting substrate inhibition **is** not particularly **misleading** at low substrate concentrations.
基質阻害を無視することについての仮定は，低濃度の基質においては特に誤りではない。

Length（Lengthy）：長さ（長々しい）；距離（→ Long）

1. **The length of** the Columbia River from the Canadian border to the Pacific Ocean is approximately 745 river miles.
カナダの国境から太平洋までのコロンビア川の長さは，およそ 745 river miles（河川マイル）である。

2. **The length of** the Walnut Creek reach from Building A to Pond B is approximately 500 m.
Walnut Creek 区間，建物 A から池 B までの距離は，およそ 500 m である。

3. The lake has a total area of 25,300 km^2, total volume of 470 km^3, **length of** 386 km, and mean width of 17 km.
この湖は，全面積が 25,300 km^2，全容積が 470 km^3，湖長が 386 km，平均湖幅が 17 km ある。

4. An initial decision faced in identifying stream populations is whether to describe the condition of streams in terms of stream segments or **the total length of** streams.
水流母集団の確認で直面する初期決定は，水流の状態を，水流区間において述べるか，水流全長において述べるかである。

- **..... in length**：長さは …

5. At this level a line of some twenty to thirty elongated depressions, each measuring about 100 m **in length**, extends from north to south.
   20 から 30 くらいの細長いくぼみからなる線が北から南にのびている。窪みの長さはそれぞれ 100 m くらいである。
6. The Hocking River, located in Southeastern Ohio, is a medium-sized river (1197mile^2 drainage area) of about 100 mi **in length**.
   Hocking 川は，オハイオ州南東部に所在し，長さ約 100 mile の中規模河川である（流域面積 1197mile^2）。

- **lengthy**：長々しい

7. This challenge resulted in a search for numerical expressions in a form simpler to understand than long species lists and well-thought but **lengthy** technical expressions of the data.
   この課題により，生物種の長大なリストや，綿密だが長々しいデータの技術的説明よりも理解しやすい形での数値表現が追求されるようになった。
8. First and foremost, we thank Dr. Robert Ozawa, USEPA for the **lengthy** discussions and his insights on the biocriteria process and appropriate bioassessment approaches.
   まず第一に，我々が感謝を捧げる相手は，USEPA の Dr. Robert Ozawa である。彼は，長時間にわたる論議や，生物クライテリア過程ならびに適切な生物アセスメント手法に関する洞察を提供してくれた。

Less (See also **Greater** and **Larger**)：
以下；あまり … ない（➡ **Greater** と **Larger**）

- **less than**：以下

1. The concentrations of iron and manganese in the water supply wells **are less than** 1 mg/L.
   給水用井戸の水にふくまれている鉄とマンガンの濃度は 1 mg/L 以下である。
2. In all samples, the group A elements occur at concentrations that **are less than** 10 mg/kg.
   すべてのサンプルで，グループ A の元素は濃度が 10 mg/kg 以下 である。
3. At concentration **less than** 2 mg/L, the rate is concentration-dependent.
   濃度が 2 mg/L 以下では，その反応速度は濃度に依存する。
4. These values are both two orders of magnitude **less than** Hazard Level of 1 for groundwater and soil.

これらの値は，地下水と土壌の両方での危険レベル1より100倍低い。
5. Field measurements are somewhat **less** accurate **than** measurements made in a laboratory but they offer the important advantage of providing immediate results to the volunteers.
現地測定は，実験室測定ほど正確でないが，ボランティアに直ちに結果を提示するという重要な利点を持つ。
6. Generally, **less** variability is expected among surface waters within the same region **than** between different regions.
一般に，異なる地域の地表水域よりも，同じ地域の地表水域の方が変動性が低いと予想される。

- **Less is known .....**：… は，あまり知られていない

7. **Less is known** regarding the inhibitory effects of nickel on nitrifying organisms.
硝化菌に対するニッケルの阻害効果は，あまり知られていない。

- **Less attention has been paid to .....**：
… には，あまり注意が払われてこなかった

8. **Less attention has been paid to** the level of other amino acids.
他のアミノ酸のレベルには，あまり注意が払われてこなかった。

Level：レベル；水準；段階；程度

1. One of the major concerns with regional reference sites is their **acceptable level** of disturbance.
地域参照地点に関わる重要事の一つは，擾乱の許容可能レベルである。
2. It has not yet been determined which **level** would be most appropriate for the water quality standard.
どのレベルが水質基準に最も適しているかは，まだ未決定である。
3. **The** benzene concentration was 0.24 mg/L, two orders of magnitude below the 10−6 **risk level**.
ベンゼンの濃度は0.24 mg/Lで，10−6リスクレベルの1/100の規模であった。
4. The biocriteria are consistent with **a good level of** community performance.
生物クライテリアは，高水準の群集能力と一貫している。
5. EPA (2003) studied the **level of agreement** between bioassay and instream biosurvey results using data collected near 43 separate facilities.
EPA (2003) は，43施設の付近での収集データを用いて，生物検定と水流内生物調査の結果の一致レベルを研究した。

6. A number of reference sites were sampled to ascertain as accurately as possible the **background levels of** all parameters under investigation.
 調査中のパラメータすべてのバックグラウンドレベルを可能な限り正確に確認するために，多くの参照地域でサンプルが採集された。
7. Potential uses of biological criteria in the total maximum daily loads(TMDL) process are also being explored at both the state and federal **levels**.
 日最大負荷量(TMDL)過程における生物クライテリアの潜在的利用も，州レベルおよび連邦レベルで追求されている。

★ Level が concentration（濃度）といった意味で使われる場合の例文を下に示します。

8. The excess nutrients are the primary cause of the reduced **oxygen levels** observed in the lake.
 この湖で観察される酸素レベルの減少は，過剰栄養が主要な原因である。
9. In the most aquatic systems these processes are capable of reducing **free copper levels** to very low values.
 ほとんどの水生システムにおいて，これらのプロセスは遊離銅レベルを非常に低い値に下げることができる。
10. There are several questions that arise from these observations of **arsenic levels** in surface waters: 1 ....., 2 .....,
 表面水のヒ素レベルの観察からいくつかの質問が生ずる。1 …，2 …，
11. The PCB issue dates back to 1973, when state fish and game officials detected unusually **high levels of** PCBs in the river's striped bass population.
 PCB 問題は 1973 年にさかのぼる。その当時，州の fish and game 当局者が川にいるストライプトバスの個体群に異常に高いレベルの PCB を検出していた。
12. This concern has arisen lately as analytical techniques for detecting water contaminants have improved to the point where it is possible to detect contaminants **at the part per-trillion level**.
 水汚染物質を検出するための分析技術が汚染物質を ppt レベルで検出可能なまでに向上したため，最近この関心が高まった。

Lie：置かれている；の状態にある

1. The potential aspects **lie** outside the scope of this paper.
 可能は側面については，この論文の範囲外である。
2. The answer to those questions **lies** in the fact that the stereospecificity of enzymes is not exact.
 それらの質問への答えは，酵素の立体特異性が厳密でないという事実に見いださ

れる。

Light : 光；解明する

- **in light of** ..... :
  … に照らし合わせて；… を考慮に入れて；… の観点から見て

1. **In light of** these facts, ...
   これらの事実を考慮に入れて
2. An evaluation of the methodology **in light of** the methodology application is presented in section 6.
   方法論の評価は，この方法論のアプリケーションを考慮に入れて第6節に掲げられている。
3. While this may seem enigmatic **in light of** current strategies to regionalize wastewater flows, the presence of water with a seemingly marginal chemical quality can successfully mitigate what otherwise would be a total community loss.
   これは，汚水流量を地域分けするという現行戦略の観点からは不可解に思われるかもしれないが，化学的に不十分な水質の水の存在が生物群集の全損をうまく緩和しうるのである。

- **〜 shed light on** ..... : … に光を当てる；… が解明される

4. Let me try to **shed some light on** a few points.
   いくつかの点に光を当ててみよう。
5. The new work **could shed** more **light on** these issues.
   この新しい研究は，これらの問題により多くの光を当てるかもしれない。
6. It is possible that more frequent observations at the boundary zones of these phases **would shed further light on** this problem.
   これらの相の境界域におけるいっそう頻繁な観察によって，この問題が解明される可能性がある。

- **〜 throw light on** ..... : … に光を投げ込む

7. It is intended to **throw light on** the physiological aspects of bacterial growth.
   バクテリア成長の生理学上の様相に光を投げ込むことを意図としている。

Like (Unlike)：と同様（と異なり）

1. However, **like** any other valued fish species, it does have specific habitat and water quality requirements.
   しかしながら，他の貴重な魚種と同様，具体的な生息地および水質の要件を有する。
2. **Unlike** the smaller rotary kilns, the waste is burned with the fuel right in the flame.
   より小型の回転灯とは異なり，廃棄物は燃料と共に炎の中で燃やされる。
3. **Unlike** other methods, bioassessments can be used to document instream improvements that result from wastewater facility upgrades and the implementation of best management practices.
   他の方法と異なり，生物アセスメントは，廃水施設向上および最善管理慣行の実施に起因した流入改善の文書化に利用できる。
4. **Unlike** the USEPA Environmental Monitoring and Assessment Program (EMAP) sampling design, the Idaho DEQ database was not collected under a statistically random design for the location of sampling sites.
   USEPAの環境監視評価プログラム（EMAP）サンプリング設計と違い，アイダホ州DEQのデータベースは，サンプリング地点の位置について統計的に無作為な計画下で収集されなかった。
5. **Unlike** chemical parameters, multimetric biological indices integrate chemical, biological, and physical impacts to aquatic systems and portray both condition and status in terms of designated use attainment/nonattainment in direct terms.
   化学的パラメータと異なり，多メトリック生物指数は，化学的・生物的・物理的影響を水生系に統合して指定使用達成／未達成における状態・状況を直接表す。

Likelihood：可能性

1. The **likelihood of** a false positive error at ST57 is relatively low since there is visual evidence of contamination.
   現場ST57において汚染が無いのに有ると間違える可能性は比較的低い。なぜなら，汚染の証拠が目に見えるからである。
2. The **likelihood of** a false negative decision is higher, due to uncertainty regarding the vertical extent of soil contamination and the potential for groundwater contamination.
   縦方向の土壌汚染と地下水汚染の可能性が不確実であるために，誤った否定の決

定になる可能性がより高い。

Likely(Unlikely)：ありそうな；多分(ありそうもない)

1. The heavy metals are another **likely** source of inhibition.
   この重金属は，多分(成長を)抑制するもう1つの原因である。
2. It **likely** represents a worst case under low flow conditions.
   それは多分，小流量状態下で最も悪いケースを表す。
3. Since external impacts are minimal, measured variability **would be most likely** due to the sampling inconsistencies.
   外部影響が最小であるから，測定された変動性は，サンプリングの不整合性に起因する見込みが高い。
4. Many facilities **will likely** have considerable data for one assessment but relatively little for another.
   多くの施設で，1つのアセスメントに関するデータベースは，豊かだが，他のアセスメントに関するデータベースは，貧弱である。
5. Not only with this result in a loss of valuable information, it **will likely** result in the continued degradation of the aquatic resource.
   これは，貴重な情報の損失を招くばかりか，水資源の継続的な劣化を招く見込みが強い。

   - **It seems likely that** ……… ：…のことはありそうに思われる
   - **It appears likely that** ……… ：…のことはありそうに見える
   - **It is likely that** ……… ：おそらく…であろう
   - **It is unlikely that** ……… ：…ということはありそうもない
   - **It is quite likely that** …… ：…ということは十分考えられる
   - **It is most likely to** …… ：…の見込みがきわめて強い

6. **It seems likely that** such a consistent pattern of contamination would result from in situ conditions rather than migration.
   このような一貫した汚染パターンは，移動条件というより現場の条件に起因するように思われる。
7. **It appears likely that** the cause is simply increased dispersion.
   この原因は，単に増強した散乱であるように見える。
8. **It is unlikely that** extreme care was taken to prevent deposition of DDT into the Black River during the 27 years of plant operation.
   27年の工場創業期間中，Black川へのDDTの堆積を防ぐために出来るだけの注意が払われという事は，ありそうもない。
9. We specifically discuss benthic macroinvertebrate monitoring since **it is**

most likely to be usable in biological criteria issue.
筆者は特に大型無脊椎動物モニタリングについて論じる，なぜなら，それは生物クライテリア問題に役立つ見込みがきわめて強いからである。

● ~ is likely to ..... : ～は … である可能性がある

10. Photolysis **is** not **likely to** be important.
    光分解は重要である可能性がない。
11. Approaches for assessing the response of stress proteins in organisms when exposed to chemicals are currently under development, but they must be cost−effective before they **are likely to** be widely used in biological assessment and criteria program.
    化学物質に曝された際の生物体内のストレス蛋白質の応答評価するための手法を目下案出中である。だが，費用効果的でなければ，これらが生物アセスメントおよび生物クライテリアプログラムにおいて多用される見込みは，低いと思われる。
12. The main goal of any risk assessment is to determine those conditions **likely to** produce harm.
    どんなリスク評価でもの主な目的は，害を作り出す可能性が高いそれらの条件を決定することである。

★ more, most とか highly など likely を強調する使い方を下に示します。

13. Long−term ecological effects due to shoreline oiling **were highly unlikely**.
    海岸線での油田事業による長期的な生態影響は，ほとんどありそうもない。
14. Thus streams draining comparable watersheds within the same region **are more likely to** have similar biological, chemical, and physical attributes than from those located in different regions.
    故に，同じ地域内の類比可能な流域を流れる水流は，異なる地域に所在する水流に比べ，類似の生物的・化学的・物理的属性を持つ見込みが高い。
15. The apparent reduction in the oxygen utilization **is most likely** the result of soil gas diffusion.
    酸素消費の明白な減少は，土壌ガス拡散の結果である可能性が最も高い。

● What is more likely is that ..... : より可能性が高いのは，… である

16. **What is more likely is that** biosurveys targeting multiple species and assemblages provide improved detection capability over a broader range as well as protection to a larger segment of the ecosystem than a single species or assemblage approach.
    複数の生物種および群がりをターゲットにした生物調査は，単一の生物種または群がり手法に比べて，より広範な探知能力向上をもたらす見込みが高いし，より

広範な生態系保護になるだろう。

Likewise：同じく；同様に

1. It is **likewise** unclear how well limestone reactors will neutralize more highly concentrated solutions.
   石灰岩反応器が，より濃厚な溶液をどれほどよく中和するかは同じく不明確である。
2. For macroinvertebrates the issue of identifying midges to the genus/species level (as opposed to the family level) proved **likewise** to be a farsighted decision given the value of this group in diagnosing impairments.
   大型無脊椎動物について，小虫を（科レベルでなく）属／種レベルで識別する問題は，同様に，損傷判定におけるこの集団の価値に鑑みて，先見的決定であることが判明した。
3. **Likewise**, if production of a site is considered high based on organism abundance and/or biomass, and high production is natural for the habitat type under study (as per reference conditions), biological condition would be considered good.
   同様に，生物の数度および／または生物体量に基づきある地点での生産が高いとみなされ，また調査対象生息地について（参照状態によって）高生産が自然であるならば，生物学的状態は良好だと考えられよう。

Limit：限界；制限する；限りがある

1. Although the effort centered on surface water quality, **limited** sampling was conducted on tailings deposits.
   努力を地表水の水質に集中させたけれども，選鉱廃堆積物を対象に行なわれたサンプリングには限りがあった。

    - only limited ..... ：ほんの限られた …
    - only a limited amount of ..... ：ほんの限られた量の …

2. **Only limited** funds are available to support research on the development of inexpensive analytical methods.
   ほんの限られた資金だけが，低価な分析方法の開発研究の援助のために利用できる。
3. In most case, **only a limited amount of** stream water quality data is available.
   ほとんどの場合，ほんの限られた量の水流水質データしか利用できない。

    ★Limitの動詞としての使用例文を下に示します。

4. Oxygen is not the only factor that can **limit** bioremediation.
   酸素が，生物処理を制限できる唯一の要因ではない。
5. Comparison with other technologies on a case-by-case basis **is limited**.
   「ケースバイケース」での他の技術との比較に限りがある。
6. Since many taxa are difficult to distinguish even with a microscope in a laboratory, the levels of taxonomic identification in the field **are limited**.
   実験室で顕微鏡を使っても多くの分類群が区別しづらいので，現地での分類識別は，限定的である。
7. Although sample surveys have been used for many years in a variety of arenas, their acceptance and use for characterizing natural resources **has been limited**.
   サンプル調査は，長年にわたり様々な場で用いられてきたが，天然資源の特性把握のためのその受容と利用は限られていた。

- **～ is limited to .....**：〜は … に限定される

8. Existing knowledge **is** too **limited to** provide a sound basis for planning to prevent accelerated eutrophication and its adverse effects.
   既存の知識は，加速する富栄養化とその悪影響の防止計画のための効果ある基礎を呈するにはあまりにも限定されている。
9. The impacts we observe in Illinois **are not limited to** Illinois or even the Midwest.
   イリノイ州で観察されたインパクトは，イリノイ州や米国中西部諸州のみに限られない。
10. This subsection **is not limited to** effects caused by the release of chemical pollutants, but rather may address effects from any cause of impairment.
    この項は，化学的汚濁物質排出による効果に限定されず，あらゆる損傷原因からの効果を扱う。

Limitations：限界；制限

1. There are two known **limitations** of the synthetic leachate as currently formulated.
   現在調合されている合成浸出物には2つの周知の限界がある。
2. The model program was modified in order to eliminate the following **limitations**: ....
   モデルプログラムは，次に示す限界をなくすために修正された。
3. As a result of these **limitations**, a suitable oxidizer is required.
   これらの制限の結果として，適した酸化剤が必要とされる。

4. This points out the **limitations** of assigning a single degradation rate coefficient for the entire soil profile.
  これは，土壌プロフィール全体に一つの分解速度係数を割り当てることには限界があることを指摘する。
5. Using standing crop alone for assessing and detecting changes in water quality, however, has some **limitations**.
  しかし，水質変化の評価・探知のために生物体量のみを用いることには，いくつか限界がある。

Line：線

1. In this figure, **the dashed line** represents how their concentrations vary with time.
  この図では，点線はそれらの濃度がどのように時間で変化するかを示している。
2. The fit between the model and data is certainly good；**the least square line** presented by Ueda et al.(1997) is of slightly steeper slope.
  モデルとデータは確かに一致している。Uedaら(1997)によって示された最少二乗法による線は，少し勾配が大きい。
3. If a relationship was observed, **a 95％ line of best fit** was determined and the area beneath trisected following the method recommended by Fukui et al.(1984).
  関係が見られたならば，ベストフィットの95％ラインが決定され，その下方の面積は，Fukuiら(1984)の推奨方法に従って三等分された。
4. This is particularly true if **baseline data** are available for the waterbody in question.
  これは，特に当該水域について基準データが利用可能な場合に当てはまる。
5. The overall effort should focus on building improved "**line of evidence**".
  努力全体は，改善された「証拠ライン」の構築に主眼を置くべきである。
6. When flying over the Southern Rocky Mountains along the Wyoming / Colorado **state line** in mid-1980, this author noticed differences in timber management practices between states.
  1980年代中期，ワイオミング州／コロラド州の境界に沿ったSouthern Rocky山脈の上空を飛んだ時に，著者は，両州での材木管理慣行の相違に気づいた。

Link（linkage）：つながり（結合）

- ～ **is linked to** …… ：～は … と結びついている

Link

1. It is well known that methemoglobinemia **is linked to** nitrate in drinking water.
   メトヘモグロビン血症が飲中の硝酸と関係があるということは，よく知られている。
2. It is believed that heart disease **is linked to** smoking.
   心臓病の原因は，喫煙と関係があると思われる。

   ● **The linkage of A and B**：AとBの結合
3. The **linkage of** dynamic chemical runoff **and** instream chemical fate and transport models is a recent accomplishment.
   ダイナミックな化学物質の流出モデルと河川内での化学物質の衰退・輸送モデルの結合は，最近の業績である。

List (Listing)：リスト；一覧表；リストする；一覧表にする（リスト）

1. The following is **a list of** the equipment used in this study.
   この研究に使用した機器を以下に，列挙する。
2. They **listed** these in a decreasing order of importance.
   かれらは，重要性が減少する順にこれらを一覧表にした。
3. The estimated rate constants **are listed** in Table 2.
   推定速度定数を表2にリストする。
4. The results of the analysis of the eight samples from the monitoring wells **are listed** in Table 3.
   モニタリング用の井戸からの8つのサンプルの分析の結果を表3にリストする。

   ● **listing**：リスト
5. A complete **listing of** the chemical results is provided by Ikeda (2004).
   化学的な結果の完全なリストが池田（2004）によって提供されている。

Literature：文献；調査報告書

1. **A literature summary** of longitudinal dispersion coefficients for streams and rivers is reported in Table 1.
   小川と河川における縦方向拡散係数の文献値が要約され，表1に報告されている。
2. **Much of the literature** concerned with waste treatment applications of nitrification has been devoted to municipal sewage.
   硝化処理の適用に関連する文献の多くが，都市下水に捧げられてきた。

3. Good agreement was obtained between experimentally determined stability constants and **available literature values**.
   実験を通して決められた安定性定数と利用可能な文献値との間で良い合意が得られた。

   - **review of the literature on ......**：… に関わる文献レビュー
   - **examination of the literature on ......**：… に関する文献の考察

4. **Reviewing the literature on** sonochemistry of organochlorine compounds as summarized above did not lead to any publications on the use of ultrasound for chemical monitoring.
   上に要約したように有機塩素化合物の超音波化学分解に関する文献レビューは，化学的モニタリングの為の超音波利用に関する出版につながらなかった。

5. **Examination of the literature** on sorption reveals a great deal of confusion regarding the time to reach the sorption equilibrium.
   吸着についての文献調査によれば，吸着均衡に達っするまでの時間に関してかなりの混同がある。

   - **in the literature**：文献に

6. It has been well established **in the research literature that** .....
   … ということは，研究文献でしっかり確立していた。

7. Table 4 reports some values found **in the literature**.
   表4には，文献にて見いだされた若干の値が報告されている。

8. The health hazards of environmental mercury are well documented **in the literature**.
   環境での水銀の健康障害は，文献に明確に記録されている。

9. Many studies found **in the literature** report the results of relatively short-duration runs.
   報告書に見いだされる多くの研究は，比較的短時間の実験結果を報告している。

10. Ecological information for many algae, particularly diatom, has been recorded **in the literature** for over a century.
    多くの藻類，特に珪藻についての生態学的情報は，1世紀にわたり文献に記録されてきた。

11. Recent advances in AOP treatment technology as well as their use in industrial applications are summarized **in the literature**.
    AOP処理技術における最近の進歩は，産業的なアプリケーションでのそれらの使用と同様にこの文献に要約されている。

12. The mechanisms constituting ABCD processes have been discussed extensively elsewhere **in the literature** (12–18).

Literature

ABCDプロセスを構成している仕組みは，文献(12–18)の至る所で広範囲に論じられてきた。

13. Horie (1976) claims that over 3000 species have ecological information **in the literature**, which can be built into an autecological database appropriate for the periphyton in a region.
Horie (1976) によれば，文献には3000種以上の生態学的情報が収められており，1地域における付着生物に適した個生態学的データベースを構築しうる。

- **some literature values** ……… : 若干の文献値
- **Based on the literature data, it is probable that** …… :
文献によれば，…のことはありそうだ。

14. Table 2 gives **some literature values** for the vertical dispersion coefficient.
表2に，縦断拡散係数の文献値を若干示す。
15. Good agreement was obtained between experimentally determined stability constants and **available literature values**.
実験によって決められた安定定数と入手可能な文献値の間にかなりの合意が得られた。
16. Attempts to calibrate the model and keep parameters in a reasonable range **based on literature data** proved unsuccessful.
モデルをキャリブレートして，パラメータを文献データに基づく妥当な範囲にとどめる試みは，不成功であることが判明した。

- **a large body of literature** ：多くの文献
- **a wealth of literature** ：豊富な文献
- **an immense literature** ：多くの文献
- **the vast majority of the literature** ：大多数の文献
- **a paucity of literature** ：不足している文献
- **much of the literature** ：文献の多く

17. There is **a large body of literature on** biodegradation of PCBs.
生物によるPCBの分解に関する文献が多く存在する。
18. There is **a wealth of** technical **literature**, but it is written for use by scientists and engineers familiar with the technical terms and background.
専門的な文献は豊富にあるが，それは専門用語・背景に精通した科学者および技術者によって使用されるように書かれている。
19. There is **an immense literature of** field experiments to show the effects of ……
…の影響を示す現場実験に関する文献が多く存在する。
20. **The vast majority of the literature** reporting toxicity studies of copper(II)

contains conclusions drawn from results based on an incomplete consideration of the chemistry of copper (II).
銅(II)の毒性研究を報告している大多数の文献では，銅(II)の化学的性質の不完全な考察に基づく結果によって結論ずけられている。

21. **A paucity of literature on** the efficiency of heavy metals removal in treatment processes now exists.
現在，処理過程での重金属の除去効率に関する文献が不足している。

22. **Much of the literature** concerned with waste treatment applications of nitrification has been devoted to municipal sewage.
硝化の廃液処理への適用に関係した文献の多くが都市下水に集中されてきた。

Little：ほとんど … ない

★ few（ほとんどない）と a few（少しある）は数えられる名詞につかわれるのに対して，little（少ししかない）と a little（少しある）は数えられない名詞に用いられます。

1. **Little** legal attention has been paid to this trend.
   この傾向には法律的な注意があまり払われていない。
2. It was concluded that ICI scores are consistent at locations where **little** man-induced **change** has occurred.
   人為的変化があまり生じていない場所では，ICI スコアは，一貫していると結論づけられた。
3. Many facilities will likely have considerable data for one assessment but relatively **little** for another.
   多くの施設で，1つのアセスメントに関するデータベースは，豊かだが，他のアセスメントに関するデータベースは，貧弱である。
4. However, as with most general solutions, these equations provide **little insight into** the solution.
   しかしながら，たいていの一般的解決と同じように，これらの式はその解決にほとんど洞察を加えていない。
5. **Little consideration** has been given to the effects of salt water on the nitrification process in an activated sludge treatment system.
   活性汚泥処理系において，塩水が硝化過程に及ぼす影響はほとんど考慮されてこなかった。
6. It should be noted that in major portions of the country, topographic watersheds either cannot be defined or their approximation has **little meaning**.

米国の大部分において，地形流域は，画定されえないか，その概算がほとんど無意味であることに留意すべきである。

7. Comparisons of toxicity data between tests conducted on-site to tests conducted off-site on samples have shown **little variation**.
サンプルについてオンサイトで行なわれたテストとオフサイトで行なわれたテストの間の毒性データの比較では，ほとんど変化を示さなかった。

8. Examination of Table 2 provides **little evidence** of any correlation between influent metal concentrations and percentage removal.
表2の考察から，流入水の金属濃度とパーセンテージ除去率との間の相互関係に関する証拠はほとんどなにも出てこなかった。

9. The problem is that **little effort** is being expended on studying ecosystems holistically and attempting to define differences in patterns of ecosystem mosaics.
問題点は，生態系の全体論的研究や，生態系モザイクのパターン差を定義づける試行にあまり力が入れられていないことである。

- **little or no**：ほとんど，または全く…しない

10. Metrics with **little or no** relationship to stressors are rejected.
ストレッサーとの関係がほとんど，あるいは全く無いメトリックは，却下される。

11. The smallmouth bass (Micropterus dolomieui) is one of the most important game species in midwestern rivers and streams. Furthermore, this is a species that requires **little or no** external support in the way of supplemental stocking.
コクチバス (Micropterus dolomieui) は，中西部の河川水流における最重要な漁獲対象種の1つである。さらに，補足的放流という外部支援をほとんどまたは全く必要としない魚種である。

12. Understanding the underlying concepts and theoretical assumptions of biocriteria may be a challenging to environmental managers that have **little or no** experience with biological assessments.
生物クライテリアの根本的概念と理論的仮定を理解することは，生物アセスメントの経験に乏しい，もしくは経験のない環境管理者にとって課題になろう。

- **there is little agreement on** …… (See **Agreement**)：
  …は，ほとんど一致していない (➡ **Agreement**)
- **there is little question that** …… (See **Question**)：
  …には，ほとんど疑問の余地がない (➡ **Question**)
- **there is little understanding of** …… (See **Understand**)：
  …が，ほとんど理解されていない (➡ **Understand**)

- **there is little doubt that** .....(See **Doubt**)：
  …には，疑問の余地がほとんどない(➡ **Doubt**)
- **Little is known**(See **Known**)：ほとんど知られていない(➡ **Known**)
- **little work has been done on**.....(See **Work**)：
  …については，ほとんど研究がされてこなかった(➡ **Work**)

Locate：位置する；所在する

1. The Iowa River Basin **is located in** central Iowa.
   アイオワ川の流域はアイオワ州の中央に位置している。
2. Well D **is located just to** the north of the power plant.
   井戸Dは発電所の北に位置している。
3. The Blair Lake facility **is located in** a remote area.
   ブレア・レーク施設は遠く離れたエリアに位置している。
4. Source Area LF04 **is located** approximately 3 miles east−northeast from the south end of the runway.
   汚染源エリアLF04は滑走路の南端からおよそ3マイル東北東に位置している。
5. Source Area SS37 **is located** approximately 300 feet east of Building 433, just east of F Avenue between Q Road and C Street.
   汚染源エリアSS37は，Q道とC通りの間のF大通りのすぐ東にある433号棟の東，およそ300フィートに位置している。
6. The long−term fixed station **is located on** the mainstream approximately 1 mi (1.6 km) downstream from the confluence of the East and West Branches and upstream of the Elyria wastewater treatment plant.
   East小流とWest小流の合流点から約1mile(1.6km)下流の本流と，Elyria廃水処理場の上流に長期的な固定観測場所がある。
7. Sites from a northern ecoregion grouped with sites **located within** the northern areas of neighboring coregions.
   州北部の生態地域の地点は，近隣生態地域の北部区域に所在する地点とともにグループ分けされた。
8. The Hocking River, **located in** Southeastern Ohio, is a medium−sized river (1197mile^2 drainage area) of about 100 mi in length.
   Hocking川は，オハイオ州南東部に所在し，長さ約100mileの中規模河川である(流域面積1197mile^2)。
9. The Harn Creek−Natsu River system **is located on** the north side of the Mizudori River in Ohfuku City.
   Ohfuku Cityでは，Harn小流−Natsu川システムはMizudori川の北側に位置している。

★オーフクシティー (Ohfuku City) は，シティーが都市名の一部であるため，City は大文字になります。カンサスシティー (Kansas City)，Iowa City (アイオワシティー) も同例です。市名には，冠詞 "the" が付きませんが，シティーが都市名の一部でない場合は，the city of Richland (Richland 市) のように the city of ～ となります。

★位置に関連する例文を下に示します。

10. Lake Erie, the southernmost of the Laurentian Great Lake, is between 42°45' and 42°50' north latitude and 78°55' and 83°30' west longitude.
    Lake Erie は，Laurentian Great Lake の南端にあり，北緯 42°45' と 42°50' の間，西経 78°55' と 83°30' の間に位置する。

Long：長さ

1. A five-foot **long** split-barrel core sampler was used to collect soil samples.
   長さ5フィートのスプリットバレル (胴体割れ) コアサンプラーが土壌サンプルを収集するめるために使われた。
2. Cell 4 contains a 4 m wide by 8 m **long** pool with a capacity of about 200,000 liters.
   セル4は幅4メートルに長さ8m，容量が約 200,000 リットルのプールを内蔵している。
3. The dimensions of each basin are 125 feet **long**, 67 feet wide, and 21 feet deep.
   各ため池の寸法は長さ125フィート，幅67フィートと深さ21フィートである。

   ● **long term**：長期にわたる

4. The model is intended to compute **long term** trends.
   このモデルでは，長期の傾向が計算されるように意図されている。
5. The model represents a tool to estimate the **long-term** effort of human activities on the water environment.
   このモデルは，水環境に対する人間活動の長期にわたる努力を推定するツールに相当する。
6. This work will aid in making predictions about **long-term** buildup of organic layers on the stone.
   この研究は，石への長期にわたる有機層の蓄積を予測するのを助けるであろう。

   ● **long periods**：長期にわたる
   ● **long range**：長期にわたる
   ● **long standing**：長期にわたる

7. Repeated tests over **long periods** are more significant.
   長い期間にわたっての繰り返されたテストは，いっそう重要である。
8. The possible **long-range** effects of the disposal of trace metals in the coastal environment are matter of growing concern.
   沿岸環境において微量金属の処分による長期にわたる潜在的影響は，これからますます懸念される事態である。
9. A **long-standing** heavy metals problem in this stream occurred below an industrial effluent discharge.
   この水流における長年の重金属問題は，産業廃液排出口の下流で発生した。

   ● **a long history of** ..... ：… の長い歴史

10. Arsenic has **a long history of** use for its toxic and medicinal properties.
    ヒ素は毒として，また薬として長い歴史を持つ。

    ● **as long as** ..... ：… する限り
    ● **so long as** ..... ：… する限り

11. **As long as** we are requested to make assessments of the integrity of surface water ecosystems, a framework to rank and partition natural variability is essential.
    地表水生態系の健全性のアセスメントを我々が求められるからには，自然変動を格づけして仕分けするための枠組みが不可欠である。
12. The index comprises homogeneous variables **as long as** characteristics from a common organizational level such as fish, birds, chemicals, etc. are used.
    この指数は，魚類，鳥類，化学物質といった共通の組織レベルからの特性が用いられる限り，均質な変数からなる。
13. **So long as** biocriteria are authorized by state law, they are not prohibited by the CWA.
    生物クライテリアは，州法で公認される限り，CWAによって禁止されない。
14. **So long as** the science is sound and well supported, the findings likely will be upheld in the courts.
    科学が堅実で十分に裏づけられる限り，その所見は，法廷で是認されるだろう。
15. Traditionally, federal courts defer to reasonable administrative agency interpretations of federal statutes, **so long as** those readings are not foreclosed by the language or legislative history of the statute.
    伝統的に連邦裁判所は，連邦法への行政機関による合理的な解釈が法律の文言または立法史によってあらかじめ排除されていない限り，それらに従っている。

- **It has long been known that .....**：… は，かなり以前から知られている

16. **It has long been known that** certain groups of macroinvertebrates are more pollution-tolerant than others.
    大型無脊椎動物の一部集団が他集団よりも汚濁耐性が強いことは，かなり以前から知られている。

- **no longer .....**：もはや … ではない

17. The widespread public concern about these problems draws attention to the fact that the quality of drinking water supplies can **no longer** be taken for granted.
    これらの問題に対する広い公共の関心が，供給飲料水の水質は，もはや安全とはいえないという事実に注意が注がれている。

Look：見る；調べる

1. We must **look** more carefully **into** .....
   …をもっと念入りに調べなければならない。
2. Lakes **are looked at** more from a dynamic and mechanistic point of view, rather than a descriptive one.
   湖は，記述的な見地よりどちらかと言うと，いっそう動勢・機械論的見地から見られる。

Magnitude：大きさ；規模；重要性

1. The **magnitude of** the difference between the positive and negative charges indicates that ....
   プラス電荷とマイナス電荷の相違の大きさは，… を示唆する。
2. The **magnitude** and significance **of** chemical contamination of aquatic environments are increasingly evident.
   水生環境の化学汚染の大きさと重要性は，ますます明白である。
3. Furthermore, from Fig. 2, it can be seen that the **magnitude of** the effect is erratic.
   さらに，図2から，その効果の規模が常軌を逸したといえる。

   ● **in the magnitude of** .....：… の程度の

4. This discrepancy between the samples collected two years apart suggests that significant variations **in the magnitude of** soil contamination can occur either over very short horizontal distances or over a relatively short time period.
   2年間隔てて収集されたサンプルの間でのこの矛盾は，土壌汚染の程度の有意な相違が非常に短い水平線距離で，あるいは比較的短期間にわたって起こりえることを示唆する。

   ● ..... **order of magnitude**：… 倍；… 桁

5. Arsenic levels are more than **an order of magnitude** lower in the low flow sampling period.
   ヒ素レベルは，低流量状態でのサンプリング時においては一桁以上低い。
6. Concentrations of naphthalene and TPH increased by over **an order of magnitude** between 1988 and 1989 in well D-3.
   ナフタレンとTPHの濃度が，井戸D-3において，1988と1989の間に10倍以上増加した。
7. The same organism may have **an order of magnitude** more chlorophyll under low-light and nutrient-rich conditions than under high-light and nutrient-poor conditions.

Maintain

同一生物において，多光量・低養分状態よりも少光量・高養分状態の方が十倍ほど多くのクロロフィルを有する場合がある。

8. Benzene exceeded the 0.005 mg/L MCL by over **two orders of magnitude**.
   ベンゼンが0.005 mg/L MCLを100倍以上超えていた。
9. The benzene concentration was 0.24 mg/L, **two orders of magnitude** below the 10^{-6} risk level.
   ベンゼンの濃度は0.24 mg/Lで，10^{-6}リスクレベルの1/100の規模であった。

   ★ MCLは Maximum Comtaminant Level（最大汚染濃度）の略です。

10. These values are both **two orders of magnitude** less than Hazard Level of 1 for groundwater and soil.
    これらの値は，地下水と土壌の両方での危険レベル1より100倍低い。
11. Concentration of DDT in the Tennessee River sediments is approximately **two orders of magnitude** below those in Indian Creek.
    Tennessee川堆積物中のDDTの濃度は，Indian Creekにおけるそれらの濃度のおよそ1/100以下である。
12. Vertical profiles also exhibited **order of magnitude** differences between the upper and lower layer average Lindane concentrations.
    縦断プロフィールにおいて，リンデン濃度の上層の平均と下層の平均で桁違いの相違を示した。

    ● ～ **is order of magnitude greater than** ……：
       ～は…と比べて桁外れに大きい

13. The desorption and adsorption coefficients **are orders of magnitude greater than** the settling coefficient.
    脱着係数および吸着係数は，沈降係数と比べて桁外れに大きい。

Maintain (Maintaining)：維持する（維持すること）

1. Lobsters **were maintained** in an aquarium containing artificial seawater at 10℃.
   ロブスターは人口海水を入れた水槽の中で10℃で飼育された。
2. The problem of how to **maintain** optimal nitrification rates is not only a question of the presence of aerobic conditions.
   最適な硝化反応速度をどのように維持するかという問題は，単に好気性状態であるかないかだけではない。
3. A principal objective of the CWA (Clean Water Act) is to restore and

maintain the biological integrity of the nation's surface waters.
CWA（水質汚濁防止法）の主要目標は，米国の地表水域の生物学的健全度を回復・維持することである。

4. An underlying assumption of this approach is that the patch work of performance standards, when implemented as a whole, will be sufficient to restore and **maintain** the physical, chemical, and biological integrity of the water.
この手法の根本的な前提は，実施基準の集成が全体として実施された場合，その水の物理的・化学的・生物的な完全度を十分に回復・維持するだろうというものである。

- **maintaining**：維持すること

5. Advocates seek to persuade others of the importance of **maintaining** water resource quality.
環境擁護者（advocacy）たちは，水資源の質の維持の重要性を他者に納得させようと努めている。

6. Work over the last decade by investigators, such as Kubota (1993), have emphasized the importance of riparian areas for **maintaining** the quality and function of the hyporheic zones of streams.
Kubota (1993) などの調査者による過去10年間の研究は，河床間隙層の質・機能の維持にとっての水辺区域の重要性を強調した。

Make：…を〜にする；…に〜をさせる；作る；行う

1. Several new approaches **make** periphyton analysis a better indicator than it was 5 or 10 years ago.
いくつかの新手法のおかげで，付着生物分析は，5〜10年前よりも優れた指標となっている。

2. Compared to the Scioto River, the Ottawa River exhibits evidence that, despite some improvements, **makes** it one of the most severely impaired rivers in the state.
Scioto 川に比べ，Ottawa 川は，若干の改善にもかかわらず，同州で最もはなはだしく影響された河川の一つという証拠を示している。

- **make meaningful extrapolations from .....**：
  …からの推定を有意義にする

3. We must develop an understanding of ecosystem regionalities at all scales, in order to **make meaningful extrapolations from** site-specific data

collected from case studies.
我々は，ケーススタディで収集された地点特定データからの推定を有意義にするために，すべての縮尺での生態系地域性への理解を生じさせなければならない。

- ~ **make progress**(See **Progress**)：
  ~は進歩を成し遂げる(➡ **Progress**)
- ~ **make use of** .....(See **Use**)：~は … を活用する(➡ **Use**)
- ~ **make it clear that** .....(See **Clear**)：
  ~は … ということを明らかにする(➡ **Clear**)
- **What makes it clear is that** ..... (See **Clear**)：
  明らかなことは….ということである(➡ **Clear**)
- **make comparison between A and B**(See **Comparison**)：
  … と … を比較する(➡ **Comparison**)
- **in making comparison between**..... **and** .....(See **Comparison**)：
  … と … を比較するにあったては(➡ **Comparison**)
- ~ **make decisions on**(**about**).....(See **Decision**)：
  ~は … について決定をする(➡ **Decision**)
- **in making decisions on**(**about**).....(See **Decision**)：
  … について決定を下す際(➡ **Decision**)
- ~ **make the determination**(See **Determine**)：
  ~が決定する(➡ **Determine**)
- ~ **makes it possible to** .....(See **Possible**)：
  ~は… を可能にする(➡ **Possible**)

★ Make は受動態で多様に使用されます。下に掲げた例文を参考にして下さい。

- ~ **is made**：~がなされる；~が行われる

4. Water quality **measurements were** not **made** as part of the survey.
   水質測定が，調査の一部として行われなかった。
5. Much **progress and advancement has been made** in the past decade.
   進歩と前進の多くは，過去10年間における成果である。
6. During the course of testing, several **modifications were made** to the criteria.
   検定中にクライテリアに対しいくつかの修正が行われた。
7. Some minor **revisions were made** in 1985, but the approach remained essentially the same through 1987.
   1985年にいくつか細かな改正が施されたが，この手法は，基本的に1987年まで存続した。
8. Over the past few years, several **studies have been made on** groundwater

contamination by VOCs.
過去数年にわたって，VOCs による地下水の汚染についての研究がいくつかなされてきた。

9. In evaluating the performance of a model relative to some field measurement or another model, qualitative rather than quantitative **statements are** often **made**.
あるフィールド測定，またはもう1つのモデルとの比較からモデルの性能を評価する場合，定量より定性質的な報告がしばしばなされる。

10. This **distinction is made** necessary by the widespread degree to which macrohabitats have been altered among the headwater and wadable streams in the HELP ecoregion.
この弁別が必要になった理由は，HELP 生態地域における源流・徒渉可能河川で，マクロ生息地が大幅に改変されたことである。

11. This is similar to **observations that we have made** in the Scioto River downstream from Columbus, Ohio, and elsewhere.
この観察結果は，オハイオ州 Columbus 市の Scioto 川下流および他の場所における筆者の観察結果と似通っている。

12. With the exception of a few **experiments made** with KI, all experiments suggest that the reactor generates hydroxyl radicals.
KI を使用した少数の実験を除いて，すべての実験はこの反応器がヒドロキシル遊離基を発生させることを示唆する。

13. Field measurements are somewhat less accurate than **measurements made** in a laboratory but they offer the important advantage of providing immediate results to the volunteers.
現地測定は，実験室測定ほど正確でないが，ボランティアに直ちに結果を提示するという重要な利点を持つ。

- **～ is made of ……**：～は…で製作されている

14. Artificial substrates **can be made of** a variety of materials and can be floated just below the surface or anchored to the bottom of the stream.
人造底質は，様々な素材で製作し，水面のすぐ下に浮かべたり，水流底に留めたりできる。

- **argument can be made against** ……（See **Argument**）：
議論が … に対してされうる（➡ **Argument**）
- **attempts were made to** …..（See **Attempt**）：
… のために十分な試みがなされた（➡ **Attempt**）
- **choices is made**（See **Choices**）：選択が行われる（➡ **Choices**）

- **decision was made to** ..... (See **Decision**)：
  …することが決定された（➡ **Decision**）
- **determination is made** (See **Determination**)：
  判断が下される（➡ **Determination**）
- **examination was made of** ..... (See **Examination**) ..... ：
  について検討がなされた（➡ **Examination**）
- **improvement in** ..... **was made** (See **Improvement**)：
  …の改善が行われた（➡ **Improvement**）

Manifest：表す；明らかにする

1. Since that time, a much greater understanding has developed of how each level of treatment **manifests** itself in the aquatic environment.
   それ以降，各水準の処理が水生環境でどれほど効果を発揮するかについて，理解がかなり深まった。
2. The underlying concept of the biomarkers approach in biomonitoring is that contaminant effects occur at the lower levels of biological organization before more severe disturbances **are manifested** at the population or ecosystem level.
   生物モニタリングにおける生物標識手法の根底にある概念は，より深刻な攪乱が個体群レベルや生態系レベルで明らかになる前に，より低次の生物レベルで汚染影響が生じるというものである。

Manner (See also **Fashion**)：方法（➡ **Fashion**）

- **in this manner**：この方法で
- **in a slightly different manner**：それぞれ若干異なる形で
- **in a satisfactory manner**：満足のいく方法で
- **in a similar manner**：同様に
- **in the same manner**：同じように

1. **In this manner**, each remedial action can be readily screened to determine its effectiveness in achieving the specified objectives.
   この方法で，修復活動のひとつひとつが，特定目的の達成にあたってその有効性を決定するために，容易に選別できる。
2. Each state traditionally has conducted biomonitoring **in a slightly different manner**.
   各州は，従来それぞれ若干異なる形で生物モニタリングを行っている。

3. This model is capable of correlating growth data from many situations **in a satisfactory manner**.
   このモデルは，多様の条件下からの成長データを満足のいく方法で関連づけることができる。
4. **In a similar manner**, it may be necessary to select reference sites from another class of stream, lake, or wetland if all sites in a particular class are highly disturbed.
   同様に，ある等級における地点すべてがきわめて攪乱されていれば，別の等級の水流，湖沼，湿地から参照地点を選ぶ必要もあろう。
5. Although the probability of all ecosystems becoming unhealthy **in the same manner** is unrealistic, it is important to note that response signature are definable based on patterns of specific perturbations.
   すべての生態系が同じように不健全になる確率は，非現実的だが，具体的攪乱パターンに基づき応答サインが定義可能なことに注目することが重要である。

Many：多く

1. Such materials have **many** technological applications.
   このような材料には多くの技術的適用がある。
2. **Many** compounds have a potential for chemical modifications by both abiotic and biotic systems.
   多くの化合物が，非生物システムと生物システムの両方によって化学変形を起こす可能性を持っている。
3. **Many** regions will require increasingly efficient treatment plants.
   多くの地域で効率的な処理場をますます必要とするであろう。
4. **Many more** states are in the early stages of developing the biological survey or reference site methods needed for biological criteria.
   さらに多くの州が，生物クライテリアに必要な生物調査方法または参照地点方法の策定初期段階にある。
5. Akira Koga provided **many** hours of support in the development of the basic computer programs.
   Akira Koga は，長時間を割き基礎的なコンピュータプログラムの作成を支援してくれた。
6. **Many** are also heavily impacted by urbanization and industries, and all streams relative to watersheds of 30 mi^2 or more have been channelized at one time or another.
   多くは，都市化や産業の影響を強く受けており，30m^2 以上の流域に関わるすべての水流は，ある時期に流路制御を受けてきた。

- **many of .....** : …の多く

7. **Many of** America's largest cities use lead pipe extensively for water service connection.
   アメリカの最大都市の多くが，給水接続のために鉛のパイプを広範囲に使用している。
8. The Mississippi River project illustrates **many of** the decisions, procedures, and results involved in using HSPF.
   この Mississippi 川プロジェクトは，HSPF の使用に関与する決定，手順，結果の多くを例証している。

- **..... as many** : …ほど
- **as many as .....** : …と同じほど

9. Twice **as many** impaired waters have been discovered by using biological criteria and chemistry assessments together than were discovered using chemistry assessments alone.
   化学アセスメントの単独利用に比べて，生物クライテリアと化学アセスメントの併用によって損傷水域が2倍も多く発見された。
10. The groundwater in the alluvium beneath the tailings has a possibility of being comprised of water from **as many as** four sources.
    堆積選鉱屑の下にある沖積層の地下水は，およそ4つの水源からの水が混ざっている可能性がある。

Matter : 問題；事態

- **~ is a matter of concern** : ～は懸念される事態である
- **~ is a matter of debate** : ～は討論上の問題である
- **~ is another matter** : ～は別問題である

1. The possible long-range effects of the disposal of trace metals in the coastal environment **are matter of growing concern**.
   沿岸環境において微量金属の処分による長期にわたる潜在的影響は，これからますます懸念される事態である。
2. The existence of both mechanisms has been established, but their relative importance **is a matter of debate** and probably depends on the method by which cavitation is produced.
   両方のメカニズムの存在は確立されてきたが，それらの相対的重要性は討論上の問題であって，おそらくキャビテーションを生じさせる方法に左右されるであろう。

- **as a matter of practice**：実際問題として

3. Nevertheless, courts have noted that the implementation of water quality criteria that are not based on chemical-specific numeric criteria can be difficult **as a matter of practice**.
   それでも法廷は，化学数値クライテリアに基づいていない水質クライテリアは，実際問題として難しいと指摘してきた。

- **no matter whether** ..... ：… かどうかにかかわらず
- **no matter what** ...... ：どんなに

4. Naphthalene produced this inhibition **no matter whether** the melanin was fully aggregated or more less dispersed at the time of initial exposure.
   最初にさらされた時にメラニンが完全に凝結していたか，あるいはある程度分散していたかの状態にはかかわらず，ナフタリンはこの抑制効果を呈した。

- **..... matter**：…の物質

5. Little is known about the chemical nature of **organic matter** in groundwater.
   地下水での有機物質の化学的性質については，ほとんど知られていない。
6. There was no consistent relationship between the specific oxidation rate and the concentration of **organic matter** in the reactor.
   この反応器で，固有の酸化速度と有機物の濃度の間には一貫した関係がなかった。
7. Heavy metals may be subject to precipitation, adsorption to **particulate matter** and subsequent sedimentation.
   重金属は沈殿，粒子への吸着，それに伴う堆積作用を受ける。

Maximum (Maximize) (See also **Minimum**)：最大（最大化）(→ **Minimum**)

1. At the B site, chloride has **a maximum value of** 100 mg/L.
   B地点では，塩化物が最大値で100 mg/Lである。
2. The error bars show the minimum and **maximum** average measured soil-water contents.
   誤差バーは，測定された土壌の水分含有量の最小・最大平均値を示している。
3. The central basin is the largest in terms of area (16, 317 km^2) and has a mean depth of 25 m, while the eastern basin is the deepest, **with a maximum depth of** 64 m.
   中央水域は面積では最大 (16,317 km^2) で，平均湖深が25 mである。一方，東水

域は最も深く，最大湖深が64 m である。

- **maximize**

4 In order to **maximize** the meaningfulness of extrapolations from these studies and the use of data collected from national or international surveys, we must develop a clear understanding of ecosystem regionalities.
これら研究からの推定や，国内調査・国際調査での収集データ利用の意義を最大化するため，我々は，生態系地域性へのより明確な理解を進展させるべきである。

Means：手段；方法

1. The multivariate approach provides **a powerful means of** maximizing ecological information obtained from periphyton assemblages.
この多変量手法は，付着生物群から得られた生態学的情報を最大化するための有効手段となる。
2. The U.S. Supreme Court and lower courts have indicated repeatedly that water quality standards must be enforced through NPDES permits, and in fact, that NPDES permits are **the primary means of** translating water quality standards into enforceable requirements.
米国最高裁判所およびその下位裁判所は，これまで再三，水質基準がNPDES許可を通じて実施されねばならないこと，そして実際，NPDES許可が水質基準を実施可能な要求へと移し換える主要手段であることを主張してきた。

- **as a means of .....**：…の手段として

3. The former can be used just as much as the latter **as a means of** experimentation.
前者は後者とまったく同じぐらいよく実験手段として使用できる。
4. These animals were introduced to the caldera **as a means of** mosquito control.
これらの動物はカルデラに蚊制御の手段として導入された。

- **by means of .....**：…の方法で

5. The USGS constructed 40 monitoring wells **by means of** auger drilling.
USGSは，オーガードリルを使う方法で40のモニタリング用井戸を建設した。

Meaning (meaningful)：意味（有意義な）

1. Over the linear scale of the IBI even gross changes **have meaning** with

regard to the relative condition of the community and types of impacts that are present.
IBIの線形スケールで，大幅な変化は，生物群集の相対状態，ならびに存在する影響の種類に関して意味を有する。

- **~ have little meaning**：〜がほとんど無意味である

2. It should be noted that in major portions of the country, topographic watersheds either cannot be defined or their approximation **has little meaning**.
米国の大部分において，地形流域は，画定されえないか，その概算がほとんど無意味であることに留意すべきである。

- **meaningful**：有意義な

3. It was felt that this data would be particularly **meaningful** with respect to the .....
このデータが … に関して特に重要な意味を持つであろうと感じられた。
4. Without this knowledge, a **meaningful** analysis is very difficult.
この知識なしでは，有意義な分析は非常に困難である。
5. The results can be more **meaningful** than those that are normally released for publication.
この結果は，通常，出版物として発表されている結果よりも有意義であり得る。
6. Without a sufficient theoretical basis it would be very difficult, if not impossible, to develop **meaningful** measures and criteria for determining the condition of aquatic communities.
十分な理論的基礎がなければ，水生生物群集の状態を決定するための有意義な測度および基準(クライテリア)を設定することは，仮に不可能でなくても，非常に困難だろう。
7. We must develop an understanding of ecosystem regionalities at all scales, in order to make **meaningful** extrapolations from site-specific data collected from case studies.
我々は，ケーススタディで収集された地点固有データからの推定を有意義にするために，すべての縮尺での生態系地域性への理解を生じさせなければならない。

Measure：測定；測定する

1. The choice of groups to be identified and analyzed is related to the **measures** that will be used, as was discussed in detail above.
前述のとおり識別・分析されるべき集団の選択は，用いられる測度に関係している。

2. A **measure of** how much the quality can be improved can be derived through changing management practices in selected watersheds where associations were determined.
水質がどれほど改善されうるかを示す測度は，関連性が決定された流域における管理慣行の変更を通じて得られるだろう。

- **measured .....** ：測定された …

3. In examining the **measured** data further, it was discovered that .....
測定されたデータをさらに調べると，… が発見された。
4. A robust statistical treatment of these data often can help determine whether or not the biological response is attributable to the **measured** pollutant.
これらのデータへの確かな統計処理は，生物応答が測定された汚濁物質に帰せられるか否か決定するのに役立つことが多い。

- **to measure** ：測定すること

5. Existing multimetric approaches are robust in their ability **to measure** biological condition.
既存の多メトリック手法は，生物学的状態の測定能力において確かである。
6. The idea being brought forward in this research is **to measure** several of these general parameters.
この研究にて前進的考え方は，これらの一般的なパラメータのいくつかを測定するということである。

Measurement ：測定

1. These **measurements** show generally good agreement with expected concentration values.
これらの測定値は，予期していた濃度値と一般に良い一致を示している。
2. The writers want to show herein other difficulties which can arise in toxicity **measurements**.
著者は，毒性測定の際に発生し得る他の難問題をここに示したい。

- **measurements were obtained** ：測定が得られた
- **measurements were made using .....** ：… を使って測定された

3. Accurate flow **measurements have not been obtained**.
正確な流量測定が得られなかった。
4. The **measurements were made using** an electronic thermometer.

その測定には，電子温度計が使われた。

Meet (Meeting)：満たす（満たすこと）

1. Laboratory experiments were performed **to meet** three objectives: 1).....; 2)....., and 3)....
3つの目的を満たすために研究室で実験が行なわれた。それらは1)…，2)…，および3)…
2. Many states have used biological monitoring **to meet** these requirements for identification and reporting of waterbody condition.
多くの州が水塊状態の確認・報告についてこれらの要件を満たすため，生物モニタリングを用いてきた。
3. These people may know of sites that **meet** some or most of the site selection criteria.
これらの人々は，地点選択基準を幾分または大部分満たす地点を知っているかもしれない。
4. Vermont uses biological criteria from ambient stream data to determine whether two different types of biological standards **are being met**.
バーモント州は，環境水流データからの生物クライテリアを用いて，2種類の生物基準が満たされているか否か決定している。

- **meeting**：満たすこと

5. This report is a first step toward **meeting** this goal.
この報告はこの目的達成に向かっての第一歩である。
6. The model has been judged potentially capable of **meeting** the user's needs.
このモデルは，潜在的に使用者の必要性を満たすことが出来ると評価されてきた。
7. Activities requiring a permit under Section 404 of the CWA must be certified as **meeting** provisions of the water quality standards by the state water quality agency.
CWA第404条のもとで許可を要する諸活動は，州水質担当機関によって水質基準の規定を満たしていると認定されなければならない。

Mention (Aforementioned)：言及；言及する（前述の）

1. **Mention of** trade names or commercial products does not constitute endorsement or recommendation for use.

商標名や商品への言及は，その利用を必ずしも是認していないし，また推奨もしていない。
2. The discussion will emphasize assessments of stream environments, but aspects of monitoring ponds and lakes also **will be mentioned**.
本章の力点は，水流環境アセスメントに置かれているが，湖沼モニタリングについても言及していく。
3. Despite the **aforementioned** impoundments and flow alterations, overall habitat conditions in the Scioto River are good to excellent.
前述の人工湖と流量改変にもかかわらず，Scioto 川の全般的生息場所状態は，良ないし優である。

- **As previously mentioned**：前に述べたように
- **As was previously mentioned**：前述のとおり
- **As mentioned earlier**：以前に言及したように
- **It should be mentioned that .....**：… が言及されるべきである

4. **As previously mentioned**, industry has not been forced to address issues related to numeric criteria on a national scale.
前述のとおり，産業界は，全国規模では数量的クライテリアに関わる問題への取組みを強いられていない。
5. **As was previously mentioned**, Illinois EPA has employed a system of tiered aquatic life uses in the state water quality standards since 1978.
前述のとおり，イリノイ州 EPA は，1978 年以降，州水質基準において段層式の水生生物使用システムを用いてきた。
6. **As was mentioned earlier**, HCOO$^-$ did not originate from the photochemistry of MeOH solvent.
以前に言及されたように，HCOO$^-$ は MeOH 溶媒の光化学が由来ではない。
7. **It should be mentioned that** the drinking water criteria for manganese is more than two orders of magnitude lower than the irrigation water criteria.
マンガンの飲料水基準が潅漑水基準より二桁以上低いことは，言及されるべきである。

Method：方法

1. The following discussion presents examples of **methods** presently used in evaluating the periphyton assemblages.
以下の論考は，付着生物群がりを評価する際に現在用いられている方法の諸例である。
2. The **method** described in this paper can improve the interpretation of the

results.
この論文に記述された方法によって，この結果の解釈を改善することができる。
3. The **method** used here may be a new approach not only for gaining better results, but also for acquiring new knowledge and insights about toxic behavior.
ここで使われた方法は，もっと良い結果を得ることだけではなく，毒物の挙動について新しい知識と洞察を得るのに，新しいアプローチであるかもしれない。
4. The assessment of biological condition using a suite of metrics to define biocriteria is rapidly becoming **the method of choice** among state water resource agencies.
生物クライテリアを定めるための一連のメトリックを用いて生物学的状態アセスメントは，急速に州の水資源担当機関にとって選択手法になりつつある。
5. When used in a monitoring program, these **methods** provide an indication of what specific environmental characteristics are changing over time, and can be used to help diagnose the cause of changes in the aquatic system.
監視プログラムで用いた場合，これらは，どの環境特性が経時的に変化しているか示す指標となり，水生系における変化原因を判定するのに役立つ。

- **by the methods**：この方法で
- **using the methods**：この方法を用いて
- **following the method**：この方法に従って

6. Sampling time can be as rapid as 20 min/site, as shown **by the methods** used in Montana (Ikuda, 1993).
サンプリング時間は，モンタナ州での方法 (Ikuda, 1993) で示されたとおり，1地点当り20分間で済む。
7. Biosurveys were conducted near Coralville Station from 1988 to 1991 **using methods** that conformed to Iowa DEQ protocols or with minor deviations.
Iowa DEQ プロトコルに準拠した，もしくはやや偏向した方法を用いて，1989年から1991年まで Coralville 発電所付近で生物調査が行われた。
8. If a relationship was observed, a 95% line of best fit was determined and the area beneath trisected **following the method** recommended by Fukui et al. (1984).
関係が見られたならば，ベストフィットの95％ラインが決定され，その下方の面積は，Fukui ら (1984) が推奨する方法に従って三等分された。

Methodology：方法論

1. The study was intended to demonstrate a **methodology**.

この研究は方法論を実証するように意図された。
2. An evaluation of the **methodology** in light of the methodology application is presented in section 6.
方法論の評価は，この方法論の適用を考慮に入れて第6節に掲げられている。
3. Because of a lack of extensive knowledge on pesticide toxicity, the **methodology** was not fully verified.
殺虫剤毒性に関する広い知識が欠如していたために，この方法論は十分に実証されなかった。

- **The methodology described herein**：ここに記述された方法論

4. **The methodology described herein** makes it possible to gain insight into water quality changes to be expected over the life of a groundwater recharge project.
ここに記述された方法論は，地下水涵養プロジェクト期間中に生じるであろうと思われる水質変化の洞察を可能にする。

Mind：心

- **keep in mind**：心に留めておく；念頭におく
- **bear in mind**：心に留めておく

1. It is necessary to **keep in mind** that .....
… ということを心に留めておくことが必要である。
2. In developing a TMDL it is important to **keep in mind** certain constraints on the WLA portion.
TMDLを策定において，WLA部分における特定の制約を念頭におくことが重要である。

★ TMDL は Total Maximum Daily Loads（日最大許可負荷量），WLA は Waste Load Allocation（汚濁負荷割当）の略語です。

3. Whatever the reasons for the differences, it is important to **keep in mind** that .....
この相違の理由が何であるとしても，… を念頭におくことは重要である。
4. We hope the reader will **bear in mind that** the real world is somewhat different.
我々は，読者が実世界がいくぶん異なっているということを心に留めておくであろうことを期待する。

Minimal (Minimally)：最小の（最小に）

1. Runoff and volatilization losses were predicted to be **minimal**.
   流去水と揮発による減少が最小であると予測された。
2. Since external impacts **are minimal**, measured variability would be most likely due to the sampling inconsistencies.
   外部影響が最小であるから，測定された変動性はサンプリングの不整合性に起因する見込みが高い。
3. The method utilizes the qualitative, natural substrate collection procedure, which necessitates one site visit and **minimal** laboratory analysis.
   この方法は，定性的な自然底質採集手順を活用しており，それは1回の地点訪問と最小限の実験室分析を必要とする。
4. Whether or not petroleum entering the marine environment will have substantial or **minimal** impact depends on interactions among complex variables that are only now beginning to be understood.
   海の環境に流入している石油が本質的な影響，または最小影響を持つであろうかどうかは，複雑な変数の間での相互作用に依存が，これらの変数は今日やっと理解され始めてきているにすぎない。

　　● minimally：最小に

5. When selecting these sites one must account for the fact that **minimally** disturbed conditions often vary considerably from one region to another.
   これらの地点を選ぶ際には，最小攪乱状態が地域ごとに大幅に異なりうるという事実を考慮しなければならない。
6. The focus at this step is twofold: to locate and reject disturbed areas, and to seek and retain **minimally** disturbed areas.
   このステップでの主眼は2つある。それは攪乱区域を位置特定し排除すること，それに，最小攪乱区域を探して確保することである。

Minimize：最小にする

1. Considerable effort is being made **to minimize** the release of contaminants to receiving water bodies.
   受水域への，汚濁物質の流出を最小に抑えるために，かなり多くの努力がなされている。
2. The standard procedure outlined above was established so that the problems which often obscure interpretation of results **are minimized**.

結果の解釈をしばしば不明瞭にする問題点が最小になるように，上に概説された標準的な手順は確立された。
3. A lack of adverse effects indicated the importance of dilution **in minimizing** impacts to microbial communities.
悪影響が無いということは，微生物群集への影響を最小化するのに希釈が重要であることを示した。

Minimum(See also **Maximum**)：最小(→ **Maximum**)

1. The error bars show the **minimum** and maximum average measured soil-water contents.
エラーバーは，測定された土壌の水分含有量の最小・最大平均値を示している。
2. The WWH use represents the **minimum** goal of the CWA(Clean Water Act).
WWH 使用は，CWA(水質汚濁防止法)の最低目標を表している。

- **as a minimum**：最低限でも

3. Idaho DEQ, **as a minimum**, should have tested the hypothesis of no differences in biological performance for reference sites on impounded and free-flowing sections.
アイダホ州 DEQ は，最低限でも貯水部および流水部の参照地点について生物能力差がない，という仮説を検証すべきだった。

Minutes(See also **Hour**, **Day**)：分(→ **Hour**, **Day**)

1. Ishida(1993)identified a rapid metal removal phase occurring in the first 30-60 **min**.
Ishida(1993)は，急速な金属除去が最初の 30-60 分に生じることを見いだした。
2. A 10-mL sample of calibration standards was contacted with 1 mL of hexane in conical bottom test tubes followed by shaking for 2-5 **minutes** by hand.
コニカルボトム試験管で，10 mL の較正標準溶液サンプルに 1 mL のヘキサンを加えてから，2-5 分間手で振った。
3. Sampling time can be as rapid as 20 **min/site**, as shown by the methods used in Montana(Eda, 1993).
サンプリング時間は，モンタナ州での方法(Eda, 1993)で示されたとおり，1 地点当り 20 分間で済む。
4. Since the flow of gas through each 10-L reactor is 0.85 L/min, the residence time is approximately 12 **min**.

それぞれの 10 リットルの反応器を通過するガスの流れが 0.85 L / 分であるので，滞留時間は約 12 分である。

Mislead（See Lead）：誤りに導く；誤りをもたらす（→ **Lead**）

Mixture：混合；混合物（水）

1. The incremental reactivity of the nine-component **mixture** was also measured.
   9つの成分の混合物の増加的な反応性もまた測定された。
2. We have used this approach to study the reactivities of individual hydrocarbons and hydrocarbon **mixtures**.
   我々は，個別の炭化水素と炭化水素混合物の反応性を研究するために，このアプローチを使った。
3. The reactivity of **a mixture of** VOCs can be calculated from the reactivity of the individual components.
   VOCs の混合物の反応性は，それらの個々の成分の反応性から計算することができる。

Mode：方式

- **in batch mode**：バッチ方式で
- **in either batch or continuous mode**：
  バッチ方式または連続フロー方式で

1. In the present work, 2 L of aqueous formic acid has been used and the operation was **in batch mode**.
   ここで述べている実験では，2 リットルのギ酸溶液を使用し，バッチ方式でおこなった。
2. The reactor has a total capacity of 7.0 L and can be operated **in either batch or continuous mode**.
   この反応器は，全容量が 7.0 リットルであり，バッチ方式または連続フロー方式で操作されうる。

Model：モデル

1. Using the ABC **model**, we investigated .....

Model

ABC モデルを使うことによって … を調べた。
2. By employing the ABC **model**, we examine .....
ABC モデルを使って … を調べている。
3. The **model** we developed below is an attempt to .....
私たちが構築した下記のモデルは，… する試みがある。
4. Monod's **model** is the most widely accepted mathematical model of microbial growth.
Monod モデルは，微生物の成長を説明するのに最も広く受け入れられている数学モデルである。
5. The **model** program was modified in order to eliminate the following limitations: ....
モデルプログラムは次に示す制約をなくすために修正された。
6. In many instances, mathematical **models** are essential to generalize the results of laboratory growth studies.
多くの場合，数学モデルは実験室系細菌増殖研究結果を一般化するために不可欠である。
7. Such detailed understanding of the interactions is impossible from steady-state **models**.
このような相互作用の詳細な理解は定常モデルからは不可能である。
8. Candidate **models** are selected based on knowledge of aquatic systems, flora and fauna, literature review, and historical data.
候補のモデルは，水生系に関する知見，植物相と動物相，文献査読，史的データに基づいて選ばれる。

- **model indicates that** ..... ：モデルは … を示す
- **model represents** ..... ：モデルは … に相当する；モデルは … を表す
- **model emphasizes** ..... ：モデルは … を強調する
- **model provides** ...... ：モデルは … を提供する
- **model yields** ..... ：モデルは … を生み出す

9. The **model** developed in this paper **indicates that** ......
この論文で構築されたモデルは … を示している。
10. The **model represents** a tool to estimate the long-term effort of human activities on the water environment.
このモデルは，水環境に対する人間活動の長期にわたる努力を推定するツールに相当する。
11. SOIL **model** does not adequately **represent** the aldicarb behavior in this application.
SOIL モデルは，aldicarb の挙動を十分に説明しない。

12. The **model emphasizes** the water and sediment transport in aqueous systems.
　　このモデルは，水系での水と堆積物の移送を強調する。
13. Mathematical **models provide** a basis for quantifying the inter-relationships among the various toxic chemicals.
　　数学モデルは，種々の有毒化学物質間の相互関係を定量化する基礎を提供する。

　　● model has been structured to ……：
　　　モデルは … するために構成された
　　● model is intended to …….：このモデルは … するように意図している
　　● model is applied to …..：モデルは … に用いられている
　　● model is successful in …..：モデルは … において好結果をだす
　　● model is needed：モデルが必要とされる
　　● model makes it clear that …..：このモデルは … を明らかにする

14. **A model has been structured to** account for the relevant factors.
　　モデルは，関連要因を説明するために構成された。
15. Since **the model is intended to** compute long-term trends, average annual values are simulated and variability within the year is ignored.
　　このモデルは，長期傾向を計算するように意図されるので，年平均の値がシミュレートされ，年内の可変性が無視される。
16. The **model is applied to** Clear Lake, a large and relatively shallow lake in Northern California.
　　このモデルは，Clear 湖に用いられている。この湖は北カリフォルニアにあり，広くて比較的浅い湖である。
17. The **model is less successful in** predicting the degree of lateral dispersion.
　　このモデルは，横断方向の拡散の度合いを予測する点で好結果をださない。
18. In the past, both empirical and theoretical **models have been needed** in this field.
　　過去には，経験的なモデルと理論的なモデルの両方がこの分野では必要とされた。

　　★ Model の仮定に関連する例文を下に示します。

19. **A basic assumption for the model was that** biofilm diffusion limitations were not significant.
　　バイオフィルム拡散限界は重要ではないということが，このモデルの基本的な仮定であった。
20. It is important to note that most **models represent** an idealization of actual field conditions and must be used with caution to ensure that the

underlying model assumptions hold for the site-specific situation being modeled.
ほとんどのモデルは，実際の現地状態を理想化しており，モデル化されている地域特定状況の基礎をなしているモデル仮定を確証するよう注意を払って使われなくてはならない，ということを指摘することが重要である。

★Modelの較正に関連する例文を下に示します。

21. The **model has been calibrated with** data collected at Woods Lake.
このモデルは，Woods湖で収集されたデータを用いて較正された。
22. Attempts to **calibrate the model** and keep parameters in a reasonable range based on literature data proved unsuccessful.
モデルを較正して，パラメータを文献データに基づく妥当な範囲にとどめる試みは，不成功であることが判明した。
23. By using realistic values for these parameters, the **calibration** process at least ensures that the concentrations predicted by the model are within an order of magnitude of the actual values.
これらのパラメータに現実的な値を使うことによって，較正過程は，モデルによって予測された濃度が少なくとも実際値と同じ桁内にあることを保証する。

★Modelの機能に関連する例文を下に示します。

24. This **model is capable of** correlating growth data from many situations in a satisfactory manner.
このモデルは，多様の条件下からの成長データを満足のいく方法で関連づけることができる。
25. The **model has been judged** potentially capable of meeting the user's needs.
このモデルは，潜在的に使用者の要求を満たすことが出来ると評価されてきた。
26. The **model computations predict** the equilibrium distribution.
モデルによる計算から平衡分布が予測される。

★Modelの評価に関連する例文を下に示します。

27. **In evaluating the performance of a model** relative to some field measurement or another model, qualitative rather than quantitative statements are often made.
あるフィールド測定，またはもう1つのモデルとの比較からモデルの性能を評価する場合，定量的というより定性的な報告がしばしばなされる。
28. The fit between the **model** and data is certainly good ; the least square line presented by Wada et al. (1997) is of slightly steeper slope.

モデルとデータは確かに一致している。Wada ら (1997) によって示された最少二乗法による線は，少し勾配が大きい。

29. The **model** values correspond reasonably well with the experimental data.
モデル値は実験データとかなりよく一致している。
30. The **model** results and experimental values are given in Figure 5.
モデルの結果と実験値を図5に示す。
31. This report should serve to give the reader an idea of what is involved in testing **a model**.
この報告書は，モデル試験に何が関与するのか，その考え方を読者に提示するのに役立つはずである。

★ Model の特徴に関連する例文を下に示します。

32. A key distinguishing feature of our **model** is .....
我々のモデルの重要で際立った特徴は，… である。
33. Some known features of the behavior of iodine are not included in the **model**, because they are not relevant to the problem of interest.
ヨードの挙動に関してすでに知られている特徴の一部はモデルに含められていない。なぜなら，それらは，ここで取り上げている問題点とは直接に関係が無いからである。
34. The **model** incorporates the effects of mass transport, chemical interactions, hydrodynamic dispersion, and radioactive decay.
このモデルには，物質輸送，化学的な相互作用，流体力学的拡散，放射性崩壊の影響が盛り込まれている。
35. Chemical interactions of the **model** are depicted in Figure 1.
このモデルの化学的な相互作用が図1に描写されている。
36. The linkage of dynamic chemical runoff and instream chemical fate and transport **models** is a recent accomplishment.
動的な化学物質の流出モデルと河川内での化学物質の衰退・輸送モデルの結合は，最近の業績である。
37. The **model** formulation incorporates chemical decay and transport mechanisms of particulate and diffusive exchange between water column and sediment.
モデル式は化学的分解，微粒子の輸送メカニズム，水柱と堆積物の間の拡散的交換を含んでいる。

★ Model の記述に関連する例文を下に示します。

38. It was found that the observed processes are described satisfactory by this **model**.

Model

このモデルは観察される過程を満足に記述する，ということがわかった。

39. The description of fate of materials discharged into an estuary requires a **model** of considerable generality and complexity.
河口に放出される物質の衰退を説明するには，かなり一般性，ならびに複雑さを持つモデルを必要とする。

40. These simplified **models** contain two types of terms describing the transport of suspended or dissolved material through the estuary.
これらの単純化されたモデルは，2種の項を含み，それらは河口での浮遊物質と溶解性物質の輸送を記述している。

★Modelの速度論に関連する例文を下に示します。

41. Because the dechlorination reaction rates were relatively slow as compared to CP desorption kinetics, solid−liquid partitioning equilibrium was assumed, and partitioning rates were not incorporated into the **model**.
CPの脱着速度と比較して，脱塩素反応速度が比較的遅かったので，固体−液体間の分配平衡が仮定され，分配速度はモデルに取り入れられなかった。

42. Kinetic coefficients from the single−Cp tests were incorporated into a Michaelis−Menten competitive inhibition **model**.
単一Cpテストからの動力学係数がMichaelis−Menten競合抑制モデルに取り入れられた。

43. From time to time, attempts to develop mathematical **models** of bacterial growth have appeared and in many instances mathematical **models** are essential to generalize the results of laboratory growth studies.
時折，バクテリアの成長を説明する数学モデルの構築が試みられてきた。そして多くの場合，数学モデルは実験室での成長の研究結果を一般化するために不可欠である。

★Modelによる計算に関連する例文を下に示します。

44. According to the **model calculations**, ...
モデルによる計算によれば，…

45. **Model calculations** using the program MINEQL show that ....
プログラムMINEQLを使ってのモデル計算は，…を示す。

★Modelによる予測に関連する例文を下に示します。

46. **Model predictions** show reasonable agreement as can be seen from both Figures 2 and 3.
図2と3の両方から見られるように，モデルによる予測は合理的な一致を示している。

47. For such extreme operating conditions, deviations between **model prediction** and experiments can be expected.
このような過激な操作上の条件のために，モデル予測と実験の間のずれが予想されうる。

48. The kinetic **model is** less successful **in predicting** the initial pollutant concentration ....
動力学モデルは，初期の汚濁物濃度を予測することにおいて，成功には至らない。

★ Model シミュレーションに関連する例文を下に示します。

49. Existing pesticide transport **simulation models** can be classified into analytical and numerical models.
既存の殺虫剤輸送をシミュレーションするモデルは，分析モデルと数値モデルに分類できる。

50. Since the **model** is intended to compute long-term trends, average annual values are simulated and variability within the year is ignored.
このモデルは長期傾向を計算するように意図されるので，年平均の値がシミュレートされ，年内の変動性が無視される。

51. EXAMS is specifically designed to simulate the fate and persistence of organic chemicals in aqueous ecosystems.
EXAMS は特に水の生態系で有機化学物質の衰退と持続性をシミュレーションするように設計されている。

★ Model の結果に関連する例文を下に示します。

52. PESTAN, PISTON, and PRZM give comparable results.
PESTAN，PISTON ならびに PRZM は類似の結果を出す。

53. The **models** presented in this paper are intended to supersede the preliminary results given in a recent report (Sasaki, 2002).
この論文に記載されているモデルは，最近の報告書(Sasaki,2002)に記載された予備結果に取って代わることを目的としている。

★ Model 予測値に関連する例文を下に示します。

54. The results presented here reflect recent modifications of the global transport **model**, and, thus, the estimates are somewhat different than previously reported.
ここで報告された結果は，地球上輸送モデルの最近の修正を反映している。それゆえに，予測値は前に報告された値より，よりいくぶん異なっている。

Modify（Modification）：修正する（修正；改修）

1. The model program **was modified** in order to eliminate the following limitations: .....
   このモデルプログラムは，次に示す制約を排除するために修正された。
2. The kinetic expression for growth **is modified** in order to account for environmental effects.
   成長の速度表現は，環境の効果を説明するために修正される。
3. The TMDL determines the allowable loads and provides the basis for establishing or **modifying** controls on pollutant sources.
   TMDLでは，許容負荷量が決定され，そして汚染物源制御を確立または修正するための基礎が提示される。

   ● **modification**：修正；改修

4. The results presented here reflect recent **modifications** of the global transport model.
   ここで提出された結果は，地球規模の輸送モデルの最近の修正を反映している。
5. During the course of testing, several **modifications** were made to the criteria.
   検定中にクライテリアに対しいくつかの修正が行われた。
6. Many compounds have a potential for **chemical modifications** by both abiotic and biotic systems.
   多くの化合物が，非生物システムと生物システムの両方によって化学変化を起こす可能性を持っている。
7. New York DEP acknowledges, albeit briefly, that site-specific **modifications** may be appropriate in some circumstances.
   ニューヨーク州DEPは，いくつかの状況では地点特定の改修が適切であることを一時的にだが認めている。
8. Stream **habitat modification** was identified as the third leading cause of aquatic life impairment in the 1992 Wisconsin Water Resource Inventory.
   水流生息地改変は，1992年のウィスコンシン州水資源調査一覧において，水生生物損傷の第三の主因と確認された。
9. Regional scale **modifications** in management practices may be feasible, allowing for the significant recovery of impaired aquatic resources.
   管理慣行における地域スケール修正も実行可能であり，これは傷ついた水生資源をかなり回復させる。

   ● **with minor modification**：少しの修正で

10. Although these indices have been regionally developed, they are typically appropriate over wide geographic areas **with minor modification**.
これらの指数は，地域的に設定されてきたものだが，一般に，やや修正すれば広い地理的範囲に適している。
11. The stability constant was determined by the method of Miyamoto and Kobayashi (1986) **with minor modifications**.
安定性定数は，MiyamotoとKobayashi (1986) の方法を少しの修正した方法で決定された。

● river modification ：河川改修工事

12. This paper examines the ecological damages caused by the **river modification**.
この論文では，河川改修工事が生態に及ぼした損傷.を検討する。
13. The ecological damages caused by the **river modification** needs to be examined in detail.
河川改修工事が生態に及ぼした損傷について詳しく調査する必要がある。
14. The HELP ecoregion is affected by significant and widespread historical land use and **stream channel modifications** dating from the 19th century.
HELP生態地域は，19世紀から，甚大かつ広範な史的土地利用および水流路改変の影響を受けている。

Monitor (Monitoring)：モニター；監視する；観測する（モニタリング）

1. Many local and state watershed management decisions have been based, at least in part, on data collected by volunteer **monitors**.
多くの地方および州の分水界管理上の決定は，少なくとも部分的にボランティアモニターの収集データに基づいている。
2. The water **was monitored** continuously for pH and temperature.
この水のpHと温度が連続的に観測された。
3. Remote, difficult to reach areas **have also been monitored** by citizen volunteers when they were too difficult for state personnel to reach on a timely basis.
カバーしづらい遠隔地も，州職員が時間通りに到着しかねる場合には，市民ボランティアによってモニターされてきた。
4. The study focused on data from 56 sites on streams that were part of Georgia's designated Scenic Rivers Programs and which **were monitored** by both Georgia DNR volunteers and by Georgia EPA biologists.
この研究の主眼が置かれたのは，ジョージア州の指定するScenic Rivers Programs

(景観河川プログラム)の一部でジョージア州のDNRボランティアとジョージア州EPA所属生物学者の両方がモニターした水流の56地点でのデータである。

- **monitoring**：モニタリング

5. **Volunteer monitoring** can help provide that data.
   ボランティアモニタリングは，そのようなデータを提供するのに役立ちうる。
6. **Volunteer monitoring programs** have been developed to provide education and advocacy opportunities as well as basic data collection.
   ボランティアモニタリングプログラムが設けられた目的は，啓発機会や擁護機会を提供して，基礎データ収集を可能にすることであった。
7. The discussion will emphasize assessments of stream environments, but aspects of **monitoring** ponds and lakes also will be mentioned.
   本章の力点は，水流環境アセスメントに置かれているが，湖沼モニタリングについても言及していく。
8. Marui (1993) described the use of rapid assessment approaches **in water quality monitoring** as somewhat analogous to the use of thermometers in assessing human health.
   Marui (1993)は，水質モニタリングにおける迅速アセスメント手法の利用について，人間健康を評価する際の体温計に若干似たものと評した。
9. It is important to note that the four-tiered system has been widely used **in biomonitoring** because it recognizes the limitations of refinement and accuracy that can be ascribed to this science.
   この四段層方式が生物モニタリングで多用されている理由は，この科学に帰されるる改良および正確性の限界を認識している点にあることを念頭におくべきである。
10. The State of Montana Environmental Protection Agency **has been** intensively **monitoring** the condition of Montana's surface waters since the late 1970s.
    モンタナ州環境保護局は，1970年後半以降，モンタナ州の地表水域状態の徹底的なモニタリングを行ってきた。

Month：月

- **in late September**：9月下旬に
- **every month over a 12-month period**：12カ月間にわたって毎月
- **during the summer and fall months**：夏から秋にかけて

1. Devices for sampling of gas were installed **in late September**.
   ガスをサンプリングのための装置が9月下旬に設置された。
2. Soil samples at the field sites were collected **every month over a 12-**

month period.
土壌サンプルが，野外地点においての12カ月間にわたって毎月収集された。
3. The majority of fish species encountered in warmwater rivers and streams are essentially sedentary **during the summer and fall months**.
温水の河川・水流で見られる魚種の大半は，基本的に夏から秋にかけて定住性である。
4. Lysimeters were sampled on three occasions, **in mid June**, **late August** and **late September**, 2001.
ライシメータでサンプルが，2001年の6月中旬，8月下旬と9月下旬の3回収集された。

More (See also Less)：もっと；以上（→ Less）

1. The test offers the advantage of being **more** sensitive and much faster.
このテストには，いっそう敏感で，より迅速であるという利点がある。
2. To help us understand the earlier experiments, we are now running **more** controlled experiments.
以前の実験を理解するために，我々は今より制御された実験を行なっている。
3. Further research into sediment transport is needed to develop **more** effective tools to better address these issues.
もっと良くこれらの問題を扱うためのより効果的なツールを開発するには，堆積物輸送に対するいっそうの研究が必要です。
4. The most critical cleanup issue facing DOE is the need to adopt a **more** practical policy on the program's objectives.
DOEが直面している最も重要な浄化問題は，プログラムの目標のために，いっそう実務的な政策を採用する必要性です。
5. It was clear that a better tool was needed to **more** consistently and accurately characterize the aquatic communities.
水生生物群集をより整合的かつ正確に特性把握するための優れたツールが必要なことは，明らかであった。
6. The EPA conducted a **more** extensive review of the contaminant content of the existing deposits.
EPAは，既存の堆積物での汚染物質含有量のいっそう広範なレビューを実施した。

  ● more than ..... ：… よりも；… 以上
7. Rate enhancements of **more than** tenfold are common.
反応速度の増速は，通常10倍以上である。
8. It may raise **more** questions **than** it answers.

More

それは回答よりも多くの質問を提起するかもしれません。
9. Questions come **more** easily **than** answers.
   答えをだすより質問をだすほうがいっそう容易である。
10. Arsenic levels are **more than** an order of magnitude lower in the low flow sampling period.
    ヒ素レベルは，低流状態でのサンプリングの期間においては一桁以上低い。
11. The results can be **more** meaningful **than** those that are normally released for publication.
    この結果は，通常，出版物として発表されている結果よりも有意義であり得る。
12. To date, qualitative methods, although used for many applications where the usefulness of the results is **more** important **than** the scientific rigor of the techniques used, have not been widely accepted.
    今日までのところ，定住物方法は，技法の科学的厳密性よりも結果の有用性が重要な多くの適用例で用いられているが，まだ広く容認されてはいない。
13. In evaluating the accuracy of the estimated parameter, **more** weight should be given to the ability of the model to fit the data points near segment D **than** to its ability to fit the points near segment G.
    推定パラメータの正確さを評価することにおいて，モデルが川の切片 G の近くのデータポイントで合う能力により，切片 D の近くでデータポイントに合うことができる能力に，もっとウエートが置かれるべきである。
14. It has been known for decades that certain groups of macroinvertebrates are, in general, **more** tolerant of pollution **than** are others.
    大型無脊椎動物の一部集団が一般的に他集団に比べ汚濁耐性が強いことは，数十年前から知られている。
15. It has become increasingly evident that the environmental impact of a particular metal species may be **more** important **than** the total metal concentration.
    特定の金属種の環境に対する影響が，金属すべての濃度の影響よりいっそう重要であるかもしれないということがますます明白になってきた。
16. The same organism may have an order of magnitude **more** chlorophyll under low–light and nutrient–rich conditions **than** under high–light and nutrient–poor conditions.
    同生物が，多光量・低養分状態よりも少光量・高養分状態の方が十倍ほど多くのクロロフィルを有する場合がある。
17. It should be mentioned that the drinking water criteria for manganese is **more than** two orders of magnitude lower than the irrigation water criteria.
    マンガンの飲料水基準が潅漑水基準より二桁以上低いことは言及されるべきである。

- **in more detail**：より詳細に
- **more detailed** ..... ：より詳細な

18. The risk and benefits of this proactive strategy are discussed **in more detail** below.
    この先取的戦略の持つ危険と便益については，後ほど詳述する。
19. A **more detailed** explanation of Bailey's approach and its limitations with respect to attempts to use it to frame aquatic ecosystems are given later in this chapter.
    Baileyの手法と，水生生態系の枠組みづくりへのその利用に関する限界についてのより詳細な説明を本章の後半で述べる。
20. **More detailed** discussions of biological criteria development for the States of Maine, New York, and Texas can be found elsewhere in this volume.
    メイン州，ニューヨーク州，テキサス州での生物クライテリア設定に関するより詳細な論考は，本書のほかのところにも記されている。

- **no more than** ..... ：… 以上ではない；ただの～に過ぎない；でしかない

21. Sample processing and species proportional counts of a sample should take **no more than** 3 hr.
    サンプル処理とあるサンプルの生物種比率カウントに，3時間以上かけてはならない。

- **much more**：さらに

22. **Much more** research is needed in such basic areas as sampling effort, and indicator sensitivity and variance.
    サンプリング活動，指標の感度・分散といった基礎領域における研究はさらに必要とされる。

- **one or more of** ..... ：1つないし複数の …

23. States may choose to concentrate on **one or more of** these uses of biological criteria when developing their programs.
    諸州は，プログラム策定時に，生物クライテリアの上記用途のうち1つないし複数に専念することを選ぶ場合もある。

- **more A, more B**：よりAになるほど，よりBになる

24. **The more** toxic the whole effluent, **the more** test organisms exposed over a fixed period of time will die.
    WET (whole effluent toxicity) が強くなるほど，所定期間にわたって曝露された被

験生物の死亡数が増えるだろう。

- ～ is more likely to ……(See likely)：～は … の見込みが高い(➡ likely)

Most：大半の；大部分の

- Most ～ are ……：大半の～は … である
- at most sites：大半の地点で
- Mony, if not most, of ……：大半でなくても，多数の…
- In most areas：大半の区域において
- The most accurate way：最も正確な方法

1. **Most states are** developing bioassessment methods in advance of implementing biological criteria programs.
   大半の州は，生物クライテリアプログラムの実施に先立って生物アセスメント方法を案出中である。
2. Replicate samples were taken **at most sites**, including ecoregion reference sites.
   生態地域参照地点など大半の地点で反復サンプルが採取された。
3. **Many, if not most, of** these states are focusing their programs on the development and maintenance of biological criteria.
   これらの州のうち，大半でなくとも多数は，プログラムの主眼を生物クライテリアの設定・維持に置いている。
4. **In most areas**, the use of watersheds is an obvious necessity in defining and understanding special patterns of aquatic ecosystem quality and addressing ecological risk.
   大半の区域において，流域の利用は，水生生態系の質の空間パターンの画定・理解，ならびに生態学的リスクへの対処において明らかに必要不可欠である。
5. **The most accurate way** to decrease sample variability is to collect from only one type of habitat within a reach and to composite many samples within that habitat.
   サンプル変動性を減じるための最も正確な方法は，ある河区において1種類だけの生息場所から採集し，その生息場所における多数のサンプルを複合することである。

- Most of ……：… の大半は

6. **Most of** the early legislation was geared toward the protection of waters for human use.
   初期の法律の大半は，人間による利用（有益用途）のための水域保護を目的として

いた。
7. The study revealed that **most of** the streams possessed physical habitat that enhanced the development of communities of diverse aquatic fauna.
この研究によれば，大半の水流は，多様な水生動物相群集の発生を促す物理的な生息地を有する。
8. The significance of these higher arsenic concentrations is not completely understood for **most of** the sites.
これら比較的高濃度の砒素の重大性は，大部分の地域で完全には理解されていない。

- **The best and most .....** ：最善かつ最も … な

9. **The best and most** representative sites for each stream class are selected and represent the best set of reference sites from which the reference condition is established.
各水流等級について最善かつ最も代表的な地点が選ばれて，参照地点セットにされ，それに基づき参照状態が定められる。

- **one of the most** ：最も … の一つは

10. **One of the most** important attributes is molecular weight.
最も重要な特質の1つは分子量である。
11. **One of the most** common spatial frameworks used for water quality management and the assessment of ecological risk and nonpoint source pollution has been that of hydrologic units (or watersheds).
水質管理と生態系リスクおよび面源汚濁のアセスメントに用いられる最も一般的な空間枠組みは，水文単位（すなわち流域）であった。
12. The smallmouth bass (Micropterus dolomieui) is **one of the most** important game species in midwestern rivers and streams. Furthermore, this is a species that requires little or no external support in the way of supplemental stocking.
コクチバス（Micropterus dolomieui）は，中西部の河川水流における最重要な漁獲対象種の1つである。さらに，補足的放流という外部支援をほとんどまたは全く必要としない魚種である。

- **~ is the most widely accepted .....** (See **Accept**)：
  ~は最も広く受け入れられている … である（➡ **Accept**）
- **~ is most affected** (See **Affect**)：~ は最も影響される（➡ **Affect**）
- **~ is most appropriate** (See **Appropriate**)：
  ~ が最も適している（➡ **Appropriate**）
- **it is almost certain that .....** (See **Certain**)：

- …は，ほぼ確実である（➡ **Certain**）
- ~ **play the most important role**（See **Role**）：
  ~は最も重要な役割を果たしている（➡ **Role**）
- **it is most likely to** .....（See **Likely**）：
  …の見込みがきわめて強い（➡ **Likely**）
- ~ **is most important**（See **Important**）：
  ~は最も重要である（➡ **Important**）
- **most importantly**（See **Importantly**）：
  最も重要なことに（➡ **Importantly**）

Much：多くの；かなりの

- **Much of** .....：かなりの…
- **Much has been written about** .....：…については多く書かれてきた
- **Not much is known concerning** .....：…に関してはあまり多くは知られていない

1. **Much of** this speculation has centered on arsenic.
   かなりの疑いがヒ素に集中してた。
2. **Much has been written about** the effects of petroleum on marine life, but few experiments have dealt.
   原油が海洋生物に与える影響については多く書かれてきたが，それに対する実験はほとんどされていない。
3. **Not much is known concerning** the nature and behavior of natural systems.
   自然系の性質と挙動に関しては，あまり多くは知られていない。

- **Much work has been done on** .....：
  …に関する研究がかなりされてきた
- **Much less work has been done on** .....：
  …に関する研究はあまりされていない
- **Much progress and advancement has been made**：
  多くの進歩と前進がなされてきた
- **Much more research is needed in** .....：
  …における研究がさらに必要とされる

4. Clearly, **much work needs to be done** to evaluate and polish the proposed technique.
   明らかにかなりの研究が，提案された技巧を評価して，そしてそれに磨きをかけ

5. **Much less work has been done** on the activation of less reactive metals.
   反応性がより低い金属の励起に関する研究は，あまりされていない。
6. **Much progress and advancement has been made** in the past decade.
   多くの進歩と前進が，これまでの10年間になされてきた。
7. **Much more research is needed** in such basic areas as sampling effort, and indicator sensitivity and variance.
   サンプリング活動，指標の感度・分散といった基礎領域における研究はさらに必要とされる。

　　● **as much as .....**：…と比べて同じぐらい

8. The former can be used just **as much as** the latter as a means of experimentation.
   前者は後者とまったく同じぐらい実験手段として使用される。
9. High-intensity ultrasound dramatically increases the rates of reaction by **as much as** 200-fold.
   強度の超音波は，反応速度を200倍くらい劇的に増過する。

Multiply (Multiple)：乗じる（複数の）

1. The number of taxa in each group **is multiplied by** the group's tolerance value (intolerant =3, facultative = 2, tolerant = 1).
   各集団における分類群数にその集団の耐性値（不耐性 = 3，任意性 = 2，耐性 = 1）を乗じる。

　　● **multiple**：複数の

2. This observation could be explained if **multiple** dechlorinating microorganisms were present in the filter.
   もし多種の脱塩素微生物がこのフィルターに存在していたなら，この観察は説明できるかもしれない。
3. Typically, the area of interest is an entire state, or a particular river basin or region of a state or **multiple** regions.
   通常，関心区域は，1州全体，もしくは1州の特定の河川流域，または地域や複数州地域である。
4. Design of such a repository would make use of **multiple** engineered and natural barriers.
   このような貯蔵所の設計は，技術によるバリア（遮断物）と自然なバリアを複式で活用している。

Multitude：多数の；多くの

1. **A multitude of** low marshy islands and sand bars occupy much of the central portion of the bay.
   湾の中央部の大部分は，多数の低沼地と砂州から成っている。
2. The study of potential synergistic or antagonistic effects of **a multitude of** toxic substances present in many effluents becomes virtually an impossible task.
   多くの排水中に存在している多数の毒性物質による潜在的相乗作用または拮抗作用を研究することは，事実上，不可能な作業となる。

N

Namely：つまり；すなわち

1. Feedwater conductivity is controlled for the same reason as reactor coolant conductivity, **namely** to minimize scale formation.
   供給水の電気伝導度は，反応器冷却水の電気伝導度をコントロールする理由と同じ理由でコントロールされる。すなわちスケール生成を最小限に抑えるためである。

Nature：本質；特質；性質

1. A key aspect of this approach is its **iterative nature**.
   このアプローチの重要な点は，それの持つ反復特性である。

   - **in nature**：本質的に

2. A variety of decisions must be made, not all of which are technical or scientific **in nature**.
   このためには様々な決定が必要なのだが，そのすべてが技術的決定や科学的決定というわけではない。
3. Reality is dynamic **in its nature**.
   現実は本質的にダイナミックである。

   - **the nature of .....**：… の性質
   - **The nature of ～ is.....**：〜の本質は … である

4. Little is known about the chemical **nature of** organic matter in groundwater.
   地下水での有機物質の化学的性質については，ほとんど知られていない。
5. Not much is known concerning **the nature** and behavior of natural systems.
   自然界のシステムの性質と挙動に関しては，あまり多くは知られていない。
6. Because of **the nature of** ecoregions, the ideal way of evaluating them would be through use of an ecological index of integrity.
   生態地域の性質故に，その理想的な評価方法は，生態学的な健全性指標を用いることだろう。
7. Resource managers and scientists have come to realize that **the nature of**

these resources varies in an infinite number of ways, from one place to another and from one time to another.
資源管理者や科学者は，これらの資源の性格が，場所ごと，時間ごとに無限な様相で異なることを理解するようになっている。

8. This is probably attributable to **the variable nature of** activated sludge.
これは，おそらく活性汚泥の可変的性質に帰することが出来る。

9. Satisfying this requirement involves understanding **the nature of variability** that originates with sampling frequency and/or seasonal influences.
この要件を満たすには，サンプリング頻度および／または季節的影響に由来した可変性を理解する必要がある。

Near (Nearly)：近く；近くに（ほとんど；ほぼ）

1. The sampling site is at the downstream end of the reach, **near** the confluence of Cherry Creek and the Portneuf River.
サンプリング地点は，Cherry 小流と Portneuf 川の合流近く，その下流区域の終わりにある。

2. The arsenic concentrations will not change very much **in the near future** regardless of any channel shifting.
水路が移動したとしても，ヒ素の濃度は近い将来ではそれほど変化しないであろう。

3. Biosurveys were conducted **near** Conesville Station from 1988 to 1991 using methods that conformed to USEPA protocols or with minor deviations.
USEPA プロトコルに準拠した，もしくはやや逸脱した方法を用いて，1989 年から 1991 年まで Conesville 発電所付近で生物調査が行われた。

● **nearly**：ほとんど；ほぼ

4. New York DEP has conducted biosurveys throughout the state for **nearly** 20 years.
ニューヨーク州環境保護局 (DEP) は，約 20 年間にわたり州全域で生物調査を行ってきた。

5. The modeling of certain constituents will not be **nearly** as critical to answering study question.
ある特定成分をモデル化することは，研究疑問に答えることほど重要ではないであろう。

6. Response to the question in **nearly** every case has been that ecoregions are the desired regional framework.
この質問への返答は，ほとんどの場合，生態地域が望ましい地域枠組だという

ものだった。
7. Carbon dioxide concentrations in atmosphere have increased **nearly** 30%.
大気中での二酸化炭素の濃度がおおよそ30％増加した。
8. Some of the original models have been changed in **nearly** every subsequent version, whereas others have been largely retained.
オリジナルモデルのいくつかは，その後のバージョンほぼすべてで変更されたが，そのまま保たれたモデルも多い。

- **nearly all of** ..... (See **All**)：ほぼすべての … (➡ **All**)

Necessary：必要な

1. These data provided the **necessary** information to develop and calibrate a model of this system.
これらのデータは，このシステムのモデルを構築しキャリブレートするために必要な情報を提供した。
2. Thus, a weight-of-evidence approach can become increasingly obvious when hypothesis testing demonstrates the **unnecessary** usage of independent application
それ故，証拠加重値手法は，仮説検定が独立適用の不要さを実証した場合には，ますます明瞭になりうる。

- **it is necessary to** ..... ：…することが必要である

3. **It is necessary to** recognize that the proposed model depends on its intended purpose.
提案されたモデルが，意図した目的に依存することを認識することが必要である。
4. In regard to the dynamic behavior of systems, **it is necessary to** consider the concepts of ....
システムのダイナミックな挙動に関して，…の概念を考慮することが必要である。
5. Whatever the reasons for the differences, **it is necessary to** keep in mind that .....
この相違の理由が何であるとしても，…を念頭におくことが必要である。
6. Seasonality is a well-understood concept, therefore, **it is not necessary to** sample in multiple seasons for the sake of data redundancy.
季節性は，十分に理解されている概念であるから，データ冗長性のために複数季節でサンプリングする必要がない。
7. **It may 6e necessary to** select reference sites from another class of stream, lake, or wetland if all sites in a particular class are highly disturbed.

ある等級における地点すべてがきわめて攪乱されていれば，別の等級の水流，湖沼，湿地から参照地点を選ぶ必要もあろう。

- **〜 is necessary**：〜が必要である
- **〜 is made necessary**：〜が必要になる

8. Modeling **is necessary** in making decisions about water resources.
モデルが，水資源についての決断をくだすために必要である。
9. More extensive and detailed investigations **are necessary to** gain a further understanding of these processes.
いっそう大規模かつ詳細な調査が，これらのプロセスをいっそう理解するために必要である。
10. As we increase our awareness that a holistic ecosystem approach to environmental resource assessment and management **is necessary**, we must also develop a clearer understanding of ecosystems and their regional patterns.
我々は，環境資源の評価・管理への全体論的生態系手法が必要だとより深く認識するにつれて，生態系とその地域パターンもより明確に理解しなければならない。
11. This distinction **is made necessary** by the widespread degree to which macrohabitats have been altered among the headwater and wadable streams in the HELP ecoregion.
この区別が必要になった理由は，HELP 生態地域における源流・徒渉可能水流で，マクロ生息地が大幅に改変されたことである。

- **if necessary**：もし必要なら

12. The latter can be dealt with on a case-by-case or site-specific basis **if necessary**.
後者は，必要ならケースバイケースもしくは地点特定ベースで対処されうる。

Necessity：必要

- **〜 is a necessity**：〜 は必要不可欠である

1. In most areas, the use of watersheds **is an obvious necessity** in defining and understanding special patterns of aquatic ecosystem quality and addressing ecological risk.
大半の区域において，流域の利用は，水生生態系の質での空間パターンの画定・理解，ならびに生態学的リスクへの対処において明らかに必要不可欠である。

2. Watershed studies **are a necessity**, but equally important is the development of an understanding of the spatial nature of ecosystems, their components, and the stress we humans put upon them.
   流域研究は必要であるが，同じく重要なのは，生態系の空間的性質，その構成要素，そして我々人間が生態系に与えるストレスへの理解を育むことである。

   ● **out of necessity**：必要に迫られて

3. **Out of necessity**, these sites needed to be high quality and located on streams well removed from major anthropogenic pollution sources.
   必要に迫られ（必然的に），これらの地点は，質が高くなければならず，また，主要な人為的汚濁源が十分に除かれた水流に所在していなけれならない。

Need：必要性；必要である

1. The results of this assessment are used to help determine **the need for** remedial action, to select the remedial action, and to determine risk-based cleanup levels for the remedial action.
   このアセスメントの結果は，修復活動の必要性を決定するのを手伝うため，修復活動を選択するため，そして修復活動のリスクベースの浄化レベルを決定するため，に使われる。
2. They saw **the need for** research to better understand the process by which the regions are defined, and how quantitative procedures could be incorporated with the currently used, mostly qualitative methods to increase replicability.
   彼らは，各地域が定義される過程や，反復可能性を増すため目下用いられているおおむね質的な方法に量的手順がどれほど盛り込まれるか，を一層理解するための研究の必要性を認めた。

   ● **needs**：必要性；ニーズ

3. The model has been judged potentially capable of meeting the **user's needs**.
   このモデルは，潜在的に使用者の必要性を満たすことが出来ると評価されてきた。
4. The purpose of this article is to briefly describe some critical research and **information needs** in surface water quality.
   この論文の目的は，表面水の水質における重要な研究のいくつかと情報の必要性を簡潔に説明することである。
5. Each of these types of criteria was developed to address specific **regulatory needs** of managements as related to the permitting of point source

discharges.
これらの種類のクライテリアそれぞれは，点源排出の許可に関連した管理の具体的な規制ニーズに取り組むために設定された。

- **Needs for 〜 exist**：〜の必要性が存在する
- **A need exists for .....**：…のための一つの必要性が存在する

6. **Needs for** ecoregional frameworks **exist** al all scales.
   生態地域枠組みの必要性は，あらゆる縮尺で存在する。
7. **A need exists for** prediction of future trends and control in contaminated areas.
   汚染された地域に対し，これからの傾向の予測と管理の一つの必要性が存在する。

- **needs include.....**：必要性は … を含む。
- **needs have been identified**：必要性が識別されてきた
- **the need for 〜 was established**：〜の必要性が確立された

8. **Needs include** various types of information concerning toxic substances in water supplies.
   必要とするものの中には，供給水中の毒性物質に関する種々のタイプの情報が含まれる。
9. Based on the existing data and the conceptual model, the following data **needs have been identified**: 1) ....., 2) ......, and 3) ....
   既存のデータと概念モデルに基づいて，データの必要性が次のように識別されてきた。1)…，2)…，3)…である。
10. The **need** for a cost−effective "rapid" biological assessment method **was established**.
    費用効果的で「迅速な」生物アセスメント方法の必要性が確立された。

- **Need for 〜 becomes evident in .....**：
  〜の必要性は… において顕著となる

11. **Need for** establishment of modified biocriteria due to limitations in habitat quality **will become evident in** applications of habitat assessment routines.
12. 生息地質における限界故に，改善された生物クライテリアを定める必要性は，生息地アセスメントルーティンの適用において顕著となろう。

- **There is a need for .....**：…のために必要である
- **There is a need to .....**：…することが必要である

13. **There is a need for** an alternate tool such as a water quality model.
    水質モデルのような代わりの手段が必要である。

14. **There is** also **a need to** develop models which can be used to predict water quality changes within distribution systems.
    また，給水システム内での水質変化を予測するために使うことができるモデル構築の必要性もある。

    - ~ **underscore the need to** ..... : ～は … する必要性を強調する
    - ~ **demonstrate the need to** ..... : ～は … する必要性があることを示す

15. It also **underscores the need to** develop both technologies and management strategies to protect drinking water supplies.
    それはまた，飲料水供給を保護するための技術戦略と管理戦略の両方を開発する必要性を強調する。

16. This analysis **demonstrates the need to** access and interpret community information beyond the ICI and individual metrics.
    この分析は，ICIおよび各メトリックを超えて生物群集情報にアクセスし解釈する必要性があることを示している。

    - **with the increased need for** ..... : … の必要性が増加するとともに

17. **With the increased need for** predicting the movement and distribution of chemicals in an aquatic environment, knowledge of the rate constants governing the uptake and clearance of chemical in fish is required.
    水生環境での化学物質の移動・分配を予測する必要性が増加するとともに，魚体内での化学物質の摂取・除去を左右している速度定数の知識が必要とされる。

    ★ Need を動詞として使う場合の例文を下に示します。

18. Pilot studies or small-scale research **may be needed** to define, calibrate, and evaluate models.
    モデルを確定し，較正し，評価するには，予備的研究または小規模研究が必要だろう。

19. As suggested above, much more research **is needed** in such basic areas as sampling effort, and indicator sensitivity and variance.
    上に示唆されるように，サンプリング活動，指標の感度・分散といった基礎領域における研究はさらに必要とされる。

    - ~ **need to be done to** ..... : ～が … するためには必要である
    - ~ **need to be evaluated to** ..... :
      ～が … するためには判断されるべきである
    - ~ **need to be examined in detail** :
      ～について詳しく調査する必要がある

20. Clearly, much work **needs to be done to** evaluate and polish the proposed technique.
明らかにかなりの研究が，提案された技巧を評価して，そしてそれに磨きをかけるためには必要である。
21. The condition of the pavement in the area **needs to be evaluated to** determine the potential for continued infiltration and leaching.
その区域の舗装面の状態を評価し，継続的な浸入・浸出の可能性が判断されるべきである。
22. The ecological damages caused by the river modification **needs to be examined in detail**.
河川修正工事が生態におよぼした損傷について詳しく調査する必要がある。

Neither：どちらも … ではない

1. Classification forms the foundation of science and management because **neither** can deal with all objects and events as individuals.
分類は，科学と管理の基礎をなす。なぜなら，どちらもすべての対象や事象を個別にさえ扱うことはできないからである。

- **Neither A nor B**：AでもなければBでもない

2. **Neither** TCE disappearance **nor** chloride appearance was noted during 2 hr illumination of a catalyst−free reactor.
TCEの消失と塩化物の生成のいずれも，無触媒反応器での2時間の照射中に確認されなかった。

Nevertheless：それにもかかわらず

1. **Nevertheless**, courts have noted that the implementation of water quality criteria that are not based on chemical−specific numeric criteria can be difficult as a matter of practice.
それでも法廷は，化学数値クライテリアに基づいていない水質クライテリアは，実際問題として難しいと指摘してきた。

No：なかった

- **No A, nor B**：AがなかったしBもなかった

1. **No** reaction products were identified, **nor** was dehalogenation actually

demonstrated.
反応生成物がみつけられなかったし，脱ハロゲンも事実上証明されなかった。
2. **No** gas evolution was observed, **nor** were any other products detected by GC in either the gas or liquid phase.
ガスの発生が観察されなかったし，また，他のいかなる生成物も GC によっては気体相または液体相で検出されなかった。

Nonetheless：それでも

1. At present, few states can characterize even a small fraction of their water resources with biological survey data. **Nonetheless**, some states are instituting comprehensive biological monitoring networks based on a rotational basin approach, wherein waterbody assessments rotate among watersheds on regular intervals.
現在，生物調査データで各自の水資源のごく一部でも特性把握できる州は，きわめて少数である。それでもいくつかの州は，交代流域手法に基づき総合的な生物モニタリングネットワークの構築に取り組んでいる。この場合，水域アセスメントは，定期間隔で対象分水界が交代するのである。
2. **Nonetheless**, to guard against potential legal attacks on water quality criteria based on chemistry or toxicity, agencies should be prepared to explain fully the limitations as well as the capabilities of bacteria.
それでも，化学的特性または毒性に基づく水質クライテリアへの法的攻撃に対し防護するため，諸機関は，生物クライテリアの限界と能力を十分に説明できるよう備えるべきである。

Notable：顕著な

1. The most **notable** effect of copper inhibition is shown by the 0.4 mg/L copper experiment.
銅による抑制の最も顕著な効果は，0.4 mg/L の銅による実験で示される。
2. One **notable** exception is cadmium which was detected at higher than commonly expected values throughout the study area.
1つの顕著な例外はカドミウムで，一般的予想値より高い値が調査地域全体で検出された。

Note：留意する

- note that ......：… ということに留意してください

Note

1. Hunt (2001) reviewed the inhibitory effects of many compounds and **noted that** .....
   Hunt (2001) は多くの化合物の抑制効果をレビューして，.....と指摘した。
2. **Note that** all coefficients indicated in the figure are used for the analysis.
   数字で示されている係数すべてが，この分析に使われることに留意してください。

   - **It is important to note that** ..... ：… を留意することは重要である
   - **It must be noted that** ..... ：… ということを留意すべきである
   - **It should be noted that** ..... ：… ということは指摘されるべきである
   - **It is interesting to note that** ..... ：… に注目すれば，興味深い
   - **It is of interest to note that** ...... ：… を指摘することは興味がある
   - **It is interesting to note that** .... ：… を指摘することは興味がある

3. **It is important to note that** the four-tiered system has been widely used in biomonitoring.
   この四段層方式が生物モニタリングで多用されていることを念頭におくべきである。
4. **It is important to note that** most models represent an idealization of actual field conditions and must be used with caution to ensure that the underlying model assumptions hold for the site-specific situation being modeled.
   ほとんどのモデルは，実際の現地状態を理想化していること，そしてモデルの対象となっている特定な状況に対してモデルの基礎仮定が有効であることを保証するため注意して使われなくてはならないということ，を指摘することは重要である。
5. Although the probability of all ecosystems becoming unhealthy in the same manner is unrealistic, **it is important to note that** response signature are definable based on patterns of specific perturbations.
   すべての生態系が同じように不健全になる確率は，非現実的だが，具体的擾乱パターンに基づき応答サインが定義可能なことに注目することが重要である。
6. **It must be noted that** numerous sites along a stream are examined before making a use decision.
   使用決定を下す前に水流沿いの多数地点が検討されることを留意すべきである。
7. **It should be noted that** the drinking water criteria for manganese is more than two orders of magnitude lower than the irrigation water criteria.
   マンガンの飲料水基準が潅漑水基準より二桁以上低いことは，指摘されるべきである。
8. **It is interesting to note that** the USEPA database also includes sites sampled over multiple years in areas of lesser quality and impacted by human activities.

USEPAデータベースが質の劣る区域で複数年度にわたりサンプリングされ，人為的活動に影響された地点も含んでいることに注目すれば，興味深い。

- **unless otherwise noted**(See **Unless**)：
  注意書きが無い場合は(➡ **Unless**)：
- **as noted above**(See **Above**)：上に指摘したように(➡ **Above**)

Noteworthy：注目すべき

1. **Noteworthy is** the substantial removal of lead from the water.
   この水からの鉛の本質的な除去は，注目に値する。

   - **It is noteworthy that** ‥‥‥：… は注目すべきである

2. **It is noteworthy that** the drinking water criteria for manganese is more than two orders of magnitude lower than the irrigation water criteria.
   マンガンの飲料水基準が潅漑水基準より二桁以上低いことは，注目すべきである。

Notice：気付く；通告

1. When flying over the Southern Rocky Mountains along the Wyoming/Colorado state line in mid-1980, this author **noticed** differences in timber management practices between states.
   1980年代中期，ワイオミング州／コロラド州の境界に沿ったSouthern Rocky 山脈の上空を飛んだ時に，著者は，両州での材木管理慣行の相違に気づいた。
2. One legal critic, for example, questioned whether biocriteria are reliable, scientifically repeatable, provide dischargers with adequate **notice of** acceptable conduct, and can reliably determine cause-and-effect relationships.
   例えば，ある法律批評家は，生物クライテリアが信頼でき，科学的に反復でき，容認可能な行為の十分な通告を排出者へ提供でき，因果関係を確かに決定できるか否か，疑問視した。

Notion：概念

- **The notion that** ‥‥‥：… という観念
- **There is general acceptance of the notion that** ‥‥‥：
  … であるという概念が一般的に受け入れられている

1. **The notion that** fish are too mobile to use effectively as an indicator group

is perhaps the most often raised liability of using this group.
魚類は，指標集団として適切に用いるには移動性が高すぎるという観念は，おそらく最もしばしばあげられるこの集団の利用を阻む欠点だろう。

2. **There is general acceptance of the notion that** organic compounds may play a major role in changing speciation of trace metals.
有機化合物が，微量金属の化学種の変化に主要な役割をしているであろうという概念が一般的に受け入れられている。

Notwithstanding：それにもかかわらず

1. **Notwithstanding**, few can tolerate heavy pollutional stress and, as such, can be good indicators of environmental conditions.
それにもかかわらず，強度な汚濁ストレスに耐えるものは，ごく少数なので，環境状態の優れた指標となる。

Number：数

1. There was a significant correlation between the calculated zinc concentration and **the number of** cells present.
計算された亜鉛濃度と存在しているバクテリア数の間に有意な相互関係があった。

- a number of ‥‥‥：多数の…
- a considerable number of ‥‥‥：かなりの数の…
- an infinite number of ‥‥‥：無数の…
- total number of ‥‥‥：…の総数

2. As a result of **a number of** experimental studies, ‥‥‥
多くの実験研究の結果として，…

3. **A number of** kinetic models have been prepared.
多数の動力学モデルが(今まで)用意されてきた。

4. **A number of** factors contribute to the differences in measured values of DDTR concentrations in biota.
多くの要因が，生物体内DDTRの測定濃度の相違をもたらしている。

5. **A number of** these documents were reviewed during the course of this study and the review document is appended as Appendix A.
この研究過程で多くのこれらの文章がレビューされ，そのレビュー文章が付録Aとして付加されている。

6. There are **a number of** other federal programs that involve research and monitoring activities relevant to water criteria development.

他にも，水質クライテリア設定に関わる研究・監視活動に関連する多数の連邦プログラムが存在する。

7. This application can be useful in **a number of** ways.
   この適用は，多くの方法において有用となりうる。
8. While **a number of** HSPF applications are in progress, few studies are complete.
   多数の HSPF アプリケーションが進行中である一方，ほとんど研究が完成していない。
9. Data were gathered from **a number of** experiments using a pilot-scale reactor.
   試験規模の反応槽を使った多数の実験から，データが収集された。

- **a considerable number of** .....：かなりの数の…
- **a limited number of** .....：限られた数の…
- **an infinite number of** .....：無限な数の…
- **an increasing number of** .....：数が増えつつある…

10. Over the past few decades, **a considerable number of** studies have been conducted on the effects of temperature on the growth of nitrifying bacteria.
    ここ数十年にわたり，温度が硝化菌の成長に及ぼす影響に関してかなりの数の研究が行なわれてきた。
11. Ecological studies typically focus on **a limited number of** parameters that might include one or more of the following: (a), (b), and (c).
    生態学研究が主眼を置くのは，通常，限られた数のパラメータであり，これには次のパラメータのうち1つ以上が含まれる。(a), (b), および (c)。
12. Resource managers and scientists have come to realize that the nature of this resource varies **in an infinite number of** ways, from one place to another and from one time to another.
    資源管理者や科学者は，これらの資源の性格が，場所ごと，時間ごとに無限な様相で異なることを理解するようになっている。
13. **An increasing number of** multistate and multiagency cooperative efforts are focusing on the development of biological criteria.
    生物クライテリア設定に力点を置いた多州・多機関の協力活動が数を増やしている。

- **in number**：数において

14. Their populations have been severely reduced **in number** in the last few years.
    これまでの数年で，それらの個体数がひどく減少していた。
15. The report revision represents a significant increase **in the number of** Solid

Waste Management Units at LANL.
この修正された報告書は，LANLにおける廃棄物処理ユニット数の重要な増加を表している。

Objective：目的

- **The objective of** ～ **is to** .....：～の目的は … である
- **One objective of** ～ **is to** .....：～の1つの目的は … である
- **The key objective of** ～ **is to** .....：～の主要目的は … である
- **The overall objective of** ～ **is to** .....：～の全体的な目的は … である
- **A principal objective of** ～ **is to** .....：～の主要目標は … である

1. **The objectives of** the research **were to** gather all possible information available to define potential contaminant source area.
   この研究の目的は，汚染源の可能性がある場所を明らかにするために可能な情報すべてを収集することであった。
2. **One objective of** the initial site investigation **should be to** determine the mass of contaminants adsorbed to the soil.
   最初の現地調査の1つの目的は，土壌に吸着した汚染物質量を決定すべきことである。
3. **The key objective of** this analysis **is to** determine whether or not the feedback gained from the biological community can communicate about and characterize these differences.
   この分析の主要目標は，生物群集から得られたフィードバックがこれらの差を伝達し特性把握できるか否か決定することである。
4. **The overall objective of** this research **is to** conduct basic studies into possible causes of biological nitrification process instability.
   この研究の総合目的は，生物硝化プロセス不安定性を起こしうる原因に関する基礎研究を行うことである。
5. **A principal objective of** the CWA is to restore and maintain the biological integrity of the nation's surface waters.
   CWAの主要目標は，米国の地表水域の生物健全性を回復・維持することである。

   ★ CWAはClean Water Actの略語で，水質汚濁防止法と訳されています。

- **The objectives of** ～ **is as follows** .....：～の目的は次のとおりである
- **The objectives of** ～ **are three-fold** .....：～は次の3つの目的を持つ

6. **The objectives of** the limited sampling program at ST6 **are as follows**: .....
   ST6地点での，限定されたサンプリングプログラムの目的は次の通りである。
7. **The objectives of** this research project **are three-fold**: 1) .....; 2) .....; and 3) .....
   この研究プロジェクトは，次の3つの目的を持つ。1)…，2…，そして3)…である。

   - ~ **have the following objectives** ：～は次の事を目的としている
   - **in achieving the objectives** ：これらの目的の達成にあたって
   - ~ **is part of the objective of** ..... ：～は … の目的の一部

8. The investigation reported in this paper **had the following objectives**: 1) ....., 2) ....., and 3) .....
   この論文で報告された研究は次の事を目的としていた。1)…，2…，そして3)…である。
9. Each remedial action can be readily screened to determine its effectiveness **in achieving the specified objectives**.
   修復活動のひとつひとつが，特定目的の達成にあたってその有効性を決定するために，容易に選別できる。
10. Elucidation of reaction mechanism **was not part of the objective of** the present work.
    反応機構の解明は，この研究の目的の一部ではなかった。

Observation (Observational) ：観察(観測的)

1. **Observations over** seasonal patterns indicate that .....
   季節のパターンを通しての観察は … を示す。
2. This **observation** could be explained if multiple dechlorinating microorganisms were present in the filter.
   もし多種の脱塩素微生物がこのフィルターに存在していたなら，この観察は説明できるかもしれない。
3. A comparison of the **observations with** the computed results shows good agreement.
   観察結果と計算結果の比較すると，それらは良い整合を示している。
4. Several possible mechanisms may account for the above **observations**.
   いくつかの可能なメカニズムによって，上の観察結果が説明されるかもしれない。
5. This is similar to **observations** that we have made in the Scioto River downstream from Columbus, Ohio, and elsewhere.
   この観察結果は，オハイオ州Columbus市のScioto川下流および他の場所におけ

6. The question of the stability of organomercury compounds is raised by some of the following **observations**.
有機水銀化合物の安定性に対する質問が，次のいくつかの観察によって提起される。
7. These **observations** raised several questions which we attempted to answer to some degree in the experiments described below.
これらの観察はいくつかの質問を提起した。その質問に，我々は下に記述した実験にて，ある程度答えを出そうと試みた。
8. The major **observations** regarding the fate of trace metals identified in the water column metal study by Inouye (2002) include: .....
Inouye (2002) による水コラム金属研究で明らかにされた微量金属の成り行きに関しての主要な観察は.....を含む。

- **observational** ：観測的

9. For kick sample sites, substrate particle size is obtained by making **observational** estimates to designate percentage of each of the seven EPA size categories, as listed in Weber (1993).
キックサンプル地点について，基質の粒子サイズは，Weber (1973) により列挙された7つのUSEPAサイズ区分それぞれのパーセンテージを指定するための観測的推定で得られる。

Observe：観察する

1. It was found that the **observed** processes are described satisfactory by this model.
観察されたプロセスが，このモデルによって十分に説明されることがわかった。
2. The computational results along with the **observed** data are shown in Fig. 4.
計算結果が，観察データとともに図4に示されている。
3. A comparison of the **observed** data with the computed results shows good agreement.
観察データと計算結果との比較では良い一致を示している。
4. A reaction mechanism has been proposed that accounts for the **observed** experimental data.
観察された実験データを説明する反応機構が提案された。
5. The excess nutrients are the primary cause of the reduced oxygen levels **observed** in the lake.
この湖で観察される酸素レベルの減少は，過剰栄養が主要な原因である。
6. If a relationship **was observed**, a 95％ line of best fit was determined and

the area beneath trisected following the method recommended by Fukuchi et al. (1984).
関係が見られたならば，ベストフィットの95％ラインが決定され，その下方の面積は，Fukuchiら（1984）の推奨方法に従って三等分された。

Obtain (Obtaining) (Obtainable)：得る；入手する（得ること）（入手可能な）

- **..... obtained** ：得られた…

1. The value **obtained** here is in good agreement with that of Suzuki et al. (2002).
   ここで得られた数値は Suzuki ら (2002) が報告した値と良く一致する。
2. A brief summary is included below as an aid to understanding the results **obtained** in this project.
   このプロジェクトで得られた結果を理解する手助けとして短い要約が下に含まれている。
3. The multivariate approach provides a powerful means of maximizing ecological information **obtained from** periphyton assemblages.
   この多変量手法は，付着生物群がりから得られた生態学的情報を最大化するための有効手段となる。
4. Considering all the data **obtained** in the study, the reaction appears to be zero-order, i.e., the rate is independent of the ammonia concentration.
   この研究で得られたすべてのデータを考慮すると，この反応はゼロ次であるように思われる。すなわち，その速度はアンモニアの濃度の影響を受けない。

- **〜 is obtained** ：〜が得られる

5. A comparison between SESOIL results and those of Jones et al. (1983) **was obtained**.
   ジョーンズら (1983) の結果と SESOIL の結果との比較が得られた。
6. Good agreement **was obtained** between experimentally determined stability constants and available literature values.
   実験を通して決められた安定性定数と利用可能な文献値との間で良い一致が得られた。
7. The temperature dependence **was obtained by** plotting the rate constant vs. temperature.
   この速度定数を温度に対してプロットすることによって，温度依存性が得られた。
8. Additional insight into the fate of DDE **can be obtained by** considering the magnitudes of the parameters affecting chemical fate.

化学物質の推移に影響を与えるパラメターの規模を考慮することによって，DDEの推移について追加の洞察を得ることができる。

9. Four kick sample sites, substrate particle size **is obtained by** making observational estimates to designate percentage of each of the seven EPA size categories, as listed in Weber (1993).
四つのキックサンプル地点について，基質の粒子サイズは，Weber (1973) により列挙された7つのUSEPAサイズ区分それぞれのパーセンテージを指定するための観測的推定で得られる。

   ● **to obtain** ..... ：…を得るための

10. Sufficient attempts were made **to obtain** water samples.
水サンプルを得る十分な試みがされた。

11. A series of experiments were initiated in an attempt **to obtain** results similar to run #1.
実験#1に類似する結果を得る試みで一連の実験が始められた。

   ● **obtaining** ..... ：…を得ること

12. The protocol provided the means for **obtaining** such an answer in a defensible way, and the results have been used to make regulation decisions.
このプロトコルは，そうした回答を弁護可能な形で導くための手段となった。そして，結果を用いて規制決定が下された。

13. One common source of error arises from the difficulty **of obtaining** a representative soil sample and the lack of reproducibility of organics analysis of soils.
エラーを生じさせる一般的原因1つは，代表的な土壌サンプル採集の困難さ，それに土壌の有機分析の再現性の欠如である。

   ● **obtainable** ：入手可能な

14. The unit description includes all current, reasonably **obtainable** information on the unit.
このユニットの記述は，このユニットについての現在の，ある程度入手可能な情報すべてを含む。

Obvious (Obviously)：明白な；当然な(明白に；明らかに)

1. The **obvious** question concerns the discrepancies between this and earlier reports.

この当然な質問は，この報告書と前の報告書の間の矛盾に関係している。

2. The **obvious** need for every organism group to have a dispersal mechanism is critical if that species is to be sustained.
その生物種が持続されるべきならば，あらゆる生物集団が分散メカニズムを持つことの明らかな必要性がきわめて大事である。

3. Knowledge of the factors affecting the movement and transformations of these substances are of **obvious** importance in understanding and controlling the hazard.
これらの物質の移動・変化に影響を与えている要因に対する知識は，危険を理解し，またコントロールするうえで明らかに重要である。

- ～ **is obvious**：～は明瞭である；明白である
- ～ **is less obvious**：～は明瞭さに欠ける；明白さに欠ける

4. While the mobility of fishes compared to the more immotile groups **is obvious**, this alone does not disqualify fish as a valid indicator.
あまり動けない集団に比べて魚類の移動性は，明瞭だが，この点のみでは，魚類が有効な指標として不適格になるわけではない。

5. The trend for whole body game fish **is less obvious** and not necessarily declining.
ゲーム魚一体による（魚の体内濃度）傾向は，明白さに欠け，そして必ずしも低下していない。

- **obviously**：明らかに

6. The approach in using sonication is **obviously** applicable to organic compounds which contain halides.
ソニケーションを使うアプローチは，ハロゲンを含む有機化合物に対し明らかに適用できる。

7. **Obviously**, these choices must be made with considerable care and documented so that they do not include fundamentally different communities.
明らかに，こうした選択をする際，根本的に異なる群集が含まれないようにするため，かなり入念に行って文書化しなければならない。

Occupy：成る；占める

1. A multitude of low marshy islands and sand bars **occupy** much of the central portion of the bay.
湾の中央部の大部分は，多数の低沼地と砂州から成っている。

2. Jamaica Bay **occupies** approximately 52 km^2.
   ジャマイカ湾の面積は，およそ 52 km2 である。
3. Different classes of algae have different proportions of internal structure **occupied by** vacuoles, which are important sources of variation in measurements of biovolume.
   異なる綱の藻類は，液胞に占められた内部構造の比率が異なっており，それは，生物体積測定における変動の主因となる。

Occur (Occurring) (Occurrence)：起こる；発生する (生じること) (発生)

1. A long-standing problem **occurred**.
   長年の問題が発生した。
2. Potassium **occurs** in the range of 10 to 38 mg/L.
   カリウムが 10 から 38 mg/L の範囲で発生する。
3. A long-standing heavy metals problem in this stream **occurred** below an industrial effluent discharge.
   この水流における長年の重金属問題は，産業廃液排出口の下流で発生した。
4. In areas where large meanders **occur** in the existing channel, ...
   既存の水路で大きい蛇行がおこる地域で，…
5. Bank erosion problems often **occur** hand-in-hand with riparian vegetation disturbance.
   河岸浸食 (bank erosion) 問題は，しばしば水辺植生攪乱と一緒に発生する。
6. Related phenomena **occur** with cavitation in liquid-solid system.
   関連した現象が液体-固体システムでのキャビテーション (空洞現象) によって起こります。
7. Determination as to whether a release **has occurred** is generally made at the time a waste line is decommissioned.
   放出が起こったかどうかについての判断は，一般的に廃物ラインが使用済みになる時に下される。
8. A consistently high percentage of agreement **occurred** only when both the volunteer and professional analyses rated the site in the fair/poor ranges.
   高率の合致が一貫して生じる事例は，ボランティアと専門家の両方が可／不可の値域にある地点を対象とした場合に限られていた。

   ● **occurring**：生じること

9. The difficulty of dealing with a large mass of information is that of the many interactions **occurring** within the community, some may be related to water quality while others may not.

大量の情報を取り扱う場合の困難は，生物群集において多くの相互作用が生じることであり，それらは，水質に関係したり，関係しなかったりする。

10. Photocatalytic oxidation processes have been shown capable of mineralizing a wide range of naturally **occurring** and anthropogenic organic compounds.
触媒光反応による酸化プロセスによって，広範囲の自然態物質や人造有機化合物を無機物に変化させることができることが示されている。

- **occurrence**：発生

11. This **occurrence** indicates that nickel did not remain within the treatment units.
この発生は，処理ユニット内にニッケルが残っていなかったことを示唆する。

12. ....., depending on the **occurrence** of rainfall events.
降雨事象の発生に依存して，…

Of .....：…のうち

1. **Of** the 13 VOCs detected, TCE was detected most frequently.
検出された13のVOCsのうち，TCEが最も頻繁に検出された。
2. **Of** these 45 states, about half have documentation (mostly in draft form) supporting the methods and analyses and providing program rationale.
これら45州のうち，ほぼ半数は。方法・分析を裏づけ，プログラムの根拠を提示する文書（主に草稿）を備えている。
3. **Of** these states, 25 states also have fish assemblage monitoring program.
これらの州のうち，25州もまた魚類群がりサンプリングプログラムを備えている。

- **Of all .....**：…すべてのうち

4. **Of all** the ions removed, several are particularly important.
除去されるイオンすべてのうち，いくつかは特に重要である。
5. **Of all** the BTEX chemicals contained in the various fuels used at Eielson, benzene has the greatest mobility in the groundwater system.
Eielsonで使われている種々の燃料に含まれるBTEX化学物質すべてのうち，ベンゼンが地下水系で最も大きい可動性を持っている。

Offer：提供する

1. This method **offers** similar information.
この方法は，似たような情報を提供する。
2. The project **offers** good insight into the anaerobic treatment of wastewater.

プロジェクトは，廃水の嫌気性処理に良い洞察を加えている。
3. Beside cost-effectiveness, rapid assessments **offer** standardization of analysis and accuracy of habitat classification.
費用効果性に加え，迅速アセスメントは，分析の標準化や，生息地分類の正確性をもたらす。
4. This often provides insight about the factors (s) responsible for degradation and **offers** a diagnostic capability.
これは，しばしば劣化の要因に関する洞察をもたらし，判別能力をもたらす。

- ～ offer an advantage (See Advantage)：
  ～は利点をもたらす(➡ Advantage)
- by offering .....：… を提供することによって

5. **By offering** a more robust framework based on multiple and regionally calibrated reference sites, the chance for deriving an inappropriate biocriterion is greatly reduced.
地域的にキャリブレーションされた複数参照地点に基づいたより確かな枠組みを提供することによって不適切な生物クライテリアが導出される確率は，大幅に低下する。

Once：いったん…すると

1. **Once** field data are collected, processed, and finalized, the next step is to reduce the data to scientifically and managerially useful information.
現地データが収集・処理・最終確定されたならば，次のステップは，データをまとめて科学的・管理的に有用な情報にすることである。

One：1つ；の一つは；人；もの；事

1. **One** objective of the site investigation.....
現地調査の1つの目的は…
2. **One** major factor in determining .....
…を決定することにおいての1つの重要な要因は…
3. **One** objective of the initial site investigation should be to determine the mass of contaminants adsorbed to the soil.
初現地調査の1つの目的は，土壌に吸着した汚染物質の量を決定するべきことである。
4. **One** common source of error arises from the difficulty of obtaining a representative soil sample and the lack of reproducibility of organics

analysis of soils.
通常起こりやすいエラー源の1つは，代表的な土壌サンプルを得ることが困難で，かつ土壌有機分析の再現性の欠如から生じる。

5. **One** subregion consists of disjunct areas at or above timberline with heavy snowpack and most of the alpine glacial lakes in the ecoregion.
ある小地域は，高木限界以上で雪塊氷原を持つ離散区域からなり，高山氷結湖沼の大半がこの生態地域にある。

- **one or more** ：1つ以上

6. Ecological studies typically focus on a limited number of parameters that might include **one or more of** the following:
生態学研究が主眼を置くのは，通常，限られた数のパラメータであり，これには次のパラメータのうち1つ以上が含まれる。

- **one such .....** ：そのような…の1つ

7. **One such** waste type is radioactive waste.
そのような廃棄物タイプの1つが放射性廃棄物である。

- **~ such as this one** ：このような~

8. Certainly streams **such as this one** should be protected and not be allowed to degrade to standards and expectations set for streams typical of most of the region.
確かにこのような水流は，保護されるべきであり，その地域の大半の典型となる一連の水流についての基準や期待値を下げることは許されない。

- **from one ..... to another** ：ごとに
- **from one ..... to the next** ：それぞれ
- **one another** ：お互いに；相対して

9. When selecting these sites one must account for the fact that minimally disturbed conditions often vary considerably **from one** subregion **to another**.
これらの地点を選ぶ際には，最小擾乱状態が小地域ごとに大幅に異なりうるという事実を考慮しなければならない。

10. The affinity of a particular resin for different ions varies considerably **from one** ion **to the next**.
異なったイオンに対する特定樹脂の相性は，一種のイオンから他イオンで，それぞれかなり違う。

11. Streams/watersheds within each set should be similar to **one another**

regarding "relative disturbance" and should reflect higher water quality than streams with watershed with similar proportions in each subregion but with greater human impact.
各セットにおける水流／流域は，「相対擾乱」に関して似ているべきであり，各小地域における類似部だが，人為的影響がより強い水流よりも高い水質を反映すべきである．

12. Relationships among species are complex and changes in **one** can lead to unexpected changes in **another**.
生物種間の関係は，複雑なもので，ある生物種での変化が他の生物種での予測不能な変化に通じうる．

- **one of .....**：…の一つは

13. Finally, **one of** the major concerns with regional reference sites is their acceptable level of disturbance.
最後に，地域参照地点に関わる重要事の一つは，擾乱の許容可能レベルである．

- **one of the best**（See **Best**）：最も…の一つは（➡ **Best**）
- **one of the most**（See **Most**）：最も…の一つは（➡ **Most**）
- **One**：人；もの；事

14. **One** must take into consideration the season, as well as precipitation and temperature deviations (both long and short term) and how they may have affected vegetation and land cover patterns.
まず，季節や降水量・温度の偏差（長期的／短期的）を考慮に入れたうえで，それらが植生および土地被度パターンにどれほど影響してきたかを検討すべきである．

15. Instead of simply counting the number of cells per unit area, **one** can determine cell biovolume and use it to account for the differences in sizes of cells that are enumerated.
単位面積当り細胞数を単に数える代わりに，細胞の生物体積を決定し，それを用いて列挙される細胞サイズの差を説明しうる．

16. A primary release is **one** from the primary contaminant source, a secondary release is one that occurs, for example, from the contaminated soil to the groundwater.
主要な流出は第一汚染源から生ずるもので，第二の流出は例えば，汚染土壌に生じ地下水まで浸透するものである．

Ongoing：進行中である

1. Laboratory work addressing some aspects of these problems and related

modeling efforts **are ongoing** at the Idaho State University.
これらの問題のある局面を扱う実験研究と関連したモデリングの努力がアイダホ州立大学で進行中である。

2. Little work has been published for lakes and reservoirs, but significant efforts **are ongoing** as part of the USEPA effort to develop consistent bioassessment protocols.
湖沼および堰止め湖に関する研究は，あまり発表されていないが，整合的な生物アセスメントプロトコルを設定するUSEPA活動の一環としてかなりの努力が払われつつある。

3. This project was designed to involve the county's citizens in **ongoing** waterways education and restoration and to foster the relationship between citizens and government to ensure that both are more responsive to the needs of the environment.
このプロジェクトの意図は，郡民が進行中の水路啓発・再生活動に関与するよう促すこと，市民と政府の関係を育んで，両者が環境ニーズにより対応的であるようにすることであった。

Only：ほんの…だけ；たった…だけ；ごく；たんに

1. For the most part, **only** very small streams have watersheds completely within any one of these subregions.
大部分で，ごく小さな水流だけがこれらの小地域のいずれかにすべて属する流域を持つ。

2. This issue can be addressed **only** briefly.
この問題にはごく手短に触れる。

3. Oxygen is not the **only** factor that can limit bioremediation.
酸素が，生物学的修復を制限することができる唯一の要因ではない。

● **Only a few** ..... ：ほんの数個の…

4. **Only a few** species of fish can survive in this river.
ほんの数種の魚類だけが，この川では生息できる。

5. **Only a few** rocks need to be sampled, because each has a periphyton assemblage with hundreds of thousands of individuals.
サンプリングに必要なのは，数個の岩石だけである。なぜなら，1個の岩石に個体数が数百ないし数千もの付着生物群がりがあるからである。

● **not only** ..... ：…ばかりか
● **not only** ..... , but ..... ：…ばかりなく，…である

- **not only ....., but also .....**：… だけではなく，… である

6. **Not only** with this result in a loss of valuable information, it will likely result in the continued degradation of the aquatic resource.
これは，貴重な情報の損失を招くばかりか，水資源の継続的な劣化を招く見込みが強い。
7. This survey has shown that a number of states are using **not only** the multimetric approach, **but** are often using multiple assemblages.
この調査によれば，多数の州が多メトリック手法ばかりでなく，しばしば複数群がりも用いている。
8. The 1972 act specifically required states **not only** to identify impaired waters, **but** to rank them in order of priority for cleanup and restoration.
1972年法は，具体的に諸州が被損傷水域を確認することに加え，それらを浄化・回復の優先順位で格づけすることも要求した。
9. Biocriteria could be used **not only** to demonstrate the degree of impairment downstream of a violator, **but also** help prove the likely source of the harm through comparative analysis.
生物クライテリアは，ある違反者よりも下流での損傷度を示すことのみならず，比較分析を通じて害の有望な発生源を立証することにも利用できる。
10. The mature females were **not only** smaller **but also** produced fewer eggs, resulting in fewer young in the next generation.
成熟した雌は，より小さいだけではなく，より少数の卵を産卵するため，次の世代で幼魚がより少数となる。

- **only limited .....** (See **limited**)：ほんの限られた … (➡ **limited**)
- **only when .....** (See **When**)：… のときにだけ (➡ **When**)

Onset：発端；始まり；開始；着手

- **at the onset of .....**：… の始めに

1. TCP was introduced **at the onset of** the test or 1 hr after spiking the reactor with TeCP.
実験の開始時または TeCP を反応器に加えてから1時間後に，TCP が投入された。
2. **At the onset of** each project, and at the initial idea and data gathering meetings, the question of whether ecoregions or special purpose regions are desired is always asked.
各プロジェクトの発端で，また，アイデアやデータ収集のための初回会合で，常に生態地域と特殊目的地域のいずれが望ましいか，という質問が発せられる。

Order：順；順序；順番

1. **The order of** increasing toxicity on a basis of the bacterial growth rate was copper, cadmium, and nickel.
バクテリアの成長速度によって決めた毒性が上昇する順は銅，カドミウムそしてニッケルだった。
2. Among the metal studied, **the preferred order of** uptake by activated sludge was found to be in the sequence of lead > copper > cadmium > nickel.
研究された金属の中で，活性汚泥による吸着の優先オーダーは鉛＞銅＞カドミウム＞ニッケルの順であることが判明した。
3. In a comparison of the photocatalytic degradation of acetic acid with that of the chloroacetic acids in oxygenated aqueous dispersions of TiO_2, **the order of** the initial rates of CO_2 release was dichloroacetic acid > chloroacetic acid > acetic acid.
酸素を投入し TiO_2 拡散した水において，酢酸の光触媒分解反応と塩化酢酸の光触媒分解反応の比較では，CO_2 放出の初期速度はジクロロ酢酸＞クロロ酢酸＞酢酸の順であった。

　● **in a ..... order**：…の順に
　● **in order of .....**：…の順位に

4. The sites are addressed **in a downstream order**.
これらの地点は下流に行く順で扱われる。
5. They listed these **in a decreasing order of** importance.
かれらは，重要性が減少する順にこれらを一覧表にした。
6. The 1972 act specifically required states not only to identify impaired waters, but to rank them **in order of** priority for cleanup and restoration.
1972年法は，具体的に諸州が被損傷水域を確認することに加え，それらを浄化・回復の優先順位で格づけすることも要求した。

　● **on the order of .....**：…のオーダーで

7. Atmospheric half-lives for PCBs may be **on the order of** weeks.
PCBの大気中での半減期は，数週間のオーダーであろう。
8. Recent estimates seem to indicate residence times **on the order of** 2–10 years.
最近の推定では，滞留時間が2年から10年のオーダーであるように思われる。

　● **in order to .....**：…するために

9. The model program was modified **in order to** eliminate the following limitations: ....
   このモデルプログラムは，次に示す限界をなくすために修正された。それらは.....である。
10. The kinetic expression for growth is modified **in order to** account for environmental effects.
    この成長速度の表現は，環境影響を説明するために修正されている。
11. **In order to** fully understand the potential impact of local resource management decisions, data should be available on the status of all the waterbodies within a watershed.
    地方の資源管理決定による潜在的影響を十分に理解するため，データは，ある流域におけるすべての水域について利用可能とされるべきである。
12. The field investigations were focused on the local study areas **in order to** obtain sufficient data to develop an understanding of the hydrogeochemical conditions.
    水理地質学化学的状態を理解するのに十分なデータを得るために，この野外調査は近郊の研究地域に絞られた。
13. **In order to** maximize the meaningfulness of extrapolations from these studies, we must develop a clear understanding of ecosystem regionalities.
    これら研究からの推定の意義を最大化するため，我々は，生態系の地域性へのより明確な理解を進展させるべきである。

- order of magnitude ......(See **Magnitude**)：
  … 倍；… の規模（➡ **Magnitude**）

Originate (Originating)：根元である；由来する（に始まる）

1. Satisfying this requirement involves understanding the nature of variability that **originates** with sampling frequency and/or seasonal influences.
   この要件を満たすには，サンプリング頻度および／または季節的影響に由来した可変性を理解する必要がある。

- **originate from** ...... ：… に由来する

2. The present high levels of acid deposition **originate** predominantly **from** human activities.
   現状の高濃度の降酸は主に人間活動による。
3. As was mentioned earlier, HCOO⁻ did not **originate from** the photochemistry of MeOH solvent.

以前に言及されたように，HCOO⁻ は MeOH 溶媒の光化学が根元ではない。

- **originating** ..... ：… に始まる

4. For example, the proportion of sites with IBI scores <30 is illustrated by following the vertical line **originating** at 30 on the x axis to its intersection with the CDF, the following the horizontal line until it intersects the y axis.
例えば，IBI スコアが 30 点未満の地点の比率は，x 軸の 30 点で始まる垂直線を CDF との交点まで辿った後，それが y 軸を横切るまで水平線を辿れば示される。

Other：他の

1. **Other** sources of information include interviews and review of internal correspondence.
他の情報源はインタビュー，そして内部通信のレビューである。
2. **Other** staff who also made contributions to the process includes Hiroaki Nakajima, Hideki Morita, and Hiromasa Sakamoto.
本章に貢献してくれた他職員は，Hiroaki Nakajima, Hideki Morita, そして Hiromasa Sakamoto である。
3. The result is in accordance with **other** studies on Crustacea.
この結果は，Crustacea についての他の研究のとおりになっている。
4. Each phase is taking into account the interaction with **the other**.
それぞれの段階が他との相互作用を考慮に入れている。
5. Just because a model has been found valid for one use does not mean it is appropriate for some **other** use.
ただモデルが 1 つの使用で有効であることが見いだされたからといって，それが何か他の使用において適切であることを意味しない。
6. The present model structure may apply to **other** compounds as well as to **other** species.
現在のモデル構造は，他種と同様に他の化合物に当てはまるかもしれない。

- **These and other** ..... ：これらおよび他の …

7. While we have dealt with most of these issues in Michigan, **these and other** issues will arise elsewhere, thus regionally consistency in achieving a resolution of these issues will be needed.
筆者は，ミシガン州におけるこれらの問題の大半を扱ってきたが，これらおよび他の問題は，他地域でも生じることで，これらの問題の解決における地域的整合性が必要だろう。

- **other than .....**：以外の

8. Ambient sites **other than** reference sites should be surveyed as part of the database.
参照地点以外の環境地点は，データベースの一部として調査されるべきである。

- **On the other hand**：他方

9. **On the other hand**, their low stiffness and tendency to creep are disadvantages that must be carefully weighed against their advantages.
他方，それらの剛度の低さおよびクリープの傾向は不利点であり，それらの利点と慎重に比較考察されなくてはならない。

- **like any other .....**：他の… と同様

10. However, **like any other** valued fish species, it does have specific habitat and water quality requirements.
しかしながら，他の貴重な魚種と同様，具体的な生息地および水質の要件を有する。

- **In other cases**：他のケースでは
- **In others**：他の例では

★ In other cases は In some cases また In other は In some と組でよく使われます。

11. **In some cases**, TMDL development can be straight-forward and relatively simple. **In other cases**, a phase approach may be more appropriate.
ある場合には，TMDL 開発は単純で，そして比較的簡単であり得る。他のケースでは，段階的アプローチがいっそう適切であるかもしれない。

12. **In most cases**, both anions and cation exchange is required to achieve the desired degree of water purification. **In some** installations, the anion and cation resins are contained in separate tanks. **In others**, they are mixed together.
ほとんどの場合，望まれる度合いの水清浄化を達成するには，アニオン交換とカチオン交換の両方が必要とされる。一部の施設では，アニオン・カチオン交換樹脂が別々のタンクに詰められており，また他の施設では，それらが混合されている。

- **In all other respects**：他のすべての点で
- **All other .....**：すべての他の … は
- **from one ..... to the other**：ある… から他の…へ

13. **In all other respects**, the autopsy revealed that theses animals were healthy at the time of collection.

他のすべての点で，検死は，これらの動物が採取時には健康であったことを明らかにした。
14. **All other** values are at concentrations well below this risk level.
すべての他の値は，このリスクレベルをずっと下まわる濃度にある。
15. In some cases the site-specific habitat attributes are used to help separate where the transition **from one** ecoregion **to the other** takes place.
一部の場合，地点特定生息地属性を用いて，ある生態地域から他の生態地域への推移発生場所が分けられる。

Others：他のもの

1. All things are somewhat different, yet some things are more similar than **others**.
あらゆる事物は，それぞれ若干異なるが，あるものは他のものよりも類似性が強い。
2. Some of the original models have been changed in nearly every subsequent version, whereas **others** have been largely retained.
オリジナルモデルのいくつかは，その後のバージョンほぼすべてで変更されたが，そのまま保たれたモデルも多い。
3. The difficulty of dealing with a large mass of information is that of the many interactions occurring within the community, some may be related to water quality while **others** may not.
大量の情報を取り扱う場合の困難は，生物群集において多くの相互作用が生じることであり，それらは，水質に関係したり，関係しなかったりする。

Otherwise：さもなければ

1. Leopold (1949) stated that "A thing is right when it tends to preserve the integrity, stability, and beauty of the biotic community. It is wrong when it tends to **otherwise**."
Leopold (1949) 曰く，「生物群集の完全性，安定性および美を保つ傾向にあれば，その参照地点は正しい。そうでない傾向にあれば，それは誤っている」。
2. While this may seem enigmatic in light of current strategies to regionalize wastewater flows, the presence of water with a seemingly marginal chemical quality can successfully mitigate what **otherwise** would be a total community loss.
これは，汚水流量を地域分けするという現行戦略の観点からは不可解に思われるかもしれないが，化学的に不十分な水質の水の存在が生物群集の全損をうまく緩

和しうるのである。

- **unless otherwise noted**：注意書きが無い場合は
- **unless otherwise stated**：述べられていないなら

3. **Unless otherwise noted**, all information and procedures for working with worksheet apply to macro sheets.
注意書きが無い場合は，ワークシートで作業するためのすべての情報・手順がマクロシートに適用される。

Out：のうち；延びて；広げて

1. Our thanks **go out to** the many state agency biologists who have applied and refined the multimetric approach and have strengthened the concept through their interactions and testing.
さらに，多数の州機関所属の生物学者にも感謝したい。彼らは，多メトリック手法を応用・改善し，対話や試験を通じて，この概念を補強してくれた。
2. Insofar as past state biological criteria efforts **grew out of** biological monitoring efforts, future efforts will likely build on current monitoring programs.
過去の州の生物クライテリア努力が生物モニタリング努力から発している限り，将来の努力は，現行のモニタリングプログラムに立脚するであろう。

- **Out of .....**：…のうち

3. **Out of** 645 waterbody segments analyzed, biological impairment was evident in 49.8 ％ of the cases where no impairment of chemical water quality criteria was observed.
分析対象の水域区分645のうち，化学的な水質クライテリアでの損傷が観察されなかった事例の49.8 ％で生物損傷が顕著だった。
4. **Out of** necessity, these sites needed to be high quality and located on streams well removed from major anthropogenic pollution sources.
必要に迫られ，これらの地点は，質が高くなければならず，また，主要な人為的汚濁源が十分に除かれた水流に所在していなけれならない。

- **carried out**（See **Carry**）：行われる（➡ **Carry**）
- **point out .....**（See **point**）：…を指摘する（➡ **point**）
- **turned out to .....**（See **turn**）：結果的に…となった（➡ **turn**）

Outcome：結果；成果

1. Case histories are presented below showing the **outcome of** impairment criteria testing in a variety of situations.
   さまざまな状況での損傷クライテリア検定の結果を示した事例史を下に述べる。
2. The **outcome of** these effort has been helpful to design engineers and treatment plant operators in solving a treatment problem resulting from heavy metal ion toxicity.
   重金属イオン毒性に起因している処理問題を解決するにおいて，これらの努力の結果は，デザインエンジニアと処理場オペレーターに役立った。
3. Insufficient attention is directed toward assessing cumulative impact of these policies in terms of costs and **outcomes**.
   費用と成果の面でこれらの政策の累積影響評価に十分な注意が払われていない。

Outline：概説する；輪郭を描く

1. This paper **outlines**.....
   この論文では，…について大要を述べている。
2. Figure 1 **outlines** the investigation and remedial process which consists of six basic components: ....
   図1が，6つの基本的な構成要素から成る調査・修復プロセスを概説する。それらは…
3. The standard procedure **outlined above** was established so that the problems which often obscure interpretation of results are minimized.
   結果の解釈をしばしば不明瞭にする問題点が最小になるように，上に概説された標準的な手順は確立された。

Outweigh：まさる；はるかに上回る

1. The advantages **outweigh** the disadvantage that .....
   …という利点が不利点よりもまさっている。
2. In contrast, other states did not undertake substantial bioassessment programs because they felt the cost of bioassessment **outweighed** the benefits.
   対照的に，他の多くの州は，生物アセスメントの費用が便益をはるかに上回ると考えたせいで，実質的な生物アセスメントプログラムを行わなかった。

Over：について；に関する；にわたり；にわたって

1. This allows for the analysis of incremental changes in aquatic community performance **over** space and time.
   これは，水生生物群集実績における時間的・空間的な漸増変化の分析を可能にする。
2. For most part, ICI scores were consistent and different by no more than 6 points **over** the sampling intervals which spanned 7 to 8 years and included 3 to 6 sampling events.
   たいていの場合，ICIスコアは，7〜8年間で3回ないし6回のサンプリングが行われるサンプリング間隔で，かなり一貫しており，6点以上の差がなかった。
3. Nishikawa (1976) claims that **over** 3000 species have ecological information in the literature, which can be built into an autecological database appropriate for the periphyton in a region.
   Nishikawa(1976)によれば，文献には3000種以上の生態学的情報が収められており，1地域における付着生物に適した個生態学的データベースを構築しうる。

★ Over を時間に関して使用する場合の例文を下に示します。

4. There are no continuous records of ecological change **over** the last 2000 years.
   過去2000年にわたっての生態学的変更の連続した記録が存在しない。
5. Soil samples at the field sites were collected every month **over** a 12-month period.
   野外地点で土壌サンプルが12カ月間にわたって毎月収集された。
6. Repeated tests **over** long periods are more significant.
   長い期間にわたっての繰り返されたテストは，いっそう重要である。
7. For each index, there has been a significant positive change **over** time at most sites.
   各指数について，大半の地点で，経年的に有意なプラス変化が見られた。
8. **Over** a 2-year period, the criteria were tested, focusing on the following questions: (1)…, (2)….
   2年間にわたってクライテリアの検定が行われ，以下の諸問題に焦点が置かれていた。(1)…,(2)….
9. Ecological information for many algae, particularly diatom, has been recorded in the literature for **over** a century.
   多くの藻類，特に珪藻についての生態学的情報は，1世紀にわたり文献に記録されてきた。
10. A 40-mile segment has been sampled repeatedly **over** multiple years, which

is a prerequisite for trend analysis.
40mileの区間で，複数年度にわたり繰り返しサンプリングされた。これは，傾向分析の必須条件である。

★ Over を空間に関して使用する場合の例文を下に示します。

11. Reported concentrations are the result of composite sampling **over** 30 to 60 cm intervals.
    報告された濃度は30から60cmセンチメートル間隔の複合サンプリングの結果である。
12. Although these indices have been regionally developed, they are typically appropriate **over** wide geographic areas with minor modification.
    これらの指数は，地域的に設定されてきたものだが，一般に，やや修正すれば広い地理的範囲に適している。
13. When flying **over** the Southern Rocky Mountains along the Wyoming/Colorado state line in mid-1980, this author noticed differences in timber management practices between states.
    1980年代中期，ワイオミング州／コロラド州の境界に沿った Southern Rocky 山脈の上空を飛んだ時に，著者は，両州での材木管理慣行の相違に気づいた。

    - **over a range of** ……：…の範囲で
    - **over a wide range of** ……：広範囲の…
    - **over a** …… **range**：…の範囲で

14. The rate constants for oxidation of carboxylic acids are strongly correlated with pH **over a range of** 4-8.
    カルボン酸の酸化速度定数は4から8の範囲でpHと強い相関関係がある。
15. Naoki (1975) observed that the quantity of mercury taken up by the sludge did not alter **over a wide range of** cincentrations.
    Naoki (1975) の観察では，スラッジによって吸着された水銀の量は広範囲の濃度で変化がなかった。
16. Laboratory-scale activated sludge units were studied **over a wide range of** process operating conditions, that is, MCRT.
    実験室スケールの活性汚泥施設で，広範囲のプロセスの操作条件，すなわち，MCRTが研究された。
17. On the basis of estimated rate constants, the oxidation rate of formic acid is faster than oxalic acid **over** a pH **range of** 4-8.
    推定された速度定数によれば，formic酸の酸化速度はpH4-8の範囲でoxalic酸より速い。
18. Thus, biosurveys targeting multiple species and assemblages and more

likely to provide improved detection capability **over a broader range** as well as protection to a larger segment of the ecosystem than a single species or assemblage approach.
それ故，複数の生物種および群がりをターゲットにした生物調査は，単一の生物種または群がり手法に比べて，より広範な探知能力向上をもたらす見込みが高いし，より広範な生態系保護になるだろう。

- **over which .....** ：… に当たり；… の範囲の

19. It is better to identify large areas **over which** calibrations will be performed rather than small areas.
キャリブレーションを行うに当たり，狭い地域よりも広い地域を確認した方が良い。
20. Examining the variance structure can give insight into the extent **over which** particular biocriteria might be applicable.
変動構造を検討すれば，ある生物クライテリアが適用可能な度合に関する洞察が得られよう。

- **controversy over .....** ：… に関する論争

21. There has been considerably **controversy over** the form of the equation.
この式の形に関してかなり論争があった。

- **advantage over.....** (See **Advantage**)：
… に対しての利点 (➡ **Advantage**)
- **concern over .....** (See **Concern**)：… についての関心 (➡ **Concern**)

Overestimate (See Estimate)：過大に見積もる (➡ Estimate)

Overall：全体的

1. Activated carbon provided the best treatment **overall**.
全体として活性炭が最良処理であった。
2. The **overall** removal rate is seen to be .....
全体的な除去速度は … に見える。
3. The **overall** removal of total chromium from the plant agreed in principal with earlier studies.
このプラントからの全クロムの総体的除去量は，以前の研究のそれと基本的に一致した。
4. The **overall** effort should focus on building improved "line of evidence".

努力全体は，改善された「証拠ライン」の構築に主眼を置くべきである。
5. This study determines the **overall** quality both from an ecological point of view and also from a user's point of view.
この研究では，生態的な見地とユーザーの見地両方から総体的な質が決められる。
6. There is little doubt that much of the gypsum has formed in-situ as a result of this **overall** process of pyrite oxidation.
石膏の多くが，黄鉄鉱の総合酸化プロセスの結果として，現地で生成したことにほとんど疑いがない。
7. This is a critical juncture in the process since these initial decisions will determine the **overall** effectiveness of the effort well into the future.
これは，過程における重大時である。これらの初期決定が労力の効果性をずっと将来まで左右するからである。
8. With clearly defined goals for the maintenance of different levels of integrity through impact standards, the **overall** success of water resource management program can be evaluated.
影響標準を通じて様々な水準の健全度を保つという明確な目標によって，水資源管理プログラムの全般的成功が評価されうる。

● overall objective of …..（See Objective）：…の総合目的（→ Objective）

Overlook：看過；見過ごす；見落とす

1. We should not **overlook** the fact that …..
…という事実を見過ごすべきではない。
2. Such information can influence decisions to control certain substances or processes that might have been **overlooked** or underrated in an evaluation based on only one group.
そのような情報は，1集団のみに基づく評価において看過または過小評価されてきた物質や過程を制御する決定に影響を及ぼしうる。

Overview：全体像；概観；概略

1. **This overview is developed** from available documentation on state bioassessment and monitoring programs and an informal survey of state staff biologists.
ここで示す全体像は，州の生物アセスメント，生物モニタリングプログラム，ならびに州所属生物学者による非公式調査に関する入手可能な文書から導いたものである。

2. U.S. EPA (1990) has published comprehensive **overviews** of biocriteria rationale, whereas the **overview** by Kumagai et al. (1986) is generally regarded as the primary publication advocating using biological integrity within water quality management programs.
U.S.EPA (1990) は，生物クライテリアの根拠に関する総合的概論を刊行した。一方，Kumagai ら (1986) による概論は，一般に水質管理プログラムにおける生物完全性の利用を提唱した主要文献とみなされている。

   - ~ **offer an overview of** ..... ：～では…の概略を示す
   - ~ **provides a good overview of** ..... ：
     ～は…に関する適切な概観を提供している

3. In this chapter, we **offer an overview of** volunteer monitoring program objectives and data collection.
本章では，ボランティアモニタリングプログラムの目標およびデータ収集の概略を示すことにする。
4. EPA's User's Manual (USEPA, 1988) **provides a good overview of** how reference site data were translated into regional water quality criteria.
U.S. EPA のユーザーマニュアル (U.S. EPA, 1988) は，参照地点データが地域的水質クライテリアへ変換される様相に関する適切な概観を提供している。

Owe：負う

   - **owing to** ..... ：…であるため

1. **Owing to** the great extent of the fish industry in the Black River, it is essential that the condition of the river be kept as good as possible.
Black 川では漁業が盛んなため，この川の状態をできるだけ良く保つ必要がある。
2. Results for lower depths showed far less variation, **owing to** the influence of the constant water table elevation.
地下水面の高さが一定であるため，その影響で深いところの結果はより変化が少ない。

P

Paper：論文

- **This paper describes ……**：この論文では … について記述している
- **This paper examines ……**：この論文では … を検討する
- **This paper addresses ……**：この論文では … に取り組んでいる
- **This paper is addressed to ……**：
  この論文は … するために書かれている
- **This paper analyzes ……**：この論文では … を分析する
- **This paper concerns ……**：この論文では … を扱う
- **This paper details ……**：この論文では … について詳細に記述している
- **This paper discusses ……**：この論文は … について議論する
- **This paper outlines ……**：この論文では, … について大要を述べている
- **This paper provides ……**：この論文では… を提示する
- **This paper focuses on ……**：この論文では… に焦点を当てる
- **This paper attempt to ……**：この論文は… .をしようとしたものである
- **This paper is intended to ……**：この論文の狙いは… である

1. **This paper describes** the construction and application of a numerical model representing mercury transformations and bioavailability in the aquatic environment.
   この論文は，水生環境での水銀の変化と生物摂取を説明する数値モデルの構築と応用を記述している。
2. **This paper examines** the ecological damages caused by the river modification.
   この論文では河川修正工事が生態におよぼした損傷を検討する。
3. **This paper is addressed to** answering these questions.
   この論文はこれらの質問に答えるのに書かれている。

- **in this paper**：この論文での
- **In this paper, we examine ……**：この論文では，我々は … を検討する
- **In this paper, we explore ……**：この論文では，我々は … を探索する
- **In this paper, we compare A to B**：
  この論文では，我々はAをBと比較している

4. The method described **in this paper** can improve the interpretation of the results.
   この論文に記述された方法で，この結果の解釈を進展させることができる。
5. **In this paper, we examine** physical transport and chemical transformation processes
   この論文では，我々は物理的輸送と化学変化プロセスを検討している。
6. **In this paper, we explore** the possible use of mussels in estimating the distribution of benzo (a) pyrene in coastal waters.
   この論文では，我々は，沿岸水域での benzo (a) pyrene の分布を推定するためにイシガイの使用の可能性を探索する。
7. **In this paper, we compare** laboratory results **to** previously reported field results.
   この論文では，我々は研究室での結果と前に報告された野外調査の結果とを比較している。
   - **the purpose of this paper is** ….. (See **Purpose**)：
     この論文の目的は … である (➡ **Purpose**)
   - **scope of this paper** (See **Scope**)：
     この論文の範囲 (➡ **Scope**)

Part：部分；一部

1. Reportedly, **part of** the NBS mission is the coordination of federal monitoring activities, which should lead to better and more efficient use of biological data and assessments.
   伝聞によれば，NBS の任務の一部は，連邦モニタリング活動の調整であり，それは，生物データおよび生物アセスメントのより優良かつ効率的な利用に通じる。
   - **〜 is part of** …..：〜 は … の一部である
2. AFDM values **are part of** the assessment methodology developed by Kentucky.
   AFDM 値は，ケンタッキー州が案出したアセスメント方法論の一部である。
3. This need **was part of** a larger concern for a framework to structure the management of aquatic resources in general.
   この必要性は，水生資源全般の管理を構成する枠組みへの大きな関心の一部である。
   - **the first part of** …..：… の最初の部分
   - **other parts of** …..：… の他の部分

- **a major part of** …..：…の主要部分
- **the central part of** …..：…の中部
- **an important part of** …..：…の重要な一部
- **other part of** …..：…の他の部分

4. **The first part of** this paper provides …..
   この論文の最初の部分は…を示している。
5. Protocol II is based on comparisons with reference conditions from **other parts of** the stream that are not impaired, and classifies streams as unimpaired, moderately impaired, or severely impaired.
   プロトコルIIは，その水流で無損傷の他の部分からの参照状態との比較に基づいており，水流を無損傷，中度損傷，重度損傷に分類する。
6. When the program expands to all or **a major part of** a watershed, the effects on one area of adding pollution in another become more evident.
   そのプログラムが流域の全部または主要部分へと拡大された場合，他での汚濁追加が一区域に及ぼす効果は，一層顕著になる。
7. Regional modifications using this data set may raise this MPS expectation by approximately one third of one point for the spring dominated streams **in the central part of** the state and one sixth of one point for the eastern ecoregions.
   このデータセットを用いた地域修正は，MPS期待値を同州中部の春季優勢水流について3分の1点ほど上昇させ，州東部生態地域について6分の1点ほど上昇させるだろう。
8. Public participation in the form of public comment and public hearing is also **an important part of** Section 404 process, which addresses wetlands use and protection.
   一般市民による論評や公聴会などの形での公衆参加も404条の重要な一部であり，同条は，湿地利用や湿地保護を扱っている。

- **for the most part**：大部分で；たいていの場合

9. **For the most part**, only very small streams have watersheds completely within any one of these subregions.
   大部分で，ごく小さな水流がこれらの小地域のいずれかにすべて属する流域を持つ。
10. **For most part**, ICI scores were consistent and different by no more than 6 points over the sampling intervals which spanned 7 to 8 years and included 3 to 6 sampling events.
    たいていの場合，ICIスコアは，7～8年間で3回ないし6回のサンプリングが行われるサンプリング間隔で，かなり一貫しており，6点以上の差がなかった。

- **in part**：部分的に；一部
- **in large part**：多くの部分で

11. This is due to **in part to** the fact that .....
    これは，一部 … という事実による。
12. These projects are funded, **in part**, through the AAS program.
    これらのプロジェクト資金の一部は，AASプログラムを通じて供給されている。
13. The absence of substantial cadmium removal may be due **in part** to the high solubility of cadmium ion.
    本質的なカドミウムの除去がない一因は，カドミウムイオンの高い溶解性のためであるかもしれない。
14. Spatial variability in population abundance is well known for periphyton, and is caused **in part** by substrate, rate of flow, and light intensity.
    個体群数度の空間的可変性は，付着生物について十分に知られており，その一因は，底質，流速，光度である。
15. Many local and state watershed management decisions have been based, at least **in part**, on data collected by volunteer monitors.
    多くの地方および州の流域管理上の決定は，少なくとも部分的にボランティアモニターの収集データに基づいている。
16. Significant improvement in the biological indices has been observed during the past twenty years, which has been due **in large part** to significant reductions in point source loadings.
    生物指数における大幅な改善が過去20年間に観察されており，その主因は，点源負荷の大幅な削減であった。
17. The electric output of the plant depends **in large part** on the amount of heat which can be transferred in the steam generator and the condensers.
    発電所の電気出力は，スチーム発生・濃縮装置においての伝導可能な熱量にかなり依存する。

- **as part of .....**：… の一部として

18. Water quality measurements were not made **as part of** the survey.
    水質測定が調査の一部として行われなかった。
19. A baseline risk assessment is performed **as part of** the final Remedial Investigation (RI).
    基盤的（ベースライン）アセスメントが，最終の改善調査(RI)の一部として実施される。
20. The procedure by which this is done can be viewed **as part of** the regionalization process.

このための手順は，地域分け過程の一部とみなしうる。
21. Ambient sites other than reference sites should be surveyed **as part of** the database.
参照地点以外の環境地点は，データベースの一部として調査されるべきである。
22. Idaho DEQ now uses three biological indices **as part of** the numeric biological criteria.
アイダホ州DEQは，目下，数量的生物クライテリアの一部として3つの生物学的指数を用いている。
23. Other chemical, toxicological, physical, and source information must be used **as part of** the overall assessment process.
他の化学的，毒物学的，物理学的および源泉情報は，アセスメント過程全般の一部として用いられなければならない。
24. **As part of** its National Surface Water Survey, the USEPA focused attention on regions of the United States where acidic or acid-sensitive lakes and streams were expected to be found.
National Surface Water Survey（全米地表水調査）の一部として，USEPAは，酸性または酸に敏感な湖沼・水流が見出されると予想される米国各地に注意を注いだ。

Past：過去；これまでの

1. **The past decade** has witnessed a dramatic increase in the development, testing, and application of mathematical modeling for analysis of water resources problems.
これまでの10年間に，水資源問題の解析のための数学的モデリングの構築，試験，そしてその適用の急激な増加を目撃することになった。

- **in the past decade**：これまでの10年間
- **in the past couple of decades**：ここ数十年間
- **during the past decade**：これまでの10年間に
- **over the past two centuries**：これまでの2世紀にわたって
- **over the past few decades**：ここ数十年にわたって

2. A variety models have been developed **in the past decade**.
これまでの10年間，多種多様なモデルが構築されてきた。
3. Much progress and advancement has been made **in the past decade**.
多くの進歩と前進がこれまでの10年間になされてきた。
4. The idea that conditions were pristine in North America prior to Europian settlement has been convincingly challenged **in the past couple of decades**.

欧州人の入植前の北米では無傷状態だったという思考にたいして，ここ数十年，説得力ある異議が申し立てられてきた。

5. **During the past three weeks**, staff has participated in detailed data gathering sessions.
過去3週間に，スタッフが詳細なデータ収集セッションに参加した。
6. **During the past decade**, this process has attracted extensive attention as a modern technology for groundwater purification.
これまでの10年間に，このプロセスは，地下水清浄化の近代的な技術として広く注意を引いてきた。
7. **Over the past two centuries**, vast quantities of energy have been used.
これまでの2世紀にわたって，膨大な量のエネルギーが使われてきた。
8. **Over the past few decades**, a considerable number of studies have been conducted on the effects of temperature on the growth of nitrifying bacteria.
ここ数十年にわたって，温度が硝化菌の成長に及ぼす影響に関する研究がかなり行われてきた。

Particular (Particularly)：特定の（特に）

1. Are those in poor condition associated with any **particular** cities?
劣等状態の湖沼は，特定の都市に関連しているのか。
2. Heavy metals have been the cause of **particular** environmental concern.
重金属は特定の環境に関する懸念の原因となってきた。
3. **Particular** emphasis will be given in this review to natural water samples.
このレビュゥーでは，自然水サンプルが特に重視されている。
4. It has become increasingly evident that the environmental impact of a **particular** metal species may be more important than the total metal concentration.
特定の金属種の環境に対する影響が，金属すべての濃度の影響よりいっそう重要であるかもしれないということがますます明白になってきた。

● **In particular**：特に

5. **In particular**, the investigation was focused on evaluating the influence of physical hydrology on .....
特に，この調査は … に対する物理水文的影響を評価することに集中した。
6. The goal of biological criteria is to provide additional support for the state's water quality standards. **In particular**, biological criteria provide a mechanism for assessing aquatic life attainment based on the actual biological conditions of waterbodies.

生物クライテリアの目標は，州の水質基準に追加サポートを提供することである。特に，生物クライテリアは，流域の実際の生物状態に基づき水生生物達成度を評価するための仕組みをもたらす。

- **In a particular case**：特定の場合

7. **In theire particular case**, the exposed population grew slower.
   それら特定の場合，（毒物に）さらされた個体数は繁殖が遅かった。

   - **Of particular importance is** ..... (See Importance)：
     特に重要なのは … である（➡ **Importance**）
   - **particular attention should be directed to** ..... (See Attention)：
     特定の注意が … に向けられるべきである（➡ **Attention**）
   - **particular attention should be placed on** ..... (See Attention)：
     … に特定の注意が置かれるべきである（➡ **Attention**）
   - **particularly**：特に

8. This data set would be **particularly** meaningful with respect to the .....
   このデータセットは… に関して特に重要な意味を持つであろう。
9. The assumption of neglecting substrate inhibition is not **particularly** misleading at low substrate concentrations.
   基質阻害を無視することについての仮定は，低濃度の基質においては特に誤りではない。
10. Autecological information for many algae, **particularly** diatom, has been recorded in the literature for over a century.
    多くの藻類，特に珪藻についての個生態学的情報は，1世紀にわたり文献に記録されてきた。

Pattern：パターン

1. The ecoregion **patterns became clear**.
   生態地域パターンが明確になった。
2. **No simple pattern could be perceived** which related nitrification.
   硝化反応に関連する単純なパターンは認知され無いであろう。
3. In doing so, the complicated **pattern may be avoided**.
   そうすることにおいて，複雑なパターンが避けられるかもしれない。
4. The **pattern may vary** according to the type of impact and discharge.
   そのパターンは，影響および排出のタイプに応じ異なる。
5. **A consistent pattern of** contamination **would result from** in situ conditions rather than migration.

一貫した汚染パターンは，移動条件というより現場の条件に起因する。

● **pattern of .....**：…のパターン

6. North of the line in Wyoming, **patterns of** logging activity were apparent, whereas south of the line they were not.
ワイオミング州境の北では伐採活動パターンが顕著だったのに対し，州境の南ではパターンが認められなかった。
7. The data show a similar **pattern of** distribution of heavy metals at each site.
これらのデータは，各地点において類似の重金属分布パターンを示す。
8. Little effort is being expended on attempting to define differences in **patterns of** ecosystem mosaics.
生態系モザイクのパターンの相違を定義づける試みにあまり力が入れられていない。
9. Levin (1992) stressed that to gain an understanding of the **patterns of** ecosystems in time and space and the causes and consequences **of patterns**, we must develop the appropriate measures and quantify these **patterns**.
Levin(1992)は，時間・空間における生態系パターンや，パターンの原因・結果を理解するために，我々が適切な測度を定め，これらのパターンを数量化しなければならないことを強調した。

● **patterns in .....**：…のパターン

10. Definite **patterns in** biological community data exist.
生物群集データの明確なパターンが存在する。
11. **Patterns in** human activities must be considered as well, and these often vary as a function of ownership or political unit, as well as ecoregion.
人間の活動におけるパターンも考慮すべきであり，それは，しばしば所有単位や政治単位，そして生態地域に相関して異なる。

Perceived：認知される

1. This is especially true in such cases where a waterbody **may be perceived as** being at risk due to new dischargers.
ある水域が，新しく生じた排出者のために危険な状態にあるものと見なされるかもしれない場合，これは特に事実である。
2. No simple pattern **could be perceived** which related nitrification.
硝化反応に関連する単純なパターンは認知され無いであろう。

Percent (Percentage)：パーセント；％；割合（パーセンテージ；百分率）

1. Nitrous oxide concentrations have risen by about 15 ％.
   亜酸化窒素の濃度が約 15 ％上昇した。
2. Carbon dioxide concentrations in atmosphere have increased nearly 30 ％.
   大気中での二酸化炭素の濃度がおおよそ 30 ％増加した。
3. Only about 5 ％ of the VOC has been converted to product in the steady-state condition.
   定常状態で，ほんのおよそ 5 ％の VOC "プロダクト" を "生成物" にする。
4. Under steady-state conditions, CTEX removal rates were greater than 98 ％ in both reactors.
   定常状態の条件下では，CTEX の除去率は両方の反応器で 98 ％以上であった。
5. USEPA (1990b) found good agreement between results in about 20 ％ of cases when effluent (end-of-pipe) toxicity was assessed, and in about 30 ％ of cases when mixing zone toxicity was measured.
   USEPA は，流出（管末）毒性を評価した際には約 20 ％の事例で，また，混合区域毒性を測定した際には約 30 ％の事例で，結果でのかなりの一致を認めた。
6. In the same Minnesota results cited above, chemical and biosurvey results agreed 58 ％ of the time, and impairment was indicated by chemical but not biological criteria at only 6 ％ of the sites.
   前記のミネソタ州でのアセスメント結果において，化学的調査結果と生物的調査結果は，58 ％で一致した。また，生物クライテリアでなく化学クライテリアで損傷が示された地点は，わずか 6 ％にすぎなかった。

● **Percentage**：パーセンテージ；百分率

7. When considered on the basis of agencywide water programs this **percentage** is approximately 6 ％.
   機関全体の水関連プログラムベースで考えた場合，この割合は，約 6 ％である。
8. A consistently **high percentage** of agreement occurred only when both the volunteer and professional analyses rated the site in the fair/poor ranges.
   高率の合致が一貫して生じる事例は，ボランティアと専門家の両方が可／不可の値域にある地点を対評価した場合に限られていた。

Percentile：パーセンタイル

1. Other statistics can be calculated and represented on the graphical summary, such as means, medians, and **percentiles**.

他の統計量は，平均，中央値，パーセンタイルなどグラフ要約で計算・表現されうる。
2. These plots contain sample size, medians, ranges with outliers, and 25th and 75th **percentiles**.
これらの図には，サンプル規模，中央値，外れ値，25パーセンタイルおよび75パーセンタイルの値域が含まれている。
3. The selection of the 25th **percentile** value is analogous to the use of safety factors, which is commonplace in chemical water quality criteria applications.
25パーセンタイル値の選定は，化学的水質クライテリア適用において，ごく一般的な安全因子の利用に似ている。
4. It might be more reasonable to argue for differing **percentiles** for different regions depending upon the nature of regional-scale degradation.
地域規模劣化の性質に応じ，異なる地域についてパーセンタイルを違えることを主唱した方が合理的だろう。
5. A recent article describes some approaches for defining how well various percentiles can be estimated and notes pitfalls in the use of extreme **percentiles** for these purposes.
最近の論文では，様々なパーセンタイルがどれほど十分に定められるか決定する手法を述べており，これらの目的のための極値パーセンタイルの利用における陥穽を記している。

Perform：実行する；行う

1. Laboratory experiments **were performed** to meet three objectives: 1).....; 2)....., and 3)....
3つの目的を満たすために研究室で実験が行なわれた。それらは1)…，2)…，および3)…
2. It is better to identify large areas over which calibrations **will be performed** rather than small areas.
較正を行うに当たり，狭い地域よりも広い地域を確認した方が良い。
3. The longitudinal examination of biological sampling results **is** also **performed** in an attempt to interpret and describe the magnitude and severity of departure from the numerical biological criteria.
生物サンプリング結果の縦断的検討も，数量的生物クライテリアからの逸脱の規模・程度を解釈し記述するために行われる。
4. The goal is to achieve a suitable fit between the planned modeling effort and the data, time, and money available to **perform** the study.

ゴールは，計画されたモデリングの努力とデータ，時間，それに研究を実行するためにアクセス可能な資金の間で，適当な合致に達することです。

Period：期間

1. The very early **time periods** were suggestive of rapid loss of a small amount of PCP ; however, additional work is needed to confirm this.
   非常に早い時間の期間は小量の PCP が速い速度で分解することを意図する。しかしながら，更に追加の研究がこれを確証するために必要である。

   ● **for a period of .....**：… の間

2. **For a period of** approximately 100 years, from the original discovery of gold at Deadwood Gulch in 1875 until the late 1970's, huge volumes of mining and milling wastes were discharged into Whitewood Creek.
   1875 年に Deadwood Gulch での最初の金の発見から 1970 年代後期までおよそ 100 年の間，大量の鉱山採掘廃棄物と粉砕廃棄物が Whitewood Creek に放出された。

   ● **during the period .....**：… の期間に

3. **During the period of** June 14 to July 25, 1999, ...
   1999 年の 6 月 14 日から 7 月 25 日の期間に，

4. **During period 2**, average influent and effluent CF concentrations were 0.30 and 0.11 mg/L, respectively.
   期間 2 では，入水および出水中での平均の CF 濃度は，それぞれ 0.30 および 0.11mg/L であった。

5. These streams were all sampled **during the summer period** (June through September), to correspond to critical low flow and elevated temperature conditions.
   これらの水流は，臨界低流量および高温状態に対応するため，すべて夏季(6～9月)にサンプリングされた。

   ● **over a ..... period**：… の間にわたって

6. Soil samples at the field sites were collected every month **over a 12-month period**.
   土壌サンプルが，野外地点においての 12 カ月間にわたって毎月収集された。

7. Dissolved oxygen, pH, temperature, and specific conductance were measured in the field hourly **over a 24-h period** with either a Hydrolab(R) Surveyor Ⅱ or Data Sonde.
   溶存酸素，pH，水温，および導電率が Hydrolab(R) Surveyor Ⅱ または Data Sonde

のいずれかで，24時間にわたり，1時間ごとに現地で測定された。
8. Repeated tests **over long periods** are more significant.
長い期間にわたっての繰り返しのテストは，いっそう重要である。

Permit（Permitting）：許す；可能にする；できる（を可能にする；ができる）

1. The modeling framework WASTOX **permits** the user to examine the transport of a toxic chemical.
モデリング枠組みWASTOXをつかって，ユーザーはある有毒化学物質の輸送を調べることができる。
2. Examination of local gaseous and liquid concentrations **permitted** an explanation of the complex events occurring under transient state operating conditions.
局地的なガスと液体の濃度の検討から，瞬間状態運転条件下で生じる複雑なイベントの説明ができる。
3. Research is needed to develop inexpensive detection methods which will be **permit** small utilities and consumers to monitor effectively for the presence of harmful microbial agents in drinking water.
小規模水道事業者，および消費者が飲料水に存在する有害な微生物を監視できる安価な検出方法を開発するために研究が必要とされる。

● **permitting**：…を可能にする；…ができる

4. Predictive models can forecast future water quality, thus **permitting** trials of possible control strategies at very reasonable expense.
予測モデルは将来の水質を予測することができるため，かなり妥当な費用で可能なコントロール戦略を試すことができる。

per se：それ自体

1. State courts have not yet ruled on the validity of water quality criteria **per se** or as applied.
同州の裁判所は，水質クライテリア自体またはその適用の有効性をまだ裁定していない。
2. Nothing in the CWA mandates state adoption of biocriteria **per se**.
CWAの文言は，いずれも生物クライテリア自体の州による採用を命じていない。
3. Description of these buildings, **per se**, has not been included in this report, as they are not considered wastes.
これらの建物それ自体の記述は，この報告に含められなかった。なぜなら，これ

ら(建物)は廃棄物とはみなされないからである。

Perspective：見地；観点

- **with a ..... perspective**：…を視野に入れて
- **from a ..... perspective**：…の視点から

1. In recent years there has been an increasing awareness that effective research, inventory, and management of environmental resources must be undertaken **with an** ecosystem **perspective**.
   近年，環境資源の効果的な研究・調査一覧作成・管理は，生態系を視野に入れて行うべきだ，との認識が強まってきた。
2. A specific aspect of transport and transformation is examined **from** a systems **perspective**.
   輸送・変換の特定の側面が，システムの視点から検定される。
3. Actually, **from** a biological **perspective**, the differences in index score parameters are not unexpected.
   実際，生物学上の視点から，指数評点パラメータにおける差異は，予想外ではない。
4. **From a** regulatory **perspective**, these data provide good technical justification for a modified biological expectation.
   規制上の視点から，これらのデータは，生物学的期待値修正のための適切な技術的正当化事由となる。

- **in perspective**：大局的に；総体的に

5. Radioactive wastes from various nuclear processes, one source of harmful radiation, should be viewed **in perspective**.
   種々の核プロセス(有害な放射能源の1つである)からの放射性廃棄物は，総体的にとらえるべきである。

Place (See also **emphasis**)：置く(→ **emphasis**)

- ～ **is placed on** .....：～が(…の上に)置かれる
- ～ **is placed into** .....：～が(…の中に)置かれる

1. Research emphasis **should be placed** preferentially **on** problems of risk management.
   研究の主眼点は，リスクマネージメントの問題に優先的に置かれるべきである。
2. Particular attention **should be placed on** the development of treatment

methods for removing nitrates from groundwater supplies.
特定の注意が，給水用地下水から硝酸塩を取り除くための処理方法の開発に置かれるべきである。

3. It is apparent from Table 3 that a significant inhibitory effect **was placed on** the specific oxidation rate as a result of extreme winter temperatures.
極度な冬の温度が原因で相当な抑制効果が特定の酸化速度に及ぼしたことは，表3から明白である。

4. The design of the tubes allowed them to **be placed into** the sonicator horn at the same depth.
このチューブの設計によって，それらはソニケータホーンの中に同じ深さで置くことができた。

5. A sample solution (4 mL) **was placed** in a batch system.
サンプル溶液(4 mL)がバッチシステムに移された。

- ~ **is in place** ：～を設置する
- ~ **is held in place** ：～を設置する

6. No program to monitor tanks for leaks **was in place** prior to that date.
その日以前には，タンク漏れを監視するプログラムが使用されていなかった。

7. For bedrock samples, gear usually includes a 3- to 4-in. diameter PVC pipe with a rubber seal that **can be held in place** where the rock is brushed clean and the sample removed by suction.
床岩サンプルの場合，サンプリング装置には，通常，直径3～4in のラバーシールドPVC(ポリ塩化ビニル)管が含まれており，これを設置すれば，岩石を磨いてサンプルを吸引除去できる。

- ~ **take place** ：～が行われる;～が生じる
- ~ **is taking place** ：～が行われている;～が起きている

8. It is evident that a marked improvement **took place**.
顕著な改善が生じたことは，明らかである。

9. In some cases the site-specific habitat attributes are used to help separate where the transition from one ecoregion to the other **takes place**.
一部の場合，地点特定生息地属性を用いて，ある生態地域から他の生態地域への推移発生場所が分けられる。

10. If the proper construction of the IBI (or similar evaluations) **takes place** up front, the eclipsing and other problems can be abated.
適切なIBI(または，類似評価)の構築がなされれば，隠蔽問題および他の問題が緩和されうる。

11. Pyrite oxidation **is taking place** in the tailings.

黄鉄鉱の酸化が選鉱廃物の中で起きている。

Play：演じる

- ~ **come into play**：～は演じるようになる; 有用になる

1. The hydraulic conductivity **comes into play** only when quantitative discharge calculations are made.
   流体の伝達速度は，排出量を計算するときにだけ，有用になる。

    - ~ **play a role in** .....：
      ～は … において役割を演じる；～は … において役割を果たす
    - ~ **played a major role in** .....：～は … において重要な役割を演じた
    - ~ **have played a major role in** .....：
      ～は … において重要な役割を演じてきた
    - ~ **play a significant role in** .....：～は … に重要な役割を演ずる
    - ~ **are known to play an important role in** .....：
      ～が … において重要な役割を演ずるということが知られている
    - ~ **play no more than an ancillary role in** .....：
      ～は … では従属的役割しか演じない
    - ~ **has always played a central role in** .....：
      ～は常に … において中心的な役割を演じてきた

2. These **played a major role in** assigning and evaluating use designations, water quality management plans, and advanced treatment justifications.
   これらは，使用指定の割当・評価，水質管理プラン，および高度処理正当化において重要な役割を演じた。
3. Humans have probably **played a major role in** shaping landscape pattern and molding ecosystem mosaics for thousands of years.
   人間は，おそらく数千年にわたり，景観パターンの形成と生態系モザイクの生成において重要な役割を演じてきた。
4. Microorganisms are believed **to play a significant role in** the disappearance of trifluralin in soil.
   微生物が，土壌でのtrifluralinの消失に重要な役割を演ずると考えられる。
5. Organic mercury compounds **are known to play an important role in** the toxicological behavior of mercury.
   有機水銀化合物が，水銀の毒物学的挙動において重要な役割を演ずるということが知られている。
6. This program **has already played an important role in** litigation and

enforcement.
このプログラムは，既に訴訟と法執行において重要な役割を演じている。
7. They **play no more than an ancillary role in** human health risk assessment.
それらは，人間健康リスクアセスメントでは，従属的役割しか演じない。
8. Biological data **has always played a central role in** the Illinois water quality standards, particularly for the determination of appropriate and attainable aquatic life use designations.
生物学的データは，常に，適切かつ達成可能な水生生物使用指定の決定について，特に，オハイオ州の水質基準において中心的な役割を演じてきた。
9. Statistical sample surveys **can play an important role in** characterizing the biological condition of lakes and streams.
統計的サンプル調査は，湖沼・水流の生物学的状態を特性把握するうえで重要な役割を演じうる。
10. To date, a few states have developed comprehensive and sophisticated biological criteria programs that **play a critical role in** protecting water resource quality.
今日までのところ，水資源の質保護において重要な役割を演じるような総合的で精巧な生物クライテリアプログラムを策定した州は，ごく少数である。

Plot (See also **Figure**)：図示する；作図する；プロットする（→ **Figure**）

1. The reference site results were pooled on a statewide basis prior to constructing the drainage area scatter plots.
参照地点結果が全州ベースでプールされた後，流域面積の散点図が作製された。
2. **From an analysis of these plots**, two general conclusions may be drawn. First, ……
これらのプロットの解析から2つの一般的な結論が引き出されるかもしれない。第一に，…

- ~ **Typical plots of ~ are presented in ……**：
  ~の典型的なプロットが…に提示されている
- **Plots of ~ are shown in ……**：~のプロットを…に示す。
- ~ **is shown in …… by a three-dimensional plot of ……**：
  ~は…を表す3次元プロットで…に示されている
- **The plots for ~ were examined**：
  ~についてのプロットが検討された
- **These plots contain ……**：これらの図は…を含んでいる

3. **Typical plots of** the nitrate concentration versus time are presented in Fig. 4.
硝酸濃度対時間の典型的なプロットが図4に提示されている。
4. **Plots of** the glucose concentrations against the sorbose concentrations are shown in Figure 6.
グルコースの濃度対ソルボースの濃度のプロットを図6に示す。
5. Copper toxicity profile **is shown in** Figure 3 **by a three-dimensional plot of** the specific growth rates as a function of the two variables, total ammonia concentration and total copper concentration.
銅毒性プロフィールを全アンモニア濃度と全銅濃度の二変数を関数として成長速度定数を表す3次元のプロットにより図3に示す。
6. **The plots for** each parameter **were examined** to determine if any visual relationship with drainage area existed.
各パラメータについての図が検討され，流域面積との目測関係が存在するか否か決定された。
7. **These plots contain** sample size, medians, ranges with outliers, and 25th and 75th percentiles.
これらの図は，サンプル規模，中央値，外れ値，25パーセンタイル値および75パーセンタイル値の領域を含んでいる。
    - **Figure N shows a plot of ......**：図Nに…のプロットを示す
    - **Fig. N is a plot of ......**：図Nは…のプロットである
8. **Figure 6 shows a plot of** changing nitrification rates versus time.
図6に，硝化速度の変化・対・時間のプロットを示す。
9. **Figure 2 is a plot of** five growth curves, which shows both the experimental and computed results.
図2は5つの（バクテリア）成長カーブのプロットであり，実験結果と計算結果の両方を示している。
    - **...... plotted**：プロットされた…；作図された…
10. The results **plotted** in Figure 6 may be seen to be in general agreement with the theory.
図6にプロットされた結果はその理論と概して一致するように見られるかもしれない。
11. Figure 6 is a display of the data **plotted along with** this hypothetical line.
図6は，この仮説線と一緒に作図したデータの図示である。
    - **～ is plotted**：～がプロットされる；～が作図される

12. Using all the estimated rate constants, the calculated accumulation of CO_2 **is plotted** in Figure 2.
    すべての推定速度定数を使って計算された CO_2 の蓄積量を図2に図示する。
13. In this figure the A concentration **is plotted** versus the B concentration.
    この図には，Aの濃度対Bの濃度が図示されている。
14. Ordinate values in Figure 2 **are plotted** logarithmically to show linear correspondence to logarithmic increase of bacterial concentration.
    図2の縦座標には，バクテリア濃度の増加が対数直線関係にあることを示すために対数値がプロットされている。

   ● **by plotting**：プロットすることによって

15. The temperature dependence was obtained **by plotting** the rate constant vs. temperature.
    この速度定数を温度に対してプロットすることによって，温度依存性が得られた。
16. This is done **by plotting** the biological index results by river mile for the subject survey area.
    このため，調査対象区域について，河川距離(マイル)ごとに生物学的指数の結果が作図される。
17. It appears that there is nothing gained **by plotting** the data in this fashion.
    データをこの様式で作図することによっては，何も得られないように思われる。

Point：点

1. Let me try to shed some light on **a few points**.
   いくつかの点に光を当ててみよう。
2. The total pollutant load to a waterbody is derived from **point**, **nonpoint**, and background sources.
   水域への全汚濁負荷量は点源と非点源，それにバックグラウンド汚濁源からなる。

   ● **data points**：データポイント

3. **All data points** in the individual experiments appeared to become linear over the range of contact times studied.
   個別の実験でのすべてのデータポイントが，調査された接触時間範囲では直線になるように見えた。

4. This correlation coefficient was 0.99 indicating a surprising good fit to the **data points** employed.
この相関係数は 0.99 で，使用したデータポイントに驚くほど一致することを示唆している。
5. The generalized equations to be used for design procedures must be developed considering the entire range of operating parameters and must contain **a large number of data points** to enhance the confidence in the design.
操作パラメータの全範囲を考慮しながら，設計過程に使うための一般化した式を開発をしなければならない。また，設計の信頼度を上げるために多くのデータがなくてはならない。
6. In evaluating the accuracy of the estimated parameter, more weight should be given to the ability of the model to fit **the data points** near segment D than to its ability to fit the points near segment G.
推定パラメータの正確さを評価することにおいて，モデルが川の切片 G の近くのデータポイントで合う能力により，切片 D の近くでデータポイントに合うことができる能力に，もっとウエートが置かれるべきである。

- **The important point is that .....**：重要な点は…である

7. **The important point is that** the same technique should be used at each site that is to be compared.
重要な点は，比較対象の各地点で同一技法を用いることである。

- **At this point**：この時点では

8. **At this point** there is insufficient information on source area ST5 to draw a conclusion on probable risk.
この時点では，確かなリスクについて結論を引き出すには，汚染源 ST5 についての情報が不十分である。

- **to the point**：まで

9. Stream and most riverine fish species are not excessively mobile **to the point** where they are unusable as indicators.
水流および大半の河川の魚種は，指標に適さないほど移動性が高くない。
10. Because so few states have progressed **to the point** of proposing legally binding narrative or numeric criteria at this time, industry involvement has been relatively limited on a national scale.
現時点では，法的拘束力を持つ記述的または数量的なクライテリアまで前進している州は，ごく少数なので，業界関与は，全国規模で比較的限られている。

11. This concern has arisen lately as analytical techniques for detecting water contaminants have improved **to the point** where it is possible to detect contaminants at the part per-trillion level.
水汚染物質を検出するための分析技術が汚染物質を ppt レベルで検出可能なまでに向上したため，最近この関心が高まった。

- **~ point out .....** ：～は … を指摘する
- **it is important to point out .....** ：… を指摘することが重要である
- **it is of interest to point out .....** ：… を指摘することは重要である
- **it must be pointed out .....** ：… は指摘されなくてはならない
- **~ have been pointed out** ：～が指摘されてきた
- **as ~ have pointed out** ：～が指摘しているように

12. This **points out** the limitations of assigning a single degradation rate coefficient for the entire soil profile.
これは土壌プロフィール全体に一つの分解速度係数を当てはめることには限界があることを指摘している。

13. Barbour et al. (1991) **point out that** Cricotopus (Nostocldius) spp. is an important component of pristine, western mountain streams and is atypical of other members of this genus with regard to pollution tolerance.
Barbour ら(1991)は，Cricotopus (Nostcladius) spp.が米国西部の原始山地水流の重要な構成要素であり，汚濁耐性に関してこの属の他種と異なることを指摘した。

14. In day-to-day activities of a regulatory agency **it is important to point out** to client (e.g., dischargers) that short stretches of modified stream do not preclude application of stringent water quality rules.
規制機関の日常的活動において，（排出者など）クライアントに対し，改変水流の短区間は，厳しい水質ルールの適用を阻まないことを指摘することが重要である。

15. **It is of interest to point out** the fundamental difference between A and B.
AとBの間の基本的な相違を指摘することは重要である。

16. **It must be pointed out** at this stage that .....
… は，この段階で指摘されなくてはならない。

- **point of view** (See **View**)：見地(➡ **View**)

Portion：部分

1. Our interest centers on the land-based **portion of** the N cycle.
我々の関心は，窒素循環の地上ベースの部分に集中している。

- **a significant portion of .....** ： … の重要な部分
- **the central portion of .....** ： … の中央部
- **in major portions of .....** ： … の大部分において

2. **A significant portion of** the world's PCB burden is transported through the atmosphere.
世界のPCB負荷の大部分が大気中で輸送される。
3. A multitude of low marshy islands and sand bars occupy much of **the central portion of** the bay.
湾の中央部の大部分は，多数の低沼地と砂州から成っている。
4. It should be noted that in **major portions of** the country, topographic watersheds either cannot be defined or their approximation has little meaning.
この国の大部分において，地形流域は，境界が定まらない，その概算がほとんど無意味であることに留意すべきである。

Pose：もたらす；提起する

1. These three sites **pose** significantly lower **risks** to human health and the environment than the other four.
これらの3つの地域は，他の4地域より人間の健康と環境にいちじるしく低いリスクを提する。
2. The development and adoption of biocriteria into state water quality standards **poses challenges** as well as opportunities.
州の水質基準における生物クライテリアの設定・採択は，課題ばかりでなく，機会をもたらす。
3. Improved detection techniques have now shifted concern to the threat **posed by** toxic chemicals.
検出技術の進歩により，有害な化学物質による脅威が心配事となった。

- **～ pose problem**（See **Problem**）：
～が問題をもたらす；～が問題を提起する（➡ **Problem**）
- **question poses .....**（See **Question**）：
質問は … を提起する（➡ **Question**）：
- **question is posed**（See **Question**）：質問が提出される（➡ **Question**）

Possibility：可能性

1. The **possibility of** ..... cannot be entirely ruled out.

…の可能性をまったく排除するわけにはいかない。
2. Concern arose about the **possibility of** an explosion.
爆発の可能性について懸念が高まった。

- ～ **address the possibility of** ..... ：～は…の可能性を強調する
- ～ **examine the possibility** ：可能性を検討する
- ～ **have a possibility of** ..... ：…は…の可能性がある

3. This report **will address the possibility of** contamination at any of the three sites.
この報告書は，3地点のいずれにおいても汚染の可能性があることを強調するであろう。
4. Data are not yet available to **examine this possibility** quantitatively.
この可能性を定量的に検討するためのデータがまだ得られていない。
5. The groundwater in the alluvium beneath the tailings **has a possibility of** being comprised of water from as many as four sources.
堆積選鉱屑の下にある沖積層の地下水は，4つもの水源からの水が混ざっている可能性がある。

- **There is a possibility that** ..... ：…という可能性がある
- **Possibilities exist for** ..... ：…といった可能性が存在する

6. **There is a possibility that** surface runoff and sediment wash influences the quality of water in these wells.
流去水と堆積物洗浄水がこれらの井戸の水質に影響を与えるという可能性がある。
7. **Possibilities exist for** benthic macroinvertebrates, such as insect larval head capsule abnormalities or aberrant net-spinning activities of certain caddisflies, but these metrics are currently cost-prohibitive.
大型底生無脊椎動物について，幼生昆虫の頭部膜異常や，トビケラの異常な糸紡ぎといった可能性が存在するものの，これらのメトリックは，目下，高費用すぎる。

- **in the possibility of** ..... ：…の可能性に
- **in view of the possibility of** ..... ：…の可能性から判断して

8. Interest has also emerged recently **in the possibility of** using sonication to remediate groundwater.
地下水修復のために超音波処理を使う可能性に，関心が最近また高まってきた。

Possible (Impossible)：可能な(不可能な)

1. One **possible** conclusion from our analyses is that .....

我々の分析から導かれる結論の一つはは … ということである。
2. Several **possible** mechanisms may account for the above observations.
いくつか考えられるメカニズムによって，上の観察結果が説明されるかもしれない。
3. Persons involved in **possible** field tests must make this determination on a case-by-case basis.
実施する可能性のある現場での実験に関係している人々は，状況によってこの決定をしなくてはならない。
4. The **possible** long-range effects of the disposal of trace metals in the coastal environment are matter of growing concern.
沿岸環境における微量金属の処分による長期にわたる潜在的影響は，これからますます懸念される事項である。
5. There exist three **possible** levels at which the biological impairment criteria could be instituted in New York State:
ニューヨーク州で制度化される生物損傷クライテリアには，3つの段階が考えられる。
6. All things are somewhat different, yet some things are more similar than others, offering a **possible** solution to the apparent chaos.
あらゆる事物は，それぞれ若干異なるが，あるものは他のものよりも類似性が強く，混沌と見えるものに対する解決策となりうる。
7. The overall objective of this research is to conduct basic studies into **possible** causes of biological nitrification process instability.
この研究の最終的な目標は，生物硝化プロセスを不安定にさせる原因に関する基礎研究を行うことである。

● as ..... as possible ： できる限り…

8. Ideally, reference sites should be **as** undisturbed **as possible**.
理想的には，参照地点は，できる限り無攪乱であるべきである。
9. Any discrepancies should be resolved before actually drawing a sample to prevent **as** much confusion **as possible**.
いかなる不一致も，混乱をできる限り防ぐため，実際のサンプリング前に解決されるべきである。
10. The primary concern in selecting the two sites is to assure that the physical characteristics are **as** similar **as possible**.
2地点の選定における主な関心事は，物理的特性ができる限り似るよう努めることである。
11. In contrast to the complexity of the processes being modeled, the model itself must be **as** simple to use **as possible**.
モデル化されているプロセスの複雑さとは対照に，モデルそれ自体はできる限り

簡単に使えるものでなければならない。

12. A number of reference sites were sampled to ascertain **as** accurately **as possible** the background levels of all parameters under investigation.
調査中のパラメータすべてのバックグラウンドレベルを可能な限り正確に確認するために，多くの参照地域でサンプルを採集した。

- **it is possible .....：…が可能である**

13. In the best of situation, **it is possible to** achieve an annual water balance within five percent.
最良の状態は，年度の水収支を5パーセント内で達成することが可能である。

14. While **it is possible for** reference sites to double as upstream control sites, the reverse is not always true.
参照地点を上流の対照地点の2倍にすることが可能な場合でも，逆は必ずしも真でない。

15. This concern has arisen lately as analytical techniques for detecting water contaminants have improved to the point where **it is possible to** detect contaminants at the part per-trillion level.
水汚染物質を検出するための分析技術が汚染物質をpptレベルで検出可能なまでに向上したため，最近この関心が高まった。

16. **It is possible that** more frequent observations at the boundary zones of these phases would shed further light on this problem.
これらの相の境界域におけるいっそう頻繁な観察によって，この問題が解明される可能性がある。

- **～ make it possible to .....：～は…を可能にする**

17. The methodology described herein **makes it possible to** gain insight into water quality changes to be expected over the life of a groundwater recharge project.
ここに記述された方法論は，地下水涵養プロジェクト期間中に生じるであろうと思われる水質変化の洞察を可能にする。

- **～ would have been possible：～がしえただろう**

18. None of this **would have been possible** without the excellent data management and processing skills of Dave Oda.
Dave Odaの優れたデータ管理・処理能力がなければ，何事もなしえなかったろう。

19. This chapter **would not have been possible** without the many years of field work, laboratory analysis, and data assessment and interpretation by

members of the USEPA, Ecological Assessment Section.
本章は，USEPA Ecological Assessment Section の職員による長年にわたる現地調査，実験分析，およびデータ評価・解釈がなければ，存在しえなかった。

- **whenever possible**：可能なら

20. **Whenever possible**, base the area on natural vs. political, boundaries to maximize applicability and the probability of selecting the least disturbed sites possible.
可能なら，常に最小擾乱地点選定の可能性および確率を最大限に高めるため，自然的境界および政治的境界に基づくべきである。

- **impossible**：不可能な

21. Such detailed understanding of the interactions is **impossible** from steady-state models.
このような相互作用の詳細な理解は，定常モデルからは不可能である。

Postulate：仮定する

1. **It is postulated that** TCE degradation involve the following sequential reactions.
TCE の分解が，次の連続的な反応を伴うと仮定される。
2. TCE degradation **was postulated** to involve the following sequential reactions.
TCE の分解が，次の連続的な反応を伴うと仮定された。

Potential (Potentially)：可能性；電位（潜在的に）

1. Leachates from waste sludge pose **potential** environmental problems.
廃棄汚泥からの浸出水は潜在的な環境問題を提起する。
2. Its **potential** impact on the scientific community is large and still developing.
科学の世界への潜在的影響は大きく，そしてまだ進展している。
3. The condition of the pavement in the area needs to be evaluated to determine **the potential for** continued infiltration and leaching.
その区域の舗装面の状態を評価し，継続的な浸入・浸出の可能性が判断されるべきである。
4. The regulated community is concerned about the **potential for** more stringent permits and other restrictions that may be leveraged by biocriteria.

規制の対象となる業界は，生物クライテリアにてこ入れされる，より厳しい許可限度および他の制限の可能性について懸念している。

5. The increasing threat of contamination and the **potential** hazard have focused a great deal of attention on the fate of mercury in the environment.
汚染と潜在的な危険性に対する脅威の増加は，環境での水銀の衰退に多くの関心を集めた。

6. These **potential** release sites do not fall under the definition of Solid Waste Management Units ; however, they are areas of environmental concern.
これら流出可能地点は，廃棄物管理ユニットの定義下ではない。しかしながら，それらは環境にとって関心のある地点である。

7. **Potential** uses of biological criteria in the total maximum daily loads (TMDL) process are also being explored at both the state and federal levels.
日最大負荷量（TMDL）過程における生物クライテリアの利用可能も，州レベルおよび連邦レベルで追求されている。

- ～ **provide a potential** ：～は可能性がある
- ～ **have a potential** ：～は潜在性を持っている

8. The tailings **provide a potential** to cause degradation of groundwater quality.
この選鉱廃物は，地下水の水質劣化を起こす可能性がある。

9. Many compounds **have a potential** for chemical modifications by both abiotic and biotic systems.
多くの化合物が，非生物システムと生物システムの両方によって化学変形を起こす潜在性を持っている。

- potential ：電位

10. These processes are affected by the concentration, size, and valence of metal ions on one hand and the redox **potential**, ionic strength, and pH of the solution on the other.
これらのプロセスは，金属イオンの濃度，サイズ，結合価の影響を受ける一方，他方では溶液の酸化還元電位，イオン強度，pHの影響を受ける。

- potentially ：潜在的に

11. The following is a list of his criticisms and a response to a **potentially** limited viewpoint of the IBI.
以下に，彼の批判と，IBIに限られているかもしれない観点への反応を列挙していく。

12. Contamination of groundwaters with synthetic chemicals poses new and

potentially severe problems.
合成化学物質による地下水汚染が新しい，そして厳しいかもしれない問題ををもたらす。

13. The model has been judged **potentially** capable of meeting the user's needs.
このモデルは，使用者の要求を満たすことが出来るだろうと評価されてきた。

Practice：実行；慣行

1. Useful information on farming **practices** had been gathered for the Missouri River study.
農業作業に関する有用な情報がMissouri川の研究のために集められてきた。
2. This may not have been a difference in state **practice**. It may have reflected differences in ownership, say, between federal and state or federal and private.
これは，各州慣行の相違ではなかったかもしれない。所有における差，つまり，連邦所有地と州有地の差，連邦所有地と私有地の差を反映していたのかもしれない。
3. Regional scale modifications in management **practices** may be feasible, allowing for the significant recovery of impaired aquatic resources.
管理手法における地域単位での修正が可能であり，これは傷ついた水生資源をかなり回復させることができる。
4. A measure of how much the quality can be improved can be derived through changing management **practices** in selected watersheds where associations were determined.
水質がどれほど改善されうるかを示す尺度は，関連性がみられる流域における管理手法の変更を通じて得られるだろう。

● **a questionable practice**：疑問の余地のある慣行

5. Using synthetic chelator chemicals is **a questionable practice** from an environmental standpoint.
合成のキーレーター化学物質を使うことは，環境の見地からは疑問の余地のある方法である。

● **as a matter of practice**：実際問題として

6. Courts have noted that the implementation of water quality criteria that are not based on chemical-specific numeric criteria can be difficult **as a matter of practice**.

法廷は，化学的に決められた数値クライテリアに基づいていない水質クライテリアの実施は，実際問題として難しいと指摘してきた。

- **In practice**：実際問題として；実のところ

7. **In practice**, the criteria as informal guidance are moving towards achieving the de facto status of standards, and are able to be effectively applied in their current form.
実のところ，非公式指針たる損傷クライテリアは，事実上の標準へ向かっており，現行の形で効果的に適用されうる。
8. **In practice**, however, the use of biocriteria and biological assessments to establish new permit requirements raises interesting and challenging questions.
実際問題として，新規許可要件を定めるための生物クライテリアおよび生物アセスメントの利用は，非常に興味深く，難しい問題を提起する。

Preceding：すぐ前の；先立つ

1. **In the years preceding** these recommendations, Kaga et al. (1981) developed an operational definition of biological integrity.
これらの勧告の数年前，Kaga ら(1981)が生物完全性の実用的定義を定めた。
2. Another reason for using single-purpose frameworks, as mentioned **in the preceding section**, stemmed from the belief that …,
前章で述べたとおり，単一目的枠組みを用いる別の理由は … という思考にあった。
3. The **preceding** results demonstrate that discernable patterns in biological community information can be used to determine the probable cause of certain types of impairments.
前述の結果は，生物群集情報における区別可能なパターンは，ある種の損傷の潜在的原因を決定するため用いうることを実証している。

Precise (Precisely)：正確な；明確な（正確に；明確に）

1. **To be more precise**, ...
より正確に言うと，…
2. Additional work is needed to more **precisely** define and set expectations for these two metrics, and to test their usefulness in field applications.
これら2つのメトリックについて目標をより正確に設定し，現地適用におけるその有用性を試すため，さらなる研究が必要とされる。

Predict (Predictive) (Prediction) (Unpredictable):
予測する（予測的）（予測）（予測不能な）

1. The model computations **predict** the equilibrium distribution.
   モデルによる計算から均衡分布が予測される。
2. Runoff and volatilization losses **were predicted** to be minimal.
   流去水と揮発による減少が最小であると予測された。
3. By using realistic values for these parameters, the calibration process at least ensures that the concentrations **predicted** by the model are within an order of magnitude of the actual values.
   これらのパラメータに現実的な値を使うことによって，キャリブレーション過程では，モデルによって予測された濃度が少なくとも実際値と同じ桁内にあることを保証する。

   ● **to predict** ：予測すること

4. The failure of the model **to predict** it may be attributed to errors in the model input data.
   それを予測するモデルの失敗は，モデル入力データのエラーに帰されるかもしれない。
5. **To predict** whether a behavioral response to a chemical pollutant will occur, one must ask whether the organism can detect the pollutant at concentrations likely to be encountered in field situations.
   化学汚染物質に対する挙動反応が起こるかどうか予測するには，現場の状況で起こりえる濃度で生物が汚染物を検知できるかどうか知らなくてはならない。

   ● **predicted** ..... ：予測される …

6. The **predicted** CO_2 concentration is higher than the observed concentration at later time.
   予測された CO_2 濃度は，後に観察された濃度より高い。
7. A comparison of **predicted** and measured Alachlor residues in the soil is given in Figure 3.
   土壌における Alachlor 残余の予測値と測定値の比較を図3に示す。
8. Figure 3 shows good agreement between the **predicted** and observed suspended solids concentrations for the first 150 meters downstream.
   図3が，最初の150メートル下流で浮遊固体濃度の予測値と観察値が合っていることを示す。

   ● **predicting** ：予測すること

9. The model is less successful in **predicting** the degree of lateral dispersion.
 このモデルは，横断方向の拡散の度合いを予測するにはあまりよくない。
10. Understanding the reactions of PCBs with OHC is crucial to **predicting** the future persistence of PCBs and the hazards associated with their presence.
 PCB 持続性および PCB の存在による危険性を予測するのに，PCB と OHC の反応を理解することが不可欠である。
11. With the increased need for **predicting** the movement and distribution of chemicals in an aquatic environment, knowledge of the rate constants governing the uptake and clearance of chemical in fish is required.
 水生環境での化学物質の移動・分配を予測する必要性が増加するとともに，魚体内での化学物質の摂取・除去を左右している速度定数の知識が必要とされる。
12. Design models that do not include competition between CPs risk over predicting CP dechlorination rates in CP mixtures and under **predicting** the time for remediation.
 塩化フェノール（CP）の間の競合反応を含まない設計モデルは，CP 混合物において CP 脱塩速度を過剰に予測し，さらに改善に要する時間を不十分に予測する恐れがある。

- **predictive**：予測的
- **unpredictable**：予測不能な

13. **Predictive** models can be used to help define the relationships between water pollutants and their sources.
 予測モデルは，水環境の汚染物とそれらの流入源との関係を定義するのをに使うことができる。
14. Work has shown that highly variable and **unpredictable** flow regimes can have strong influences on fish assemblages.
 研究によれば，可変性が強くて予測不能な流量状況は，魚類の群がりに強い影響を及ぼしうる。

- **prediction**：予測

15. A need exists for **prediction** of future trends and control in contaminated areas.
 汚染された地域に対し，これからの傾向の予測と管理の必要性がある。
16. The validity of these **predictions** could be determined by an examination of the biotic condition of the stream.
 これらの予測の正当性は，水流の生物状態の観察結果によって決定されえる。
17. For such extreme operating conditions, deviations between model **prediction** and experiments can be expected.

このような極端な操作条件では，モデル予測値と実験値の間の逸脱が予想される。

18. Model **predictions** show reasonable agreement as can be seen from both Figures 2 and 3.
図2と図3の両方に見られるように，モデルによる予測が妥当であることを示している。

19. A greater understanding of kinetic behavior **should lead to better predictions** of the behavior of PCP during anaerobic bioremediation.
速度論的挙動をより良く理解することは，嫌気性生物による修復中のPCPの挙動をもっと良く予測することにつながる。

20. This work will aid **in making predictions** about long-term buildup of organic layers on the stone.
この研究は，石への長期にわたる有機層の蓄積を予測するのを助けるであろう。

Preferred（Preferentially）：優先の（優先的に）

1. Among the metal studied, the **preferred** order of uptake by activated sludge was found to be in the sequence of lead > copper > cadmium > nickel.
研究された金属の中で，活性汚泥による吸着の優先オーダーは鉛＞銅＞カドミウム＞ニッケルの順であることが判明した。

2. Research emphasis should be placed **preferentially** on problems of risk management.
研究の主眼点は，リスクマネージメントの問題に優先的に置かれるべきである。

Premise：前提

1. Many of Saito's concerns are **based on the premise that** they are heterogeneous.
Saitoの懸念の多くは，それらが異質だという前提に基づいている。

Preparation：準備

● ～ **in preparation**：準備中

1. Based on the information gathered and the publication **in preparation**, a first approximation of reference ecoregions and subregions is compiled.
収集情報に基づき，また，準備中の刊行物の概説に基づき，改良された生態地域および小地域の初の概要が作成される。

Presence (See also Absence)：存在（→ Absence）

1. There is very limited published data on **the presence of** mercury in the Purple River.
   Purple 川において，水銀の存在を発表したデータがほとんどない。
2. These data clearly demonstrate **the presence of** two different fish communities on the same waterbody.
   これらのデータは，同一水域に2種類の魚類群集が存在することを明らかに示している。
3. The recent death of horses is not irrelevant to **the presence of** selenium in the grass.
   最近の馬の死は牧草にセレンが存在していることとは無関係ではない。
4. The analysis is based solely on the **presence or absence of** a group and does not take into consideration the abundance of that group.
   この分析は，ある集団の有無のみに基づいており，その集団の数度を考慮していない。
5. Understanding the reactions of PCBs with OHC is crucial to predicting the future persistence of PCBs and the hazards associated with **their presence**.
   PCB の将来の持続性および PCB の存在による危険性を予測するのに，PCB と OHC の反応を理解することが不可欠である。
6. An examination by Toda et al. (2002) of oysters and other shellfish exposed to an oil spill revealed **the presence of** large quantities of petroleum compounds in the body tissues.
   カキと他の貝への流出した石油の影響をみた Toda ら (2002) の研究によれば，貝の体組織に大量の石油化合物が存在することが明らかになった。

● **in the presence of** ..... ： … が存在するときの

7. Possible explanation for the strong PMPA response **in the presence of** the pesticides include: 1) ......, 2) .....
   殺虫剤が存在するときの PMPA の強い反応については，次のような説明が可能である。
8. This approach assumes that the control and study site differ only **in the presence of** pollution.
   この手法は，対照地点と研究地点が汚濁の有無においてのみ異なると仮定している。

Present (Presently) ：現在；存在する；提示する（現在）

- **At present time** ：現在
- **At the present time** ：現在
- **from 1980 to the present** ： 1980から現在まで

1. **At the present time**, there are several uncertainties.
   現在，いくつかの不確実性がある。
2. **At present**, few states can characterize even a small fraction of their water resources with biological survey data.
   現在，生物調査データで各自の水資源のごく一部でも特性把握できる州は，きわめて少数である。
3. Because of methodological uncertainties, time-trend analysis has been limited to the period **from 1940 to the present**.
   方法論的不確実性のため，時傾列分析は1940から現在までの間に限定されてきた。
4. Data are available **from the late 1800s to the present**.
   1800年代後期から現在まで，データが得られる。

- **present .....** ：現在の…

5. The aim of the **present** study was to examine aspects of elimination of PAH by Mytilus edulis.
   現在の研究の目的はMytilus edulisによるPAHの除去の状況を調べることであった。
6. The **present** model structure may apply to other compounds as well as to other species.
   現在のモデル構造は，他種と同様に他の化合物にも当てはまるかもしれない。
7. Under the conditions of the **present** experiments, chloride was the major ionic product, and small amounts of $HCOO^-$ were detected as well.
   現在の実験条件下では，塩化物が主要なイオン生成物であった，そして同様に小量の$HCOO^-$も検出された。

- **～ is present** ：～が存在する

8. Each area can be viewed as a discrete system which has resulted from the mesh and interplay of the geologic, landform, soil, vegetative, climatic, wildlife, water and human factors which **may be present**.
   各領域は，個別的なシステムとみなされ，地質，地形，土壌，植生，気候，野生生物，水，そして潜在的な人間因子の調和および相互作用から生じてきた。

9. The study of potential synergistic or antagonistic effects of a multitude of toxic substances **present** in many effluents becomes virtually an impossible task.
多くの排水中に存在している多数の毒性物質による潜在的相乗作用または拮抗作用を研究することは，事実上，不可能な作業となる。
10. The periphyton assemblage may change up to 30％ in taxa **present** from year to year within the spring-summer period.
付着生物群がりは，年ごとに30％もの春夏間の部類群変化が生じうる。

● ～ **present** ……：～は…を提示する

11. Table 7 **presents** a summary of this comparison.
表7が，この比較の要約を提示する。
12. Fig. 4 **presents** a typical reaction profile observed during a test run.
試運転の間に観察された典型的な反応プロフィールを図4に提示する。
13. The following discussion **presents** examples of methods presently used in evaluating the periphyton assemblages.
以下の論考は，付着生物群がりを評価する際に現在用いられている方法の諸例である。
14. The purpose of this section is **to present** a description of some of these modifications and to illustrate the different directions many programs have taken.
本項の目的は，これらの修正の一部を述べることと，多くのプログラムがとった異なる方向性を示すことである。

● ～ **present a challenge**（See **Challenge**）：
～は課題をもたらす（➡ **Challenge**）
● ～ **present a problem**（See **Problem**）：
～は問題を生じうる（➡ **Problem**）
● ～ **present a risk**（See **Risk**）：～は危険性を引き起こす（➡ **Risk**）
● ～ **is presented**：～が提出される；～が発表される

15. In the beginning, case histories **are presented**.
まず初めに，事例史を下に述べる。
16. Comparison of surface water analyses with previous sample **are presented** in Table 1.
表流水の分析値と以前のサンプルの分析値との比較を表1に示す。
17. Summaries of the more progressive of these interstate cooperative efforts **are presented** below.
これらの州間協力活動の前進について以下に要約する。

18. The curve shown in this figure agrees well with the curve **presented** recently by Ito et al. (2002).
 この図に示された曲線は，最近 Ito ら (2002) によって発表された曲線とよく一致している。
19. Data on the kinetics of bioaccumulation of mercury by zooplankton **will be presented**.
 動物プランクトンによる水銀の生物濃縮速度に関するデータが発表される。
20. Results **are presented** in several forms, but are normalized to allow direct comparison with other published work.
 結果がいくつかの形式で提出されているが，標準化されているために発表されている他の研究結果との直接比較を可能にしている。
21. The model **presented** in this paper is intended to supersede the preliminary results given in a recent report (Sasaki, 2002).
 この論文に記載されているモデルは，最近の報告書 (Sasaki,2002) に記載された予備結果に取って代わることを意図している。
22. The results **presented** here reflect recent modifications of the global transport model, and, thus, the estimates are somewhat different than previously reported.
 ここで報告された結果は，地球上輸送モデルの最近の修正を反映している。それゆえに，予測値は前に報告された値より，いくぶん異なっている。

- **~ is presented elsewhere**： ～は，ほかの様々なところに提出されている

23. The details of the studies and results **have been presented elsewhere** (5–9) and are only summarized here.
 その研究の詳細と結果は，ほかの様々なところに発表されているので (5-9)，ここではただ要約するだけにする。
24. Procedures for these calculations **are also presented elsewhere** (2, 3).
 これらの計算のための手順はほかのところにも発表されている (2,3)。

- **in presenting** .....： … の提出にあたって

25. For the sake of clarity **in presenting** the test results, most of the details concerning the theoretical background of the tests and the analytical procedures are provided in the Appendix B.
 テスト結果の提出にあたって，それを明快にするために，試験と分析手法の理論的な背景の細部の多くを付録 B に示す。

- **presently**： 現在

26. **Presently**, there is little understanding of the interaction of surfactants and these compounds.
現在，界面活性剤とこれらの化合物の相互作用はほとんど理解されていない。
27. It is **presently** insufficient to stress that oil is a toxic substance and fish can be affected by exposure.
原油が有毒物質であり，そして魚がそれに接触することによって影響を受けることを強調することは現在不適当である。

Previous（Previously）：前の（前に）

1. **Previous** investigation have examined .....
以前の調査で … が検査された。
2. These conclusions are in contrast to those of **previous** researchers.
これらの結論は前の研究者たち（の結論）と対照的である。
3. Improvements in the treatment process were made in the year before this sampling, and macroinvertebrate changes were not as great as **in previous years**.
サンプリング前年に処理過程の改善が行われ，大型無脊椎動物の変化は，それまで数年間ほど大幅でなかった。
4. As discussed **in the previous section**, the status of biological criteria programs across the United States can be thought of as the sum of all the states in various stages of developing and implementing biological criteria.
前節で述べたとおり，全米での生物クライテリアプログラムの現状は，生物クライテリア設定・実施の様々な段階における諸州すべての総和として考えうる。

   ● **previously**：前に

5. In this paper, we compare laboratory results to **previously** reported field results.
この論文では，我々は研究室での結果と前に報告された野外調査の結果とを比較している。
6. These data provide additional information on **previously** identified Solid Waste Management Units.
これらのデータは，前に確認された廃棄物処理施設についての追加情報を提供する。
7. All known underground storage tanks **previously** or currently used for the storage of wastes have been included.
前に，あるいは現在廃棄物の貯蔵のために使われたすべての周知の地下貯蔵タンクは含まれた。
8. As stated **previously**, the selection of the appropriate criterion depends on

the content of the database.
前述のとおり，適切なクライテリアの選定は，データベースの内容に左右される。
9. As was **previously** mentioned, Idaho DEQ has employed a system of tiered aquatic life uses in the state water quality standards since 1988.
前述のとおり，アイダホ州DEQは，1988年以降，州水質基準において段層式の水生生物使用システムを用いてきた。

Primary（Primarily）： 主な（主要に）

1. The **primary** area of interest is the Tama River.
   関心のある主要な地域はTama川である。
2. The excess nutrients are the **primary** cause of the reduced oxygen levels observed in the lake.
   この湖で観察される酸素レベルの減少は，過剰栄養が主要な原因である。
3. Development and implementation of biological criteria consists of four **primary** steps:
   生物クライテリアの設定・実施は，以下の4つの主要ステップからなる。
4. Nitrate up to 1 mg/L as N did not affect the rate of **primary** producer during 3 hr of incubation.
   3時間の培養時間では，硝酸態窒素1mg/Lまで主生産者の成長速度には影響を与えなかった。
5. A **primary** release is one from the **primary** contaminant source, a secondary release is one that occurs, for example, from the contaminated soil to the groundwater.
   主要な流出は第一汚染源から生ずるもので，第二の流出は例えば，汚染土壌に生じ地下水まで浸透するものである。
   - **primary concern is** .....（See **Concern**）：
     主な関心事は … である（➡ **Concern**）
   - **..... of primary concern**（See **Concern**）：関心の高い …（➡ **Concern**）
   - **primary emphasis**（See **Emphasis**）：主眼；重点；強調（➡ **Emphasis**）

Principle：原則

1. There is no reason why these same **principles** would not apply here.
   これらの原則がここで適用されないという理由は，見あたらない。
2. These two **principles** hold in all cases under any conditions.
   これらの2つの原則は，どんな状況であってもすべての場合に適用される。
3. In accordance with **principles** of bacterial kinetics, it was assumed that ......

バクテリアの成長速度論の原則にしたがって，…と想定された。
4. As a basic **principle of** statutory construction, general statements of statutory goals and objectives have no legal force and effect, absent specific operative provisions in the law.
法的解釈の基本原理として，法律上の目的および目標の全般的言明は，法的効力を持たず，法律における具体的な運用既定を欠いている。
5. The validity of an integrated assessment using multiple metrics is supported by the use of measurements of biological attributes firmly rooted in sound ecological **principles**.
多メトリックを用いた統合的アセスメントの妥当性は，確固たる生態学的原理に根ざした生物学的属性測定を利用することで確保される。
6. Forbes' insight and application of the **principle of** natural selection led to the establishment of a biological station on the shores of the Illinois River in 1894.
1894年にIllinois川の岸に生物監視測定所が設置されたが，Forbesの洞察と自然淘汰原理の応用が，そのきっかけとなった。

Prior：前の；以前；先立つ

1. Efforts were made to secure all **prior** existing data relevant to this subject.
この課題に関するすべての事前に既存するデータを安全に保つ努力をした。
2. No program to monitor tanks for leaks was in place **prior to** that date.
その日以前には，タンク漏れを監視するプログラムは使用されていなかった。
3. The MIwb does not require a spatial calibration **prior to** use.
MIwbは，利用に先立つ空間的較正を必要としない。
4. These samples were stored at room temperature in these bags **prior to** use in the experiments.
これらのサンプルは，実験に使用するまで，室温にてこれらの袋に保管された。
5. The reference site results were pooled on a statewide basis **prior to** constructing the drainage area scatter plots.
参照地点結果が全州ベースで集められた後，流域面積の散点図が作製された。
6. The idea that conditions were pristine in North America **prior to** Europian settlement has been convincingly challenged in the past couple of decades.
欧州人が入植する前の北米では無傷状態だったという考えにたいして，ここ数十年，説得力ある異議が申し立てられてきた。
7. Although the ICI did not meet the Warmwater Habitat (WWH) biocriterion upstream from RM 109.4 in any year **prior to** 1991, the severity of the degradation was less in the normal flow years.

1991年以前は，RM109.4上流でICIが温水生息場所(WWH)生物クライテリアを満たさなかったにもかかわらず，劣化度は，平年並み流量の年の方が小幅であった。

8. Volunteer-collected data are used for trend analysis, as screening tools **prior to** intensive investigation by water quality professionals, as a basis for making local zoning and land management decisions, and in state's 305b water resource quality reports.
ボランティアによる収集データは，傾向分析に用いられたり，水質専門家によるさらに徹底的な調査に先立つ選別ツールとして，局地的区域(localzoning)の設定および土地管理決断のための基礎として，また，諸州の305(b)水資源報告に用いられる。

Priority：優先的；プライオリティ

1. We recommend that **the highest priority for** future experimental work is to validate the existing models.
これからの実験研究の最も高い優先順位は既存のモデルを実証することにある，と我々は提言する。
2. The issue of toxics control and biological toxicity assessment has long been **a priority for** Association of Metropolitan Sewage Agencies.
毒物制御の問題と生物による毒性アセスメントは長い間，Association of Metropolitan Sewage Agenciesが優先課題であった。
3. In the view of the Task Force, the research questions of **priority importance** falls into two general categories: 1) ....., 2) ....., and 3) .....
特別委員会の観点では，研究質問の優先的重要性は2つの一般的な部門に分類される。それらは，1)…, 2)…, および3)…, である。

- **in order of priority for** .....：…の優先順位で

4. The 1972 act specifically required states not only to identify impaired waters, but to rank them **in order of priority for** cleanup and restoration.
1972年法は，具体的に諸州が被損傷水域を確認することに加え，それらを浄化・回復することを優先するよう要求した。

Probable：ありそうな

1. It seems quite **probable** that .....
…は非常にありうることである。
2. There is insufficient information on source area ST5 to draw a conclusion on

probable risk.
十分に可能性のあるリスクについて結論を引き出すには，汚染源 ST5 についての情報が不足している。

3. Discernable patterns in biological community information can be used to determine the **probable** cause of certain types of impairments.
生物群集情報における区別可能なパターンは，ある種の損傷の潜在的原因を決定するために用いうる。

4. The uncertainty associated with the definition of **probable** conditions and the consequences of a false positive or false negative error has been considered as part of the decision selection process.
予想される条件の定義と虚偽の肯定的であるか，あるいは虚偽の否定的なエラーの結果と結び付けられた不確実は決定セレクションプロセスの一部であると考えられました。

Probability：確率

1. The **probability** is high that, these animals were, at some time, introduced to the caldera as a means of mosquito control.
蚊をコントロールする手段として，あるとき，これらの動物がカルデラに放された可能性が高い。

2. The difference in TPH concentrations between the initial and four-week samples was statistically significant at the 0.025 **probability** level.
最初のサンプルと4週間後のサンプルでの TPH 濃度の相違は 0.025 の確率レベルにおいて統計的に有意であった。

3. Whenever possible, base the area on natural vs. political, boundaries to maximize applicability and the **probability of** selecting the least disturbed sites possible.
可能なら，常に最小攪乱地点選定の可能性および確率を最大限に高めるには，自然的境界および政治的境界に基づくべきである。

Problem (Problematic)：問題 (問題点の多い)

1. The magnitude and significance of the **problem has become increasingly evident**.
この問題の規模と重要性がますます明白になってきた。

2. The assessment approaches are intended to **identify** water quality **problems**.
このアセスメント手法の意図は，水質上の問題点を確認することである。

Problem

3. Aquatic organic matter **is responsible for several problems** in water supplies.
給水において，水生有機物がいくつかの問題点の原因である。

- **A major problem is** ..... ：主要な問題は … である
- **The central problem of** 〜 **is** ..... ：〜の主要な問題は … である
- **A paramount problem of** 〜 **is** ..... ：〜の主要な問題は … である
- **The problem is that** ..... ：問題点は … である
- **other problems associated with** 〜 **are** ..... ：
  〜に関わる他の問題は … である
- **Problem related to** 〜 **are** ..... ：〜に関係がある問題は … である

4. **A major problem** preventing the widespread use of ozone is the high cost of treatment.
オゾンの広範囲な使用を妨げている主要な問題は，処理コストが高いことにある。

5. **The central problem of** the stream water quality is the nonpoint sources.
この水流の主な水質問題は，面源にある。

6. **A paramount problem of** water quality is the widespread distribution of toxic substances.
主要な水質問題は，毒性物質が広範囲にわたって分布していることである。

7. **Other problems associated with** chlorophyll a **are** high temporal variability and difficulty in interpreting trends.
クロロフィルaに関わる他の問題は，時間的可変性の強さと傾向解釈の難しさである。

8. **The problem has been** in defining the regions.
問題点は，地域の定義づけであった。

9. **Problems related to** chemical contamination of surface water supplies, particularly those associated with byproducts from new treatment technologies, **are** rational in scope and receive national attention.
表流水給水における化学汚染に関連付けられ，とりわけ新しい水処理技術の副産物に結びつく問題は，その道理性を認められており，全国的な注目を集めている。特に新しい副産物と結び付けられたそれらの範囲では合理的であって，そして国家の注意を受けます。

10. **The problem is that** little effort is being expended on studying ecosystems holistically and attempting to define differences in patterns of ecosystem mosaics.
問題点は，生態系の全体論的研究や，生態系モザイクのパターン差を定義づける試行にあまり力が入れられていないことである。

11. **The problem** with using this type of framework for geographic assessment and targeting **is that** it does not depict areas that correspond to regions of similar ecosystems or even regions of similarity in the quality and quantity of water resources.
   この種の枠組みを地理アセスメントおよびターゲット設定に用いるうえでの問題点は，類似生態系の地域や，水資源の質・量において類似な均一地域に符合する区域を描写しないことである。

    - **The problem of how to .....**：…するかという問題

12. **The problem of how to** maintain optimal nitrification rates is not only a question of the presence of aerobic conditions.
    最適な硝化反応速度をどのように維持するかという問題は単に好気性状態であるか，ないかだけではない。

    - **a problem occur**：問題が発生する
    - **a long-standing problem occurred**：長年の問題が発生した

13. Bank erosion **problems often occur** hand-in-hand with riparian vegetation disturbance.
    河岸浸食問題は，しばしば水辺植生攪乱と関連して起こる。

14. **A long-standing** heavy metals **problem** in this stream **occurred** below an industrial effluent discharge.
    この水流における長年の重金属問題は，産業廃液排出口の下流で発生した。

    - **The problem arises from .....**：問題は…から起こる
    - **The problem stems from .....**：問題は…から発する

15. **The problem arises** principally **from** the discharge of the residues of human activities.
    人間活動の残余の流出から主に問題が生じる。

16. **The problem stems from** the period 1947 to 1970, when a private DDT plant discharged wastes containing DDT residues into a drainage ditch which flowed into the river.
    この問題は，1947から1970の間に民営のDDT製造プラントがDDT残余を含んでいる廃液を川に流れ出る排水溝に放流したことが発端である。

    - **A problem is encountered**：問題が生じる
    - **A problem encountered in both A and B is that .....**：
      AとBの両ケースで生じた問題は…である

17. **A similar problem is encountered** in estimating pesticide model

parameters.
殺虫剤の環境モデルのパラメータを推定する過程にて，類似の問題が生じている。

- **The problem is receiving increased attention.**：
  この問題に対する注意が増している
- **The problem of ~ receives attention**：~の問題に注意が注がれる

18. **The problem is receiving increased attention** throughout the country.
    国中で，この問題に対する注意が広がっている。

- **A problem remains**：問題は未解決である

19. **Problems do remain**, but all of the indicators point to conventional problems associated with the municipal sewer system, particularly CSOs.
    問題は，未解決だが，指標は，すべて都市下水システム，特にCSOに関わる問題点を指し示している。

    ★ここでの **do** は **remain** を強調するために使われています

- **~ detect problems**：~は問題を探知する
- **problems can be abated**：問題が緩和されうる

20. In some cases, biological impairment criteria **may detect** water quality **problems** that are undetected or underestimated by other methods. In these cases, biological impairment criteria may be used as the sole basis for regulatory action.
    一部の場合において，生物損傷クライテリアが他の方法で探知されなかった水質問題，もしくは過小評価された水質問題を探知できる。これらの場合，生物損傷クライテリアは，唯一の規制措置の根拠となるであろう。

21. If the proper construction of the IBI (or similar evaluations) takes place up front, the eclipsing and other **problems can be abated**.
    適切なIBI（または，類似評価）の構築がなされれば，隠蔽問題および他の問題を排除できる。

- **~ is a problem**：~は問題である

22. The latter **is a problem of** much practical interest.
    後者は，実用面でかなり興味のある問題である。

23. Unavailability of site-specific data for model parameter **is a major problem**.
    モデルパラメータを決めるための地域特定データが得られない状況は，主要な問題である。

- ~ present a problem：～は問題を生じうる
- ~ address the problem of ……：～は…の問題を扱う
- To address this problem：この問題点に対処するため

24. Historical data **may present a problem** because the site selection generally is not conductive to either of these processes.
史的データは，問題を生じうる。地点選定は，通常これらの過程において資するわけではないからである。
25. This report **addresses the problem** of analyzing the fate of chemicals discharged into receiving waters.
この報告書は，水域に放流された化学物質の衰退の解明に関する問題を扱っている。
26. **To address this problem**, a three-year ecoregion project was conducted.
この問題点に対処するため，3箇年の生態地域プロジェクトが行われた。
27. Laboratory work **addressing** some aspects of these **problems** and related modeling efforts are ongoing at the Idaho State University.
これらの問題の一局面を扱う室内研究とそれに関連したモデリングの努力がアイダホ州大学で進行中である。

- ~ pose a problem：～が問題をもたらす；～が問題を提起する

28. **It did not pose** a significant **problem**.
それは大きな問題にならなかった。
29. Leachates from waste sludge **pose** potential environmental **problems**.
廃棄汚泥からの浸出液は，潜在的な環境問題を提起する。
30. Contamination of groundwaters with synthetic chemicals **poses new and potentially severe problems**.
合成化学物質による地下水汚染が新たに潜在的で厳しい問題をもたらす。
31. The question of chemical speciation **poses one of the most difficult problems** to be resolved by the chemist, especially in the context of synergistic effects as encountered in natural water.
化学的種分化への質問は，特に自然水域で見られる相乗効果という文脈において，化学者による解決が最も難しい問題の1つを提出している。

- To counter this problem：この問題点に対処するため

32. **To counter this problem** we have devised a low-end scoring procedure.
この問題点に対処するため，筆者は，ローエンドスコアリング手順を考案した。

- **problematic**：問題点の多い

33. The following is a description of the **problematic** areas encountered during this survey.
次に，この調査の間に生じる問題点の多いエリアが記述されている。
34. First, the definition and delineation of reference sites **is problematic**.
第一に，参照地点の定義・画定に問題がある。

Procedure：手順

1. Details concerning the experimental procedure are given elsewhere and are described only briefly here.
実験手順に関する詳細は，ほかの様々なところで提示されているので，ここでは簡単に記述する。
2. Calibration procedures, as noted above, allow normalization of the effects of stream size.
前述のとおり，キャリブレーション手順により，水流規模の標準化を可能にする。
3. Data are tabulated in the field and laboratory, and documented via chain-of-custody procedures.
データは，現地および実験において作表され，保管連鎖手順に従って文書化される。
4. The procedure by which this is done can be viewed as part of the regionalization process.
このための手順は，地域分け過程の一部とみなされる。

- **The procedure included .....**：手順は … を含む
- **The procedures are provided in .....**：手順は … に示されている
- **The procedures for … are presented .....**：
 … のための手順は … に示されている
- **The procedure was established**：手順は確立された

5. The procedure included the following step : .....
この手順は次のステップの通りである。それは … である。
6. The analytical procedures are provided in the Appendix B.
分析手法を付録 B に示す。
7. Procedures for these calculations are also presented elsewhere (6, 7).
これらの計算のための手順はほかのところにも提出されている (6, 7)。
8. The standard procedure outlined above was established so that the problems which often obscure interpretation of results are minimized.
結果の解釈を出来るだけ明瞭にするために，上に概説された標準的な手順は確立

された。

Process：プロセス；過程；処理する

1. The investigation **process** consists of six basic components: ....
   調査プロセスは6つの基本的な構成要素から成る。それらは，…
2. The basic steps of the risk assessment **process** are the following:......
   リスクアセスメント過程の基本的な段階は次の通りである。それは，…
3. Chemical transport and transformation **processes** are examined more closely.
   化学物質の輸送・変化プロセスがより綿密に検討される。
4. We use mathematics to help understand the behavior of some physical **process** or system.
   我々は，物理的プロセス挙動あるいは物理的システム挙動の理解を深める為に数学を使う。

   - ～ **is in the process of** ..... ：～は…の過程にある
   - ～ **is processed with** ..... ：～は…で処理される

5. Tomita (1990) **is in the process of** validating his version for Lake Erie.
   Tomita (1990) は，Erie湖に関するバージョンの確証過程にある。
6. Water sample **was processed with** hydrochloric acid, then neutralized and filtered as previously described.
   水のサンプルは塩素で処理された後，前述のように中和され，そしてろ過された。

   - **in the processing** ：処理過程において

7. The plant discharges mercury used **in the processing** from early years.
   早年から，プラントは使用済みの水銀を排出している。

Progress：進歩；前進；進歩する；前進する

1. Because so few states **have progressed** to the point of proposing legally binding narrative or numeric biocriteria at this time, industry involvement has been relatively limited on a national scale.
   現時点では，法的拘束力を持つ記述的，数量的な生物クライテリアまで前進している州は，ごく少数なので，業界関与は，全国規模で比較の限られている。

   - **progress has been made** ：進歩してきた

2. Much **progress** and advancement **has been made** in the past decade.
   ほとんどの進歩ならびに前進は，過去10年間の成果である。

   - **in progress**：進行中の
   - **～ is in progress**：～は進行中である

3. The authors wish to thank the following individuals for providing information on research **in progress**: Tomoko Koga, Naomi Okada, Yoko Nakamura.
   筆者らは，以下の各氏が現在進行中の研究に関する情報を提供してくれたことについて感謝したい。Tomoko Koga, Naomi Okada, Yoko Nakamuraの各氏である。

4. While a number of HSPF applications **are in progress**, few studies are complete.
   多数のHSPFアプリケーションが進行中である一方，ほとんどの研究が未完である。

   - **～ make progress**：～は進歩を成し遂げる
   - **～ have made progress toward ……**：
     ～は…に向けて前進を遂げてきた

5. During the past few weeks, the project **has made considerable progress**.
   ここ数週の間に，プロジェクトはかなり進歩した。

6. USEPA, U.S Forest Service, and Bureau of Land Management **have made substantial progress toward** developing biological indicators for surface water assessment.
   USEPAとU.S. Forest Service（森林局）および土地管理局は，地表水アセスメントのための生物指標の設定に向けて大幅に進歩した。

Promise (Promising)：有望；見込み（有望な）

1. The bacterial luminescence bioassay **shows great promise** as it responds to a wide range of compounds.
   広範囲の化合物に対して反応するため，バクテリアの蛍光バイオアッセイは有望である。

2. Our unpublished data suggest the methods **to be of promise** for the characterization of sediments.
   我々の未発表のデータは，堆積物の特性評価のために有望な方法であろう。

   - **promising**：有望な
   - **～ seen more promising**：～がいっそう有望に思われる

3. Natural chelators of plant or microbial origin **seem more promising**.
植物あるいは微生物が起源である自然のキレートがいっそう期待できる。
4. In September 1987, USEPA published a management study entitled, Surface Water Monitoring: A Framework for Change, that strongly emphasized the need to accelerate the development and application of **promising** biological monitoring techniques in state and USEPA monitoring programs.
1987年9月にUSEPAは,『Surface Water Monitoring: A Framework for Change (地表水モニタリング：変化のための枠組み)』と題する管理研究論文を発表した。同書は,州およびUSEPAのモニタリングプログラムにおける有望な生物モニタリング技法の案出・適用を促す必要性があることを力説していた。

Propose：提案する

● ～ is proposed ：～が提案される

1. Evidence has been gathered, and solutions **have been proposed**.
証拠が集められ,そして解決策が提案されてきた。
2. A reaction mechanism **has been proposed** that accounts for the observed experimental data.
観察された実験的なデータを説明する反応機構が提案された。

● proposed by ..... ：…によって提案される；…によって提示される

3. The best known explanation among those explanations **proposed by** ～ is .....
～によって提示された説明の中で最もよく知られているのは … である。
4. Several wells exceeded the Primary Drinking Water Standards **proposed by** the EPA in arsenic and sulfate.
いくつかの井戸で,EPAによって提議されている主要な飲料水基準が,ヒ素と硫酸塩で越えていた。
5. This relationship is compliant with the "fixed demand" theory **proposed by** Wu and Yao(2001).
この関係は,Wu andYao(2001)によって提案された「不変需要」理論に従う。

● proposed ..... ：提案された

6. It is essential to recognize that the **proposed model** depends on its intended purpose.
提案されたモデルが,その意図した目的によることを認識することが不可欠である。

7. Clearly, much work needs to be done to evaluate and polish the **proposed** technique.
   明らかに多くの研究が，提案された技巧を評価して，そしてそれをより良くするためには必要である。
8. A newly **proposed** toxicity test will fit into the overall hazard management system.
   新たに提案された毒性テストが総合的な危険マネージメントシステムに組み込まれるであろう。
9. The aim of this study is also to contribute indirectly to the **proposed** data base and inventory system for the state.
   この研究の目的は，この州の為に提案されたデータベースそして目録作成システムに間接的に貢献することでもある。
10. Two of the **proposed** use designation changes were challenged in court because upgrades in designated use could possibly result in more stringent permit limitations.
    指定変更案のうち2つは，法廷で異議申立てされた。使用指定における昇格は，より厳しい許可限度をもたらす可能性があったからである。

Proportion (Proportional)：比率；割合（比例した）

1. It is relatively easy to read off the **proportion of** sites that are above or below a selected score.
   選ばれたスコアを上回る地点もしくは下回る地点の比率を読みとることは，比較的容易である。
2. Different classes of algae have different **proportions of** internal structure occupied by vacuoles.
   異なる綱の藻類は，液胞に占められた内部構造の比率が異なっている。
3. The advantage of taking this additional step is that it is now relatively easy to see the **proportion of** sites that do not meet their respective biocriterion scores.
   この追加措置による利点は，それぞれの生物クライテリアスコアを満たさない地点の比率が比較的見やすくなることである。
4. More specifically, questions such as, "What **proportion of** the stream miles (and how many stream miles) are affected by point source discharges?" can be addressed.
   より具体的には，「水流 mile 数のどれほどの比率（どれほどの水流 mile 数）が点源排出に影響されるか？」といった疑問に対処できる。
5. Streams/watersheds within each set should be similar to one another

regarding "relative disturbance" and should reflect higher water quality than streams with watershed with similar **proportions in** each subregion but with greater human impact.
各セットにおける水流／流域は、「相対攪乱」に関して似ているべきであり、各小地域における類似部だが、人為的影響がより強い水流よりも高い水質を反映すべきである。

- **proportional**：比例した

6. Sample processing and species **proportional** counts of a sample should take no more than 3 hr.
サンプル処理とあるサンプルの生物種比率カウントに、3時間以上かけてはならない。

Prove (Proven)：証明する；判明する（証明されている）

1. This **was** not **proved conclusively**, however.
しかしながら、これは断定的には証明されなかった。
2. These **have proved valuable** in assigning causes and sources to water resource impairments noted by USEPA.
これらは、USEPAにより指摘された水資源損傷の原因と出所を示す際に有益であると証明されている。
3. An attempt to use the scheme for classifying aquatic ecosystems **proved unsuccessful**.
水生生態系分類方式を用いる試みは、不成功であることが判明した。
4. Attempts to calibrate the model and keep parameters in a reasonable range based on literature data **proved unsuccessful**.
モデルをキャリブレートして、パラメータを文献データに基づく妥当な範囲にとどめる試みは、不成功であることが判明した。

- ～ **prove clearly that** ……：～は…ということを明確に証明する

5. The results **prove clearly that** the combination of UV light and hydrogen peroxide is effective at treating water contaminated with chlorinated organics.
この結果は、紫外線と過酸化水素の組み合わせが有機塩化物で汚染された水の処理に有効である、ということを明確に証明している。

- ～ **is proven**：～が証明される

6. Loss of membrane integrity **was not proven** as a cause for the observed

difference in removal between Tests 1 and 2.
薄膜の完全性の損傷が，試験1と試験2の間で観察された除去の相違の原因であるとは証明されなかった。

- ～ **have proven**：～は判明している

7. Periphyton **has proven useful for** environmental assessment.
付着生物は，環境アセスメントに有用であることが判明している。
8. Such information **has proven very useful for** assessing changes in climate, salinity, pH, and nutrients.
こうした情報は，気候，塩分濃度，pH，栄養分における変化を評価するため非常に有用だと判明している。
9. As a tool for identifying problems, advocating for mitigation and enforcement of regulations, volunteer data **have proven** very useful.
問題点を確認するためのツールとして，規制の緩和・施行を求める擁護活動や，ボランティア情報は，非常に有用だと判明している。

- **proven** .....：証明されている

10. The formulation is based upon a **proven** concept.
この公式は，証明されている概念に基づいている。

Provide：与える；提示する；提供する

1. Activated carbon **provided** the best treatment overall.
活性炭による処理が，全体的に最良であった。
2. Extensive laboratory investigations and data analysis **provided** the following conclusions.
大規模な実験調査とデータ分析が次の結論を提示した。
3. Tarou Shiota **provided** many hours of support in the development of the basic computer programs.
Tarou Shiota は，長時間を割き基礎的なコンピュータプログラムの作成を支援してくれた。
4. Metrics that are poorly defined or based on a flawed conceptual basis **provide** erroneous judgments with the potential for erroneous management decisions.
正しく規格化されていなかったり，欠陥ある概念に基づいたメトリックは，誤った判断を生じさせ，管理上の判断ミスを招く可能性がある。

- ～ **provide information on** .....：～は…についての情報を提供する

- ~ **provide a potential to** ..... ：～が … の可能性を提示する
- ~ **provide comments on** ..... ：～が … の論評を寄せる

5. These data **provide** additional **information on** previously identified Solid Waste Management Units.
   これらのデータは，前に確認された廃棄物処理施設についての追加情報を提供する。
6. It has always been assumed that the tailings **provide a potential to** cause degradation of groundwater quality.
   選鉱くずが地下水質の劣化を起こす可能性があるとは常に想定されてきた。
7. Mike Sugi **provided comments on** earlier versions of the manuscript.
   Mike Sugi が本章の草稿を論評してくれた。
8. Greg Segawa **provided** insightful **comments on** Snake River integrity.
   Greg Segawa は，Snake 川の完全性について洞察あふれる論評を寄せてくれた。

- ~ **provides a powerful means of** ..... ：～は … の有効手段となる
- ~ **provide an indication of** ..... ：～は … を示す指標となる
- ~ **provides little evidence of** ..... ：
  ～からは … の証拠がほとんどなにも出てこない

9. The multivariate approach **provides a powerful means of** maximizing ecological information obtained from periphyton assemblages.
   この多変量手法は，付着生物群から得られた生態学的情報を最大化するための有効手段となる。
10. When used in a monitoring program, these methods **provide an indication of** what specific environmental characteristics are changing over time, and can be used to help diagnose the cause of changes in the aquatic system.
    監視プログラムで用いた場合，これらは，どの環境特性が経時的に変化しているか示す指標となり，水生系における変化原因を判定するのに役立つ。
11. Examination of Table 2 **provides little evidence of** any correlation between influent metal concentrations and percentage removal.
    表2の考察から，流入水の金属濃度と除去率との間の相互関係に関する証拠はほとんどなにも出てこなかった。

- ~ **provide a basis for** ..... ：～は … の基礎を提する
- ~ **provide the means for** ..... ：～は … のための手段となる

12. Mathematical models **provide a basis for** quantifying the inter-relationships among the various toxic chemicals.
    数学モデルは，種々の毒性化学物質における相互関係を数量化するための基礎を提

13. To address this problem, a three-year ecoregion project was conducted that characterized the streams within each ecoregion, developed a classification of streams, and **provided a sound basis for** developing realistic water quality standards and beneficial uses within ecoregions.
この問題点に対処するため，3箇年の生態地域プロジェクトが行われた。それは，各生態地域内の水流を特性把握し，水流の分類を設定し，生態地域内での現実的な水質基準および有益利用の確立のための確かな基盤を提供した。
14. The protocol **provided the means for** obtaining such an answer in a defensible way, and the results have been used to make regulation decisions.
このプロトコルは，そうした回答を弁護可能な形で導くための手段となった。そして，結果を用いて規制決定が下された。

- **～ provide consistency：～は整合性をもたらす**

15. Although substantial national efforts are underway to **provide consistency** in the implementation of biological criteria, the actual development and implementation of biological criteria remains an activity of state water quality programs.
生物クライテリアの実施における整合性をもたらすために相当の全米的な努力が払われているものの，生物クライテリアの実際の設定と実施は，いまだ州の水質プログラム活動にとどまっている。
16. An approach that can use the same framework and information **provides** valuable **consistency** among the many different programs.
同じ枠組みおよび情報を用いる手段は，多くの様々なプログラムにおける貴重な整合性をもたらす。

- **..... is provided：～が出される；～が示される**

17. The analytical procedures **are provided** in the Appendix.
分析手順を付録に示す。
18. The answer to this question **is provided** by something call Dynamic Modeling.
この質問には，ダイナミックモデリングと呼ばれる方法によって答えが出される。
19. For the sake of clarity in presenting the test results, most of the details concerning the theoretical background of the tests and the analytical procedures **are provided** in the Appendix B.
テスト結果の公表において透明性を高めるため，試験と分析手法の理論的な背景の細部の多くを付録Bに示す。

Publication：出版；出版物；刊行；刊行物

1. The results can be more meaningful than those that are normally **released for publication**.
   この結果は，通常，出版物として発表されている結果よりも有意義と言える。
2. It has been subject to the agency's peer and administrative review and **approved for publication**.
   USEPAの同領域専門家および行政官による評価を経て，公刊が承認された。
3. Based on the information gathered and the **publication in preparation**, a first approximation of reference ecoregions and subregions is compiled.
   収集情報に基づき，また，準備中の刊行物の概説に基づき，改良された生態地域および小地域の初の概要が作成される。
4. Reviewing the literature on sonochemistry of organochlorine compounds did not lead to any **publications on** the use of ultrasound for chemical monitoring.
   有機塩素化合物の超音波化学分解に関する文献レビューは，化学的モニタリングの為の超音波利用に関するどの出版物にもならなかった。

Publish：発表する

1. In 1989, USEPA developed and **published** Rapid Bioassessment Protocols for Use in Streams and Rivers: Benthic Macroinvertebrates and Fish (Plafkin et al., 1989).
   1989年，USEPAは，『水流・河川での利用のための迅速生物アセスメント・プロトコル：底生大型無脊椎動物と魚類』(Plafkinet al., 1989)を作成し刊行した。
2. Little work **has been published** for lakes and reservoirs, but significant efforts are ongoing as part of the USEPA effort to develop consistent bioassessment protocols.
   湖沼および堰止め湖に関する研究は，あまり発表されていないが，整合的な生物アセスメントプロトコルを設定するUSEPA活動の一環としてかなりの努力が払われつつある。

   ● **published** ‥‥‥：発表されている…

3. These data are in agreement with published data.
   これらのデータは発表されているデータと一致する。
4. There is very limited **published** data on the presence of mercury in the Sakura River.

Sakura川において，水銀の存在を発表したデータがほとんどない。
5. Our **unpublished** data suggest the methods to be of promise for the characterization of sediments.
我々の未発表のデータは，堆積物の特性評価のために必要な手段となりうることを示唆する。
6. Results are presented in several forms, but are normalized to allow direct comparison with other **published** work.
結果がいくつかの形式で提出されているが，標準化されているために発表されている他の研究結果との直接比較を可能にしている。

Purpose（See also **Aim**, **Goal**, **Objective**）： 目的（➡ **Aim**, **Goal**, **Objective**）

1. Even before the conception of biological criteria, biological monitoring information was being used **for many purposes** within state water quality programs.
生物クライテリアの構想前でさえ，生物モニタリング情報は，州の水質プログラムにおいて多くの目的で用いられていた。

  ● **the purpose of .....**：…の目的は

2. **The primary purpose of** the present paper is to demonstrate its applicability to .....
この論文の主な目的は … にその適用性を証明することである。
3. **The purpose of** this paper is to present (1) a description of the chemical conditions and (2) preliminary interpretations.
この論文の目的は，(1) 化学的条件の記述と (2) 予備解釈を示すことである。
4. **The purpose of** this article is to briefly describe some critical research and information needs in surface water quality.
この論説の目的は，表面水の水質に関するある重要な研究と情報のニーズを簡潔に記述することである。
5. **The purpose of** this investigation was to study the effects of nickel on the completely mixed activated sludge process.
この調査の目的は，完全混合型活性汚泥プロセスにおよぼすニッケルの影響を研究することあった。
6. **The purpose of** this section is to present a description of some of these modifications and to illustrate the different directions many programs have taken.
本項の目的は，これらの修正の一部を述べることと，多くのプログラムがとった

異なる方向性を示すことである。

　　● **It is the purpose of** ～ **to** ..... ：…することが～の目的である。

7. **It is the purpose of** this research **to** determine the rate of ion exchange.
イオン交換速度を決定することがこの研究の目的である。

　　● **for the purpose of** ..... ：…のために

8. **For purposes of** comparison of output predictions with field data., ...
予測出力値と現場でのデータとの比較のために，

9. **For the purposes of** establishing numerical biocriteria, the two most important uses are Warmwater Habitat(WWH) and Exceptional Warmwater Habitat(EWH).
数量的生物クライテリア確定のために最重要な2つの使用指定は，温水生息地(WWH)と例外的温水生息地(EWH)である。

　　● **to serve the purposes of** ..... ：…の目的に適うために

10. To "**serve the purposes of** the Act," water quality standards must address chemical, physical, and biological integrity of the nation's waters.
「本法の目的に適う」ためには，水質基準が米国の水域の化学的・物理的・生物学的な完全度に取り組まなければならない。

Quality：質

1. Because of the copious **quality of** data gathered during the sampling program, ...
   サンプリングプログラム期間中に収集された豊富な良質のデータ故に，…
2. This study determines the overall **quality** both from an ecological point of view and also from a user's point of view.
   この研究では，生態的な見地とユーザーの見地両方から総体的な質が決められる。
3. Data **quality** can be roughly divided into three categories based on program objectives:
   データの質は，プログラム目標に基づいておおよそ以下の3つのカテゴリーに分けられる。
4. These rivers **are** generally **of moderate to high quality** and few sites were found that were rated either fair or poor.
   これらの河川は，おおむね質が中度(良)ないし高度(優)なものであり，可または不可の地点は，ごく少数である。
5. There appears to be sufficient evidence that water **quality** will differ for the upper and middle Natsui River due to geographic factors.
   水質が，地理的因子故にNatsui川の上流と中流で異なることを示すのに必要な十分の証拠が存在すると思われる。
6. The amount and **quality of** available information are not adequate to complete the studies.
   利用可能な情報の量と質は，これらの研究を完成させるのに十分ではない。

Quantify (Quantifiable)：数量化する(計量可能な)

1. There have been many attempts to explain, describe and **quantify** these effects.
   これらの影響を説明し，記述し，そして数量化する多くの試みがされてきた。
2. Levin (1992) stressed that to gain an understanding of the patterns of ecosystems in time and space and the causes and consequences of patterns, we must develop the appropriate measures and **quantify** these patterns.

Levin(1992)は，時間・空間における生態系パターンや，パターンの原因・結果を理解するために，我々が適切な測度を定め，これらのパターンを数量化しなければならないことを強調した。

3. In consequence, there are substantial data available to aid **in quantifying** the fate of contaminants entering the soil zone through surface deposition.
その結果として，地表面堆積を通して土壌層へ入る汚染物質の衰退を定量化するのに十分なデータが利用可能である。

- **quantifiable**：計量可能な

4. **Quantifiable** estimates of land use can be made with a known degree of certainty.
土地利用の計量可能な推定は，既知の確実性で得られる。

Quantitative (Quantitatively)：量的(量的に)

1. Table 1 summarizes **quantitative** data on .....
表1は…の量的なデータを要約する。
2. **Quantitative** data on byproducts analysis is lacking.
副成物分析の量的データが欠けている。
3. No attempt is made here to give a **quantitative** estimate of the error.
ここでは，誤差の量的な推定をする試みは、何もされなかった。
4. The hydraulic conductivity comes into play only when **quantitative** discharge calculations are made.
流体の伝達速度は，排出量を計算するときにだけ，有用になる。
5. In evaluating the performance of a model relative to some field measurement or another model, qualitative rather than **quantitative** statements are often made.
ある現場測定，またはもう1つのモデルとの比較からモデルの性能を評価する場合，量より質的な報告がしばしばなされる。

- **quantitatively**：量的に

6. Data are not yet available to examine this possibility **quantitatively**.
この可能性を量的に検討するためのデータがまだ得られていない。
7. **Quantitatively**, the results plotted in Figure 6 may be seen to be in general agreement with the theory.
定量的には，図6にプロットされた結果はその理論と一般的な一致するように見られるかもしれない。
8. Multivariate analyses, such as CCA and WA, are useful because they make

use of the information in assemblages of **quantitatively** infer ecological characteristics (e.g., pH and BOD).
CCA および WA といった多変量分析は，(pH, BOD など)生態特性を量的に推断するために群がり情報を活用しているので，有用である。

Quantity：量

1. Important to waste minimization efforts is knowledge of the **quantity** and characteristics of waste stream.
廃棄物減少化の努力において重要なのは，廃棄物フロー過程での量と特性の知識である。
2. Considerable **quantities of** these constituents were discharged in the tailings during much of the mining history.
採鉱史上，長年の間に，これらの成分のかなりの量が選鉱滓に排出された。
3. Over the past two centuries, **vast quantities of** energy have been used.
これまでの2世紀にわたって，膨大な量のエネルギーが使われてきた。
4. An examination by Toda et al. (2002) of oysters and other shellfish exposed to an oil spill revealed the presence of **large quantities of** petroleum compounds in the body tissues.
カキと他の貝に流出した石油をさらした Toda ら(2002)の研究によれば，貝の体組織に大量の石油化合物が存在することが明らかになった。

Question：質問；疑問；疑問視する

1. Testing of these **questions** yielded the following results: (1) ....., (2) ..・・・
これらの問題への吟味により(以下の結果が生じた。それらは(1)…，(2)…である。
2. More specifically, **questions** such as, "What proportion of the stream miles (and how many stream miles) are affected by point source discharges? Channel modifications? Riparian modifications? Stream bank instability? can be addressed.
より　具体的には，「水流 mile 数のどれほどの比率(どれほどの水流 mile 数)が点源排出に影響されるか？　流路改変は？　水辺改変は？　水流岸の不安定は？」といった疑問に対処できる。
3. One legal critic, for example, **questioned** whether biocriteria are reliable, scientifically repeatable, provide dischargers with adequate notice of acceptable conduct, and can reliably determine cause-and-effect relationships.

例えば，ある法律批評家は，生物クライテリアが信頼でき，科学的に反復でき，容認可能な行為の十分な通告を排出者へ提供でき，因果関係を確かに決定できるか否か，疑問視した。

4. **The question about** what constitutes a sufficiently comprehensive bioassessment is another key contemporary issue facing the implementation of bioassessments and biocriteria.
十分に総合的な生物アセスメントを構成するものが何かに関する疑問は，生物アセスメントおよび生物クライテリアの実施が現在直面している，もう一つの重大問題である。

- **The question is .....**：問題は… である
- **The question is how .....**：
  問題は，どのように…するかということである

5. However, the following types of **questions are** pertinent.
しかし，以下のような疑問が残る。

6. One relevant **question is**: how comparable are reference site to the mainstream Muskingum River near Conesvill Station?
一つの関連問題は，「参照地点を Conesville 発電所付近の Muskingum 川本流といかにして類比可能か？」ということである。

7. The critical **question is how** to determine water quality criteria.
きわめて重要な問題は，いかにして水質クライテリアをきめるかである。

- **There is little question that .....**：
  …ということには，ほとんど疑問の余地がない

8. **There is little question that** aquatic resources have been and continue to be degraded by a myriad of land use and resource use activities.
水生資源が無数の土地利用・資源利用活動によって劣化してきたこと，また劣化しつづけていることには，ほとんど疑問の余地がない。

- **〜 raises questions**：〜は質問を提起する
- **〜 raises the question of how ...**：
  このことは〜がどれほど…かという疑問を提起する
- **〜 raises the question of wheir ...**：
  〜は，どこで…かという質問を提起する

9. It may **raise** more **questions** than it answers.
それは回答よりも多くの質問を提起するかもしれない。

10. This now **raises the question of how** global warming can affect vegetation.
このことは，地球温暖化が植生にどれほど影響しうるかという疑問の提起をする。

11. This statement **raises the question of where** such information was obtained.
    どこでこのような情報が得られたかという質問を，この陳述は提起する。
12. These observations raised several **questions** which we attempted to answer to some degree in the experiments described below.
    これらの観察は，下に記述した実験にて我々がある程度答えようと試みたいくつかの質問を提起した。
13. Of course, application of biocriteria in the antidegradation program **will raise** interesting **questions**.
    当然ながら，劣化防止プログラムにおける生物クライテリアの適用は，興味深い問題を提起するだろう。
14. In practice, however, the use of biocriteria and biological assessments to establish new permit requirements **raises** interesting and challenging **questions**.
    実際問題として，新規許可要件を定めるための生物クライテリアおよび生物アセスメントの利用は，非常に興味深い問題を提起する。

- **question is raised**：質問が提起される

15. **The question** "Is nuclear power safe?" **is** also **raised** frequently.
    「原子力は安全か？」という質問もまた，しばしば提起される。
16. **Questions** and concerns about the suitability of the site **have been raised**.
    その地点の適合性に対しての質問と懸念が上げられてきた。
17. The **question** of the stability of organomercury compounds **is raised** by some of the following observations.
    有機水銀化合物の安定性に対する質問が，次の観察のいくつかによって提起される。
18. The **question was raised** as to how much the scores would change if the HPI were changed.
    もしHPIが変えられたなら，スコアがいくらを変わるだろうかという質問が提起された。

- **question arises**：疑問点が生じる

19. **The question arises** as to how we can believe that ......
    … ということを我々がいかに信じることができるか，という質問が起こる。
20. The **question arose** whether these individual deviations could be the result of measurable differences between the fishes.
    これらの個別の逸脱が魚の間の測定可能な相違の結果で有るのか否か，という質問が起こった。

21. Since EDTA is not produced by the fungus, the **question arose** as to what might be the physiological electron donor.
    EDTAは菌類によって作り出されないので，生理学的電子供与体は何であるのかという疑問が生じる。
22. **The question now arises**: What is the ecological reference site?
    ここに疑問が生じる。すなわち生態学的参照地域とはなんであろうか。
23. One **question that arises** here is .....
    ここで生じている一つの問題は … である。
24. There are several **questions that arise from** these observations of arsenic levels in surface waters: 1..., 2.....,
    表面水のヒ素レベルの観察からいくつかの質問が生ずる。1 …, 2 … である。

    - **question poses .....**：質問は … を提起する
    - **question is posed**：質問が提出される

25. The **question** of chemical speciation **poses** one of the most difficult problems to be resolved by the chemist, especially in the context of synergistic effects as encountered in natural water.
    化学的種分化への質問は，特に自然水域で見られる相乗効果という文脈において，化学者によって解決されるべき最も難しい問題の1つを提起している。
26. One of two regulatory **questions is generally posed**: 1) what are the maximum allowable discharges of toxic chemicals to a water system, and 2) how long will it take a contaminated natural water system to recover?
    規定に関する質問の2つのうち1つが一般に提出される。それら2つの質問とは1)毒性化学物質が水系に放出されうる最大許容量はいくらか，ならびに2)汚染された自然水系が回復するのにはどれほどの時間を要するのか，である。

    - **question is addressed**：問題を扱う
    - **~ is addressed to answering the questions**：
      ~は，この質問に答える為に取り組んでいる
    - **~ address the following questions**：
      ~は次の質問を扱う；~は次の問題に取り組む

27. **The questions** concerning drinking water **are addressed** in this article: 1) to what extent, if any, does acid precipitation cause a drinking water problem；2) if a problem is present, what can be done about it?
    この論文では，飲料水に関する問題を扱う。すなわち1)もし酸性雨が飲料水問題を引き起こすとすればどの程度か；2)もし問題であるのならば，その問題に対して何ができるか？
28. This paper **is addressed to answering these questions**.

Question

この論文は，これらの質問に答える為に取り組んでいる。
29. This integrated assessment was designed **to address the following questions**: 1 ....., 2 ....., and 3 .....
この統合したアセスメントは次の質問を扱うよう計画された。それらは，1 ⋯，2 ⋯，and 3 ⋯ である。

- **The answer to the question is** ..... (See also **Answer**)：
 この質問への答えは ⋯ である（➡ **Answer**）

30. **The answer to this question** is provided by something called Dynamic Modeling.
この質問への答えは，ダイナミックモデリングと呼ばれるものによって見いだされる。

31. **The answer to those questions** lies in the fact that the stereospecificity of enzymes is not exact.
それらの質問への答えは，酵素の立体特異性が厳密でないという事実に見いだされる。

- **question remains**：疑問が残される
- **There remain many unanswered questions**：
 数多くの疑問が応えられずに残っている

32. Many **questions remain** about the adequacy, frequency, and precision of the extensive chemical and microbiological monitoring.
多くの疑問が，広範な化学的・微生物学的モニタリングの妥当性，頻度，及び精度に関して残されている。

33. **There still remain many unanswered questions** regarding reaction mechanisms under a variety of ultrasonic conditions.
いろいろな超音波状態下での反応機構に関して数多くの質問が応えられずに残っている。

- **～ remains an open question**：～は未解決の問題のままである

34. This hypothesis has not been validated, and this phenomenon **remains an open question**.
この仮説は実証されていないし，それにこの現象は未解決の問題のままである。

35. Whether or not this is a metabolite of the 6-CB still **remains an open question**.
これが 6-CB の代謝物質であるかどうかは，いまだに未解決の問題のままである。

- **～ is open to question**：～が疑問視される

36. The extent to which the information can be used to characterize the resources statewide **is open to question**.
全州の資源を特性把握するため情報が用いられうる度合が疑問視される。

- 〜 **come into question** ：〜に疑問が投げかけられる

37. Recently, the cumulative costs associated with environmental mandates **have come into question**.
最近，環境指令に関連した累積コストに疑問が投げかけられている。

- **question of whether** ..... ：…かどうかという質問
- **question of whether or not** ..... ：…しているか否かという質問

38. This raised the **question of whether** biotic integrity had really been restored.
これは，生物学的健全性が実際に回復したか否かについて，疑問を生じさせたのである。

39. The crucial **question of whether** the water body becomes acidic depends on the magnitude of the watershed compensatory response.
水域が酸性になるかどうかという重大な疑問は，流域の中和反応の大きさにかかっている。

40. At the onset of each project, and at the initial idea and data gathering meetings, the **question of whether** ecoregions or special purpose regions are desired is always asked.
各プロジェクトの発端で，また，アイデアやデータ収集のための初回会合で，常に生態地域と特殊目的地域のいずれが望ましいか，という質問が発せられる。

41. This raises the **question of whether** these compounds might have contributed significantly to the processes of natural selection and mutation, and to the evolution of species.
これらの化合物が際立って自然淘汰と突然変異のプロセス，そして種の進化の一因になったかもしれないかどうかについての疑問が生じる。

42. The results of these types of analyses do more than simply answer the **question of whether** or not a waterbody is performing up to expectations with regards to biological integrity.
これらの分析の結果は，ある水域が生物完全度に関する期待に対し実績を示しているか否かという質問に単に答えるばかりではない。

- **question follows** ：質問は次のことである

43. Perhaps the most important **question** that we endeavor to answer with the 305b report **follows**: Is water quality improving or worsening?

Question

おそらく，305(b)報告書で筆者が答えようとする最重要な問いは，「水質が向上しているか，劣化しているか？」というものである。

- **Response to the question is that .....**：この質問への回答は…である

44. **Response to the question** in nearly every case **has been that** ecoregions are the desired regional framework.
この似通るすべての質問への回答は，生態地域が望ましい地域枠組みだというものだった。

- **in question**：当該の

45. This is particularly true if baseline data are available for the waterbody **in question**.
これは，特に当該水域について基準データが利用可能な場合に当てはまる。

Range：範囲；値域

1. It is **at the high end of the range** reported by Ikeda et al (1999).
   それは，Ikeda ら(1999)によって報告された範囲の上限限界にある。
2. The generalized equations to be used for design procedures must be developed considering **the entire range of** operating parameters and must contain a large number of data points to enhance the confidence in the design.
   一般化されたこの式が設計に使われるためには，操作パラメータの全範囲を考慮し手順が作成されなければならない。また，設計の信頼度を上げるためには多くのデータがなくてはならない。
3. The possible **long-range** effects of the disposal of trace metals in the coastal environment are matter of growing concern.
   沿岸環境において微量金属の処分による長期にわたる潜在的影響は，これからますます懸念される事態である。
4. Interpretation of sampling data will require considerations of zoogeography, historical abundance and distribution, and historical **ranges of** variability.
   サンプルからのデータの解釈には，動物地理，史的な数度・分布，史的な可変値域を考慮する必要があろう。

● **a wide range of .....**：広範囲の …

5. The bacterial luminescence bioassay shows great promise as it responds to **a wide range of** compounds.
   広範囲の化合物に対して反応するため，バクテリアの蛍光バイオアッセイは大変期待を持てる。
6. States have undertaken **a wide range of** efforts to improve bioassessment methods.
   諸州は，生物アセスメント方法を改良するため広範にわたる努力を払ってきた。
7. Photocatalytic oxidation processes have been shown capable of mineralizing **a wide range of** naturally occurring and anthropogenic organic compounds.
   触媒光反応による酸化プロセスによって，広範囲の自然態物質や人為的な有機化合物を無機物に変化させることができることが示されている。

8. During the past decade, this process has attracted extensive attention as a modern technology for groundwater purification and has been studied for **a wide range of** hazardous organic compounds.
これまでの10年間に，このプロセスは，地下水清浄化の近代的な技術として大きく注目を集めてきた，そして広範囲の危険な有機化合物のために研究されてきた。
9. A total of 27 stations were selected to encompass **a wide range** in sedimentary and trace metal gradients.
堆積性で微量な金属濃度の勾配にて広範な範囲をカバーするように，合計27の地点が選ばれた。

- **a broad range of .....** ：広範な …

10. For **the broad range of** human impacts, comprehensive, multiple metric approach is most appropriate.
広範な人為的影響については，総合的な複数基準手法が最も適している。
11. As described by Kita et al. (1989), **the range of** pollution sensitivity exhibited by each metric differs among metrics; some are sensitive across **a broad range of** biological conditions, others only to **part of the range**.
Kitaら(1986)が述べたとおり，各メトリックによって示された汚濁感度値域は，それぞれ異なる。広範な生物学的状態について敏感なものもあれば，ごく一部の値域のみに敏感なものもある。

- **in the range of .....** ：… の範囲で
- **in the ..... range** ：… の範囲で

12. Potassium occurs **in the range of** 10 to 38 mg/L.
カリウムは10から38 mg/Lの範囲で生じる。
13. Iron is generally **in the range of** 1 to 20 mg/L.
鉄は一般に1から20 mg/Lの範囲にある。
14. Sulfate occurs **in the hundreds to thousands of milligram per liter range**.
硫酸塩は数百mg/Lから数千mg/Lといった範囲で生じる。
15. Attempts to calibrate the model and keep parameters **in a reasonable range** based on literature data proved unsuccessful.
モデルをキャリブレートして，パラメータを文献データに基づく妥当な範囲にとどめる試みは，不成功であることが判明した。

- **over a range of .....** ：… の範囲で
- **over a wide range of .....** ：広範囲の …

16. The rate constants for oxidation of carboxylic acids are strongly correlated with pH **over a range of** 4-8.
カルボン酸の酸化速度定数は4から8の範囲でpHと強い相関関係がある。
17. On the basis of estimated rate constants, the oxidation rate of formic acid is faster than oxalic acid **over a pH range of** 4-8.
推定される速度定数によれば，ギ酸の酸化速度はpH4-8の範囲でシュウ酸より速い。
18. All data points in the individual experiments appeared to become linear **over the range of** contact times studied.
個別の実験でのすべてのデータポイントが，調査された接触時間範囲では直線になるように見えた。
19. Naoki (1975) observed that the quantity of mercury taken up by the sludge did not alter **over a wide range of** concentrations.
Naoki (1975) の観察では，スラッジによって吸着された水銀の量は広範囲の濃度で変化がなかった。
20. Laboratory-scale activated sludge units were studied **over a wide range of** process operating conditions, that is, MCRT.
研究室スケールの活性汚泥施設で，広範囲のプロセスの操作条件，すなわち，MCRTが研究された。

● **with a range of** ..... ：…の範囲に

21. The average concentration actually detected was 0.08 ppm **with a range of** 0.05 to 0.1 ppm.
実際に検出された平均濃度は0.08 ppmで，0.05から0.1 ppm範囲にあった。
22. The average would increase to 0.18 ppm **with a range of** 0.16 to 0.19 ppm.
平均して0.18 ppmまで上昇し，その範囲は0.16から0.19 ppmであろう。

● **within the range** ：範囲内で

23. The values fell **within the range** encompassed by sampling variation.
この値はサンプリング変動による範囲内で低下した。
24. The pH of reduced sediments generally falls **within a fairly narrow range** around neutrality.
還元状態にある堆積物のpHは，一般に中性付近のかなり狭い範囲内にある。

● ～ **range from A to B** ：～はAからBまでである

25. The pH levels studied **range from** moderately acid **to** weakly alkaline.
「中酸性」から「弱アルカリ性」までのpHレベルが検討された。
26. Scores for each metric **range from 1 to 5** and these scores can then be

translated into descriptive site bioassessments such as excellent, good, fair, or poor.
各メトリックのスコアは，1点から5点までであり，これらのスコアが次に記述的な生物アセスメント(優，良，可，劣)へ変換される．

Rank：格付けする

1. The 1972 act specifically required states not only to identify impaired waters, but to **rank** them in order of priority for cleanup and restoration.
1972年法は，具体的に諸州が被損傷水域を確認することに加え，それらを浄化・回復の優先順位で格付けすることも要求した．
2. As long as we are requested to make assessments of the integrity of surface water ecosystems, a framework to **rank** and partition natural variability is essential.
地表水生態系の完全性のアセスメントを我々が求められるからには，自然的可変性を格付けして仕分けするための枠組みが不可欠である．
3. Matsuda (1990) briefly discussed several studies that have shown biomarker responses correlate with predicted levels of contamination and with site **rankings** based on community level measures of ecological integrity.
Matsuda(1990)は，予測された汚染レベルや，生態完全性の生物群集レベル測度に基づく地点ランキングと相関する生物標識応答を明らかにした諸研究について簡潔に論じた．

Rate：位置付けする；格付けする；速度

1. Volunteers generally **rated** water quality higher at most sites than did state biologists.
一般的に，ボランティアは，州の生物学者よりもほとんどの地点で水質を高めに位置付けた．
2. A consistently high percentage of agreement occurred only when both the volunteer and professional analyses **rated** the site in the fair/poor ranges.
一貫して，高い確率で一致を得られる事例は，ボランティアと専門家の両方が可／不可の値域にある地点を対象とした場合に限られていた．
3. When the SQM analysis was used as a pass/fail screen to **rate** attainment or nonattainment of streams with Maine's Warmwater Habitat (WWH) use degradation, agreement improved.
SQM分析がメイン州の温水生息場所(WWH)使用指定の達成または未達成を格づけする合否選別基準として用いられた場合，その一致する頻度は上昇した．

- **the reaction rate**：反応速度
- **the rates of reaction**：反応速度

4. The influence of water vapor on **the reaction rate** derived from the low adsorption of ethylene due to its low adsorption affinity relative to water.
反応速度に対する水蒸気の影響は，水と比較して低い吸着性に起因するエチレンの低い吸着性から生じた。
5. High-intensity ultrasound dramatically increases **the rates of reaction** by as much as 200-fold.
高強度超音波は，反応速度を200倍くらい劇的に増過する。

Rather：かなり；むしろ

- **rather than .....**：…よりもむしろ

1. Emphasis is on explanation and intellectual stimulation, **rather than** on comprehensive documentation.
包括的な文書化よりもむしろ，説明と知的な刺激が重要視される。
2. Lakes are looked at more from a dynamic and mechanistic point of view, **rather than** a descriptive one.
湖が，記述的な見地よりどちらかと言うと，いっそう動勢・機械論的見地から見られる。
3. It is better to identify large areas over which calibrations will be performed **rather than** small areas.
キャリブレーションを行うに当たり，狭い地域よりも広い地域を確認した方が良い。
4. What seems to be lacking is a consideration of the ecological background **rather than** water chemistry background.
欠けていると思われることは，水化学の背景より生態学の背景を考慮することである。
5. It seems likely that such a consistent pattern of contamination would result from in-situ conditions **rather than** migration.
このような一貫した汚染パターンは，移動条件というより現場の条件に起因するように思われる。
6. In evaluating the performance of a model relative to some field measurement or another model, qualitative **rather than** quantitative statements are often made.
あるフィールド測定，またはもう1つのモデルとの比較からモデルの性能を評価する場合，量より質的な報告がしばしばなされる。
7. For example, over half of the 400 lakes that are monitored in Illinois and

included in the state's 305(b) reports are sampled by citizen volunteers **rather than** state water quality specialists.
例えば，イリノイ州では，400のモニタリング対象湖沼のうち，半数以上が州の水質専門家でなく，市民ボランティアの手でモニターされ，同州の305(b)報告に収められている。

Rational：合理的

1. The use of stability constant provides a **rational** basis for predicting levels of metal accumulation.
安定定数の使用は，金属の蓄積レベルを予測するための合理的基礎となる。
2. Problems related to chemical contamination of surface water supplies, particularly those associated with byproducts from new treatment technologies, are **rational** in scope and receive national attention.
表流水給水における科学汚染に関連付けられ，とりわけ新しい水処理技術の副産物に結び付く問題は，その道理性を認められており，全国的な注目を集めている。

Rationale：根拠

1. Of these 45 states, about half have documentation (mostly in draft form) supporting the methods and analyses and providing program **rationale**.
これら45州のうち，ほぼ半数は，方法・分析を支持しており，プログラムの根拠を提示する文書(主に草稿)を備えている。
2. U.S. EPA (1990) has published comprehensive overviews of biocriteria **rationale**, whereas the overview by Kumagai et al. (1986) is generally regarded as the primary publication advocating using biological integrity within water quality management programs.
U.S. EPA (1990) は，生物クライテリアの根拠に関する総合的概論を刊行した。一方，Kumagaiら(1986)による概論は，一般に水質管理プログラムにおける生物完全性の利用を提唱した主要文献とみなされている。

Reaction：反応

1. A **reaction** mechanism has been proposed that accounts for the observed experimental data.
観察された実験データを説明する反応機構が提案された。
2. Chlorobenzene and all intermediate products disappeared within the first 2

hr of the **reaction**.
クロロベンゼンとすべての中間生産物は分解反応し，最初2時間以内に消失した。

3. Catalytic **reactions** are of enormous importance in both laboratory and industrial applications.
触媒反応は，研究的および産業への適用の両方で非常に重要である。

4. Each of these **reaction** pathways will be discussed in greater detail below.
これら反応過程のそれぞれが下により詳細に記述される。

5. No **reaction** products were identified, nor was dehalogenation actually demonstrated.
反応生成物がみつけられなかったし，脱ハロゲンも事実上証明されなかった。

6. TCE degradation was postulated to involve the following sequential **reactions**.
TCE分解が，次の継続的な反応を伴うと仮定された。

7. Elucidation of **reaction** mechanism was not part of the objective of the present work.
反応機構の解明は，この研究の目的の一部ではなかった。

● **the reaction of A with B**：AとBの反応

8. **The reaction of** gold **with** hemoglobin is of interest, for example, because studies have shown that gold accumulates in the red blood cells of proteins.
例えば，ヘモグロビンと金の反応は興味がある。なぜなら，研究が金がタンパク質の赤血球に蓄積することを示したからである。

● **the reaction rate**(See **Rate**)：反応速度(➡ **Rate**)
● **the rates of reaction**(See **Rate**)：反応速度(➡ **Rate**)

Reactor：反応槽

1. The BBB **reactor** has been described elsewhere (3–7).
このBBB反応器は，ほかのいくつもの論文に記述されてきた(3–7)。

2. The use of sonochemical **reactors** for potential wastewater treatment applications has not been addressed to date.
音響科学の反応装置を用いた応用が，下水処理方として利用可能性があるとは，今までには考えられませんでした。

3. The 1116 commercial **reactors** in the U. S. are all of the light-water type.
米国での1116の商業用原子炉のすべては，軽水タイプである。

4. Effluent TCA concentrations from **the reactor** CF-2 were consistently

lower than influent TCA concentrations.
反応器 CF-2 からの廃水中の TCA 濃度は，一貫して入水の TCA 濃度より低かった。
5. With the exception of a few experiments made with KI, all experiments suggest that **the reactor** generates hydroxyl radicals.
KI を使用した少数の実験を除いて，すべての実験はこの反応器でヒドロキシル遊離基が発生することを示唆する。
6. Eventually a steady-state condition will be achieved **in the reactor**.
結局には，この反応器において定常状態が達成されるであろう。
7. Data were gathered from a number of experiments **using a pilot-scale reactor**.
試験規模の反応器を使った多数の実験から，データが収集された。

Read：読む

- **read off**：読みとる

1. It is relatively easy to **read off** the proportion of sites that are above or below a selected score.
選ばれたスコアを上回る地点もしくは下回る地点の比率を読みとることは，比較的容易である。

Realize：理解する

- ～ **come to realize that** .....：～は … を理解するようになる

1. Resource managers and scientists **have come to realize that** the nature of this resource varies in an infinite number of ways, from one place to another and from one time to another.
資源管理者や科学者は，これらの資源の性格が，場所ごと，時間ごとに無限な様相で異なることを理解するようになっている。

Reason：理由

- **The reason for** .....：… の理由
- **Another reason for** .....：… の別の理由
- **The main (primary) reason for this is that** .....：
  これの主な理由は … ということにある
- **It is for this reason that** .....：
  … はこの理由による；この理由により … である

1. **The reasons for** making such decisions should be documented.
   そのような決定を下した理由を文書化すべきである。
2. **Another reason for** using single-purpose frameworks stemmed from the belief that .....
   単一目的枠組みを用いる別の理由は … という思考にあった。
3. **The primary reason for** the differences in population responses is the much lower reproductive output.
   個体群返答に相違がある主な理由は，より低い生殖生産にある。
4. **Possible reasons for** these divergences are discussed below.
   これらの相違の可能な理由が下に論じられる。

   - **for assorted reasons**：さまざまな理由で
   - **for a variety of reasons**：いろいろな理由で
   - **for the reasons mentioned above**：上記の理由により

5. 1,4-Dioxane is of environmental concern **for assorted reasons**.
   1,4-Dioxane はさまざまな理由で環境上関心がある。
6. Wide and anomalous variations in the mercury background concentrations exist **for a variety of reasons**.
   水銀のバックグランドの濃度の広範な，そして異常なほどの相違にはいろいろな理由がある。

   - **There is no reason for** .....：… の理由はない
   - **There is a reason to believe that** .....：… を信じる理由がある
   - **One reason is that** .....：一つの理由は … ということである

7. **This is one reason that** the selection of only the most pristine sites as references is inadvisable.
   これは，最も無傷な地点だけの選定が勧められない理由の一つである。
8. **There is a reason to believe that** it will continue to be so well into the future.
   それが未来に向けて良くあり続けるであろうと確信する理由が存在する。

   - **There is no reason why** .....：という理由は見あたらない
   - **This is the reason why** .....：それが … の理由である
   - **The reason why ～ is that** .....：
     なぜ～かという理由は … ということである

9. There is no **reason why** these same principles would not apply here.
   これらの原則がここで適用されないという理由は，見あたらない。
10. In concept, **there is no legal reason why** NPDS permit cannot be used to

implement biocriteria.
概念上，NPDES 許可を用いて生物クライテリアを実施してはならないという法的理由は，存在しない。

- **Whatever the reasons for .....**：…の理由が何であるとしても

11. **Whatever the reasons for** the differences, it is important to keep in mind that .....
この相違の理由が何であるとしても，…を念頭におくことは重要である。

Reasonable：合理的な

- **reasonable agreement**（See also **Agreement**）：
  合理的な合意（➡ **Agreement**）

1. **Reasonable agreement** was achieved.
   合理的な一致が得られた。
2. Model predictions show **reasonable agreement** as can be seen from both Figures 2 and 3.
   図2と図3の両方に見られるように，モデル予測は合理的な一致を示す。
3. A linear regression of measurement and predicted values yielded an R^2 of 0.75, indicating a **reasonable agreement**.
   測定値と予測値の線形回帰線が，0.75 の R^2 をもたらし，合理的な一致を示した。

- **It is reasonable to expect that .....**：…と思うことは合理的である
- **It is reasonable to suppose that .....**：…と考えることは理にかなう
- **It is reasonable to surmise that .....**：
  …と推測することは合理的である
- **It is reasonable to assume .....**：…と仮定するのが道理的である
- **It might be more reasonable to .....**：
  …した方がより合理的かもしれない

4. **It is reasonable to expect that** the process of pyrite oxidation and gypsum formation will continue to occur for a long time period.
   黄鉄鉱の酸化と石膏生成プロセスが長期間起こり続けるであろうと思われる。
5. **It is reasonable to expect** some inhibition due to the elevated levels of BOD5 and COD.
   高いレベルの BOD5 と COD によれば，若干の抑制があると予想できる。
6. **It might be more reasonable to** argue for differing percentiles for different regions depending upon the nature of regional-scale degradation.

地域規模で生じている劣化の性質に応じ，異なる地域についてパーセンタイルを違えることを主唱した方が合理的かも知れない。
7. By analogy with the oxidation of formic and oxalic acids, **it is reasonable to assume** a second-order reaction for glyoxylic acid because it has a structure similar to the other two acids.
ギ酸とシュウ酸の酸化反応が類似していることから，グリオキシル酸の反応は第2次反応であると仮定するのが理にかなっている。なぜなら，それは他の2つの酸に構造が類似しているからである。

Record ： 記録；記録する

1. Even with **a record** of that length, results are inconclusive.
こんなに長い記録でさえ，結果はまだ確実ではない。
2. In any case, there are no continuous **records of** ecological change over the last 2000 years.
いずれにしても，過去2000年にわたっての生態変化の継続した記録が存在しない。
3. A decision was made **to record** this information even though the eventual importance of its use was not immediately apparent.
たとえその利用の最終的重要性が直ちに明らかでなくても，この情報を記録することが決定された。
4. Ecological information for many algae, particularly diatom, **has been recorded** in the literature for over a century.
多くの藻類，特に珪藻についての生態学的情報は，1世紀にわたり文献に記録されてきた。

Recognize ： 認識する

1. At least in the context of antidegradation, USEPA **recognized that** human and ecological use protection involves more than maintenance of chemical water quality.
少なくとも劣化防止の文脈において，USEPAは，人為的・生態的な使用保護が化学的水質維持以上のものを伴うことを認識していた。
2. It is important to note that the four-tiered system has been widely used in biomonitoring because it **recognizes** the limitations of refinement and accuracy that can be ascribed to this science.
この四段層方式が生物モニタリングで多用されている理由は，この科学に帰されうる改良および正確性の限界を認識している点にあることを念頭におくべきである。

- **It is recognized that .....**：… であることが認識される
- **It must be recognized that .....**：
  … であることを認識しなければならない
- **It is important to recognize that .....**：
  … であることを認識することが重要である
- **It is increasingly recognized that .....**：
  … であることがますます認められている

3. **While it is recognized that** individual waterbodies differ to varying degree, the basis for having regional reference sites is the similarity of watersheds within defined geographical regions.
   個々の水域が様々に異なることは，認識されるものの，地域参照地点を設けることの基礎は，所定の地理的地域内での流域の類似性にある。
4. **It must be recognized that** the concept and definition of ecoregions are in a relatively early stage of development.
   生態地域の概念・定義は，まだ発展初期段階にあることを認識しなければならない。
5. **It is important to recognize that** the proposed model depends to a significant extent on its intended purpose.
   提案されたモデルが，意図的な目的にかなり左右されることを認識することが重要である。
6. **It is increasingly recognized that** photochemical processes may be important in determining the fate of organic pollutants in aqueous environments.
   水環境において有機汚濁物の衰退を決定するのに，光化学プロセスが重要であるかもしれないことがますます認められている。

Recommend：勧告する;推奨する；推薦されている；勧告されている

1. **We recommend that** the highest priority for future experimental work is to validate the existing models.
   これからの実験研究の最も高いプライオリティが既存のモデルを実証することにある，と我々は推奨する。
2. For streams with riffle or cobble substrate, consideration of the following four measures **is recommended**：(1) ....., (2) .....,
   浅瀬または大礫基質の水流については，以下の4つの測点を考慮することが推奨される。
3. Relying exclusively on this measurement of standing crop **is not recommended**；however, in combination with other indicators, such as

DBI, AFDM can be useful.
生物体量の測定のみに頼ることは推奨できない。しかし，DBI，AFDMなど他の指標との組合せは有用であろう。

4. If a relationship was observed, a 95％ line of best fit was determined and the area beneath trisected following the method **recommended** by Fukuda et al.(1984).
関係が見られたならば，ベストフィットの95％ラインが決定され，その下方の面積は，Fukudaら(1984)の推奨方法に従って三等分された。

● **It is highly recommended that** .....：…することがきわめて望ましい

5. **It is highly recommended that** all assessments include not only biological community information but habitat information.
すべてのアセスメントが生物群集情報のみならず，生息場所情報も含むことがきわめて望ましい。

● **recommended** .....：推薦されている；勧告されている

6. Cadmium was detected above the **recommended** drinking water criteria.
カドミウムが推奨される飲料水クライテリアを超過した濃度で検出された。
7. Iron exceeded **recommended** concentrations in two wells at the Rockland sites.
鉄分がRockland地域では2つの井戸で勧告濃度を超えていた。

Recommendation：推薦；勧告

1. In the years preceding these **recommendations**, Kimura et al.(1986) developed an operational definition of biological integrity.
これらの勧告よりも数年前，Kimuraら(1986)が生物学的健全性の実際的定義を定めた。
2. Mention of trade names or commercial products does not constitute endorsement or **recommendation** for use.
商標名や商品への言及は，その利用を必ずしも是認していないし，また推奨もしていない。

Reduce：低下する；減少する

1. Their populations **have been severely reduced** in number in the last few years.
これまでの数年で，それらの個体数がひどく減少していた。

2. By offering a more robust framework based on multiple and regionally calibrated reference sites, the chance for deriving an inappropriate biocriterion **is greatly reduced**.
複数の地域較正参照地点(reference site)に基づいたより確かな枠組みを提供することによって不適切な生物クライテリアが導出される確率は，大幅に低下する。
3. Three decades of Federal controls **have sharply reduced** the vast outflows of sewage and industrial chemicals into America's rivers and streams.
30年間にわたる連邦政府による防止策は，米国の河川・水流への下水および産業化学物質の大量流出を激減させてきた。
4. Once field data are collected, processes, and finalized, the next step is **to reduce** the data to scientifically and managerially useful information.
現地データが収集・処理・最終確定されたならば，次のステップは，データをまとめて科学的・及び管理的に有用な情報にすることである。

- **reducing**：下げること

5. The major effort to control eutrophication has been directed towards **reducing** the input of .....
富栄養化をコントロールする主要な努力が…の投入量を減らす方向に向けられてきた。
6. In the most aquatic systems these processes are capable of **reducing** free copper levels to very low values.
ほとんどの水生システムにおいて，これらのプロセスは遊離銅レベルを非常に低い値に下げることができる。

- **reduced** .....：減少した…

7. The excess nutrients are the primary cause of the **reduced** oxygen levels observed in the lake.
この湖で観察される酸素レベルの減少は，富栄養が主要な原因である。
8. Nonpoint sources are now impeding further biological improvements observed in larger rivers that resulted from **reduced** point source impacts.
点源影響の減少によって大河川で観察された一層の生物学的な向上を面源が妨げている。

Reduction (Reducing)：減少；削減(下げること)

1. It is difficult to ensure, a priori, that implementing nonpoint source controls will achieve expected load **reductions**.
点源制御の実施が予想される負荷量の削減を達成するであろうことを先見的に保

証することは難しい。
2. The apparent **reduction in** the oxygen utilization is most likely the result of soil gas diffusion.
酸素消費の明白な減少は，土壌ガス拡散の結果である可能性が最も高い。
3. Significant improvement in the biological indices has been observed during the past twenty years, which has been due in large part to **significant reductions in** point source loadings.
生物指数における大幅な改善が過去20年間に観察されており，その主因は，点源負荷の大幅な削減であった。

Redundancy：余剰；冗長性

1. To avoid **redundancy** in discussing these alternatives, ...
この選択肢を論じることにおいて冗長性を避けるために，…
2. Seasonality is a well-understood concept, therefore, it is not necessary to sample in multiple seasons for the sake of data **redundancy**.
季節性は，十分に理解されている概念であるから，データ冗長性のために複数季節でサンプリングする必要がない。

Refer：を意味する；を示す；に関わる

● ～ **refers to** ..... ：～は…を示す；～は…を意味する

1. Relative abundance of taxa **refers to** the number of individuals of one taxon as compared to that of the whole community.
分類群の相対豊富度（relative abundance）とは，生物群集全体の個体数に比した一分類群の個体数を意味する。
2. Sensitivity **refers to** the numbers of pollution-tolerant and -intolerant species in the sample.
感度は，そのサンプルにおける汚濁耐性種および汚濁不耐性種の数に関わっている。
3. Simply **referring to** a waterbody as a high-quality or significant resource is inadequate given the penchant for these characterizations to have unique attributes according to the individual making the pronouncements.
個々の発表によれば，これらの特性が独自の属性を帯びがちな傾向を鑑みて，ある水域を単に高質または有効資源と評するだけでは不十分である。

Reference：参照；言及；参照文献

1. The average value in the **reference** soils was 1.9 mg/kg.
参考土壌での平均値は 1.9 mg/kg であった。
2. Arsenic was not detected in the **reference** wells.
参考井戸ではヒ素が発見されなかった。
3. The cadmium level in the C zone is high relative to the **reference** site.
C 区域のカドミウムのレベルは，参照地点と比較すると高い。
4. In the sanitary engineering field only scattered **references** on acclimation to cation toxicity are available.
衛生工学の分野では，カチオン毒性に対する順応に関する参考文献はまれにしか存在しない。

★ **Reference**（参考文献）の引用に関する例文を下に示します。

1. For kick sample sites, substrate particle size is obtained by making observational estimates to designate percentage of each of the seven EPA size categories, **as listed in Ueda（1993）**.
キックサンプル地点について，基質の粒子サイズは，Ueda (1993) により列挙された 7 つの USEPA サイズ区分それぞれのパーセンテージを指定するための観測的推定で得られる。
2. These are then converted to phi values **as in Yamada（1962）**, and mean particle size is calculated.
これらは，次に Yamada (1962) におけるような phi 値に変換され，粒子サイズ平均が計算される。

Reflect：反映する

1. The results presented here **reflect** recent modifications of the global transport model.
ここで提出された結果は，地球規模の輸送モデルの最近の修正に反映している。
2. They **reflect** practical situation and as a result present the greatest challenge with respect to modeling.
それらは実務的な状況を反映し，そして結果として，モデリングに関する最大の難問を提示している。
3. As would be expected, the fish community found at each site **reflects** the predominant habitat features.
予想どおり，各地点で見られた魚類群集は，優勢な生息地特色を反映している。

4. Streams/watersheds within each set should be similar to one another regarding "relative disturbance" and **should reflect** higher water quality than streams with watershed with similar proportions in each subregion but with greater human impact.
   各セットにおける水流／流域は，「相対攪乱」に関して似ているべきであり，各小地域における類似部だが，人為的影響がより強い水流よりも高い水質を反映すべきである。
5. Thus, metrics **reflecting** biological characteristics may be considered as appropriate in biocriteria programs if their relevance can be demonstrated, response range is verified and documented, and the potential for application in water resource programs exists.
   それ故，生物学的特性を反映するメトリクスは，その関連性が実証され，応答値域が検証・文書化され，水資源プログラムにおける応用可能性が存在する限り，生物クライテリアプログラムに適しているとみなされよう。

Regard（Regarding）：に関する；に対する（に関して；について）

- **in regard to** ……：…に関して
- **with regard to** ……：…に関して

1. **In regard to** the dynamic behavior of systems, it is necessary to consider the concepts of …..
   システムの力学的な挙動に関して，…の概念を考慮することが必要である。
2. Some of the results given in Table 3 have important implications **with regard to** the PCB distribution.
   表3に示された結果の一部は，PCBの分布に関する重要な含蓄を持っている。
3. **With regard to** taxonomy it was essential in our analysis to have the macroinvertebrates identified to the lowest possible level, particularly the midges.
   分類群に関して，大型無脊椎動物，特に小虫を可能な限り低いレベルまで識別することが筆者の分析に欠かせなかった。
4. In recent years, many researchers have been interested in the water quality of these lakes **with regard to** both acidification and nutrient enrichment.
   近年，多くの研究者が酸性化と養分濃縮に関して，これらの湖沼の水質に興味を抱いている。
5. Over the linear scale of the IBI even gross changes have meaning **with regard to** the relative condition of the community and types of impacts that are present.

IBI の線形スケールで，大幅な変化は，生物群集の相対状態，ならびに存在する影響の種類に関して意味を有する。

- **regarding** ..... : …に関して；…について

6. Less is known **regarding** the inhibitory effects of nickel on nitrifying organisms.
硝化菌に対するニッケルの阻害効果に関しては，あまり知られていない。

7. There still remain many unanswered questions **regarding** reaction mechanisms under a variety of ultrasonic conditions.
いろいろな超音波状態下での反応メカニズムに関して，まだ多くの疑問が残っている。

8. This language was not trivial and caused a great deal of concern **regarding** how to define "integrity" (especially biological integrity) and the measurements to be applied.
この文言は，些細なものでなく，「完全性」(特に生物完全性)の定義づけと適用すべき測定値について強い関心を生じさせた。

Regardless：ともあれ；いずれにしても；にもかかわらず

1. **Regardless**, such within-ecoregion differences in land use and land cover must be distinguished from ecoregional characteristics.
ともあれ，土地利用と土地被度におけるこのような生態地域内での差異を生態学的特性と区別しなければならない。

- **regardless of** ..... : …にもかかわらず

2. The arsenic concentrations will not change very much in the near future **regardless of** any channel shifting.
移動経路が変化したとしても，ヒ素の濃度は近い将来ではそれほど変化しないであろう。

3. **Regardless of** whether Fukui is right or wrong, there is much to be learned from reconstructing the cultural and ecological histories of Easter Island.
Fukui の理論が正しいか間違っているかにかかわらず，イースター島の文化的・生態的歴史を再構築することから学ぶべきことが多くある。

4. Like Metric 8, this is a negative metric and, as such, a low number (< 50 individuals) or an absence of organisms in a sample defaults to a zaro score for the metric **regardless of** the presence or absence of the specified tolerant taxa.
メトリック8と同様にこれは，負のメトリックなので，所定の耐性分類群の有無

にかかわらず，生物が少数(50匹未満)または皆無であればデフォルトとして0とスコアされる。

Region：地域

1. Many **regions** will require increasingly efficient treatment plants.
   多くの地域で効率的な処理場をますます必要とするであろう。
2. The steady flow assumption is most applicable **in this region**.
   この地域では，定常流量の仮定が最も適用可能である。
3. Health problems related to the intake of contaminated water have been encountered **in some regions of** Taiwan.
   台湾のある地域では，汚染された水の摂取に関連した健康問題に直面してきた。
4. The risk of error due to inappropriate classification **across** heterogeneous **regions** should be avoided.
   異質な地域を適切さに欠ける分類によって生じる誤差リスクを避けるべきである。

Relate（Related）：関係がある（関連した）

- ～ **is related to** ……：～は…と関係がある
- ～ **is closely related to** ……：～は…と密接に関わっている
- ～ **is strongly related to** ……：～は…と強く関係している
- ～ **is likely related to** ……：～は…に関係する見込みが強い

1. The choice of groups to be identified and analyzed **is related to** the measures that will be used, as was discussed in detail above.
   同定され・分析されるべき集団の選択は，前に詳細に述べたように，用いられる測度に関係している。
2. The results confirm the hypothesis that copper uptake by algae **is related to** the free cupric ion activity and is independent of the total copper concentration.
   この結果は，藻類による銅の摂取は遊離銅イオンの活性と関係があり，全銅濃度とは関係がないという仮説を確証する。
3. Current efforts to develop and implement biological criteria **are closely related to** independent state efforts to apply biological monitoring information to water resource quality assessment.
   生物クライテリアを設定・実施するための現在の努力は，各州が生物モニタリング情報を水資源質アセスメントに適用する努力と密接に関わっている。

4. The Biological Response Signatures can be particularly useful in demonstrating that the observed degradation **is likely related to** specific discharges, especially those involving Complex Toxicity impact type.
生物応答サインは，特に観察された劣化が複雑毒性影響に関わる具体的排出に関係する見込みが強いことを実証するうえで有用である。

- **～ may relate to .....** ：～は … と関連付けられるかもしれない

5. Their activities **may relate to** thermal changes, flow changes, sedimentation, and other impacts on the aquatic environment.
それらの活動は熱変化，流量変化，堆積，それに水生環境への影響と関連付けられるかもしれない。

- **～ has related A to B .....** ：～はAをBに関連づ付けた

6. Recent work by Egawa et al. (2000) **has related** hydrology **to** lake alkalinity.
Egawa ら (2000) の最近の研究は，水文学を湖におけるアルカリ度に関連付けた。

- **as related to .....** ： … に関連した

7. Each of these types of criteria was developed to address specific regulatory needs of managements **as related to** the permitting of point source discharges.
これらの種類のクライテリアそれぞれは，点源排出の許可に関連した管理の具体的な規制ニーズに取り組むために設定された。

- **related .....** ：関連した …

8. **Related** phenomena occur with cavitation in liquid−solid system.
関連した現象が液体−固体システムでのキャビテーション（空洞現象）によって起こります。

9. Laboratory work addressing some aspects of these problems and **related** modeling efforts are ongoing at the Idaho State University.
これらの問題の若干の局面を扱っている学術研究，および関連したモデリングの努力はアイダホ州立大学で進行中である。

Relationship (Interrelationship)：関係（相互関係）

1. One legal critic questioned whether biocriteria are reliable and can reliably determine **cause−and−effect relationships**.
ある法律批評家は，生物クライテリアが信頼でき，因果関係を確かに決定できるか否か，疑問視した。

- **A relationship was derived** ..... : 一つの関係が導かれた
- **This relationship has been identified by** ..... :
  この関係は … によって確認されている
- **A relationship was observed** ..... : 一つの関係が見られた
- **This relationshps is compliant with** ..... : この関係は … に従う
- **A simple relationshps could be established** ..... :
  一つの単純な関係が確証されえる

2. **A relationship was derived** on the basis of energy considerations.
   一つの関係がエネルギーを考慮して導かれた。
3. **This relationship has been identified by** several other workers, and fitted to Langmuir and Freundlich isotherms.
   この関係は，数人の他の研究者によって確認され，Langmuir と Freundlich 等温線に当てはめられた。
4. If **a relationship was observed**, a 95 % line of best fit was determined and the area beneath trisected following the method recommended by Furuike et al. (1984).
   関連性を見い出せれば，95% 適合線が決められ，その下方の面積は，Furuike ら (1984) の推奨方法に従って三等分された。
5. This **relationship is compliant with** the "fixed demand" theory proposed by Wu and Yao (2001).
   この関係は，Wu and Yao (2001) によって提案された「不変需要」理論に従う。
6. If a simple **relationship could be established**, it would be of great in planning the future direction of any development work on a new chemical.
   もし単純な関連を築けられるのならば，いかなる新化学物質開発研究の将来方針を計画するにあたって，それは偉大なことであろう。

- **relationship between A and B** : A と B の関係

7. Figure 2 presents **the relationship between** A **and** B.
   図2に A と B の関係を示す。
8. A number of studies indicate that there is **a relationship between** A and B.
   多くの研究が，A と B の間に関係があることを示唆している。
9. Findings to date indicate no **relationship between** the original source of DDT **and** PCBs.
   今日までの調査結果は，DDT と PCBs の最初の汚染源の関係を示していない。
10. There was no consistent **relationship between** the specific oxidation rate **and** the concentration of organic matter in the reactor.
    この反応器内で，特定な酸化速度と有機物の濃度の間には一貫した関係がなかった。
11. Predictive models can be used to help define the **relationships between**

water pollutants **and** their sources.
予測モデルは，水汚濁物とそれらの流入源との関係を定義するのを助力するために使うことができる。

12. Large variations in temperature may mask the **relationships between** water pollutants **and** their sources.
大きな温度変移によって，水汚濁物とそれらの流入源との関係が隠されてしまうかもしれない。

13. **The relationship between** cellular copper content **and** cupric ion activity shows good agreement with the hyperbolic function.
細胞の銅含有量と銅イオン活性の間の関係は，双曲線関数で示すと良く一致する。

14. Koike and Sasaki reported **a linear relationship between** the log of Ne/Nin **and** the applied UV dose as predicted by Eqn 1.
KoikeとSasakiは，式1によって予測されるlog(Ne/Nin)と使われたUV放射線量の間の線形関係を報告した。

- **relationship to .....** ：…との関係
- **little or no relationship to .....** ：
  …との関係がほとんど，あるいは全く無い
- **relationship with .....** ：…との関係
- **relationships among .....** ：…間の関係

15. Metrics with **little or no relationship to** stressors are rejected.
ストレッサーとの関係がほとんど，あるいは全く無いメトリックは，却下される。

16. Its simplicity and numeric **relationship to** the original four zones of stream pollution lead to the development of a widely used biotic index in the United States.
それは，単純で，汚濁水流の当初4つのゾーンと数値関係を持つので，米国で広く用いられる生物指数として確立された。

17. A strong inverse **relationship with** watershed size exists.
流域規模との強い反比例関係が存在する。

18. The plots for each metric were examined to determine if any visual **relationship with** drainage area existed.
各メトリックについての図が検討され，流域面積との目測関係が存在するか否か決定された。

19. Certain metrics that showed no positive **relationship with** drainage area required the use of an alternate trisection method.
流域面積と正の関係を示さないいくつかのメトリックは，代替の三等分法を必要とした。

20. **Relationships among** species are complex and changes in one can lead to

unexpected changes in another.
生物種間の関係は，複雑なもので，ある生物種での変化が他の生物種での予測不能な変化に通じうる。

- **interrelationship among .....** ： … 間の相互関係

21. Mathematical models provide a basis for quantifying the **interrelationships among** the various toxic chemicals.
数学モデルは種々の有毒化学物質間の相互関係を数量化する基礎を示す。
22. Most agree with a general definition that ecoregions comprise regions of relative homogeneity with respect to ecological systems involving **interrelationship among** organisms and their environment.
衆目の一致した一般的定義では，生態地域とは，生物およびその環境の相互関係を伴う生態系に関して比較的同質な地域からなる。

Relative (Relatively)：
比較的な；比較上の；相対的な（比較的に；相対的に）

1. They provide direct insight into the **relative** importance of the various mechanisms.
それらは，種々のメカニズムの相対的重要性について直接に洞察を加えている。
2. Most, however, agree with a general definition that ecoregions comprise regions of **relative** homogeneity with respect to ecological systems involving interrelationship among organisms and their environment.
しかしながら，一致した一般的定義では，生態地域とは，生物およびその環境の相互関係を伴う生態系に関して比較的同質な地域からなる。

- **relative to .....** ： … と比較すると
- **～ is relative to .....** ： ～は … と関連している

3. The cadmium level in the C zone is high **relative to** the reference site.
区域Cのカドミウムのレベルは，参照地点と比較すると高い。
4. The results of the arsenic analyses can be considered **relative to** the flow conditions in the creek.
ヒ素分析の結果は，小川の流れの状態と比較して考慮されえる。
5. This is somewhat surprising, in view of the order of magnitude higher water column decay rate of DDE **relative to** Lindane.
Lindaneと比較して水柱でのDDEの桁違い高い分解速度の見地からして，これは驚きに値する。
6. In evaluating the performance of a model **relative to** some field

measurement or another model, qualitative rather than quantitative statements are often made.
ある現場測定，もしくはモデルとの比較からモデルの性能を評価する場合，量より質的な報告がしばしばなされる。

7. Many are also heavily impacted by urbanization and industries, and all streams **relative to** watersheds of 30 mi^2 or more have been channelized at one time or another.
多くは，都市化や産業の影響を強く受けており，30 m^2 以上の流域に関わる水流は，すべてある時期に流路制御を受けてきた。

- **relatively**：比較的に；相対的に

8. Information about the kinetics of bioaccumulation is **relatively** scarce.
生物体内蓄積の速度論についての情報は比較的欠乏しい。
9. Its development can be straight-forward and **relatively** simple.
それの開発は単純で，そして比較的簡単であり得る。
10. Groundwater modeling is a **relatively** new field that was not extensively pursued until about 1965.
地下水モデリングは比較的新しい分野で，1965年頃までは幅広く追究追されなかった。
11. It is **relatively** easy to read off the proportion of sites that are above or below a selected score.
選ばれたスコアを上回る地点もしくは下回る地点の比率を読みとることは，比較的容易である。
12. It must be recognized that the concept and definition of ecoregions are in a **relatively** early stage of development.
生態地域の概念・定義は，まだ発展初期段階にあることを認識しなければならない。
13. These types of streams and watersheds would comprise **relatively** undisturbed references for the region.
この種の水流および流域は，その地域について比較的影響を受けないな参照地点からなる。

Relevant (Relevance)：関連している；関連がある（関連性）

- ～ **is relevant to** ……：～は … に関連している
- ～ **have relevance to** ……：～は … と関連性がある

1. For a metric to be useful, it **must be relevant to** the biological community under study and to the specified program objectives.

あるメトリックが有用たるには，研究対象の生物群集ならびに明確なプログラム目標に関連していることが必要である。
2. There are a number of other federal programs that involve research and monitoring activities **relevant to** biological criteria development.
　他にも，生物クライテリア設定に関わる研究・監視活動に関連して，多数の連邦プログラムが存在する。

Rely（Relying）：依拠する；頼る（頼ること）

- **rely on** ..... ：…に依拠する
- **rely upon** ..... ：…に依拠する

1. The bioassay **relies on** the assumption that only free copper ions are toxic.
　このバイオアッセイは，遊離銅イオンだけが毒性を持つという仮定に左右される。
2. The contemporary regulatory-intensive approach **relies heavily on** a legal style of rule setting.
　現代の規制重視手法は，主に法律スタイルの規則策定に依拠している。
3. These strategies **must rely on** assessing various combinations of chemical-specific and WET criteria in the absence of instream biosurvey data.
　これらの戦略は，水流内生物調査データがない場合，化学クライテリアとWETクライテリアの様々な組合せを評価することに頼らなければならない。
4. Contemporary water resource management **has relied upon** performance (technology-based) standards.
　現代の水資源管理は，実績（技術ベース）標準に準拠してきた。

- **relying**：頼ること

5. **Relying** exclusively on this measurement of standing crop is not recommended.
　生物体量の測定のみに頼ることは推奨できない。

Remain（Remaining）：残される；とどまる（残っている）

1. This occurrence indicates that nickel did not **remain** within the treatment units.
　この発生は，処理ユニット内にニッケルが残っていなかったことを示唆する。
2. Some minor revisions were made in 1985, but the approach **remained** essentially the same through 1987.

1985年にいくつか細かな改正が施されたが，この手法は，基本的に1987年まで存続した。
3. Arguments over the merits of narrative vs. numeric biological criteria **will likely remain**.
記述生物クライテリアと数値生物クライテリアのメリットをめぐる議論は，今後も続く見込みが強い。
4. Problems do **remain**, but all of the indicators point to conventional problems associated with the municipal sewer system, particularly CSOs.
問題は，未解決だが，指標は，すべて都市下水システム，特にCSOに関わる問題点を指し示している。
5. Although substantial national efforts are underway to provide consistency in the implementation of biological criteria, the actual development and implementation of biological criteria **remains** an activity of state water quality programs.
生物クライテリアの実施における整合性をもたらすために相当の全米的な努力が払われているものの，生物クライテリアの実際の設定・実施は，いまだ州の水質プログラム活動にとどまっている。

- **there remain .....** ：…が残っている

6. Although we have presented here a framework from within which numerical biocriteria can be developed and implemented by states, **there remain** important areas for future development and research. Some of these follow:
我々は，本章で，数量的生物クライテリアが州ごとに設定・実施されるための枠組みを示してきた。だが，今後の研究を要する重要な領域がいくつか残っており，例えば，以下のとおりである。

- **much remains to be done**：
残された課題は多い；多くの課題が残されている

7. Despite the fact that the large amount of work has completed for wadable warmwater streams, **much remains to be done**.
徒渉可能温水水流について多くの研究が終わったにもかかわらず，残された課題は多い。
8. Despite the large amount of effort that has been directed towards IBI development, **much remains to be done**, both in terms of generating new versions for different regions and habitat types, and in terms of validating existing versions.
IBI案出へ向けて多大な労力が払われてきたにもかかわらず，様々な地域や生息場所種類について新バージョンを生じること，ならびに既存バージョンを確証す

ることの両方においてまだ多くの課題が残されている。

- **~ remain unanswered**：~は解決されないまま残っている
- **There remain unanswered questions regarding ……**：
  … に関して解決されてない疑問が残っている

9. Fundamental questions **remain unanswered**.
   基本的な疑問は解決されないまま残っている。
10. **There still remain many unanswered questions regarding** reaction mechanisms under a variety of conditions.
    いろいろな状態下での反応機構に関して解決されてない疑問がまだ多く残っている。

- **question remains**(See question)：疑問が残される（➡ **question**）
- **~ remains an open question**(See question)：
  ~は未解決の問題のままである（➡ **question**）：
- **remaining**：残っている

11. Any noncombustible debris **remaining** is typically landfilled.
    残っているどんな不燃瓦礫も主として埋め立て地に処分される。

Repeat（Repeatedly）：繰り返す（繰り返して）

1. **Repeated** tests over long periods are more significant.
   長い期間にわたっての繰り返されたテストはいっそう重要である。
2. To validate the above hypothesis, the experiments **were repeated** with a catalyst concentration of 300 ppm.
   上記の仮説を実証するため300 ppmの触媒濃度で実験が再度行われた。

- **repeatedly**：繰り返して

3. A 40-mile segment has been sampled **repeatedly** over multiple years, which is a prerequisite for trend analysis.
   40mileの区間で，複数年度にわたり繰り返しサンプリングされた。これは，傾向分析の必須条件である。
4. The U.S. Supreme Court and lower courts have indicated **repeatedly** that water quality standards must be enforced through NPDES permits.
   米国最高裁判所およびその下位裁判所は，これまで再三，水質基準がNPDES許可を通じて実施されねばならないことを主張してきた。

Report(Reportedly): 報告;報告書;報告する(伝聞によれば)

1. A number of **reports indicate that** .....
   多くの報告が … を示唆する。
2. This **report is** a revision of the Solid Waste Management Units Report submitted in 1988.
   この報告書は,1988年に提出された廃棄物処理施設報告の修正版である。
3. This **report is subject to** revisions based on additional information.
   この報告書は,追加情報に基づいて修正の対象となる。
4. Some concern was expressed by the author of the **report** over the accuracy of the majority of the data.
   大多数のデータの正確さについて,若干の懸念がこの報告の著者によって表わされた。
5. The obvious question concerns the discrepancies between this and earlier **reports**.
   この当然な質問は,この報告書と前の報告書の間の矛盾に関係している。

   ● **This report addresses** ..... :
      この報告書は … を扱う;この報告書は … を強調する

6. **This report addresses** the problem of analyzing the fate of chemicals discharged into receiving waters.
   この報告書は,水域に排出された化学物質の衰退過程を解析するときに生じる問題を扱っている。
7. **This report will address** the possibility of contamination at any of the three sites.
   この報告書は,3地点のいずれにおいても汚染可能性があることを強調するであろう。

   ● **This report serves** ..... :この報告書は … に役立つ

8. **This report should serve** to give the reader an idea of what is involved in testing a model.
   この報告書は,モデルのテストには何が関与するのかといった考えを,読者に与えるのに役立つはずである。
9. **This report will serve** as a basis for RFI work plans developed by ERP.
   この報告書は,ERPによって作成されるRFIワークプランの基盤となるであろう。

   ● **in the reports**:報告書に

10. The research described **in this report** has been funded by the USEPA.
    この報告書で述べてきた研究は，USEPAから資金供給されたものである。
11. A complete discussion of the data can be found **in the HLA reports**.
    このデータの完全な論議がHLA報告書に見いだされる。
12. These descriptions are based on information contained **in the EPA reports** (EPA, 2002).
    これらの記述は，EPA報告書((EPA,2002)に含まれている情報に基づいている。

　　　● **reported** ….. ：報告されている …

13. In this paper, we compare laboratory results to previously **reported** field results.
    この論文では，我々は研究室での結果を前に報告されている野外調査の結果と比較した。

　　　● **reportedly** ：伝聞によれば

14. **Reportedly**, part of the NBS mission is the coordination of federal monitoring activities, which should lead to better and more efficient use of biological data and assessments.
    伝聞によれば，NBSの任務の一部は，連邦モニタリング活動の調整であり，それは，生物データおよび生物アセスメントのより優良かつ効率的な利用に通じる。

★**Report**の動詞として使用例文を下に示します。

15. Kimura and Sasaki (2001) **reported** a linear relationship between the log of Ne/Nin and the applied UV dose as predicted by Eqn 1.
    KimuraとSasaki(2001)は，Ne/Ninと使用したUV放射量の関係はEqn1によって予測されるように直線関係にあると報告した。
16. The results of this effort **are reported** in Band et al. (1986) and Ford et al. (1987).
    この努力の結果は，Bondら(1986)およびFordら(1987)によって報告されている。
17. The estimates are somewhat different than previously **reported**.
    この見積もりは前に報告されている値よりいくぶん異なっている。

Represent (Representing) (Representative) ：
表現する；表す；示す（代表）（表現する）

1. The thick line **represents** the overall mean measured soil concentration.
   濃く塗られた線は土壌の測定濃度の総合平均値を示す。

2. In this figure, the dashed line **represents** how their concentrations vary with time.
この図では，点線はそれらの濃度がどのように時間で変化するかを示している。
3. The model does not adequately **represent** the endocrine disruptors' behavior in this application.
このモデルは，この応用では内分泌かく乱物質の挙動を十分に表現しない。
4. Other statistics **can be** calculated and **represented** on the graphical summary, such as means, medians, and percentiles.
他の統計量は，平均，中央値，百分位数などグラフ要約で計算・表現できる。

　● **representing**：表現する

5. This paper describes the construction and application of a numerical model **representing** mercury transformations and bioavailability in the aquatic environment.
この論文は，水環境での水銀の変化と生物利用性を表現する数値のモデルの構築と応用を説明する。

　● **representative of .....**：…の代表

6. Ideally, reference sites for estimating attainable biological performance should be as undisturbed as possible and be **representative of** the watersheds for which they serve as models.
理想的には，達成可能な生物学的実績を推定するための参照地点は，できる限り無攪乱であるべきで，それがモデルを務める流域の代表となるべきである。

Require(Requiring)：必要とする(要する)

1. Decommissioning of these units **requires** removal of all stored explosives.
これらの設備の廃止には，貯蔵されている爆発物すべての撤去が必要とされる。
2. This **will require** new approaches to management with an emphasis on the assessment of a wide expression of ecological impacts.
これには，生態影響の広範な表現への評価に力点を置いて新たな管理手法を要するだろう。
3. Interpretation of sampling data **will require** considerations of zoogeography, historical abundance and distribution, and historical ranges of variability.
サンプリングデータの解釈には，動物地理，史的な数度・分布，史的な可変値域を考慮する必要があろう。
4. Many regions **will require** increasingly efficient treatment plants.

多くの地域で効率的な処理場をますます必要とするであろう。

5. The description of fate of materials discharged into an estuary **requires** a model of considerable generality and complexity.
   河口に放出される物質の減少を説明するには，かなり一般的かつ複雑さを持つモデルを必要とする。
6. The 1972 act specifically **required** states not only to identify impaired waters, but to rank them in order of priority for cleanup and restoration.
   1972年法は，具体的に諸州が汚染水域を確認することに加え，それらを浄化・回復の優先順位で格づけすることも要求した。

- **～ is required** ：～が必要とされる

7. Knowledge of the rate constants governing the uptake and clearance of chemicals in fish **is required**.
   魚体内で，化学物質の入取・排出を左右している速度定数の知識が必要とされる。
8. Two weeks **were required** for complete equilibrium to be achieved.
   均衡が完全に達成されるには2週間が必要とされた。
9. As a result of these limitations, a suitable oxidizer **is required**.
   これらの限界の結果として，適した酸化剤が必要とされる。
10. With the increased need for predicting the movement and distribution of chemicals in an aquatic environment, knowledge of the rate constants governing the uptake and clearance of chemical in fish **is required**.
    水生環境での化学物質の移動・分配を予測する必要性が増加するとともに，魚体内での化学物質の摂取・除去を左右している速度定数の知識が必要とされる。
11. The type of data **required** depends to a larger extent on the chemical being tested.
    必要とされるデータのタイプは，対象とする化学物質に大きく左右される。

- **requiring .....** ：…を要する

12. Activities **requiring** a permit under Section 404 of the CWA must be certified as meeting provisions of the water quality standards by the state water quality agency.
    CWA第404条のもとで許可を要する諸活動は，州水質担当機関によって水質基準の規定を満たしていると認定されなければならない。

Requirement：必要条件

1. The condition of the stream can be inferred from the taxa present and what

is known of their **requirements** and tolerance.
   水流状態は，常在分類群や，その要件・耐性に関する知見から推断されうる。
2. Biological impairment criteria are intended to supplement existing chemical standards and toxicity testing **requirements**.
   生物損傷クライテリアの意図は，既存の化学基準および毒性試験要件を補完することである。
3. Many states have used biological monitoring to meet these **requirements** for identification and reporting of waterbody condition.
   多くの州が水塊状態の確認・報告についてこれらの要件を満たすため，生物モニタリングを用いてきた。
4. Satisfying this **requirement** involves understanding the nature of variability that originates with sampling frequency and/or seasonal influences.
   この要件を満たすには，サンプリング頻度および／または季節的影響に由来した可変性を理解する必要がある。
5. Section 303, which addresses the establishment of water quality standards, includes **requirements** for public involvement in public hearings on proposed changes.
   303条は，水質標準の設定を扱っており，変更案に関する公聴会への一般市民の参加を要求している。

Research(See also **Study**)：研究（→ **Study**）

1. They saw the need for **research** to better understand the process by which the regions are defined, and how quantitative procedures could be incorporated with the currently used, mostly qualitative methods to increase replicability.
   彼らは，各地域が定義される過程や，反復可能性を増すため目下用いられているおおむね質的な方法に量的手順がどれほど盛り込まれるか，を一層理解するための研究の必要性を認めた。

    ● **further research**：いっそうの研究

2. This is an area for **further research** and an opportunity for interstate cooperation.
   この領域について，いっそうの研究が必要であり，州間協力の機会が存在する。
3. **Further research into** sediment transport is needed to develop more effective tools to better address these issues.
   もっと良くこれらの問題を扱うためのより効果的なツールを開発するには，堆積物輸送に対するいっそうの研究が必要である。

- **research has been undertaken to .....** :
  …するために研究が着手されてきた
- **research must be conducted to .....** :
  …するための研究が行われねばならない
- **research is being conducted to .....** :
  …するための研究が行われている

4. **Research has been undertaken to** increase the level of understanding of the effects of heavy metals.
   重金属の影響に関する理解力レベルを向上させるために研究が着手されてきた。
5. **Research must be conducted to** demonstrate how the two approaches are complementary.
   これら2つの手法がどれほど補完的かを実証するための研究が行われねばならない。
6. **Current research is being conducted to** test the efficacy of this application to lakes, reservoirs, estuaries, and large rivers.
   現在，この概念を湖沼，貯水池，河口域および大河川へ応用することの効力を試す研究が行われている。

- **research is needed** : 研究が必要とされる

7. **Intensified research may be needed to** enable proper selection of target organisms.
   対象生物の適切な選択をするためには，研究強化が必要とされるかもしれない。
8. As suggested above, **much more research is needed** in such basic areas as sampling effort, and indicator sensitivity and variance.
   上に示唆されるように，サンプリング活動，指標の感度・分散といった基礎領域における研究はさらに必要とされる。
9. **Further research into** sediment transport **is needed** to develop more effective tools to better address these issues.
   もっと良くこれらの問題を扱うためのより効果的なツールを開発するには，堆積物輸送に対するいっそうの研究が必要です。

Respect：関する

- **with respect to .....** : …に関する

1. Microorganisms differ **with respect to** their aluminum toxicity tolerance.
   微生物はアルミニウム毒性許容性において異なっている。
2. This data set would be particularly meaningful **with respect to** .....

このデータセットは … に関して特に重要な意味を持つであろう。
3. Numerous pollutant studies, especially **with respect to** heavy metals, have already been performed on the < 63−μm fractions, allowing better comparison of results.
多くの研究が，特に重金属においては，<63−μm 部分でなされてきたので，他の結果との比較をより可能にしている。
4. Numerous pollutant studies, especially **with respect to** heavy metals, have already been performed on the < 63−μm fractions, allowing better comparison of results.
多くの研究が，特に重金属においては，<63−μm 部分でなされてきたので，他の結果との比較をより可能にしている。
5. Most agree with a general definition that ecoregions comprise regions of relative homogeneity **with respect to** ecological systems involving interrelationship among organisms and their environment.
一般的定義では，生態地域とは，生物およびその環境の相互関係を伴う生態系に関して比較的同質な地域からなる。
6. A more detailed explanation of Aoki's approach and its limitations **with respect to** attempts to use it to frame aquatic ecosystems are given later in this chapter.
Aoki の手法と，水生生態系の枠組みづくりへのその利用に関する限界についてのより詳細な説明を本章の後半で述べる。

● **In all other respects**：すべての他の点で

7. **In all other respects**, the autopsy revealed that theses animals were healthy at the time of collection.
他のすべての点で，検死は，これらの動物が採取時には健康であったことを明らかにした。

Respective（Respectively）：それぞれの（それぞれ）

1. In the case illustrated here, about 82 % of the sites do not achieve their **respective** goals.
ここで示した事例において，地点の約 82％がそれぞれの目標に達成していない。
2. The advantage of taking this additional step is that it is now relatively easy to see the proportion of sites that do not meet their **respective** biocriterion scores.
この追加措置による利点は，それぞれの生物クライテリアスコアを満たさない地点の比率が比較的見やすくなることである。

- **respectively**：それぞれ

3. During period 1, average influent and effluent CF concentrations were 0.30 and 0.11 mg/L, **respectively**.
   期間1では，入水および出水中での平均のCF濃度は，それぞれ0.30および0.11mg/Lであった。
4. The concentrations of TCA in the effluent of Port-1 and Port-2 were 0.25 mg/L and 0.40 mg/L, **respectively**.
   ポート-1とポート-2からの流出水中のTCAの濃度は，それぞれ0.25 mg/Lと0.40 mg/Lであった。
5. Fourteen of nineteen (75%) and seventeen of nineteen (90%) scores were within plus or minus two and four points of the median, **respectively**.
   スコア19個のうち14個 (75%)，19個のうち17個 (90%) がそれぞれ中央値からプラスマイナス2，4点であった。

Response：応答；反応

1. The crucial question of whether the water body becomes acidic depends on the magnitude of the watershed compensatory **response**.
   水域が酸性になるかどうかという重要な質問は流域の中和反応の大きさにかかっている。
2. Approaches for assessing the **response of** stress proteins in organisms when exposed to chemicals are currently under development.
   化学物質に曝された際の生物体内ストレス蛋白質の応答を評価するための手法を目下案出中である。
3. Although this was not part of the original experimental design, the **response of** the process to an original load is critical in determining its utility for this application.
   これはオリジナルの実験計画法の一部ではなかったけれども，このオリジナルの負荷量に対するプロセスの反応は，このアプリケーションではその利用を決定することにおいてきわめて重要である。

- **in response to ......**：…に応えて

4. We hypothesize that a specific sequence of events will occur **in response to** chronic nitrogen amendments.
   長期にわたる窒素の土壌改良に応じて，複数のイベントが特定の順序で起こるであろう，と我々は仮定する。
5. **In response to** these findings, the Commission changed the manner in

which it assigned aquatic life uses to unclassified water bodies
これらの所見に応えて委員会は，未分類水域への水生生物利用の指定方法を変更した。

Responsible：原因である

● **responsible for** .....：…の要因である；…の原因である

1. This often provides insight about the factors **responsible for** degradation and offers a diagnostic capability.
   これは，しばしば劣化の要因に関する洞察をもたらし，判別能力をもたらす。
2. Sedimentation is widely held as **responsible for** degradation of fish communities in warmwater and coldwater streams, and many of the mechanisms of this degradation have been well documented.
   堆積は，温水水流ならびに冷水水流において魚類群集劣化の原因と広く認められており，劣化メカニズムの多くが十分に文書化されている。

● ～ **is responsible for** .....：～は…の原因である

3. Aquatic organic matter **is responsible for** several problems in water supplies.
   給水において，水生有機物がいくつかの問題点の原因である。
4. It was the author's conclusion that seasonal and habitat changes **were responsible for** much of the variability of species abundance at a given site.
   ある特定地点の生物種豊富性の変化を大きく左右するのは季節的変化や生息地変化である，というのが著者の結論である。

Result：結果；結果として生じる；帰着する

1. This paper presents the **results** of .....
   この論文は…の結果を提する。
2. These **results** are in agreement with the laboratory studies.
   これらの結果は実験室での研究結果と一致している。
3. The sample provided reliable **results**.
   このサンプルは信頼できる結果をもたらした。
4. These **results** are consistent with the fact that....
   これらの結果は....という事実と一貫している。
5. Consistent with the **results** of Wu and Yao (2001), ...
   Wu and Yao (2001) の結果と一貫して，…

- **results show that .....**：結果は … を示している
- **result indicates .....**：結果は … を示す
- **The results from ..... indicate that .....**：
  我々の … の結果は … ということを示している
- **Our results suggest that .....**：我々の結果は … を示唆している

6. **The result clearly shows that .....**
   その結果は … ということを明らかに示している。
7. **Results from** this study **showed** more information would be obtained by sampling more individual streams than more sites on the same stream.
   この研究結果によれば，同じ水流の多数地点でのサンプリングよりも，多数水流でのサンプリングの方がより多くの情報が得られる。
8. **Results** summarized in Table 1 **indicate that .....**
   表1で要約した結果が，… のことを示している。
9. **Our results suggest that** the reactivity is additive for the concentration range studied.
   我々の結果は，研究された濃度の範囲では反応性が加算的であることを示唆している。

- **The results demonstrate .....**：結果は … を実証している

10. **The results demonstrate** the utility of using the theoretical range of the IBI to differentiate between and interpret different types of impacts.
    結果は，異なる影響種類を差別化するために IBI の値域を利用することの有用性を実証している。
11. **The preceding results demonstrate that** discernable patterns in biological community information do exist and can be used to determine the probable cause of certain types of impairments in combination with other source and instream chemical/physical data.
    前述の結果は，生物群集情報における区別可能なパターンが存在し，特に他の源泉および水流内化学／物理データと併用すれば，ある種の損傷の潜在的原因を決定するため用いうることを実証している。

- **These results support the hypothesis that .....**：
  これらの結果は … という仮説を支持する
- **The results confirm the hypothesis that .....**：
  この結果は … という仮説を確証する

12. **These results directly support the hypothesis that** natural levels of copper in seawater can be toxic to plankton.

これらの結果は，海水中に自然に存在する銅のレベルがプランクトンにとって有毒であり得るという仮説を直接支持している。

13. **The results confirm the hypothesis that** copper uptake by algae is related to the free cupric ion activity and is independent of the total copper concentration.
    この結果は，藻類による銅の摂取は遊離銅イオン活性と関係があり，全銅濃度とは関係がないという仮説を確証する。

    - **results is provided by .....**：結果が … によって提供される
    - **results were obtained**：結果が得られた

14. A complete listing of the chemical **results is provided by** Ikeda (2001).
    化学的な結果の完全なリストが池田 (2001) によって提供される。
15. The following **results were obtained**: (a).....; (b).....; and (c).....
    次のような結果が得られた。(a)…, (b)…, それに (c)… である。

    - **The results are shown in .....**：この結果は … に表されている
    - **The results are listed in .....**：この結果は … に列挙されている

16. **The results are shown in** Figure 3.
    この結果は図3に表されている。
17. **The results** of investigations on three different heavy metals **are shown in** Fig. 9.
    3つの異なった重金属についての調査の結果は図9に表す。
18. The **results** of the analysis of the eight samples from the monitoring wells **are listed in** Table 3.
    モニタリング用の井戸からの8つのサンプルの分析の結果を表3に列挙する。

    - **The results of ～ is .....**：～の結果は … である

19. One of the **results of** modeling **is** identification of nonpoint sources.
    このモデル構築の結果の一つは面源の確定である。

    - **The results can be expressed in .....**：この結果は … で表わされうる
    - **Results are presented in .....**：結果が … で提出される

20. **The results can be expressed in** a fashion analogous to an adsorption isotherm.
    この結果は吸着等温線に類似した方法で表現されうる。
21. **Results are presented in** several forms, but are normalized to allow direct comparison with other published work.
    結果がいくつかの形式で提出される，しかしその結果を標準化することによって

他の発表された研究との直接比較を可能にする。

- **~ is the result of .....** ：~は … の結果である

22. The apparent reduction in the oxygen utilization **is most likely the result of** soil gas diffusion.
酸素消費の明白な減少は，土壌ガス拡散の結果である可能性が最も高い。
23. The contamination detected and characterized at the OU5 source area **is primarily a result of** land disposal practices.
OU5汚染源地域において検出され，特徴づけられる汚染は，主に埋立処分の結果である。

- **~ result in .....** ：~は … という結果になる

24. Insufficient knowledge about regional expectations **can result in** misinterpretations about the severity of impacts in streams.
地域期待値に関する知見不足は，川の影響の動性に関する誤った解釈を招きうる。
25. Not only with this **result in** a loss of valuable information, it **will likely result in** the continued degradation of the aquatic resource.
これは，貴重な情報の損失を招くばかりか，水資源の継続的な劣化を招く見込みが強い。
26. This loss of habitat quality **has resulted in** extinctions, local extirpations, and population reductions of fish species and other aquatic fauna in the United States.
こうした生息場所での質低下は，米国における魚種および他の水生動物相の，絶滅，局所的根絶，個体数減少をもたらした。
27. Two of the proposed use designation changes were challenged in court because upgrades in designated use **could** possibly **result in** more stringent permit limitations.
用途指定変更案のうち2つは，法廷で異議申立てされた。使用指定における昇格は，より厳しい許可限度をもたらす可能性があったからである。

- **resulting in .....** ：… をもたらす

28. The mature females were not only smaller but also produced fewer eggs, **resulting in** fewer young in the next generation.
成熟した雌は，より小さいだけではなく，より少数の卵を産卵するため，次の世代で幼魚がより少数となる。
29. A series of batch tests was carried out to test the hypothesis that the same enzymes are used to attack chlorophenols with the same position-specific dechlorination reactions, **resulting in** competition between these

compounds.
位置特定の同じ脱塩反応において同じ酵素が塩化フェノールを攻撃するために使われ，これらの化合物の間に競合反応をもたらす，という仮説をテストするために一連のバッチ試験が行われた。

- **~ result from .....** ：
~は … によって（結果として）生じる；～は … に起因する

30. It seems likely that such a consistent pattern of contamination **would result from** in-situ conditions rather than migration.
このような汚染の一貫したパターンは，移動してきたと考えるよりは，現地の条件に起因しているように思われる。
31. Each area can be viewed as a discrete system which **has resulted from** the mesh and interplay of the geologic, landform, soil, vegetative, climatic, wildlife, water and human factors which may be present.
各領域は，個別的なシステムとみなされ，地質，地形，土壌，植生，気候，野生生物，水，そして潜在的な人間因子の調和および相互作用から生じてきた。
32. Physical characteristics of individuals that may be useful for assessing chemical contaminants **would result from** microbial or viral infection, some sort of tetragenic or carcinogenic effects during development of that individual.
化学的汚染を評価するにあたり，固体の身体的特徴が有用なこともある。
33. Nonpoint sources are now impeding further biological improvements observed in larger rivers that **resulted from** reduced point source impacts.
点源影響の減少によって大河川で観察された一層の生物学的な向上を現在面源が妨げている。
34. Bioassessments can be used to document instream improvements that **result from** wastewater facility upgrades and the implementation of best management practices.
生物アセスメントは，廃水施設向上および最善管理慣行の実施に起因した流入改善の立証に利用できる。

- **as a result of .....** ：… の結果として

35. **As a result of** a number of experimental studies, ...
多くの実験的研究の結果として，…
36. **As a result of** these limitations, a suitable oxidizer is required.
これらの限界の結果として，適した酸化剤が必要とされる。
37. There is little doubt that much of the gypsum has formed in-situ **as a result of** this overall process of pyrite oxidation.

石膏の多くが，黄鉄鉱の総合酸化プロセスの結果として，現地で生成したことにほとんど疑いがない。
38. A significant inhibitory effect was placed on the specific oxidation rate **as a result of** extreme winter temperatures.
極度な冬の温度が原因で相当な抑制効果が特定の酸化速度にかかった。

Reveal：明らかにする；明らかになる

1. The analysis **revealed** the following: .....
   この分析は，以下を明らかにした。
2. In addition, it **reveals** cumulative effects of pollution from many sources on aquatic communities.
   さらに，多くの源からの汚濁が水生生物群集に及ぼす累積効果も明らかにする。
3. In an analysis of resources expended during FFY 1987 and 1988 the following **were revealed**.
   1987〜88連邦会計年度に行われた資源分析において，以下が明らかになった。
4. Examination of the literature on sorption **reveals** a great deal of confusion regarding the time to reach the sorption equilibrium.
   吸着についての文献調査によれば，吸着均衡に達っするまでの時間に関して多くの混乱がある。
5. An examination by Toda et al. (2002) of oysters and other shellfish exposed to an oil spill **revealed** the presence of large quantities of petroleum compounds in the body tissues.
   カキと他の貝に流出した石油をさらしたToda ら(2002)の研究によれば，貝の体組織に大量の石油化合物が存在することが明らかになった。

● 〜 **reveal that** ..... ：〜から … ということが明らかになる

6. An examination of this process **revealed that** the reduction process was sequential.
   このプロセスの検討から，この還元プロセスが連続的であったことが明らかになった。
7. Figures 4 and 5 **strikingly reveal that** the copper toxicity is also influenced by the substrate concentration.
   図4と図5によれば，銅の毒性が基質濃度によっても影響をうけることが極めて明らかである。
8. The study **revealed that** most of the streams possessed physical habitat that enhanced the development of communities of diverse aquatic fauna.
   この研究によれば，大半の水流は，多様な水生動物相群集の発達を促す物理的な

生息地を有する。
9. Further investigations into the time course of metal uptake in activated sludge **have revealed that** this process occurs in two stages.
汚泥中で時間とともに変わる金属摂取をさらに調査し，このプロセスが2つの段階で起こることを明らかにした。

Reverse：逆；反対

- **the reverse is true**：逆も正当である
- **the reverse is not always true**：逆は必ずしも真ではない

1. **The reverse will be true for** compressive axial stresses.
圧縮軸のストレスについては逆も真であろう。
2. While it is possible for reference sites to double as upstream control sites, **the reverse is not always true**.
参照地点を対照地点の2倍にすることが可能な場合でも，逆は必ずしも真でない。

Review(Reviewing)(See also **Literature**)：査読；レビュー；レビューする（レビューすること；再検討すること）(➡ **Literature**)

1. There is a short but good **review** by .....
…による短いが，すぐれたレビューがある。
2. **A literature review** revealed that .....
文献のレビューによって…が明らかになった。
3. **Based on a review of** the literature on acid rain, ...
酸性雨に関する文献のレビューに基づいて，…
4. Probably the most complete **review of** the chemistry of diesel fuels is the study by Akagi et al.(2004).
ディーゼル燃料の化学に関する，おそらく最も完全なレビューはAkagi ら(2004)による研究である。
5. Other sources of information include interviews and **review of** internal correspondence.
他の情報源は，インタビュー，そして内部通信のレビューである。
6. We appreciated critical **review comments** from three anonymous reviewers, which greatly improved an earlier draft of this manuscript.
筆者は，3人の匿名査読者からの批評的論評に感謝する。それは，本章の草稿を大幅に改善してくれた。
7. Some of the state program documents used to develop information for this

summary were in draft form at the time of our **review** and are representative of preliminary programs.
本論のための情報作成に用いられた州プログラム文書の一部は，我々による査読時点で草稿段階にあり，予備的プログラムを表している。

- ● 〜 **conducted a review of** ……：〜は…のレビューを実施した
- ● 〜 **have not yet gone through the review process**：
  〜は，まだ，レビュー過程を終えていない

8. The EPA **conducted a** more extensive **review of** the contaminant content of the existing deposits.
   EPAは，既存の堆積物での汚染物質含有量のいっそう広範なレビューを実施した。
9. The biological impairment criteria developed **have not yet gone through the review** and comment **process** towards implementation into the state regulations.
   設定された生物損傷クライテリアは，まだ，州規則での実施へ向けての査読・論評過程を終えていない。

- ● **in the review**：レビューでは

10. **In the excellent review** presented by Numata, ……
    Numataによって発表されたすぐれたレビューでは，
11. Particular emphasis will be given **in this review** to natural water samples.
    このレビューでは，自然水サンプルが特に重視されるであろう。

★ **Review** の動詞としての使用例文を下に示します。

12. Hushimi (2001) **reviewed** the inhibitory effects of many compounds and noted that ……
    Hushimi (2001)は多くの化合物の抑制効果をレビューして，…と指摘した。
13. Tamaki (2001) **has reviewed** proposed models for the periphyton assemblage.
    Tamakiは，付着生物群がりについて提案されたモデルのレビューを行った。
14. The literature on the nitrification process **was fully reviewed by** ……
    硝化プロセスに関する文献が…によって十分にレビューされた。
15. Some aspects of these deficiencies **have been reviewed** by Koike (2004).
    これらの欠陥の一側面がKoike (2004)によってレビューされた。
16. A number of these documents **were reviewed** during the course of this study and the review document is appended as Appendix A.
    この研究過程で多くのこれらの証拠書類がレビューされ，レビュー書類が付録Aとして付加されている。

17. Descriptions of the complicated mechanisms involved in individual contaminant oxidations using AOPs **have been extensively reviewed** in the literature.
AOPsを使っての個々の汚染物質酸化に関与する複雑なメカニズムの記述が，この文献で広範にレビューされた。
18. Because these approaches have been recommended for fish assemblages and for benthic macroinvertebrates, they **will not be reviewed** extensively here.
魚類群がりならびに底生大型無脊椎動物についてこれらの手法がこれまで勧告されてきたので，それらについてはここでは広範囲なレビューはしない。
19. The second study **reviewed** here is that of Kurama et al.(2003), who evaluated the PRID model.
ここでレビューされた2番目の研究はKuramaら(2003)によるもので，彼らはPRIDモデルを評価した。

● reviewing ：レビューすること；再検討すること

20. Wakamoto, **in reviewing** the work of Oike and Inoki, suggested that .....
Wakamotoは，OikedとInokiの研究を再検討し，…を示唆した。
21. **Reviewing** the literature on sonochemistry of organochlorine compounds did not lead to any publications on the use of ultrasound for chemical monitoring.
有機塩素化合物の超音波化学分解に関する文献レビューは，化学的モニタリングの為の超音波利用に関するどの出版物にもならなかった。
22. Environmental managers within industry can become well acquainted with basic water quality criteria concepts by **reviewing** the four documents cited above.
業界内の環境管理者は，上記4つの文書を繙けば，基礎的な水質クライテリア概念を十分に把握することができる。

Revision ：修正

1. This report is a **revision of** the Solid Waste Management Units Report submitted in 1988.
この報告書は1988年に提出された廃棄物処理施設報告の修正版である。
2. Consequently, Arkansas determined that the water quality standard driving this process needed **revision**.
その結果，アーカンソー州は，この過程を導く水質基準の改正が必要だと決定したのである。

3. New SWMUs identified after this date will be included **in the next revision** to this report.
この日付後に認識された新しいSWMUsは、この報告書の次の修正版に入れられる。
4. Upon completion of the initial **revision of** ecoregions and delineation of subregions, sets of reference sites are identified for each subregion.
生態地域の初回改良、ならびに小地域の画定の完了時に、各小地域について参照地点セットが確認される。

- **revisions were made**：改正が施された

5. Some minor **revisions were made** in 1985, but the approach remained essentially the same through 1987.
1985年にいくつか細かな改正が施されたが、この手法は、基本的に1987年まで存続した。

- **~ is subject to revisions**：
~は修正の対象となる；~は修正の適用を受ける

6. This report **is subject to revisions** based on additional information.
この報告書は追加情報に基づいた修正の対象となる。

- **with some minor revisions**：若干の修正を加えて

7. The feed solution was prepared according to the medium described in Ref 8 for nitrifying bacteria **with some minor revisions**.
この流入溶液は、硝化菌に関して参照8に記述されている溶媒に従って、若干の修正を加えてつくられた。

Risk：リスク；危険性

1. The 10^{-6} **risk level** for chrysene is 56 $\mu g/kg$.
クリセンの10^{-6}のリスクレベルは56 $\mu g/kg$である。
2. All other values are at concentrations well below this **risk level**.
すべての他の値は、このリスクレベルをずっと下まわる濃度にある。
3. The basic steps of the **risk assessment** process are the following:……
リスクアセスメント過程の基本的な段階は次のことである。それは、…
4. The results of this assessment are used to select the remedial action and to determine **risk-based** cleanup levels for the remedial action.
このアセスメントの結果は、復元活動を選択するため、そしてリスクベースの浄化レベルを決めるために使われる。

- ～ **present a risk**：～はリスクを引き起こす

5. There is insufficient information to determine whether contamination present at this source area may **present an unacceptable risk**.
   この汚染源での汚染が，容認出来ない危険性を引き起こすかもしれないかどうか決定するには情報が不十分である。

- ～ **pose risks to** ..... ：～は … にリスクを提する

6. These three sites **pose** significantly higher **risks to** human health and the environment than the other four.
   これらの3つの地域は，他の4地域より人間の健康と環境に著しく高いリスクを提する。

River (including Stream, Creek)：川 (Stream, Creek をも含む)

★ River, Stream および Creek の位置に関連する例文を下に示します。

1. The Harn Creek–Natsu River system **is located on** the north side of the Mizudori River in Ohfuku City.
   Ohfuku 市では，Harn 小流–Natsu 川システムは Mizudori 川の北側に位置している。
2. This creek **drains from** the northwestern slope of the Black Hills.
   この小流は Black Hills（ブラックヒルズ）の北西部のスロープから流れ出る。
3. The Columbia River **flows through** the northern part of the Hanford Site, and turning south, it forms part of the site's eastern boundary.
   コロンビア川は Hanford 地域の北部を通過し，南に曲がり，地域の東境界線の一部を形成する。
4. The Yakima River **runs** along part of the southern boundary and joins the Columbia River south of the city of Richland.
   Yakima River（ヤキマ）川は（Hanford 地域の）南の境界線の一部に沿って流れて，そして Richland 市の南で Columbia（コロンビア）川と合流する。
5. The Big Lost River **courses through** the Idaho National Laboratory.
   Big Lost 川はアイダホ国立研究所地区を流れる。
6. The Big Lost River **travels** the INEEL and is close proximity to facilities and waste disposal / storage areas.
   Big Lost 川は INEEL 地区に達し，そしてその施設と廃棄物処理/貯蔵地域の近くを流れている。

★ River, Stream および Creek の様々な表現例文を下に示します。

7. Eleven dams exist **along** the Columbia River in the United States.

11のダムが合衆国のコロンビア川に沿って存在する。

8. The company constructed a diversion channel and a small settling pond **along** Whitewood Creek.
その会社は，Whitewood Creek に沿って迂回水路と小さな沈澱池を建設した。

9. This creek **is fed by** several small headwater stream that enter upstream of the 18 mile segment.
この小流は，18マイル区間の上流に流入するいくつかの小さい源流によって供給される。

10. The Bell Fourche River **joins** the Cheyenne River approximately 130 miles farther downstream.
Bell Fourche 川はおよそ130マイルさらに下流で Cheyenne 川に合流する。

11. In areas where large meanders occur in the existing channel, ...
既存の水路の大きい蛇行がおこる場所で，…

12. Conesville Station is an additional 102 river miles upstream from this point.
Conesville 観測地点は，この地点から更に102マイル (River Mile) 上流にある。

13. The sampling site is at the downstream end of the reach, near the confluence of Cherry Creek and the Portneuf River.
サンプリング地点は，Cherry 小流と Portneuf 川の合流近く，その下流区域の終わりにある。

14. The length of the Columbia River from the Canadian border to the Pacific Ocean is approximately 745 river miles.
カナダの境界から太平洋までのコロンビア川の長さはおよそ745リバーマイル（河川距離）である。

15. The Big Lost River is both a gaining and losing stream that fluctuates seasonally and with respect to precipitation and snow melt.
Big Lost 川は増流と減流の両方であり，季節，そして降雨と雪溶け水によって増減する河川である。

16. The Scioto River is a tributary to the Ohio River, about 185 river miles downstream of the Muskingum River/Ohio River confluence (Fig. 2).
Scioto 川は，Ohio 川支流に当たり，Muskingum 川／Ohio 川の合流点から約185リバーマイル（河川距離）下流に存在する（図2）。

17. The Columbia River is the third largest river in the United States ; it originates in Canada, flows through eastern Washington, and forms the border between western Oregon and Washington as it flows to the Pacific Ocean.
コロンビア川は合衆国での3番目に大きい川である。それはカナダから始まって，東ワシントン州を通って流れ，それが太平洋に流れ出るとき西オレゴン州とワシントン州の間に境界線を形成する。

18. Up to 800 tons of DDTR (as DDT, DDE, and DDD) are estimated to be distributed in the sediments of Huntsvill Spring Branch and Indian Creek from upstream of the plant discharge ditch to the confluence of Indian Creek with the Tennessee River.
最高 800 トンの DDTR(DDT，DDE と DDD として)が，Huntsvill Spring Branch と工場廃水溝の上流から Indian Creek と Tennessee 川の合流地点までの Indian Creek の沈降堆積物に分布していると推測される。

19. The Hocking River, located in Southeastern Ohio, is a medium-sized river (1197 mile^2 drainage area) of about 100 mi in length. The Hocking River headwaters are in glacial deposits of Fairfield County southeast of Columbus, Ohio, and it flows southeasterly through unglaciated, rugged topography to the Ohio River.
Hocking 川は，オハイオ州南東部に所在し，長さ約 100 mile の中規模河川である(流域面積 1197 mile^2)。Hocking 川の源流は，オハイオ州 Columbus 市南東部の Fairfield 郡の氷河堆積物にあり，氷河作用を受けていない高低ある地形を抜けて南東へ流れ，Ohio 川へ達している。

Role (See also Play)：役割 (→ Play)

1. **The role of** trace metals in regulating the growth rates of marine phytoplankton is not well understood.
   海洋性植物プランクトンの増殖速度制御における微量金属の役割は，よく理解されていない。

   - ~ **have a role in** .....：~ は … において役割を演じる
   - ~ **have a greater role in** .....：
     ~ は … においてより重大な役割を演じる
   - ~ **fulfill a role**：~ は … の役割を果たす

2. Biological information can be expected to **have a greater role in** water resource decisions in the future for several reasons.
   生物学的情報は，将来，いくつかの理由で水資源決定においてより重大な役割を演じるものと予想される。

3. Regional reference sites **can fulfill a dual role** as the arbiter of regionally attainable biological performance and as an upstream reference for determining the significance of any longitudinal changes.
   地域参照地点は，地域で達成可能な生物学的実績の規範として，また，縦断的変化の有意性を決定するための上流の参照として，2つの役割を果たしうる。

- **with the roles of .....**：… としての役割がある

4. The observed heavy metal content of waterways is a function of many variables, **with the roles of** some components being only partially understood.
   水路で検出された重金属含有量は，多くの変数により変動し，ただ部分的にだけ理解されているある成分としての役割がある。

- **～ play a role in .....**：
  ～は … において役割を演じる；～は … において役割を果たす（➡ **Play**）
- **～ played a major role in .....**：
  ～は … において重要な役割を演じた（➡ **Play**）
- **～ have played a major role in .....**：
  ～は … において重要な役割を演じてきた（➡ **Play**）
- **～ play a significant role in .....**：
  ～は … に重要な役割を演ずる（➡ **Play**）
- **～ are known to play an important role in .....**：
  ～が … において重要な役割を演ずるということが知られている（➡ **Play**）
- **～ play no more than an ancillary role in .....**：
  ～は … では従属的役割しか演じない（➡ **Play**）
- **～ has always played a central role in .....**：
  ～は常に … において中心的な役割を演じてきた（➡ **Play**）

Room：余地

- **There is room for .....**：… の余地がある
- **～ leave room for .....**：～には … の余地がある

1. **There** is **room for** reconsidering their conclusions.
   彼らの結論は再考の余地がある。
2. **There is still room for** argument about introducing a new species in this river.
   この川に，新しい種を導入することについては，まだ議論の余地がある。
3. Their results **leave room for** different interpretation.
   彼らの結果には異なった解釈の余地がある。

Rule：規則；法則；規定する；支配する

1. There is an exception to this adsorption **rule**.

この吸着の法則には一つの例外がある。
2. The contemporary regulatory-intensive approach relies heavily on a legal style of **rule** setting.
現代の規制重視手法は，主に法律スタイルの規則策定に依拠している。
3. EPA has established **a rule** that attainment of biocriteria should be granted disproportionate weight in demonstrating overall use attainment.
EPA は，全般的な用途達成の実証において生物クライテリア達成が不相応な加重を与えられるべきだとのルールを確定した。
4. In fact, in finding that dams were not required to have NPDES permits, one court **rule** that the fact that dams may cause "pollution" does not necessitate an NPDES permit where there is no addition of "pollutants."
実のところ，ダムが NPDES 許可を必要としないという所見において，ある裁判所は，「汚濁物質」の追加がない場合，ダムが「汚濁」を生じうるという事実は，NPDES 許可を必要とせしめない，と裁定した。

★ **Rule** の動詞としての使用例文を下に示します。

5. State courts have not yet **ruled** on the validity of water quality criteria per se or as applied.
同州の裁判所は，水質クライテリア自体またはその適用の有効性をまだ裁定していない。

- **～ is ruled out**：～を排除する

6. The possibility of ～ cannot be entirely **ruled out**.
… の可能性をまったく排除するわけにはいかない。

Sake：目的；理由

- **for the sake of .....**：… のために

1. **For the sake of** clarity in presenting the test results, most of the details concerning the theoretical background of the tests and the analytical procedures are provided in the Appendix B.
テスト結果の提出にあたってそれを明快にするために，試験と分析手法の理論的な背景の細部の多くを付録Bに示す。
2. Seasonality is a well-understood concept, therefore, it is not necessary to sample in multiple seasons **for the sake of** data redundancy.
季節性は，十分に理解されている概念であるから，データ冗長性のために複数季節でサンプリングする必要がない。

Same：同じ

1. PCE, TCE, and DCE are grouped into the **same** VOC group.
PCE，TCE，DCEは，同じVOCの部類に分類される。
2. Some minor revisions were made in 1985, but the approach remained essentially **the same** through 1987.
1985年にいくつか細かな改正が施されたが，この手法は，基本的に1987年まで存続した。
3. The agency grouped together these minor waters under the **same** preliminary aquatic life designation.
TNRCCは，これらの小水域を同じ予備的な水生生物使用指定のもとでグループ分けした。
4. These data clearly demonstrate the presence of two different fish communities on the **same** waterbody.
これらのデータは，明らかに同一水域に2種類の魚類群集が存在することを実証している。
5. The design of the tubes allowed them to be placed into the sonicator horn **at the same depth**.

このチューブの設計によって，それらをソニケーターホーンの中に同じ深さで設置することができた．

6. Differences among sites would be much greater than year-to-year variability **at the same site**.
地点間の差の方が同一地点での年次変化よりも大きいのである．

7. Generally, less variability is expected among surface waters **within the same region** than between different regions.
一般に，異なる地域の地表水域よりも，同じ地域の地表水域の方が可変性が低いと予想される．

- **At the same, ...**：同時に，…

8. **At the same**, however, the fundamental bioassessment approaches being developed are applicable to all waterbody types.
しかし同時に，設定中の基本的な生物アセスメント手法は，すべての水域型に適用可能である．

- **A and B are the same**：AとBは同じである

9. TCE **and** trichloroethylene **are the same** compounds.
TCEとtrichloroethyleneは同じ化合物である．

- **The same is true of .....**：…についても同様である

10. **The same is true of** whole effluent toxicity criteria.
これは，全流出毒性(WET)クライテリアについても同様である．

- **in the same way**（See **Way**）：同様に（→ **Way**）

Sample(Sampling)：サンプル；サンプリングする（サンプリング）

1. **Sample processing** and species proportional counts of a sample should take no more than 3 hr.
サンプル処理とあるサンプルの生物種比率カウントに，3時間以上かけてはならない．

2. A **sample solution** of 65 mL was sonicated in a cylindrical glass vessel of 50 mm i.d. with a total volume of 150 mL.
内径50 mm，全容量150 mLの円筒状のガラス容器内で，65 mLのサンプル溶液をソニケートした．

- **samples were taken at .....**：…でサンプルが採取された
- **samples were taken from .....**：サンプルは…から採取された

3. Replicate **samples were taken at** most sites, including ecoregion reference sites.
生態地域参照地点など大半の地点で反復サンプルが採取された。
4. **Samples taken from** riffles or runs are sufficient to characterize the stream.
瀬または急流から採取されたサンプルで，十分にその河川の特性を把握できる。
5. This section deals with soil **samples taken** during the drilling of monitoring wells.
この節は，モニタリング用の井戸を掘る時に採集された土壌サンプルを取り扱っている。

- **samples were collected**：サンプルは集められた
- **samples were stored**：サンプルは保管された
- **samples were frozen at −X℃**：サンプルは −X℃で冷凍された
- **samples have been analyzed for .....**：サンプルは … 分析された
- **samples were compiled for .....**：… についてサンプルが集められた
- **samples are sufficient to .....**：サンプルは … するのに十分である

6. **Samples were collected** and stored at 4 ℃ until use.
サンプルは集められ，使うまで4℃で保管された。
7. Soil **samples** at the field sites **were collected** every month over a 12-month period.
野外地点における土壌サンプルが，12カ月間にわたって毎月収集された。
8. These **samples were stored** at room temperature in these bags prior to use in the experiments.
これらのサンプルは，この実験に使用するまで，室温にてこれらの袋に保管された。
9. The collected **samples were frozen** at −50 ℃, transported to the laboratory on dry ice, and stored at −20 ℃.
採集されたサンプルは−50℃で冷凍され，研究室までドライアイス詰めで輸送され，−20℃で保管された。
10. **Samples** from each of these sites **have been analyzed for** total arsenic.
これらの各サイトからのサンプルで，全ヒ素分析をおこなった。
11. A total of 27 individual **samples were compiled for** sites just upstream of Conesville Station；51 samples were taken within the entire WAP ecoregion.
Conesville 発電所のすぐ上流の地点について合計27のサンプルが集められ，WAP 生態地域全体では51サンプルが採取された。

★**Sample** の動詞としての使用例文を下に示します。

12. It is necessary to **sample** from multiple locations.

複数地点でサンプリングする必要がある。
13. During 1988 to 1990, TPWD joined TNRCC **to sample** an additional 66 streams.
1988年から1990年にかけて，TPWDは，TNRCCと共同で，さらに66水流でサンプリングを行った。

● **～ was sampled**：～はサンプリングされた

14. The total of 60 study units **will have been sampled** after nine years.
合計60の研究単位がすべてサンプリングされるのは，9年後になる。
15. **These streams were all sampled** during the summer period (June through September), to correspond to critical low flow and elevated temperature conditions.
これらの河川は，臨界低流量および高温状態に対応するため，すべて夏季(6～9月)にサンプリングされた。
16. A 40-mile **segment has been sampled** repeatedly over multiple years, which is a prerequisite for trend analysis.
40mileの区間で，複数年にわたり繰り返しサンプリングされた。これは，傾向分析の必須条件である。
17. In Year 1 of MAHA, 266 **sites were sampled** throughout the mid-Atlantic highlands stretching from Virginia, through West Virginia and Maryland, into Pennsylvania.
MAHAの初年度に，バージニア州，ウェストバージニア州，メリーランド州からペンシルバニア州へ至る中部大西洋岸諸州高地全域の266地点でサンプリングが行われた。
18. Only a few rocks **need to be sampled**, because each has a periphyton assemblage with hundreds of thousands of individuals.
サンプリングに必要なのは，数個の岩石だけである。なぜなら，1個の岩石に付着生物群がりがあるからである。

● **sampling**：サンプリング

19. This **sampling** did not encounter evidence of excessive concentrations of arsenic.
このサンプリングでは，ヒ素の濃度が過剰であるという証拠はあがらなかった。
20. EPA conducted a more extensive **sampling** of potable wells.
EPAは，飲料水用井戸のより以上の大規模なサンプリングを実施した。
21. **Sampling** time can be as rapid as 20 min/site, as shown by the methods used in Montana (Matsuo, 1993).
サンプリング時間は，モンタナ州での方法で示されたとおり(Matsuo, 1993)，1

地点当り 20 分間で済む。
22. The **sampling site** is about 2 miles upstream of the Town of Whitewood.
サンプリング地点は Whitewood 町のおよそ 2 マイル上流にある。
23. **Resampling** of these wells by EPA confirmed the existence of arsenic concentrations elevated above background concentration.
これらの井戸での EPA による再度のサンプリングによって，ヒ素の濃度がバックグランド濃度以上であることが確認された。
24. Improvements in the treatment process were made in the year before this **sampling**, and macroinvertebrate changes were not as great as in previous years.
サンプリング前年に処理過程の改善が行われ，大型無脊椎動物の変化は，それまで数年間ほど大幅でなかった。

Satisfactory (Unsatisfactory)：満足のいく；十分に（満足のいかない）

1. It was found that the observed processes are described **satisfactory** by this model.
観察されたプロセスがこのモデルによって十分に説明されることがわかった。
2. These constant values were found to have a **satisfactory** fit to all the experimental data.
これらの一定値は，実験データすべてに対し十分に一致することが判明した。
3. This model is capable of correlating growth data from many situations **in a satisfactory manner**.
このモデルは，多様な条件下からの成長データを満足のいく方法で関連づけることができる。
4. Abundant evidence eventually accumulated that this approach **was unsatisfactory**.
結局，この手法が満足のいかないものであったという数多くの証拠が蓄積した。

Satisfy：満たす

1. For millennia, people could **satisfy** their need for water by locating permanent settlements next to rivers, lakes, or oceans.
数千年間にわたり，人間は，河川，湖沼，海洋の近くに定住することによって，水の必要性を満たすことができた。
2. **Satisfying** this requirement involves understanding the nature of variability that originates with sampling frequency and/or seasonal influences.
この要件を満たすには，サンプリング頻度および／または季節的影響に由来した

可変性を理解する必要がある。

Say：つまり

1. This may not have been a difference in state practice. It may have reflected differences in ownership, **say**, between federal and state or federal and private.
   これは，各州慣行の相違ではなかったかもしれない。所有における差，つまり，連邦所有地と州有地の差，連邦所有地と私有地の差を反映していたのかもしれない。

Scale：スケール；規模

1. With these time and **space scale**, ...
   これらの時間スケールと空間スケールで，…
2. It has also become widely accepted in **large-scale** applications.
   それは同じく大規模なアプリケーションで広く受け入れられた。
3. **Laboratory-scale** activated sludge units were studied over a wide range of process operating conditions.
   実験室規模の活性汚泥処理ユニットが広範囲のプロセス操業条件で研究された。
4. The following conclusions were established from the process emissions survey and **bench-scale** testing: 1) ....., 2), ...
   プロセス排気テストの調査，そしてベンチスケール実験から次の様な結論が確立された。1)…, 2), … である。

   - **on a ..... scale**：…のスケール(規模)で
   - **on a scale of A to B**：AからBまでのスケール(規模)で

5. Figure 3 shows how water is used **on a global scale**.
   地球規模で水がどのように使用されているかを図3に示す。
6. Because so few states have progressed to the point of proposing legally binding narrative or numeric biocriteria at this time, industry involvement has been relatively limited **on a national scale**.
   現時点では，法的拘束力を持つ記述的または数量的な生物クライテリアまで進展している州はごく少数なので，業界関与は全国規模で比較的限られている。
7. Tolerance values are assigned **on a scale of** 0 **to** 10, with 0 being the least tolerant and 10 being the most tolerant.
   耐性値は，0から10までのスケールで指定され，0が最低耐性を，10が最高耐性を表す。

- **over the time of scale**：時間のスケールにおいて
8. **Over the time of scale** of interest, this assumption is reasonably valid and may be used as a basis for a preliminary analysis of the problem.
   我々が関心ある時間のスケールにおいて，この仮定は適度に有効でり，この問題の予備解析の基盤として使用されるかもしれない。

Schematic (See also **Figure**)：図式の (→ **Figure**)

1. A **schematic** of pesticide fate and transport within a reservoir is presented in Figure 1.
   貯水池中における殺虫剤の衰退と輸送の概略図が図1に提示されている。
2. A **schematic** of phytoplankton kinetics is presented in Figure 1.
   植物プランクトンの速度論の概略図が図1に提示されている。

Scope：範囲

1. While model applications may differ greatly **in scope** and purpose, it is hoped that the representative data derived from pilot studies will be useful in this process.
   モデルのアプリケーションが応用範囲と目的で大いに異なるかもしれない一方，予備試験的研究から得られる代表的データはこのプロセスに有用あろうことが期待される。
2. Problems related to chemical contamination of surface water supplies, particularly those associated with byproducts from new treatment technologies, are rational **in scope** and receive national attention.
   表流水給水における科学汚染に関連付けられ，とりわけ新しい水処理技術の副産物に結び付く問題は，その道理性を認められており，全国的な注目を集めている。

- **outside the scope of .....**：…の範囲外
- **beyond the scope of .....**：…の範囲外
- **within the scope of .....**：…の範囲内

3. The potential aspects lie **outside the scope of** this paper.
   可能的側面については，この論文の範囲外である。
4. The potential aspects lie **beyond the scope of** this paper.
   可能的側面については，この論文の範囲外である。
5. Although it was not **within the scope of** this research to ascertain the

factors which inhibit nitrite production, the results of the study have incidentally seemed to substantiate the research of others.
亜硝酸の生成を抑制する要因を確認することはこの研究の範囲外であったけれども，この研究結果は偶然にも他の研究についても実証したようである。

- **broaden the scope of .....**：…の範囲を拡大する

6. Additionally, some programs are beginning **to broaden the scope of** sampling from a single type of waterbody such as a lake or stream to entire watersheds, including land use monitoring.
さらに，一部のプログラムは，湖沼や河川といった１種類の水域から土地利用モニタリングといった流域全域までサンプリング範囲を拡大している。

Section：区間；部；セクション；節；条

- **A section of .....**：…の区間

1. **A 26-mile (42-km) section of** the river was chosen for the studies.
この川の 26 マイル (42-km) 区間がこの研究のために選ばれた。
2. EPA should have tested the hypothesis of no differences in biological performance for reference sites on impounded and free-flowing **sections**.
EPA は，貯水部および流水部の参照地点について生物実績差に関する仮説をテストすべきだった。

- **This section deals with .....**：この節は…を取り扱う
- **The section illustrates .....**：この章では…を示す
- **The section describes .....**：この章では…を述べる
- **This section is subject to .....**：このセクションは…の適用を受ける
- **Section X recordified .....**：X 条は…を成文化した
- **Section X expanded .....**：X 条は…を拡大した
- **Section X outlines .....**：セクション X では…を概説する
- **In Section X, we consider .....**：
  セクション X では，我々は…について考察する
- **Section X will proceed to .....**：セクション X では，続けて…する

3. **This section deals with** soil samples taken during the drilling of monitoring wells.
この節は，モニタリング用の井戸を掘る時に採集された土壌サンプルを取り扱っている。
4. **The following section illustrates** both the various differences and the

many similarities of existing and emerging biological criteria programs.
次の章では，既存・新規の生物クライテリアプログラムにおける様々な相違点と多くの類似点を示すことにする。

5. **The following section describes** current state efforts in biological criteria development and illustrates both the various differences and the many similarities of existing and emerging biological criteria programs.
次に生物クライテリアにおける現行の諸州での取り組みを述べ，既存・新規の生物クライテリアプログラムにおける様々な相違点と多くの類似点を示すことにする。

6. **This section is subject to** revisions based on additional information.
このセクションは追加情報に基づいて修正の適用を受ける。

7. **Section** 303 of the 1972 act **recordified** and expanded the water quality standards provisions of the 1965 law.
1972年法の303条は，1965年法の水質基準規定を再び成文化して拡大した。

8. **The purpose of this section is to** present a description of some of these modifications and to illustrate the different directions many programs have taken.
本項の目的は，これらの修正の一部を述べることと，多くのプログラムがとった異なる方向性を示すことである。

- ~ **is discussed in subsequent section** ：~は次節で論じられる
- ~ **is given in section X** ：~を第X節に掲げる
- ~ **is presented in section X** ：~を第X節に掲げる
- ~ **is stated in Section X** ：~は第X節に述べられる

9. As will be discussed **in subsequent section**, ...
次節で論じられるように，…

10. As discussed **in the previous section**, the status of biological criteria programs across the United States can be thought of as the sum of all the states in various stages of developing and implementing biological criteria.
前節で述べたとおり，全米での生物クライテリアプログラムの現状は，生物クライテリア設定・実施の様々な段階における諸州すべての総和として考えうる。

11. A detailed discussion of these results **is given in Section** 2.3.
これらの結果について，後ほど2.3節で詳述する。

12. An evaluation of the methodology in light of the methodology application **is presented in section** 6.
方法論の評価は，この方法論のアプリケーションを考慮に入れて第6節に掲げられている。

13. The objective of the CWA, **stated in Section** 101 (a), is "to restore and

maintain the chemical, physical, and biological integrity of the Nation's waters."
101条(a)で述べられているCWAの目標は，「米国の水域の化学的・物理的・生物学的な生物完全度を回復・維持すること」である。

- **as mentioned in the preceding section**：前章で述べたとおり

14. Another reason for using single-purpose frameworks, **as mentioned in the preceding section**, stemmed from the belief that .....
前章で述べたとおり，単一目的枠組みを用いる別の理由は … という思考にあった。

- **under Section X**：第X条のもとで

15. Criteria developed **under Section** 304(a)(1) could be viewed as criteria to address the effects of pollutants, while criteria under Section(a)(2) address the broader effects of pollution.
304条(a)(1)のもとで設定されたクライテリアは，汚濁物質の影響を扱うものとみなされ，304条(a)(2)のもとでのクライテリアは，汚濁の影響を扱うものとみなされる。

16. Activities requiring a permit **under Section** 404 of the CWA must be certified as meeting provisions of the water quality standards by the state water quality agency.
CWA第404条のもとで許可を要する諸活動は，州水質担当機関によって水質基準の規定を満たしていると認定されなければならない。

Seem (Seemingly)：
のように思われる；であるらしい（かのよう；見たところでは）

1. Natural chelators of plant or microbial origin **seem** more promising.
植物や微生物が起源の自然のキレート化剤がいっそう有望に思われる。
2. The general principles used in defining metrics **seem** consistent over wide geographic areas.
メトリック確定に用いられる一般原理は，広範な地区に共通なようである。

- **There seems to be** .....：… があると思われる
- **It seems likely that** .....：… はありそうに思われる
- **it seems wise to** .....：… する方が賢明だろう
- **What seems to be lacking is** .....：欠けていると思われることは … である

3. **There seems to be** no established theory to explain this phenomenon.
この現象を説明する定説はないとおもわれる。

4. **There seems to be** no similarity between the trends in arsenic profiles and the trends in solid-phase sulfate profiles.
ヒ素プロフィールの傾向と固相の硫酸塩プロフィールの傾向との間には類似性があるように思われない。
5. **It seems likely that** such a consistent pattern of contamination would result from in situ conditions rather than migration.
このような一貫した汚染パターンは，移動条件というより現場の条件に起因するように思われる。
6. If conditions are similar among regions, or if the differences can be associated with management practices that have a chance of being altered, **it seems wise to** combine the regions for the purposes of establishing biocriteria.
仮に，各地域における状態が似ている場合や，その差違が可変性のある管理慣行に関連している場合には，生物クライテリア設定のために諸地域を組み合わせた方が賢明だろう。
7. **What seems to be lacking is** a consideration of the ecological background rather than water chemistry background.
欠けていると思われることは，水化学の背景より生態学の背景を考慮することである。

- **seemingly**：…かのよう

8. The pH dramatically increases with depth with a **seemingly** corresponding increase in calcium and magnesium.
pHは深さとともに劇的に上昇し，カルシウムとマグネシウムの増加と対応するかのようである。

Seen：見える

1. Two exceptions with cadmium and sodium **are seen** in the data.
データに見られる2つの例外はカドミウムとナトリウムである。

- ～ **is seen to be** .....：～は…に見える
- **As we have seen**：上述のとおり

2. The overall removal rate **is seen to be** .....
全体的な除去速度は…に見える。
3. The results plotted in Fig. 6 **may be seen to be** in general agreement with the theory.
図6での作図の結果は，一般的には理論と一致しているかのように見える。

4. **As we have seen**, most states are developing bioassessment methods in advance of implementing biological criteria programs.
上述のとおり，大半の州は，生物クライテリアプログラムの実施に先立って生物アセスメント方法を案出中である。

- **it can be seen that .....**：…であることがわかる

5. From Fig. 2, **it can be seen that** the magnitude of the effect is erratic.
図2から，この影響の大きさが不規則であることがわかる。

6. From these reactions, **it can be seen that** trisodium phosphate yields more sodium hydroxide than disodium phosphate.
これらの化学反応から，リン酸三ナトリウムはリン酸二ナトリウムよりも多くの水酸化ナトリウムを生じさせることがわかる。

- **as can be seen from .....**：…から見ることができるように
- **as seen from Figure N**：図Nから見られるように

7. Model predictions show reasonable agreement **as can be seen from** both Figures 2 and 3.
図2と3の両方から見られるように，モデルによる予測は妥当な合意を示している。

8. **As can be seen from** the figure, the slopes of the lines are parallel and constant.
この図から見られるように，線の勾配は平行ならびに一定である。

9. **As seen from** Table 4, the BIOMODEL estimates of biotic DDTR concentration correspond fairly closely to measured concentrations.
表4から見られるように，BIOMODELによって推定された生体内のDDTRの濃度は測定された濃度にかなり近い。

Select (Selecting)：選択する；選ぶ（選定すること）

- **〜 is selected**：〜は選ばれる

1. A total of 27 stations **were selected** to encompass a wide range in sedimentary and trace metal gradients.
堆積性で微量な金属濃度の勾配にて広範な範囲をカバーするように，合計27の地点が選ばれた。

2. Candidate models **are selected** based on knowledge of aquatic systems, flora and fauna, literature review, and historical data.
候補のモデルは，水生系に関する知見，植物相と動物相，文献査読，史的データ

に基づいて選ばれる。
3. The local study areas **were selected** on the basis of evidence suggesting that they are areas of relatively intense geochemical activity.
局地的研究エリアは，比較的激しい地質化学的活動のエリアであることを示唆している根拠をもとにして選択された。
4. The best and most representative sites for each stream class **are selected** and represent the best set of reference sites from which the reference condition is established.
各河川等級について最善かつ最も代表的な地点が選ばれて，参照地点セットにされ，それに基づき参照状態が定められる。

 ● **to select** ..... : … を選択するため

5. The results of this assessment are used **to select** the remedial action and to determine risk-based cleanup levels for the remedial action.
このアセスメントの結果は，復元活動を選択するため，そしてリスクにもとづいて浄化レベルを決めるために使われる。
6. It may be necessary **to select** reference sites from another class of stream, lake, or wetland if all sites in a particular class are highly disturbed.
ある等級における地点すべてがきわめて攪乱されていれば，別の等級の水流，湖沼，湿地から参照地点を選ぶ必要もあろう。

 ● **selected** ..... : 選定の …

7. Core samples were collected at **selected** locations throughout the study area.
コアサンプルが，研究エリア全域の選定地点にて収集された。
8. This study is an attempt to identify the effect of **selected** inorganic ions on a fixed-bed catalyst.
この研究は，選定の無機イオンが固定ベッド触媒に及ぼす影響を確認する試みである。
9. The experiments described in this article concentrated on the leaching characteristics of **selected** arsenic-bearing wastes.
この論文に記述された実験は，選ばれたヒ素を含む廃棄物の浸出特性に集中された。
10. It is relatively easy to read off the proportion of sites that are above or below a **selected** score.
選ばれたスコアを上回る地点もしくは下回る地点の比率を読みとることは，比較的容易である。
11. A measure of how much the quality can be improved can be derived through changing management practices in **selected** watersheds where associations

were determined.
水質がどれほど改善されうるかを示す尺度は，関連性が決定された流域における管理慣行の変更を通じて得られるだろう。

- **selecting**：選定すること

12. The primary concern **in selecting** the two sites is to assure that the physical characteristics are as similar as possible.
2 地点の選定における主な関心事は，物理的特性ができる限り似るよう努めることである。
13. When **selecting** these sites one must account for the fact that minimally disturbed conditions often vary considerably from one region to another.
これらの地点を選ぶ際には，最小擾乱状態が地域ごとに大幅に異なりうるという事実を考慮しなければならない。
14. Whenever possible, base the area on natural vs. political, boundaries to maximize applicability and the probability of **selecting** the least disturbed sites possible.
可能なら，常に最小擾乱地点選定の可能性および確率を最大限に高めるため，自然的境界および政治的境界に基づくべきである。

Selection：選択

1. These people may know of sites that meet some or most of the site **selection** criteria.
これらの人々は，サイト選択基準を幾分または大部分満たすサイトを知っているかもしれない。
2. Forbes' insight and application of **the principle of natural selection** led to the establishment of a biological station on the shores of the Illinois River in 1894.
1894 年に Illinois 川の岸に生物監視測定所が設置されたが，Forbes の洞察と自然淘汰原理の応用が，その導因となった。
3. In extensively disturbed regions and uniquely undisturbed regions, **the method of reference site selection** will likely be less of an issue because of relatively homogeneous conditions.
はなはだしく擾乱された地域や，珍しく擾乱されなかった地点では，比較的同質な状態の故に，参照地点選定方法がさほど問題にならないだろう。
4. The uncertainty associated with the definition of probable conditions and the consequences of a false positive or false negative error has been considered as part of the decision **selection process**.

ありそうな条件の定義と虚偽の肯定的であるか，あるいは虚偽の否定的なエラーの結果と結び付けられた不確実性は決定セレクションプロセスの一部であると考えられました。

● **selection of** ..... ： … の選択

5. The **selection of** the appropriate criterion depends on the content of the database.
適切なクライテリアの選定は，データベースの内容に左右される。
6. The criteria for **selection of** sites included the following: 1) ....., 2) ....., 3) .....
サイトの選択についてのクライテリアは，次のことを含んでいた。1)…, 2)…, 3)…。
7. Intensified research may be needed to enable **proper selection of** target organisms.
目標生物の適切な選択を可能にするためには，強化研究が必要とされるかもしれない。
8. This is one reason that the **selection of** only the most pristine sites as references is inadvisable.
これは，最も無傷な地点だけの選定が勧められない理由の一つである。
9. The **selection of** reference sites from which attainable biological performance can be defined is a key component in deriving numerical biological criteria.
達成可能な生物学的実績が定義されうるような参照地点の選定は，数量的生物クライテリアを導出する際の重大な構成要素である。

Sense：観念；認識力；センス

1. This is probably true **in the sense that** .....
これは … という意味でおそらく本当であろう。
2. Among all the techniques of waste management, waste reduction is the **common sense** solution to the prevention of future hazardous waste problems.
廃棄物マネージメントのすべてのテクニックのなかで，廃棄物削減は将来の危険廃棄物問題の防止の常識的な解決策である。

Separate：分ける；隔離する

1. If **separate** habitats are sampled, it is important to keep them separate for subsequent analysis.

隔離生息場所でサンプリングするならば，その後の分析でも隔離することが肝要である。
2. In some cases the site-specific habitat attributes are used to help **separate** where the transition from one ecoregion to the other takes place.
一部の場合，地点特定生息地属性を用いて，ある生態地域から他の生態地域への推移発生場所が分けられる。

Sequence (Sequential)：連続（一連の；連続して起こる）

- **a sequence of .....**：…の順序
- **in the sequence of .....**：…の順

1. We hypothesize that **a** specific **sequence of** events will occur in response to chronic nitrogen amendments.
長期にわたる窒素の土壌改良に応じて，複数のイベントが特定の順序で起こるであろう，と我々は仮定する。
2. Among the metal studied, the preferred order of uptake by activated sludge was found to be **in the sequence of** lead > copper > cadmium > nickel.
研究された金属の中で，活性汚泥による吸着の優先オーダーは鉛＞銅＞カドミウム＞ニッケルの順であることが判明した。

- **sequential**：一連の

3. Exposures A–C represent three **sequential** 5-week exposures.
曝露Aから曝露Cまでは，一連の5週間曝露を表している。

Series：一連；連続

- **a series of .....**：一連の…

1. **A series of** laboratory experiments were performed to assess factors that .....
…の要因を評価するために一連の室内実験が行なわれた。
2. **A series of** experiments were initiated in an attempt to obtain results similar to run #1.
実験#1に類似する結果を得る試みで一連の実験が始められた。
3. **A series of** batch tests was carried out to test the hypothesis that the same enzymes are used to attack chlorophenols with the same position-specific dechlorination reactions, resulting in competition between these compounds.

位置特定の同じ脱塩素化反応において，同じ酵素が塩化フェノールを攻撃するために使われ，これらの化合物の間に競合反応をもたらす，という仮説をテストするために一連のバッチ試験が行われた。

4. Virtually all of the petroleum industry is based on **a series of** catalytic transformations.
石油工業のほとんどすべては，一連の触媒変換に基づいている。

5. The surveying of ambient biota takes the form of either coordinated monitoring networks or **a series of** special studies.
環境生物相に対する調査は，調整された監視網もしくは一連の特別研究の形をとる。

6. In 1970–1971, **a series of** studies by the U.S. EPA (1973) were undertaken to document and characterize the discharge of tailings to Whitewood Creek.
1970から1971までに，Whitewood Creekへの選鉱廃物の排出を文書化し，特徴づけするための一連の研究がU.S.EPA(1973)によって着手された。

Serve：役となる；役目をする

1. This report **should serve to** give the reader an idea of what is involved in testing a model.
この報告書は，モデルをテストするのに何が関与するか，という考えを，読者に与えるのに役立つはずである。

2. Although it can be argued that full protection of aquatic life requires the use of biological criteria for all five activities, the implementation of biological criteria in any one area **serves to** enhance the state's water quality program.
水生生物の完全保護には，5つの活動すべてについての生物クライテリアが必要だとの主張もありうるが，いずれか1領域での生物クライテリアの実施でもその州の水質プログラムを強化できる。

3. To "**serve** the purposes of the Act," water quality standards must address chemical, physical, and biological integrity of the nation's waters.
「本法の目的に適う」ためには，水質基準が米国の水域の化学的・物理的・生物学的な完全度に取り組まなければならない。

● ～ **serve as** ……：～は…として役立つ

4. This chapter is not intended to **serve as** a comprehensive synopsis of state activities.
本章は，州活動の総合的概要を述べることを意図していない。

5. The information **could** also **serve as** a basis for developing numerical models.

情報は，数値モデル設定の基礎にもなる。
6. This report **will serve as** a basis for RFI work plans developed by ERP.
この報告書は，ERPによって作成されるRFIワークプランの基盤となるであろう。
7. Ideally, reference sites for estimating attainable biological performance should be as undisturbed as possible and be representative of the watersheds for which they **serve as** a basis models.
理想的には，達成可能な生物学的実績を推定するための参照地点は，できる限り無攪乱であるべきで，それがモデルを務める流域の代表となるべきである。

Set（Setting）：セット；設定する（背景）

1. The two **data sets** are generally in good agreement.
この2つのデータセットは，一般に良い合意にある。
2. Streams within each **set** should be similar to one another regarding "relative disturbance".
各セットにおける河川は，「相対攪乱」に関して似ているべきである。

- **set of .....**：…のセット

3. A second decision is to define the **set of** sites to which a criterion applies.
第二の決定は，クライテリアが適用される地点セットを定めることである。
4. Upon completion of the initial revision of ecoregions and delineation of subregions, **sets of** reference sites are identified for each subregion.
生態地域の初回改良，ならびに小地域の画定の完了時に，各小地域について参照地点セットが確認される。
5. The best and most representative sites for each stream class are selected and represent the best **set of** reference sites from which the reference condition is established.
各水流等級について最善かつ最も代表的な地点が選ばれて，参照地点セットにされ，それに基づき参照状態が定められる。

- **setting**：背景

6. Key information about the environmental **setting**, characteristics of the receiving waterbody, and insights into the chemical/physical dynamics of the discharge(s) are either directly or indirectly reflected by the biota.
環境背景に関する主要情報，受入水域の特性，そして排出の化学的／物理的動態に関する洞察は，生物相によって直接的または間接的に反映される。

★ Setの動詞としての使用例文を下に示します。

7. Additional work is needed to more precisely define and **set** expectations for these two metrics, and to test their usefulness in field applications.
これら2つのメトリックについて期待値をより正確に設定し，現地適用におけるその有用性を試すため，さらなる研究が必要とされる。

- **set forth**：定める

8. Sample surveys are an efficient way to meet many of the reporting requirements about the condition of lakes and streams **set forth** in the Clean Water Act.
サンプル調査は，Clean Water Act（水質汚濁防止法）で定められた湖沼・河川の状態に関する報告要件の多くを満たすための効率的な方法である。

Several：いくつかの

1. **Several** new approaches make periphyton analysis a better indicator than it was 5 or 10 years ago.
いくつかの新手法のおかげで，付着生物分析は，5～10年前よりも優れた指標となっている。

2. **Several** possible mechanisms may account for the above observations.
いくつかの可能なメカニズムによって上の観察結果が説明されるかもしれない。

3. **Several** wells exceeded the Primary Drinking Water Standards proposed by the EPA in arsenic and sulfate.
いくつかの井戸で，EPAによって提案されている主要な飲料水基準が，ヒ素と硫酸塩で越えていた。

4. Over the past few years, **several** studies have been made on groundwater contamination by VOCs.
過去数年にわたって，VOCsによる地下水の汚染についての研究がいくつかなされてきた。

5. It was considered appropriate for this study for **several** reasons.
それはいくつかの理由で，この研究に適切であると思われた。

6. There are **several** questions that arise from these observations of arsenic levels in surface waters: 1....., 2.....,
表流水のヒ濃度の観察からいくつかの質問が生ずる。1 … 2 …

7. Performance standards offer **several** advantages that will perpetuate their use in water resource management.
実績基準には，水資源管理においてそれらの利用を永続化させるといった利点がいくつかある。

8. Results are presented in **several** forms, but are normalized to allow direct

comparison with other published work.
結果がいくつかの形式で提出されているが，標準化されているために発表されている他の研究結果との直接比較を可能にしている。

- **several of .....** ：…のいくつか

9. The idea being brought forward in this research is to measure **several of** these general parameters.
この研究における前進的考え方は，これらの一般的なパラメータのいくつかを測定するということである。
10. The preceding discussion of the analysis of the capabilities of biological survey data to discriminate different types of impacts is contrary to **several of** the assertions of Suzuki(1999)in a critique of community indices.
生物学的調査データが様々な影響種類を弁別できる能力への分析に関する以前の論考は，生物群種指数を批判した Suzuki(1999)の主張の一部に反している。

- **several years** ：数年
- **several decays** ：数十年

11. Data exist from **several years** of studies that support these conclusions.
数年間の研究からなるデータが存在し，この結論を支持している。
12. The environmental concern with endocrine disruptors has been growing for the last **several years**.
ここ数年の間，内分泌かく乱物質について環境的関心が高まってきている。
13. At this rate, low-pH conditions will reach the bottom of the Nordic Main tailings within **several decays**.
この速度では，pH が低い条件が数十年以内に Nordic Main 選鉱屑層の底に達するであろう。
14. High iron low-pH water will eventually occupy the entire tailings mass, probably within **several decades** or a century.
おそらく数十年あるいは 1 世紀以内に，高濃度の鉄分を含有し pH が低い水が結局は選鉱廃棄物中全部を占めることになるでしょう。
15. Under these conditions, **several thousand years** would be required for removal of entire gypsum content of the tailings.
これらの条件の下では，選鉱くずに含まれる石膏の全部を撤去するには，数千年が必要とされるであろう。
16. In contrast to the relatively low public concern about periphyton, they have been used extensively in the analysis of water quality for **several decades**.
付着生物は，一般市民からの関心が比較的低いのに比して，数十年にわたり水質分析で盛んに用いられてきた。

Short：短い；短期の

1. There is a **short** but good review by .....
   … による短いが，すぐれたレビューがある。
2. **Short** term studies should be undertaken to determine whether constituents are being sampled adequately.
   構成要素が適切にサンプルされているかどうかを決めるため，短期の研究がおこなわれるべきである。
3. This discrepancy between the samples collected two years apart suggests that significant variations in the magnitude of soil contamination can occur either over very short horizontal distances or over a relatively **short** time period.
   2年間隔てて収集されたサンプルの間でのこの矛盾は，土壌汚染度の有意な差違が非常に短い水平距離で，あるいは比較的短期間にわたって起こりえることを示唆する。
4. In day-to-day activities of a regulatory agency it is important to point out to client (e.g., dischargers) that short stretches of modified stream do not preclude application of stringent water quality rules.
   規制機関の日常的活動において，（排出者など）クライアントに対し，改変河川の短区間は，厳しい水質ルールの適用を阻まないことを指摘することが重要である。

   ● **In short**：簡潔に言えば；要するに

5. **In short**, the discovery has considerable environmental significance.
   簡潔に，この発見はかなり環境的な重要性をもつ。

Show (Showing)：表す；示す (提示すること；示すこと；表すこと)

1. COD **showed** an increasing tendency with increasing TSS.
   TSS が増えるとともに，COD が増加する傾向を示した。
2. The relationship between cellular copper content and cupric ion activity **shows** good agreement with the hyperbolic function.
   細胞の銅含有量と銅イオン活性の間の関係は，双曲線関数で示すと良く一致する。

   ● ～ **show that** .....：～は… ということ表している

3. This clearly **shows that** .....
   これは … ということを明らかに表している。
4. The calculation of the t-test **showed that** this change was not statistically

significant at the p = 0.05 level
t検定の計算によれば，この変化は，p = 0.05 レベルで，統計的に有意でない。
5. The comparison of the two different time period **showed that** the increased index scores for the later period were highly significant ($p < 0.0001$).
これら2つの時期の比較は，最新時期の指数スコア増大がきわめて有意なことを示した ($p < 0.0001$)。

● **This study has shown that** .....：この研究は … を示した

6. This research **has shown that** .....
この研究は … を示した
7. Work **has shown that** ....
この研究は … を示した

● **Figure N shows** .....：図 N は … を表す

8. **Table 2 shows** in part that .....
表2は一部 … のことを示す。
9. **Figure 3 shows** the variation of oxalic acid concentration with irradiation time.
図3は，照射時間でのシュウ酸の濃度変化を表す。

● **results show that** .....：結果が … を示す

10. The **result clearly shows that** .....
その結果は … ということを明らかに示している。
11. **Our results show that** the reactivity is additive for the concentration range studied.
我々の結果は，研究された濃度の範囲では反応性が加算的であることを示している。
12. **Results** summarized in Table 1 **show** that .....
表1で要約した結果が，… のことを示している。
13. **Results** from this study **showed** more information would be obtained by sampling more individual streams than more sites on the same stream.
この研究結果によれば，同じ河川の多数地点でのサンプリングよりも，多数の河川でのサンプリングの方がより多くの情報が得られる。

● ～ **is shown**：が示されている

14. The results **are shown** in Figure 3.
この結果は図3に表す。
15. The results of investigations on three different heavy metals **are shown** in Fig. 9.

3つの異なった重金属についての調査の結果は，図9に表す。
16. The derivation of the impairment for species richness **is shown** as an example in Table 1.
生物種豊富度についての損傷クライテリアの導出は，表−8.1に例示している。
17. The variability inherent to each of the three biological indices used by USEPA **has been shown** to be quite low and within acceptable limits at relatively undisturbed sites.
USEPAが用いている3つの生物学的指数それぞれに内在する可変性は，比較的無攪乱な地点ではきわめて低く，許容限度内であることが示されている。

- **as shown by .....**：…によって示されたように

18. Sampling time can be as rapid as 20 min/site, **as shown by** the methods used in Montana (Sano, 1993).
サンプリング時間は，モンタナ州での方法で示されたとおり(Sano, 1993)，1地点当り20分間で済む。

- **showing**：提示すること；示すこと；表すこと

19. Case histories are presented below **showing** the outcome of impairment criteria testing in a variety of situations.
次に，様々な状況での損傷クライテリア検定の結果を示した事例史を述べる。

Significance：重要性

1. The discovery has considerable environmental **significance**.
この発見は，かなり環境において重要性をもつ。
2. The magnitude and **significance of** chemical contamination of aquatic environments are increasingly evident.
水生環境の化学汚染の規模と重要性がますます明かになっている。
3. The model results underscore the **significance of** chemical partitioning on chemical fate and highlight the importance.
モデルの結果は，化学物質の衰退におけるその物質分配性の意義を強調し，そしてその重要性を重視する。

Significant (Significantly)：重要な(際立って)

1. A basic assumption for the model was that biofilm diffusion limitations **were** not **significant**.
このモデルの基本的な仮定は，生物膜拡散限界が重要ではないということであっ

2. All results have been rounded to no more than three **significant** figures.
すべての結果は，有効数字三桁より端数を無くした。
3. Competitive adsorption between water vapor and a probe contaminant can have a **significant** influence on the oxidation rate of the contaminant.
水蒸気と調査汚染物質の間の競合性吸着が，汚染物質の酸化速度に重要な影響を与え得る。

- **significantly**：際立って

4. The flow did not increase **significantly** in this segment of Blackwood Creek.
Blackwood Creek のこの区間では流れはそれほど増加しなかった。
5. The six-week soil TPH concentrations were **significantly** lower than the initial soil TPH concentrations at the 0.025 probability level.
6 週間後の土壌 TPH 濃度は，0.025 の確率レベルにおいて最初の土壌 TPH 濃度より有意に低かった。

Similar：類似の；よく似た

1. Fish metrics offer **similar** information.
魚類メトリックは，似たような情報を提供する。
2. A **similar** problem is encountered in estimating pesticide model parameters.
殺虫剤の環境モデルのパラメータを推定する過程にて，類似の問題が生じた。
3. **Similar** concentration profiles were obtained for the other sands that were tested.
他の被試験砂では，類似の濃度プロフィールが得られた。
4. The data show a **similar** pattern of distribution of heavy metals at each site.
これらのデータは，各地点において類似の重金属分布パターンを示す。
5. Benthic metrics have undergone **similar** evolutionary developments.
底生生物メトリックは，似たような展開をたどってきている。
6. Thus streams draining comparable watersheds within the same region are more likely to have **similar** biological, chemical, and physical attributes than from those located in different regions.
故に，同じ地域内の類比可能な流域を流れる河川は，異なる地域に所在する水流に比べ，類似の生物的・化学的・物理的属性を持つ見込みが高い。

- ～ **is similar to** …..：～は … に似ている；～は … に類似している

7. This **is similar to** observations that we have made in the Scioto River

downstream from Columbus, Ohio, and elsewhere.
この観察結果は，オハイオ州 Columbus 市の Scioto 川下流および他の場所における筆者の観察結果と似通っている。

8. A series of experiments were initiated in an attempt to obtain results **similar to** run #1.
実験 #1 に類似する結果を得る試みで一連の実験が始められた。

9. Ash-free dry mass (AFDM) of periphyton samples provides an estimate of standing crop **similar to** chlorophyll a.
付着生物サンプルの無灰乾重量 (AFDM) は，クロロフィル a に似た生物体量推定値をもたらす。

10. Reference streams draining watersheds that are completely within a particular region tend to **be similar to** one another when compared to reference streams in adjacent regions.
特定地域内に存在する参照河川流域は，隣接地域における参照河川に比べ，類似傾向にある。

11. Students' samples were found to **be very similar to** those collected and analyzed at the same sites by professionals at the Alberta Environment, Water Quality Branch.
高校生によるサンプルは，その地点でアルバータ州環境・水質部の専門家が採集・分析したものとよく似ていた。

12. By analogy with the oxidation of formic and oxalic acids, it is reasonable to assume a second-order reaction for glyoxylic acid because it has a structure **similar to** the other two acids.
ギ酸とシュウ酸の酸化反応が類似していることから，グリオキシル酸の反応は第 2 次反応であると仮定するのが道理的である。なぜなら，それは他の 2 つの酸に構造が類似しているからである。

● **~ is similar among .....** ：～は … において似ている

13. If conditions **are similar among** regions, or if the differences can be associated with management practices that have a chance of being altered, it seems wise to combine the regions for the purposes of establishing biocriteria.
仮に，各地域における状態が似ている場合や，その差違が変化可能性ある管理慣行に関連している場合には，生物クライテリア設定のために諸地域を組み合わせた方が賢明であろう。

● **as similar as possible** ：できる限り似る

14. The primary concern in selecting the two sites is to assure that the physical

characteristics are **as similar as possible**.
2地点の選定における主な関心事は，物理的特性ができる限り似るよう努めることである。

- **~ is more similar than …..**：～は…よりも類似している

15. All things are somewhat different, yet some things **are more similar than** others, offering a possible solution to the apparent chaos.
あらゆる事物は，それぞれ若干異なるが，あるものは他のものよりも類似性が強く，見かけの混沌に対する潜在的解決策となる。

- **In a similar manner**：同様に

16. **In a similar manner**, it may be necessary to select reference sites from another class of stream, lake, or wetland if all sites in a particular class are highly disturbed.
同様に，ある等級における地点すべてがきわめて擾乱されていれば，別の等級の河川，湖沼，湿地から参照地点を選ぶ必要もあろう。

Similarity：類似性

1. Patterns in Major Basins and USGS (USGS, 1982) **have no similarity to** patterns of ecoregions.
大流域とUSGS水文単位（USGS，1982）におけるパターンは，生態系地域パターンと類似していない。

2. These regions generally **exhibit similarities** in the mosaic of environmental resources, ecosystems, and effects of humans and can therefore be termed ecological regions or ecoregions.
これらの地域は，一般的に環境資源，生態系および人為的影響のモザイクにおいて類似性を示すので，生態学的地域，もしくは生態地域と呼ばれる。

3. The problem with using this type of framework for geographic assessment and targeting is that it does not depict areas that correspond to regions of similar ecosystems or even regions **of similarity** in the quality and quantity of water resources.
この種の枠組みを地理アセスメントおよびターゲット設定に用いるうえでの問題点は，類似生態系の地域や，水資源の質・量において類似な均一地域に符合する区域を描かないことである。

4. While it is recognized that individual waterbodies differ to varying degree, the basis for having regional reference sites is the **similarity of** watersheds within defined geographical regions.

個々の水域が様々に異なることは認識されるものの，地域参照地点を設けることの基礎は，所定の地理的地域内での流域の類似性にある。

● similarity between ..... : … の間の類似性

5. There seems to be no **similarity between** the trends in arsenic profiles and the trends in solid−phase sulfate profiles.
 ヒ素プのロフィールの傾向と固体層の硫酸塩プロフィールの傾向の間には類似性がないようには思われる。

Simple : 単純な；簡単な

1. The model development approach can be straight−forward and relatively **simple**.
 このモデルの構築手法は単純で，そして比較的簡単であり得る。
2. Although the concept **appears simple** on the surface, it can be deceptively complex in application.
 この概念は，一見，単純に思えるが，適用の際には意外に複雑であるかもしれない。

● simple ..... : 単純な …

3. **No simple** pattern could be perceived which related nitrification.
 硝化反応に関連する単純なパターンは，認知され無いであろう。
4. These models of **simple** aqueous systems cannot approach the complexity of natural sediment−water systems.
 これらの単純な水性システムモデルは，自然の沈降堆積物−水システムの複雑さに迫ることが出来ないない。
5. If a **simple** relationship could be established, it would be of great in planning the future direction of any development work on a new chemical.
 もし単純な関係が確証されえるなら，いかなる新化学物質開発研究の将来方針を計画するにあたって，それは偉大なことであろう。
6. The present research combines sonication with commercially available probes and offers **a simple approach** toward field monitoring.
 現在の研究はソニケーションを商業的に利用可能なプローブと組み合わせ，現地モニタリング用の単純な手法を提供する。
7. This challenge resulted in a search for numerical expressions in a form **simpler** to understand than long species lists and well−thought but lengthy technical expressions of the data.
 この課題により，生物種の長大なリストや，綿密だが長々しいデータの技術的説明よりも理解しやすい形での数値表現が追求されるようになった。

- **~ is as simple to use as possible**：～の使用ができる限り簡単である

8. In contrast to the complexity of the processes being modeled, the model itself **must be as simple to use as possible**.
モデル化されているプロセスの複雑さとは対照的に，モデルそれ自身の使用ができる限り簡単でなければならない。

Situate (See also **Locate**)：位置する（→ **Locate**）

1. The Blue River study site area **is situated** on the northeastern periphery of the Green Hills uplift.
Blue River 研究地点は，Green Hills 隆起地域の北東周囲に位置している。
2. The station **is situated** on a hill approximately four miles south of a small town (population: about 3,600) and roughly 1,500 feet from the coastline of a sound.
この施設は小さな町（人口：およそ 3,600 人）のおよそ 4 マイル南にある丘の上に，そして入り江の海岸線からおよそ 1,500 フィートに位置している。

Situation：状態

1. In our experience the following are the **situations** where conflicts have arisen in New Mexico.
筆者の経験によれば，ニューメキシコ州で抵触が生じた事態は以下のとおりである。
2. They reflect practical **situation** and as a result present the greatest challenge with respect to modeling.
それらは実務的な状況を反映し，そして結果として，モデリングに関する最大の難問を提示している。

- **in the best of situation**：最良の状態で
- **in a variety of situations**：様々な状況で
- **in field situations**：現場の状況で
- **from many situations**：多様の条件下から
- **for many situations**：多くの事態について

3. **In the best of situation**, it is possible to achieve an annual water balance within five percent.
最良の状態で，年度の水の収支を 5 パーセント中で達成することは可能です。
4. Case histories are presented below showing the outcome of impairment criteria testing **in a variety of situations**.

様々な状況での損傷クライテリア検定の結果を示した事例史が下に述べられている。
5. To predict whether a behavioral response to a chemical pollutant will occur, one must ask whether the organism can detect the pollutant at concentrations likely to be encountered **in field situations**.
化学汚染物質に対する挙動反応が起こるかどうか予測するには，現場の状況で起こりえる濃度で生物が汚染物を検知できるかどうか尋ねなくてはならない。
6. This model is capable of correlating growth data **from many situations** in a satisfactory manner.
このモデルは，多様な条件下からの成長データを満足のいく方法で関連づけることができる。
7. **For many situations** this will only become evident through an iterative process.
多くの事態について，これは，反復過程でもって初めて顕著になるだろう。

- **There are certain situations** ：特定の状態がある

8. **There are certain situations** in which each type is appropriate for use.
それぞれのタイプが使用に適切である特定の状態がある。

Size：大きさ；寸法；サイズ

1. Index scores can be compared among streams of **different sizes**.
指数スコアは異なる規模の河川において比較されうる。
2. Instead of simply counting the number of cells per unit area, one can determine cell biovolume and use it to account for the **differences in sizes** of cells that are enumerated.
単位面積当り細胞数を単に数える代わりに，細胞の生物体積を決定し，列挙される細胞サイズの差をそれを用いて説明しうる。
3. These are then converted to phi values as in Kudo (1962), and mean **particle size** is calculated.
これらは，次にKudo (1962) におけるようなphi値に変換され，粒子サイズ平均が計算される。
4. The coefficient is a function of the **size**, shape, and density of the suspended particles.
この係数は浮遊粒子の大きさ，形，密度の関数で示される。
5. These plots contain **sample size**, medians, ranges with outliers, and 25th and 75th percentiles.
これらの図は，サンプルサイズ，中央値，外れ値，25パーセンタイルおよび75パーセンタイルの値域を含んでいる。

6. The Hocking River, located in Southeastern Ohio, is a **medium-sized** river (1197mile^2 drainage area) of about 100 mi in length.
Hocking 川は，オハイオ州南東部に所在し，長さ約 100mile の中規模河川 (流域面積 1197mile^2) である。

★ **Size** に関連する例文を下に示します。

7. The extent of the contamination is estimated to be approximately 200 feet by 100 feet.
汚染地帯の広さは，およそ 200 フィートかける 100 フィートであると推定される。
8. For bedrock samples, gear usually includes a 3- to 4-in. diameter PVC pipe with a rubber seal that can be held in place where the rock is brushed clean and the sample removed by suction.
床岩サンプルの場合，サンプリング装置には，通常，直径 3～4in のラバーシールド PVC (ポリ塩化ビニル) 管が含まれており，これを設置すれば，岩石を磨いてサンプルを吸引除去できる。

So：そのように；このように

● **so that**：それで；そうすることによって

1. These choices must be made with considerable care and documented **so that** they do not include fundamentally different communities.
こうした選択をする際，根本的に異なる群集が含まれないようにするため，かなり入念に行って文書化しなければならない。
2. In some cases scales are collected **so that** growth rates as well as general size can be calculated and compared.
一部の場合，鱗を採集して成長速度や一般的体長が計算・比較される。
3. Basic laboratory studies are being conducted **so that** from the field data, general conclusions may be drawn and applied elsewhere.
基本的な室内研究が行われている。そうすることによってフィールドデータから，一般的な結論が引き出され，ほかのところにも応用されるであろう。
4. The standard procedure outlined above was established **so that** the problems which often obscure interpretation of results are minimized.
上に概説された標準的な手順は，結果の解釈をしばしば不明瞭にする問題点が最小になるように確立された。

● **so ~ that .....**：あまりに～のため … となる

5. From a biological standpoint, the range of any given class is **so great that** a

criterion established for an expected biological community would be so broad as to be ineffectual.
生物学的な見地から，ある等級の値域があまりに広いため，予想生物群集について定められたクライテリアも広範すぎて無効となる。
6. The HELP ecoregion of Massachusetts has been **so** extensively ditched and drained **that** many of the small streams in this region are incapable of supporting a WWH use.
マサチューセッツ州のHELP生態地域は，徹底的に溝掘りと排水が行われてきたので，この地域の小河川の多くがWWH使用を支えられない。

Sole (solely)：唯一の (のみに；だけで)

1. Biological impairment criteria may be used as the **sole** basis for regulatory action.
生物損傷クライテリアは，規制措置の唯一の根拠となろう。
2. The analysis is based **solely** on the presence or absence of a group and does not take into consideration the abundance of that group.
この分析は，ある集団の有無のみに基づいており，その集団の存在量を考慮していない。

Solution：解；解決策；溶液

1. The focus of this effort is to analyze exhaustively the **solutions**.
この努力の焦点は，徹底的に解決策を分析することである。
2. Equation 3 is the general **solution** describing dispersion and settling in a river.
式3は，河川における拡散と沈降を説明する一般的な解である。
3. Tamaki and Ohkura (2002) have derived the following **solution**.
TamakiとOhkura (2002) は，次の解を得た。
4. Evidence has been gathered, and **solutions** have been proposed.
証拠が集められ，そして解決策が提案されてきた。
5. The acid rain problem in this region is far from **a solution**.
この地域の酸性雨問題は，解決にはほど遠い。
6. The first term on the right-hand side of the equal sign is the general **solution** and the second term is the particular **solution**.
等号の右側の最初の項は一般解で，そして2番目の項は特解である。
7. However, as with most general **solutions**, these equations provide little insight into the solution.

しかしながら，たいていの一般解と同じように，これらの式はその解にほとんど洞察を加えていない．

8. The **solution of** these equations is accomplished using finite difference approximations in identical fashion to the Water Quality Analysis Simulation Program(WASP).
   これらの方程式は，水質解析シミュレーションプログラム(WASP)と同一方法，有限差分近似を使って解かれている．

● **solution to .....**：…に対する解決策

9. All things are somewhat different, yet some things are more similar than others, offering a possible **solution to** the apparent chaos.
   あらゆる事物は，それぞれ若干異なるが，あるものは他のものよりも類似性が強く，見かけの混沌に対する潜在的解決策となる．
10. Among all the techniques of waste management, waste reduction is the common sense **solution to** the prevention of future hazardous waste problems.
    廃棄物マネージメントのすべてのテクニックのなかで，将来の危険廃棄物問題防止の常識的な解決策は廃棄物削減である．

★ **solution**(溶液)に関連する例文を下に示します．

11. **A sample solution** (4 mL) was placed in a batch system.
    サンプル溶液(4 mL)がバッチシステムに移された．
12. **A sample solution of** 65 mL was sonicated in a cylindrical glass vessel of 50 mm i.d. with a total volume of 150 mL.
    内径50 mm，全容量150 mLの円筒状のガラス容器内で，65 mLのサンプル液をソニケートした．
13. Sonoluminescence from **aqueous solutions** has been studied in some detail.
    水溶液からのSonoluminescenceは，いくぶん詳細に研究されてきた．
14. In this study, we investigated the effects of buffers and influence of Na and Ca concentrations of the **test solution**.
    この研究では，我々は緩衝液の効果と試験液のNaとCa濃度の影響を調査した．
15. It is likewise unclear how well limestone reactors will neutralize more highly **concentrated solutions**.
    石灰岩反応器が，高濃度の溶液をどれほどよく中和するかは同じく不明確である．
16. We are currently exploring the use of ultrasound in destroying chlorinated hydrocarbons **in dilute aqueous solutions**.
    現在，我々は，希釈水溶液中での塩素化炭化水素を分解するのに，超音波の使用

を探究している。

17. The **feed solution** was prepared according to the medium described in Ref 8 for nitrifying bacteria with some minor revisions.
この流入溶液は，硝化菌に関して参照 8 に記述されている培地に従って，若干の修正を加えてつくられた。

18. In the second part of this study, the effects of buffers and influence of Na and Ca concentrations of the **test solution** were investigated.
この研究の 2 番目の部分で，緩衝液の効果と試験液の Na と Ca 濃度の影響が調査された。

19. These processes are affected by the concentration, size, and valence of metal ions on one hand and the redox potential, ionic strength, and pH **of the solution** on the other.
これらのプロセスは，金属イオンの濃度，サイズ，結合価の影響を受ける一方，他方では溶液の酸化還元電位，イオン強度，pH の影響を受ける。．

★ solution（溶液）には，"a solution"，"an aqueous solution"，"the solution" と冠詞が付くか，または "solutions" と複数型で使われます。

Some：いくつか；ある

1. **Some** minor revisions were made in 1985, but the approach remained essentially the same through 1987.
1985 年にいくつか細かな改正が施されたが，この手法は，基本的に 1987 年まで存続した。

2. Nonpoint sources probably have **some** effect on biological performance at many ecoregion reference sites.
面源は，おそらく，多くの生態地域参照地点での生物行動に何らかの影響を及ぼしてきた。

3. **Some** volunteer programs focus on environmental advocacy or on enforcement of environmental permit and pollution deterrence.
一部のボランティアプログラムは，環境擁護または環境許可および汚濁阻止の施行に主眼をおいている。

● **some of .....**：… の一部

4. Although we have presented here a framework from within which numerical criteria can be developed and implemented by states, there remain important areas for future development and research. **Some of** these follow:
筆者は，本章で，数量的クライテリアが州ごとに設定・実施されるための枠組み

を示してきた。だが，今後の研究を要する重要な領域がいくつか残っており，例えば，以下のとおりである。

5. Although every program provides **some** form **of** educational benefits, the extent of educational opportunities varies among the programs, **some of which** exist almost entirely to provide education, **some for which** education is secondary to action or data collection, and others where the objectives of education and data collection are equally important.
   どのプログラムも何らかの啓発効果を有するものの，啓発機会の度合は，プログラムごとに異なっている。例えば，現存プログラムの一部は，もっぱら啓発を提供していたり，啓発が行動やデータ収集の従属的なものだったり，啓発目標とデータ収集目標が同等に重要だったりする。

   ● **in some circumstances**：いくつかの状況では

6. USEPA acknowledges, albeit briefly, that site-specific modifications may be appropriate **in some circumstances**.
   USEPAは，いくつかの状況では地点特定修正が適切であることをごく簡単にだが認めている。

   ● **some are A, others B**：Aなものもあれば，Bなものもある

7. The range of pollution sensitivity exhibited by each metric differs among metrics ; **some are** sensitive across a broad range of biological conditions, **others** only to part of the range.
   各メトリックによって示された汚濁感度値域は，それぞれ異なる。広範な生物学的状態について敏感なものもあれば，ごく一部の値域のみに敏感なものもある。

   ● **to some degree**：ある程度

8. These observations raised several questions which we attempted to answer **to some degree** in the experiments described below.
   これらの観察記録は，下に記述した実験にて我々がある程度答えようと試みたいくつかの質問を提起した。

Somewhat：いくぶん；若干；やや

1. Results from these calculations are **somewhat** uncertain.
   これらの計算からの結果は，いくぶん不確実です。
2. We hope the reader will bear in mind that the real world is **somewhat** different.
   我々は，読者が実世界がいくぶん異なっているということを心に留めておくであ

ろうことを希望する。

3. This is **somewhat** surprising, in view of the order of magnitude higher water column decay rate of DDE relative to PCB.
PCB と比較して，水柱での DDE の桁違いに高い分解速度の見地から，これはいくぶん驚きである。

4. All things are **somewhat** different, yet some things are more similar than others, offering a possible solution to the apparent chaos.
あらゆる事物は，それぞれ若干異なるが，あるものは他のものよりも類似性が強く，見かけの混沌に対する潜在的解決策となる。

5. Field measurements are **somewhat** less accurate than measurements made in a laboratory but they offer the important advantage of providing immediate results to the volunteers.
現地測定は，実験室測定ほど正確でないが，ボランティアに直ちに結果を提示するという重要な利点を持つ。

6. Since the study was intended to demonstrate a methodology, its goals were **somewhat** different than those of most engineering applications.
この研究は方法論を実証するように意図されたため，その目的はたいていのエンジニアリング適用という目的とはいくぶん異なっていた。

7. The use of these two groups is **somewhat** analogous to the use of a fish species and an invertebrate species as standard bioassay test organisms.
これら 2 集団の利用は，標準的バイオアッセイ試験生物としての魚種および無脊椎動物種の利用にやや似ている。

8. The extent of spatial overlap between different impact types throughout the database is **somewhat** variable ranging from a clear predominance of a single impact type to the overlapping influence of two or three impact types.
データベース全体での様々な影響種類の空間的重複の程度は，1 種類の影響の明確な優勢から，2, 3 種類の影響の重複まで，やや異なっている。

Sort：種類；タイプ

● **some sort of .....**：何らかの …

1. Physical characteristics of individuals that may be useful for assessing chemical contaminants would result from microbial or viral infection, **some sort of** tetragenic or carcinogenic effects during development of that individual.
化学的汚染アセスメントに有用だと思われる個体の身体特性は，細菌・ウイルス感染や，その個体の発生中における何らかの催奇形影響または発癌影響に起因する。

Sound ：確かな；信頼できる

1. Our first concern was to test where EXAMS was theoretically **sound**.
   我々の最初の関心は，どのようなサイトでEXAMSが理論的に信頼できたか，テストすることだった。
2. Together, they form the foundation for a **sound**, integrated analysis of the biotic condition.
   これらは，共に生物状態の確かな統合的分析の基礎となる。
3. This chapter introduces the elements of a **sound** survey design and illustrates the techniques for surface waters (lakes and streams) by describing several examples highlighting various components of sample surveys.
   本章では，サンプル調査の様々な構成要素を強調した事例を述べることによって健全な調査設計の要素を紹介し，地表水域（湖沼・河川）についての技法を示すことにする。

● a sound basis ：確かな基盤；効果ある基礎

4. Existing knowledge is too limited to provide **a sound basis** for planning to prevent accelerated eutrophication and its adverse effects.
   既存の知識は，加速する富栄養化とその悪影響の防止計画のための効果ある基礎を提供するにはあまりにも限定されている。
5. To address this problem, a three-year ecoregion project was conducted that characterized the streams within each ecoregion, developed a classification of streams, and provided **a sound basis** for developing realistic water quality standards and beneficial uses within ecoregions.
   この問題点に対処するため，3箇年の生態地域プロジェクトが行われた。それは，各生態地域内の河川の特性を把握し，河川の分類を設定し，生態地域内での現実的な水質基準および有益利用の確立のための確かな基盤を提供した。

Source ：源；出所

1. The heavy metals are another likely **source** of inhibition.
   重金属は，もう1つの抑制源である可能性がある。
2. Mercury found in the environment comes from two major **sources**.
   環境において見いだされる水銀には2つの主要な出所がある。
3. Toxic metals enter waste waters from a variety of **sources**.
   毒性金属がいろいろな源から下水に流入する。

4. A mass balance is constructed taking into account the inflow and outflow and the various **sources** and sinks of constituent.
物質収支は，流入と流出それに種々の構成要素の入源と出源を考慮に入れて構成される。
5. Other **sources** of information include interviews and review of internal correspondence.
他の情報源はインタビュー，そして内部通信のレビューである。
6. Our decision to approach criteria from the standpoint of measuring impairment from a pollution **source** was influenced by two major factors.
汚濁源からの損傷測定の見地でクライテリアに対処するという我々の決定は，2つの要因に影響されていた．
7. Different classes of algae have different proportions of internal structure occupied by vacuoles, which are important **sources of** variation in measurements of biovolume.
異なる綱の藻類は，液胞に占められた内部構造の比率が異なっており，それは，生物体積測定における変動の主因となる。
8. There is insufficient information to determine whether contamination present **at this source area** may present an unacceptable risk.
この汚染源での汚染が，容認出来ない危険性を引き起こすかもしれないかどうか決定するには情報が不十分である。

Speaking：言う；話す

1. **Strictly speaking**, the target subpopulation is the set of streams with bridge crossings.
厳密にいえば，ターゲット部分母集団は，橋梁交差点を持つ河川セットである。

Specific：特定の

1. A **specific** aspect of transport and transformation is examined from a systems perspective.
輸送・変換の特定の側面が，システムの視点から検定される。
2. The few studies that have been done suggest that volunteers are capable of providing good-quality data for **specific** levels of use.
ボランティアが特定の使用レベルについて高質なデータを提供できることを示す研究は，まだあまり存在しない。
3. Unavailability of **site-specific** data for model parameter is a major problem.
モデルパラメータを決めるための地域特定データが得られない状況は，主要な問

題である。
4. In some cases the **site-specific** habitat attributes are used to help separate where the transition from one ecoregion to the other takes place.
一部の場合，地点特定生息地属性を用いて，ある生態地域から他の生態地域への推移発生場所が分けられる。

Specifically：特に；具体的に

1. One area that was **specifically** excluded from this study was human health effects.
この研究から特に除外された1つのエリアは，人間健康への影響であった。
2. EXAMS is **specifically** designed to simulate the fate and persistence of organic chemicals in aqueous ecosystems.
EXAMSは，特に水性生態系での有機化学物質の衰退と持続性をシミュレートするために構成されている。
3. The 1972 act **specifically** required states not only to identify impaired waters, but to rank them in order of priority for cleanup and restoration.
1972年法は，具体的に諸州が被損傷水域を確認することに加え，それらを浄化・回復の優先順位で格づけすることも要求した。
4. We **specifically** discuss benthic macroinvertebrate monitoring since it is most likely to be usable in biological criteria issue.
我々は，特に大型無脊椎動物モニタリングについて論じる。なぜなら，それは，生物クライテリア問題に役立つ見込みがきわめて強いからである。

● **More specifically**：より具体的には

5. **More specifically**, questions such as, "What proportion of the stream miles (and how many stream miles) are affected by point source discharges? Channel modifications? Riparian modifications? Stream bank instability? can be addressed.
より具体的には，「河川距離のどれほどの比率（どれほどの水流 mile 数）が点源排出に影響されるか？ 流路改変は？ 水辺改変は？ 水流岸の不安定は？」といった疑問に対処できる。

Spite (See also Despite)：悪意；恨み (→ Despite)

● **in spite of** .....：…にもかかわらず

1. **In spite of** the extensive application of ultrasound to chemical synthesis,

the mechanism of rate enhancements in both stoichiometric and catalytic reactions of metals remained largely unexplored.
化学合成への超音波の大規模な適用にもかかわらず，反応速度が上昇するメカニズムは，金属の化学量論と触媒反応での両方にて，主として未研究のままである。

Sponsor：後援する

1. This study **was sponsored by** the Office of Water Research and Technology, under grant A–00
   この研究は，水研究技術所からの交付金 A – 007C を受けて成されました。
2. Since that time, USEPA **has sponsored** over 40 workshops on approaches to biological assessments and frameworks for developing assessment approaches.
   それ以降，USEPAは，アセスメント手法を定めるための生物アセスメントおよび枠組みに関して 40 以上ものワークショップを後援してきた。
3. Environ Company **sponsored** a laboratory study of the geochemical behavior of tailings from the Marsh Creek area, with emphasis on the effects of oxidation and the controls on arsenic behavior.
   Environ 社は，Marsh Creek 域からの選鉱くずがおよぼす地球化学的挙動に対する室内研究に出資した。その研究は，ヒ素の挙動に対する酸化効果とコントロールに力点を置いていた。

Stage：段階

- **at this stage**：この段階で
- **in two stages**：2つの段階で
- **in the early stages of .....**：…の初期段階に
- **in various stages of .....**：…の様々な段階において

1. It must be pointed out **at this stage** that .....
   … は，この段階で指摘されなくてはならない。
2. Further investigations into the time course of metal uptake in activated sludge have revealed that this process occurs **in two stages**.
   活性汚泥中で時間とともに変わる金属摂取をさらに調査し，このプロセスが2つの段階で起こることを明らかにした。
3. Many more states are **in the early stages of** developing the biological survey or reference site methods needed for biological criteria.
   さらに多くの州が，生物クライテリアに必要な生物調査方法，または参照地点方

4. It must be recognized that the concept and definition of ecoregions are **in a relatively early stage of** development.
生態地域の概念・定義は，まだ発展初期段階にあることを認識しなければならない。
5. As discussed in the previous section, the status of biological criteria programs across the United States can be thought of as the sum of all the states **in various stages of** developing and implementing biological criteria.
前章で述べたとおり，全米での生物クライテリアプログラムの現状は，生物クライテリア設定と実施の様々な段階における諸州すべての総和として考えうる。

Standpoint：見地；立場

- **from a stand point of .....**：…の見地から；…の立場からみると
- **from a ..... standpoint**：…の見地から；…の立場からみると

1. It would be desirable **from the stand point of** treatment.
それは，処理の見地から望ましいであろう。
2. This reaction is especially important **from a global standpoint**.
この反応は，世界的な見地から特に重要です。
3. Using synthetic chelator chemicals is a questionable practice **from an environmental standpoint**.
合成のキーレーター化学物質を使うことは，環境の見地から疑問の余地のある方法である。
4. Our decision to approach criteria **from the standpoint of** measuring impairment from a pollution source was influenced by two major factors.
汚濁源からの損傷測定の見地でクライテリアに対処するという我々の決定は，2つの要因に影響されていた。
5. **From a biological standpoint**, the range of any given class is so great that a criterion established for an expected biological community would be so broad as to be ineffectual.
生物学的な見地から，ある等級の値域があまりに広いため，予想生物群集について定められたクライテリアも広範すぎて無効となる。

State：述べる；曰く

1. In the last paragraph **we stated that** .....
前段落で我々は…と述べた。
2. Based on their analysis, they conclusively **stated** the following.

彼らの分析に基づいて，彼らは結論として次のことを述べた。
3. Leopold (1949) **stated** that "A thing is right when it tends to preserve the integrity, stability, and beauty of the biotic community. It is wrong when it tends to otherwise."
Leopold (1949)曰く，「生物群集の完全性，安定性および美を保つ傾向にあれば，その参照地点は正しい。そうでない傾向にあれば，それは誤っている」。

- **As stated**：述べられているように
- **As stated above**：前述のように
- **As stated previously**：前述のように

4. **As stated** in the introduction, ...
序論で述べられているように，…
5. **As they have stated**, ...
彼らが述べているように，…
6. These data will be analyzed to assess the status of biological condition and to address the goals **as stated above**.
これらのデータは，生物学的状態の現状を評価し，上述の目標に取り組むために分析されるであろう。
7. **As stated previously**, the selection of the appropriate criterion depends on the content of the database.
前述のとおり，適切なクライテリアの選定は，データベースの内容に左右される。

- **unless otherwise stated** .....：述べられていない場合は

8. **Unless otherwise stated**, all information and procedures for working with worksheet apply to macro sheets.
述べられていない場合は，ワークシートで作業するためのすべての情報・手順がマクロシートに適用される。

- **State differently**, ...：言い換えれば，

9. **State differently**, criteria developed under Section 304 (a) (1) could be viewed as criteria to address the effects of pollutants, while criteria under Section (a) (2) address the broader effects of pollution.
言い換えれば，304条(a)(1)のもとで設定されたクライテリアは，汚濁物質の効果を扱うものとみなされ，304条(a)(2)のもとでのクライテリアは，汚濁の影響を扱うものとみなされる。

- **state-of-the-art**：最新技術

10. The paper begins with a review of selected **state-of-the-art** lake water

quality models.
この論文，選ばれた最先端の湖沼水質モデルのレビューから始まっている．

Statement：陳述；言明；報告

1. This **statement** raises the question of where such information was obtained.
   どこでこのような情報が得られたかという質問を，この陳述は提起する．
2. As a basic principle of statutory construction, general **statements of** statutory goals and objectives have no legal force and effect, absent specific operative provisions in the law.
   法的解釈の基本原理として，法律上の目的および目標の全般的言明は法的効力を持たず，法律における具体的な運用既定を欠いている．
3. In evaluating the performance of a model relative to some field measurement or another model, qualitative rather than quantitative **statements are** often **made**.
   あるフィールド測定，またはもう1つのモデルとの比較からモデルの実績を評価する場合，量より質的な報告がしばしばなされる．

Statistics (Statistical；Statistically)：統計学（統計上の；統計的に）

1. From a consideration of microbial **statistics**, argument can be made against that .....
   微生物統計学の考慮から，… という議論がそれに対してされうる．
2. Other **statistics** can be calculated and represented on the graphical summary, such as means, medians, and percentiles.
   他の統計量は，平均，中央値，パーセンタイルなどグラフ要約で計算・表現されうる．

● **Statistical**：統計上の

3. **Statistical** consideration shows that .....
   統計上の考慮が … であることを示す．
4. Additionally, a t-test was added to provide **statistical** strength to a determination of significant biological impairment.
   さらに，t検定が追加され，有意の生物損傷の決定に統計的効力を与えた．

● **Statistically**：統計的に

5. Due to variability in microbial densities between samples, these changes

were not **statistically** significant.
微生物の密度はサンプル間で変わりうるため，これらの変動は統計的に有意ではない。

6. Unlike the USEPA Environmental Monitoring and Assessment Program (EMAP) sampling design, the Idaho DEQ database was not collected under a **statistically** random design for the location of sampling sites.
USEPA の Environmental Monitoring and Assessment Program (環境監視評価プログラム；EMAP) サンプリング設計と違い，アイダホ州 DEQ のデータベースは，サンプリング地点の位置について統計的に無作為な計画下で収集されなかった。

7. The difference in TPH concentrations between the initial and four-week samples was **statistically** significant at the 0.025 probability level.
最初のサンプルと 4 週間後のサンプルでの TPH 濃度の相違は 0.025 の確率レベルにおいて統計学的に有意であった。

8. The calculation of the t-test showed that this change was not **statistically** significant at the p = 0.05 level, due to high variability in species richness at the downstream site.
t 検定の計算によれば，下流地点での生物種豊富度における高い可変性故に，この変化は，p = 0.05 レベルで統計的に有意でなかった。

★統計に関連する例文を下に示します。

● **distribution**：分布

9. Although the index scores at a site are not **distributed normally**, the usage of a two-sample t-test and ANOVA model could be cautiously applied for hypothesis testing.
ある地点での指数評点は正規分布でないが，2 サンプルの t 検定および ANOVA モデルの利用が仮説の検討に適用可能である。

10. Figure 1 compares the sample and population characteristics via **a cumulative distribution function** (CDF)；a CDF describes the overall population structure, containing information about the shape of the distribution and the range of scores.
図 1 では，累積分布関数 (CDF) を用いてサンプルと母集団特性を比較している。CDF は，母集団構造全体を述べており，分布の形とスコア値域に関する情報を含んでいる。

● **fit to .....**：… に一致する

11. This correlation coefficient was 0.99 indicating a surprising good **fit to** the data points employed.

この相関係数は0.99で，使用したデータポイントに驚くほど一致することを示唆している。

- **random**：無作為な

12. Unlike the USEPA Environmental Monitoring and Assessment Program (EMAP) sampling design, the Idaho DEQ database was not collected under a statistically **random** design for the location of sampling sites.
USEPAの環境監視評価プログラム (EMAP) サンプリング設計と違い，アイダホ州DEQのデータベースは，サンプリング地点の位置について統計的に無作為な計画下で収集されなかった。

- **significant**：有意な

13. Calculation of the t-test confirmed that the impairment was **significant**.
この損傷は有意であったことがt検定の計算によって確証された。
14. For each index there has been a **significant** positive change over time at most sites.
各指数について，大半の地点で，経年的に有意なプラス変化が見られた。
15. The comparison of the two different time period showed that the increased index scores for the later period were highly **significant** ($p < 0.0001$).
これら2つの時期の比較は，最新時期に対する指数スコアの増大がきわめて有意なことを示した ($p < 0.0001$)。

- **uncertainty**：不確実性

16. The 95% confidence interval is included to display the **uncertainty** of the sample estimates.
サンプル推定の不確実性を反映するため，95%信頼区間が含まれている。

Stem：由来する；生じる；起こる

- ～ **stem from** ……：～は…から発している

1. Some of this disagreement **stems from** differences in individual perceptions of ecosystems, the uses of ecoregions, and where humans fit into the picture.
意見相違の一因は，生態系に対する各自の認識の差，生態地域の利用における人間の位置などである。
2. This authority **stems** primarily **from** Sections 303 and 304, which dictates the manner in which water quality standards are adopted.

権限は，主に303条および304条から発しており，それは，水質基準が採用される様相を指示している。

3. Another reason for using single-purpose frameworks, as mentioned in the preceding section, **stemmed from** the belief that a scientifically rigorous method for defining ecological regions must address the processes that cause ecosystem components to differ from one place to another and from one scale to another.
前章で述べたとおり，単一目的枠組みを用いる別の理由は，生態学的地域を定義づけるための科学的に厳密な方法が場所ごと，規模ごとに異なる生態系構成要素を生じさせる過程に取り組むべきだという思考にあった。

Step：段階；ステップ

1. **The basic steps** of the risk assessment process are the following : .....
リスクアセスメント過程の基本的な段階は，次のことである。それは，…
2. **The key steps** in this process are summarized in Figure 1.
この過程における主要ステップは，図1で要約されている。
3. This systematic process involves **discrete steps**, which are described as follows:
この体系的過程は，非連関的な段階を伴うものであり，それについて以下に述べる。
4. Development and implementation of biological criteria **consists** of four **primary steps** : 1) ....., 2) ....., 3) .....
生物クライテリアの設定・実施は，以下の4つの主要段階からなる。1)…，2)…，3)…
5. Additional issues concerning level-of-detail are critical to **every step** of the simulation process.
「細部のレベル」に関する追加の問題は，シミュレーションプロセスのすべてのステップに重要である。
6. Once field data are collected, processes, and finalized, **the next step** is to reduce the data to scientifically and managerially useful information.
現地データが収集・処理・最終確定されたならば，次のステップは，データをまとめて科学的・管理的に有用な情報にすることである。
7. The advantage of **taking** this **additional step** is that it is now relatively easy to see the proportion of sites that do not meet their respective biocriterion scores.
この追加措置による利点は，それぞれの生物クライテリアスコアを満たさない地点の比率が比較的見やすくなることである。

- **at each step**：それぞれのステップにおいて

8. **At each step** in the application process, we will first explain what need to be done.
アプリケーション過程のそれぞれのステップにおいて，何がされるべきかを最初に説明するであろう。

- **following steps**：以下のステップ

9. The SCI is calculated using **the following steps**: 1)....., 2)....., 3).....
SCIは，以下のステップを用いて計算される。1)…，2)…，3)…
10. **The following steps** summarize the application of the biological impairment criteria: 1)....., 2)....., 3).....
以下のステップは，生物損傷クライテリア適用の要約に相当する。1)…，2)…，3)…
11. The framework within which water quality criteria were established and used to evaluate Texas rivers and streams includes **the following major steps**：
テキサス州の河川・水流を評価するために水質クライテリアが設定・利用された枠組みには，以下の主要ステップが含まれる。

- **step–by–step**：一歩一歩

12. The guidelines are followed **step–by–step** to demonstrate how the guidelines can aid in design of a field study.
この指針書は，これがどのように現場研究の計画に役立つのかを明示するために一歩一歩段階を追って説明している。

- **stepwise**：段階的

13. The model for development and aggregation of metrics follows a **stepwise** process, which includes: (1)....., (2)....., and (3).....
メトリックの設定と集計のためのモデルは，以下のような段階的過程をたどる。(1)…，(2)…，(3)…

Straight：一直線の

- **straight–forward**：単純な

1. TMDL development can be **straight–forward** and relatively simple.
TMDLの構築は単純で，そして比較的簡単であり得る。
2. The differentiation on the basis of experimental results may not be as

straight forward as implied by these definitions.
実験結果を基にした差別化は，これらの定義によって暗示されるほど単純ではないかもしれない。

Stream (See River)：細流；小川 (➡ River)；河川

Strengthen：強化；補強

1. The planning process **can be strengthened** by interaction with other programs, allowing joint utilization of reference database.
   立案過程は，参照データベースの共同活用を可能にし，他のプログラムとの相互作用で強化されうる。
2. Kimoto et al (2004) **have strengthened** the concept through their interactions and testing.
   Kimoto ら (2004) は，対話や試験を通じてこの概念を補強してくれた。
3. Our thanks go out to the many state agency biologists who have applied and refined the multimetric approach and **have strengthened** the concept through their interactions and testing.
   さらに，多数の州機関所属の生物学者にも感謝したい。彼らは，多メトリック手法を応用・改善したうえで，対話や試験を通じて，この概念を補強してくれた。

Stress：強調する；ストレス

1. The reverse will be true for compressive axial **stresses**.
   圧縮軸のストレスについては，逆も正当であろう。
2. Notwithstanding, few can tolerate heavy pollutional **stress** and, as such, can be good indicators of environmental conditions.
   それにもかかわらず，強度な汚濁ストレスに耐えるものはごく少数なので，環境状態の優れた指標となる。
3. It is presently insufficient to **stress** that oil is a toxic substance and fish can be affected by exposure.
   原油が有毒物質であり，そして魚がそれに暴露される影響を受けることを強調することが現在不十分である。
4. Approaches for assessing the response of **stress** proteins in organisms when exposed to chemicals are currently under development.
   化学物質に曝された際の生体内ストレス蛋白質の応答を評価するための手法を目下案出中である。

5. Watershed studies are a necessity, but equally important is the development of an understanding of the spatial nature of ecosystems, their components, and the **stress** we humans put upon them.
流域の研究は必要であるが，同じく重要なのは，生態系の空間的性質，その構成要素，そして我々人間が生態系に与えるストレスへの理解を育むことである。

★ **Stress** の動詞としての使用例を下に示します。

6. Levin (1992) **stressed that** to gain an understanding of the patterns of ecosystems in time and space and the causes and consequences of patterns, we must develop the appropriate measures and quantify these patterns.
Levin (1992) は，時間・空間における生態系パターンや，パターンの原因・結果を理解するために，我々が適切な測度を定め，これらのパターンを定量化しなければならないことを強調した。

Study (See also Research)：研究；研究する（➡ Research）

● **Study is needed to .....**：研究が … するために必要である

1. **Studies are needed to** identify current prevention effort and to assess how well they are working.
現在の防止の努力を確認し，そしてそれらがどれほど稼働しているか査定するための研究が必要である。

● **study is done to .....**：研究が … のために実行される
● **study is undertaken**：研究が着手される
● **study has been conducted on .....**：… に関して研究が行われてきた
● **no study of ..... had been conducted**：
　… に関しての研究が行なわれてこなかった

2. In 1983 and early part of 1984, **the first major study was done** by EPA to develop a data base for the assessment of the existing water quality.
1983 と 1984 の早期に，最初の主要な研究が，既存の水質査定のためのデータベースを作成するために EPA（合衆国環境庁）によってなされた。

3. **Short term studies should be undertaken to** evaluate existing source water monitoring programs.
短期の研究が，既存の源水モニタリングプログラムを評価するために着手されるべきである。

4. **Few studies have undertaken to** compare data collected by volunteers with that of data collected by professionals.

ボランティアの収集データと専門家の収集データを比較する研究は，ごく少数である。

5. **The first systematic studies undertaken** by EPA in 1964 quantified cyanide loading to Rock Creek.
   1964年にEPAによって着手された最初の体系的な研究によって，Rock Creekへのシアン化物の負荷量が定量化された。
6. In 1970–1971, **a series of studies** by the U.S. EPA (1973) **were undertaken** to document and characterize the discharge of tailings to Whitewood Creek.
   1970から1971までに，Whitewood Creekへの選鉱廃物の排出を文書化し，特徴づけるための一連の研究がU.S. EPA (1973) によって着手された。
7. Various studies have been conducted on .....
   …に関してさまざまな研究が行われてきた。
8. Over the past few decades, a considerable number of studies have been conducted on the effects of temperature on the growth of nitrifying bacteria.
   ここ数十年にわたり，硝化菌の成長に及ぼす温度の影響に関する研究がかなり行なわれてきた。
9. Until very recently, no comprehensive biological study of Rock Creek had been conducted.
   つい最近まで，包括的な生物学的研究がRock Creekで行なわれてこなかった。

   - **study has been made on .....** ：…に関して研究がなされてきた
   - **study has concentrated on .....** ：研究は…に集中してきた
   - **study has focused on .....** ：研究は…に焦点を合てきた
   - **future studies will focus on .....** ：
     将来の研究は…することに集中するであろう

10. Over the past few years, several studies have been made on groundwater contamination by VOCs.
    過去数年にわたって，VOCsによる地下水汚染についての研究がいくつかなされてきた。
11. Many studies have concentrated on groundwater contamination by VOCs.
    多くの研究はVOCsによる地下水の汚染に集中してきた。
12. The study focused on data from 56 sites on streams that were part of Wisconsin's designated Scenic Rivers Programs and which were monitored by both Wisconsin DNR volunteers and by Wisconsin EPA biologists.
    この研究の主眼が置かれたのは，ウィスコンシン州の指定するScenic Rivers Programs（景観河川プログラム）の一部で，ウィスコンシン州のDNRボランティ

アとウィスコンシン州 EPA 所属生物学者の両方がモニターした河川の 56 地点でのデータである。

- **the study provided .....**：この研究から … が得られた
- **these studies demonstrated that .....**：
  これらの研究は … であることを実証した
- **previous studies of ～ have produced .....**：
  ～に関するこれまでの研究は … をもたらしている

13. The study provided detailed knowledge of the site history.
    この研究から，この地域の歴史に関する詳細な知識が得られた。
14. These studies demonstrated that middle school-aged students could be used to sample water quality with reliability nearly equivalent to trained water quality specialists.
    これらの研究は，中学生が水質サンプリングをした場合，訓練を受けた水質専門家とほぼ同等の信頼性をもたらすことを実証した。

- **few studies have examined .....**：… がほとんど研究されていない
- **our studies indicate that .....**：
  我々の研究は … ということを示している
- **more recent studies have been .....**：ごく最近の研究は … である

15. Data exist from several years of studies that support these conclusions.
    これらの結論を支持するような数年間にわたる研究のデータが存在している。
16. Despite our growing knowledge of CP anaerobic dechlorination pathways and bacteria, few studies have rigorously examined the dechlorination kinetics of CPs.
    CP の嫌気的脱塩反応過程およびバクテリアに対する知識が増えているにもかかわらず，CPs の脱塩反応速度論が徹底的に研究がされてきていない。
17. The few studies that have been done suggest that volunteers are capable of providing good-quality data for specific levels of use.
    ボランティアが特定の使用レベルについて高質なデータを提供できることを示す研究は，まだあまり存在しない。

- **in this study**：この研究で
- **in their study**：かれらの研究では
- **in the study by .....**：… による研究では
- **in the present study**：この研究では

18. In this study, we investigated .....
    この研究では，我々は … を調査した。

19. This analysis has been extended in their study.
この分析は，かれらの研究で進展してきた。
20. In the study by Takeda et al. (2002),
武田ら (2002) の研究では,
21. Considering all the data obtained in the study, the reaction appears to be zero-order, i.e., the rate is independent of the ammonia concentration.
この研究で得られたすべてのデータを考慮すると，この反応はゼロ次であるように思われる。すなわち，その速度はアンモニア濃度の影響を受けない。
22. It is possible that oxidation could have been responsible for rapid phenol degradation in the present study.
この研究では，酸化反応が，迅速なフェノール分解の原因であった可能性がある。

- **..... under study**：研究対象の…

23. For a metric to be useful, it must be relevant to the biological community under study and to the specified program objectives.
あるメトリックが有用たるには，研究対象の生物群集ならびに明確なプログラム目標に関連していることが必要である。
24. Likewise, if production of a site is considered high based on organism abundance and/or biomass, and high production is natural for the habitat type under study (as per reference conditions), biological condition would be considered good.
同様に，生物の数度および／または生物体量に基づいてある地点での生産が高いとみなされ，また調査対象生息地について (参照状態によって) 高生産が自然であるならば，生物学的状態は良好だと考えられよう。

- **～ has been studied**：～についてはこれまで研究されてきた
- **～ has never been studied**：
  ～についてはこれまで研究されなかった

25. It has been extensively studied on the basis of current knowledge of ......
それは … の最新の知識をベースにして広範に調査されてきた。
26. Sonoluminescence from aqueous solutions has been studied in some detail.
水溶液からの Sonoluminescence は，若干詳細に研究されてきた。
27. During the past decade, this process has attracted extensive attention as a modern technology for groundwater purification and has been studied for a wide range of hazardous organic compounds.
これまでの 10 年間に，このプロセスは，地下水浄化の近代的な技術として広く注意を引いてきた。そして広範囲の危険な有機化合物のために研究されてきた。

- **The objective of this study is…..** (See also **Objective** and **Purpose**)：この研究の目的は … である(➡ Objective ; Purpose)

28. The objective of this study was to examine the potential of combining sonication with existing measurement technologies for monitoring specific classes of organic pollutants in water.
この研究の目的は、特定の水の有機物汚染クラスをモニターする既存の測定技術と超音波破砕を組み合わせることについての可能性を調べることであった。

29. The overall objective of this research is to conduct basic studies into possible causes of biological nitrification process instability.
この研究の総合目的は，生物 硝化プロセスの不安定性を起こしうる原因に関する基礎研究を行うことである。

- **～ initiated a study**：～は研究に着手した

30. In response to these findings, the Commission initiated a three-year study to determine if the regional patterns would correspond to the ecoregions of Texas mapped by Omernik and Gallant (1987).
これらの所見に応えて委員会は，地域パターンがOmernikとGallant(1987)により作製されたテキサスの生態地域地図に符合するか否か決定するため、3箇年研究に着手した。

★ study(研究)の数を示す表現を下にリストしました。

- **a few studies**：2, 3の研究
- **no study**：研究はない
- **many studies**：多くの研究
- **a considerable number of studies**：かなりの研究
- **several studies**：いくつかの研究
- **various studies**：まざまな研究

31. A few studies have targeted …..
…に的を絞った研究は2,3あるだけである。

32. There has been no study that …..
…についての研究は今までなかった。

33. While a number of HSPF applications are in progress, few studies are complete.
HSPFが多く適用されつつある一方，ほとんどの研究が完全ではない。

★ study(研究)の数以外の表現を下にリストしました。

- **a comparative study of …..**：… の比較研究

- **a more recent study**：より最近の研究
- **a case study on** .....：…の事例研究
- **a detailed study of** .....：…の詳細な研究
- **systematic studies of** .....：…の組織的な研究

34. This is a comparative study of the Rock Creek watershed.
これは Rock Creek 流域の比較研究である。
35. A more recent study in Minnesota compared biological assessments made by volunteers in the Scenic Rivers Stream Quality Monitoring Program with Index of Biotic Integrity (IBI) and Invertebrate Community Index (ICI) assessments made by the Environmental Protection Agency water quality specialists.
ミネソタ州での最近の研究は，景観河川水流質モニタリングプログラムのボランティアによる生物アセスメントと，環境保護局の水質専門家による生物完全度指数 (IBI) および無脊椎動物群集指数 (ICI) アセスメントを比較した。
36. We also discuss uses of biological data collected by volunteer monitors as illustrated by a case study of the Maryland Save Our Stream (SOS) biological monitoring program.
また，ボランティアモニターにより収集された生物データの利用についても論じ，メリーランド州「Save Our Streams (SOS)」生物モニタリングプログラムをケーススタディとして示す。

Subject：課題；主題；対象

- **subject to** .....：…の対象となる；…に左右される

1. This report **is subject to** revisions based on additional information.
この報告書は，追加情報に基づいて修正の対象となる。
2. The value obtained **may be subject to** considerable error.
得られた値は，重大なエラーに左右されるかもしれない。
3. Heavy metals **may be subject to** precipitation, adsorption to particulate matter and subsequent sedimentation.
重金属は沈降，粒子への吸着，それに次ぐ堆積に左右される。
4. It **has been subject to** the agency's peer and administrative review and approved for publication.
USEPA の同領域専門家および行政官による評価を経て，公刊が承認された。
5. Physical and chemical data **will be subjected to** principal components analysis (PCA), while DCA will be applied to the fish and benthos data.
物理的・化学的データは，主成分分析 (PCA) の対象となり，DCA は，魚類および

底生生物データに適用されるだろう。

Substantiate：実証する

1. **To substantiate** the applicability of equation .....
   方程式の適用性を実証するために，…
2. Although it was not within the scope of this research to ascertain the factors which inhibit nitrite production, the results of the study have incidentally seemed **to substantiate** the research of others.
   亜硝酸の生成を抑制する要因を確認することはこの研究の範囲外であったけれども，この研究結果は偶然にも他の研究についても実証したようである。

Success：成功

1. **Success with** compounds such as TCE, $CHCl_3$ and CCl_4 will serve as proof-of-principle and form a basis for expanding the research to other pollutant classes.
   TCE，$CHCl_3$，ならびに CCl_4 のような化合物での成功は，「原則の証明」の役をし，そして他の汚染物クラスに研究を拡大するための根拠となるであろう。
2. The Scioto River perhaps represents one of the best **success stories** of any river or stream in Ohio.
   Scioto 川は，おそらくオハイオ州における河川の最も成功した例の一つである。
3. With clearly defined goals for the maintenance of different levels of integrity through impact standards, **the overall success of** water resource management program can be evaluated.
   影響標準を通じて様々な水準の完全度を保つという明確な目標によって，水資源管理プログラムの全般的成功が評価されうる。

    ● **the key to future success**：将来の成功を握る鍵

4. Arguments over the merits of narrative vs. numeric biological criteria will likely remain, and **the key to future success** will continue to be the dedication and ingenuity of state biologists.
   記述生物クライテリアと数値生物クライテリアのメリットをめぐる議論は，今後も続く見込みが強い。また，将来の成功を握る鍵は，州所属生物学者の献身や才気であろう。

Successful (Successfully) (Unsuccessful)：
成功した；好結果の（成功のうちに）（不成功の）

1. The model **was successful** in computing the degree of lateral dispersion.
   このモデルは，横断方向の拡散の度合いを計算する点で好結果をだした。
2. The kinetic model **is less successful** in predicting the initial pollutant concentration ....
   速度論的モデルは，初期の汚濁物濃度を予測することにおいて，成功には至らない。

   ● **successfully**：成功のうちに

3. Such an approach has been **successfully** applied to .....
   このような手法は，成功のうちに … に適用された。
4. While this may seem enigmatic in light of current strategies to regionalize wastewater flows, the presence of water with a seemingly marginal chemical quality can **successfully** mitigate what otherwise would be a total community loss.
   これは，汚水流量を地域分けするという現行戦略の観点からは不可解に思われるかもしれないが，化学的に不十分な水質の水の存在が生物群集の全損をうまく緩和しうるのである。

   ● **unsuccessful**：不成功の

5. An attempt to use the scheme for classifying aquatic ecosystems **proved unsuccessful**.
   水生生態系分類方式を用いる試みは，不成功であることが判明した。
6. Attempts to calibrate the model and keep parameters in a reasonable range based on literature data **proved unsuccessful**.
   モデルをキャリブレートして，パラメータを文献データに基づく妥当な範囲にとどめる試みは，不成功であることが判明した。

Such：このような

1. **Such** materials have many technological applications.
   このような材料には多くの技術的応用がある。
2. **Such** detailed understanding of the interactions is impossible from steady-state models.
   このような相互作用の詳細な理解は定常モデルからは不可能である。
3. One **such** waste type is radioactive waste.

そのような廃棄物タイプの1つが放射性廃棄物である。
4. This statement raises the question of where **such** information was obtained.
どこでこのような情報が得られたかという質問を，この陳述は提起する。
5. The reasons for making **such** decisions should be documented.
そのような決定を下した理由を文書化すべきである。
6. For **such** extreme operating conditions, deviations between model prediction and experiments can be expected.
このような過激な操作上の条件のために，モデル予測と実験の間の逸脱が予想されうる。
7. Example of **such** sources is backyard trash burning, copper smelting, and dioxins in sediments.
このような出所の例としては，裏庭でのゴミの焼却，銅の製錬，堆積物のダイオキシンである。
8. There is no evidence that any **such** effects translate into impacts on fish populations.
そのような効果が，魚類集団に対する影響につながるという証拠はない。
9. The user should be aware that **such** effects may complicate attempts to extrapolate data for photolysis rates from one aquatic medium to a very different medium.
利用者は，このような影響が光反応速度データを1つの水生媒質からかなり異質の媒質まで外挿法によって推定する試みを複雑にするかもしれないということを認識すべきである。
10. Design of **such** a repository would make use of multiple engineered and natural barriers.
このような貯蔵所の設計は，技術によるバリア(遮断物)と自然なバリアを複合的に活用している。

- **in such cases where .....**：… であるような場合
- **in such application**：このようなアプリケーションで

11. This is especially true **in such cases where** a waterbody may be perceived as being at risk due to new dischargers.
ある水域が，新しく生じた排出のために危険な状態であるものと見なされるかもしれない場合，これは特にあてはまる。
12. An equation which is of considerable importance **in such application** is .....
このようなアプリケーションでかなり重要である式は … である。

- **such as.....**：… など

13. There is a need for an alternate tool **such as** a water quality model.

水質モデルのような代わりの手段が必要である。
14. Other statistics can be calculated and represented on the graphical summary, **such as** means, medians, and percentiles.
他の統計量は，平均，中央値，パーセンタイルなどグラフ要約で計算・表現されうる。
15. Much more research is needed in **such** basic areas **as** sampling effort, and indicator sensitivity and variance.
サンプリング活動，指標の感度・分散といった基礎領域における研究はさらに必要とされる。
16. Although factors **such as** geology and soils are also important, the other factors appear to be the most important in this ecoregion.
地質や土壌などの因子も重要だが，この生態地域では，他の因子が最も重要に思われる。
17. A forum will focus on identifying practical solutions to global issues **such as** greenhouse effect, depletion of the ozone layer and toxic wastes.
フォーラムは，温室効果，オゾン層の減少，毒性廃棄物といったグローバルな問題に対する実務的な解決策を確認することに集中するであろう。
18. Success with compounds **such as** TCE, CHCl$_3$ and CCl$_4$ will serve as proof-of-principle and form a basis for expanding the research to other pollutant classes.
TCE，CHCl$_3$，ならびにCCl$_4$のような化合物での成功は，「原則の証明」の役をし，そして他の汚染物クラスに研究を拡大するための根拠となるであろう。
19. Certainly streams **such as** this one should be protected and not be allowed to degrade to standards and expectations set for streams typical of most of the region.
確かにこのような河川は，保護されるべきであり，その地域の大半の典型となる河川セットについての基準や期待値を下げることは許されない。

- **as such**：かくして

20. Notwithstanding, few can tolerate heavy pollutional stress and, **as such**, can be good indicators of environmental conditions.
それにもかかわらず，強度な汚濁ストレスに耐えるものは，ごく少数なので，環境状態の優れた指標となる。
21. This allows for the analysis of incremental changes in aquatic community performance over space and time. **As such**, this has been a useful tool for impairments between different stream and river segments.
これは，水生生物群集実績における時間的・空間的な漸増変化の分析を可能にする。かくして，これは，長期的な汚濁防止努力の有効性を実証し，様々な小流河

川区間での損傷度を比較するための有用なツールであった。
22. Like Metric 8, this is a negative metric and, **as such**, a low number (<50 individuals) or an absence of organisms in a sample defaults to a zaro score for the metric regardless of the presence or absence of the specified tolerant taxa.
メトリック8と同様にこれは，負のメトリックなので，特定の耐性分類群の有無にかかわらず，生物が少数(50匹未満)または皆無であればデフォルトとして0とスコアされる。
23. Fourteen of nineteen (75%) and seventeen of nineteen (90%) scores were within plus or minus two and four points of the median, respectively. **As such**, it was determined that changes in ICI scores at test sites compared to an ecoregional biocriterion or to an upstream control station should be considered in a zone of insignificant departure if the ICI difference is four points or less.
スコア19個のうち14個(75%)，19個のうち17個(90%)がそれぞれ中央値からプラスマイナス2，4点であった。かくして，ICIスコアの差が4点未満なら，生態地域生物クライテリアまたは上流参照地点と比較した披験地点でのICIスコアにおける変化は有意でないずれの範囲内とみなされることが決定された。
24. The QCTV is envisioned as having application when a quick turnaround is needed to problem assessment or when a screening-level, less definitive technique is desired in lieu of the more complex ICI process. **As such**, the method utilizes the qualitative, natural substrate collection procedure, which necessitates one site visit and minimal laboratory analysis.
QCTVは，問題あるアセスメントに速やかな対処が必要な場合，もしくは，より複雑なICI過程の代わりに選別レベルの決定的でない技法が用いられた場合に，適用されるものとして構想されている。かくして，この方法は，質的な自然底質採集手順を活用しており，それは1回の地点訪問と最小限の実験室分析を必要とする。

Sufficient (Insufficient)：十分 (不十分)

1. **Sufficient** attempts were made to obtain water samples.
水サンプルを得るのに十分な試みがなされた。
2. Precipitation gages and evaporation pans can be utilized with **sufficient** accuracy.
降水量計器と蒸発パンが十分な正確さで利用できる。
3. The available field data **were not sufficiently detailed to** allow an assessment of spatial variations.

利用可能なフィールドデータは，空間変化のアセスメントのためには十分に詳述されていなかった。

- **～ is sufficient to ……**：～は…をするには十分である

4. It **is** presently **insufficient to** stress that oil is a toxic substance and fish can be affected by exposure.
原油が有毒物質であり，そして魚がそれに接触することによって影響を受けることを強調することが現在不十分である。
5. It can be determined if refinements and advancement in macroinvertebrate taxonomy **have been sufficient enough to** warrant further adjustments to ICI scoring categories.
大型無脊椎動物分類群における改善と前進がICIスコアリング区分のさらなる調整を保証するに足るほど十分か否か決定されうる。

- **insufficient**：不十分

6. **Insufficient** attention is directed toward assessing cumulative impact of these policies in terms of costs and outcomes.
費用と成果の面でこれらの政策の累積影響評価にあまり注意が払われていない。
7. **Insufficient** knowledge about regional expectations can result in misinterpretations about the severity of impacts in streams.
地域期待に関する知見不足は，水流中の影響度に関する誤った解釈を招きうる。
8. Very small and intermittent streams have received **insufficient** attention everywhere but Illinois and Indiana.
小規模な間欠河川には，イリノイ州およびインディアナ州を除き，十分な注目が払われていない。
9. No particle size analysis data is currently available and **insufficient** laboratory duplicate data is available.
利用可能な粒子径分析データが現在得られない，そして利用可能な研究室での重複データが不十分である。
10. At this point there is **insufficient** information on source area ST5 to draw a conclusion on probable risk.
この時点では，十分に可能性のあるリスクについて結論を引き出すには，汚染源ST5についての情報が不十分である。
11. There is **insufficient** information to determine whether contamination present at this source area may present an unacceptable risk.
この汚染源での汚染が，容認出来ない危険性を引き起こすかもしれないかどうか決定するには情報が不十分である。

Suggest (Suggesting) (Suggestive)：
示唆する（示唆すること）（示唆的な）

1. The data in Table 2 and 3 **suggest** .....
   表2と表3のデータは…を示唆している。
2. Our unpublished data **suggest** the methods to be of promise for the characterization of sediments.
   我々の未発表のデータは，堆積物の特性評価のために有望になるであろう方法を示唆している。
3. The body of data **suggest**s a possible and effective approach toward the degradation of a wide variety of harmful toxic organic pollutants.
   多くのデータが，多様な有害毒性有機物を分解するのに可能かつ効果的なアプローチを示唆している。
4. **As suggested above**, much more research is needed in such basic areas as sampling effort, and indicator sensitivity and variance.
   上に示唆されるように，サンプリング活動，指標の感度，分散といった基礎領域における研究はさらに必要とされる。

● ～ **suggest that** ..... ：～は…を示唆する

5. Watanabe, in reviewing the work of Chiba and Ohsugi, **suggested that** .....
   Watanabeは，ChibaとOhsugiの研究を再検討し，…を示唆した。
6. Our results **suggest that** the reactivity is additive for the concentration range studied.
   我々の結果は，研究された濃度の範囲では反応性が加算的であることを示唆している。
7. The few studies that have been done **suggest that** volunteers are capable of providing good-quality data for specific levels of use.
   ボランティアが具体的な使用レベルについて高質なデータを提供できることを示す研究は，まだあまり存在しない。
8. This discrepancy between the samples collected two years apart **suggests that** significant variations in the magnitude of soil contamination can occur either over very short horizontal distances or over a relatively short time period.
   2年間隔てて収集されたサンプルの間でのこの矛盾は，土壌汚染度の有意な相違が非常に短い水平距離で，あるいは比較的短期間にわたって起こりえることを示唆する。
9. With the exception of a few experiments made with KI, all experiments

suggest that the reactor generates hydroxyl radicals.
KIを使用した少数の実験を除いて，すべての実験はこのリアクターがヒドロキシルラジカルを発生させることを示唆する。

10. Later work by Taguchi et al. (1990) **suggests that** nonpoint sources are now impeding further biological improvements observed in larger rivers that resulted from reduced point source impacts.
後のTaguchiら(1990)の研究によれば，点源影響の減少によって大河川で観察された一層の生物学的な改善を面源が妨げている。

- **suggesting**：示唆すること；示唆している

11. The study areas were selected on the basis of evidence **suggesting that** they are areas of relatively intense geochemical activity.
これらの調査地域は，地質化学的活動の比較的激しい地域であることが示唆されている根拠をベースに選択された。

- **suggestive**：示唆的な

12. The very early time periods **were suggestive of** rapid loss of a small amount of PCP ; however, additional work is needed to confirm this.
非常に早い期間においては，小量のPCPが速い速度で分解されることを示唆する。しかしながら，更なる研究がこれを確証するために必要である。

Sum(Summing)：合計；総和(合計すること)

1. The measured reactivity of a mixture was compared to that calculated from **the sum of** the measured reactivity of the mixture's individual components.
測定された混合物の反応性を，測定された混合物の個々の構成物質の合計から計算された反応性と比較した。

2. As discussed in the previous section, the status of biological criteria programs across the United States can be thought of as **the sum of** all the states in various stages of developing and implementing biological criteria.
前節で述べたとおり，全米での生物クライテリアプログラムの現状は，生物クライテリア設定・実施の様々な段階における諸州すべての総和として考えうる。

- **summing**：合計すること

3. The IR of mixtures can be calculated **by summing** the reactivity contributions of the components.
構成要素の反応性貢献度を合計することによって，混合物のIRが計算ができる。

Summarize：要約する

1. The following steps **summarize** the application of the biological impairment criteria: 1)....., 2)....., 3).....
   以下のステップは，生物損傷クライテリア適用の要約に相当する。1)…, 2)…, 3)…

   - **Table N summarizes .....**：
     表Nは…を要約する；表Nは…をまとめたものである

2. Table 1 **summarizes** quantitative data on .....
   表1は…の量的なデータを要約する。

3. Tables 1 through 6 **summarize** the experimental and computed results of the five batch runs.
   表1から表6までは，5回のバッチ実験と計算の結果をまとめたものである。

4. Figure 3 **summarizes** some of the available data that illustrate the relationship between the rate of oxidation of ammonia ion and the pH of the ambient water.
   アンモニアイオンの酸化速度と周囲の水のpHとの関係を例証する入手可能なデータのいくつかを図3に要約する。

   - **～ is summarized**：～は要約されている

5. The key steps in this process **are summarized** in Figure 1.
   この過程における主要ステップは，図-1で要約されている。

6. The details of the studies and results have been presented elsewhere (2–5) and **are only summarized** here.
   その研究と結果の詳細は，ほかの様々なところに提出されているので (2–5)，こではただ要約するだけにする。

7. Recent advances in AOP treatment technology as well as their use in industrial applications **are summarized** in the literature.
   AOP処理技術における最近の進歩は，産業的応用におけるそれらの使用と同様にこの文献に要約されている。

8. Results **summarized** in Table 1 show that .....
   表1で要約した結果が，…のことを示している。

9. The data **summarized** in Fig. 3 demonstrate that .....
   図3に要約されているデータによって，…が明示された。

   - **as summarized above**：上に要約されるように

10. Reviewing the literature on sonochemistry of organochlorine compounds **as summarized above** did not lead to any publications on the use of ultrasound for chemical monitoring.
    上に要約したように，有機塩素化合物の超音波化学分解に関する文献のレビューでは，化学的モニタリングとしての超音波利用に関する出版物は見あたらなかった。

Summary：要約

1. Table 7 presents **a summary** of this comparison.
   表7が，この比較の要約を提示する。
2. **Summaries** of the more progressive of these interstate cooperative efforts are presented below.
   これらの州間協力活動の前進について以下に要約する。
3. **A brief summary** is included below as an aid to understanding the results obtained in this project.
   短い要約が，このプロジェクトで得られた結果を理解する助けとして下に含まれている。
4. **A literature summary** of longitudinal dispersion coefficients for streams and rivers is reported in Table 1.
   小流と河川における縦方向拡散係数の文献値が要約され，表1に報告されている。
5. A few chemicals were selected from each group to build **summary** tables for each process.
   それぞれのプロセスに要約表を作成するために，少数の化学物質が各グループから選ばれた。
6. Other statistics can be calculated and represented on the graphical **summary**, such as means, medians, and percentiles.
   他の統計量は，平均，中央値，パーセンタイルなどグラフ要約で計算・表現されうる。
7. Some of the state program documents used to develop information for this **summary** were in draft form at the time of our review and are representative of preliminary programs.
   本論のための情報作成に用いられた州プログラム文書の一部は，我々による査読時点で草稿段階にあり，予備的プログラムを表している。
8. **In summary**, ...
   要約すると，…

Suitable：適した；ふさわしい

1. A **suitable** oxidizer is required.
   適した酸化剤が必要とされる。
2. The goal is to achieve a **suitable** fit between the planned modeling effort and the data, time, and money available to perform the study.
   目標は，計画されたモデリングの努力とデータ，時間，それに研究を実行するためにアクセス可能な資金の間で，適切な合致に達することである。

Supersede：取って代わる

1. The model presented in this paper is intended to **supersede** the preliminary results given in a recent report (Sasaki, 2002).
   この論文に記載されているモデルは，意図的に，最近の報告書 (Sasaki, 2002) に記載された予備結果に取って代わる。

Supplement (supplemental)：補足（補足的）

1. **To supplement** the major field study, ...
   主要な野外研究を補足するために，…
2. Biological impairment criteria are intended **to supplement** existing chemical standards and toxicity testing requirements.
   生物損傷クライテリアの意図は，既存の化学基準および毒性試験要件を補完することである。
3. This is a species that requires little or no external support in the way of **supplemental** stocking.
   これは，補足的放流という外部支援をほとんどまたは全く必要としない魚種である。

Support (Supporting)：支持；支援；支持する；裏付ける（裏付けすること）

1. This is a species that requires little or no external **support** in the way of supplemental stocking.
   これは，補足的放流という外部支援をほとんどまたは全く必要としない魚種である。

   - 〜 offer support for .....：〜は … に対する支持を提供する
   - 〜 provide support for .....：〜は … に対する支援をしてくれた
   - support for 〜 was provided by .....：

～ついて支援してくれたのは … である

2. The goal of biological criteria is to **offer** additional **support** for the state's water quality standards.
生物クライテリアの目標は，州の水質基準に追加支援を提供することである。
3. Tarou Shiota **provided** many hours of **support** in the development of the basic computer programs.
Tarou Shiota は，長時間を割き基礎的なコンピュータプログラムの作成を支援してくれた。
4. The goal of biological criteria is to **provide** additional **support for** the state's water quality standards.
生物クライテリアの目標は，州の水質基準に追加サポートを提供することである。
5. **Support for** Yuka Yoshida on this chapter **was provided by** the Montana Department of Natural Resources Bureau of Research, Study MTRS 386.
本章において Yuka Yoshida を支援してくれたのは，モンタナ州の天然資源局，研究部，研究 MTRS 386 である。

- **in support of .....** : … を支える

6. For states, this would most commonly entail trend assessments and monitoring **in support of** various water quality management and regulatory programs such as water quality standards, NPDES permitting, and nonpoint source management and assessment.
諸州にとって，これは，水質基準，NPDES 許可発出，面源管理評価といった様々な水質管理・規制プログラムを支えるための傾向アセスメントおよびモニタリングを最も一般的に含む。

★ Support の動詞としての使用例文を下に示します。

- **～ support .....** : ～は … を裏付ける；～は … を支持する

7. These results **support** the conclusion that chloride is the major ionic product.
これらの結果は，塩化物が主要なイオン生成物であるという結論を支持する。
8. Data exist from several years of studies that **support** these conclusions.
数年間の研究からなるデータが存在し，この結論を支持している。
9. These results directly **support** the hypothesis that natural levels of copper in seawater can be toxic to plankton.
これらの結果は，海水中に自然に存在する銅のレベルがプランクトンにとって有毒であり得るという仮説を直接に支持している。

- **～ is supported by .....**：～は … によって裏付けられる

10. This hypothesis **is supported by** studies of pure cultures.
    この仮説は，純菌の研究によって裏付けられる。
11. This designation of impairment **was supported by** toxicity testing results, and subsequently resulted in a reevaluation of the treatment process of this facility.
    この損傷指定は，毒性試験結果によって裏付けられ，後日，この施設の処理過程への再評価が行われた。
12. The validity of an integrated assessment using multiple metrics **is supported by** the use of measurements of biological attributes firmly rooted in sound ecological principles.
    多メトリックを用いた統合的アセスメントの妥当性は，確固たる生態学的原理に根ざした生物学的属性測定を利用することで確保される。
13. So long as the science **is** sound and **well supported**, the findings likely will be upheld in the courts.
    科学が堅実で十分に裏付けられる限り，その所見は，法廷で是認されるだろう。

- **to support .....**：… を支持するため

14. There are very few experimental data available **to support** .....
    … を支持する実験データがほとんどない。
15. There are many examples **to support** this theory.
    この理論を支持する例が多くある。
16. Only limited funds are available **to support** research on the development of inexpensive analytical methods.
    ほんの限られた資金が，低価な分析方法の開発研究の援助のために利用できる。

- **supporting**：裏づけすること

17. Of these 45 states, about half have documentation (mostly in draft form) **supporting** the methods and analyses and providing program rationale.
    これら45州のうち，ほぼ半数は，方法・分析を裏付け，プログラムの根拠を提示する文書(主に草稿)を備えている。

Surprising：驚くべき；驚くほど

1. This correlation coefficient was 0.99 indicating a **surprising** good fit to the data points employed.
   この相関係数は0.99で，使用したデータポイントに驚くほど一致することを示

唆している。

- **What is so surprising is** ..... ：驚くべきことは…である
- **This is somewhat surprising** ：これはいくぶん驚きである
- **It is not surprising that** ..... ：…は驚くに当たらない
- **It should not be surprising that** ..... ：…は驚くべきことではない

2. **What is so surprising** on Easter Island is the high abundance of corals.
   イースター島についてとても驚くべきことは，サンゴチュウが豊富なことである。
3. **This is somewhat surprising**, in view of the order of magnitude higher water column decay rate of DDE relative to PCB.
   PCBと比較して，水中でのDDEの桁違いに高い分解速度の見地から，これはいくぶん驚きである。
4. **It should not be surprising that** chemical kinetics is an integral component of models of natural systems.
   化学反応速度論が自然系モデルの必須の構成要素であるということは，驚くべきことではない。
5. This overlap **is not surprising** given that many of the CSO Toxic impacted segments were in the same streams and rivers as some of the Complex Toxic impacted segments.
   この重複は，CSO毒性影響区間の多くが複雑毒性影響区間の一部と同じ小流・河川にあったことを考えれば，驚くに当たらない。

Survey (Surveying)：調査；調査する（調査すること）

1. Results of the 1989 **fishery surveys** are given in Table 2.
   1989年の漁業調査結果を表2に示した。
2. This **survey** has shown that a number of states are using not only the multimetric approach, but are often using multiple assemblages.
   この調査によれば，多数の州が多メトリック手法ばかりでなく，しばしば複数集団も用いている。
3. Statistical **sample surveys** can play an important role in characterizing the biological condition of lakes and streams.
   統計的サンプル調査は，湖沼・河川の生物学的状態の特性を把握するうえで重要な役割を演じうる。
4. **Biosurveys** were conducted near Conesville Station from 1988 to 1991 using methods that conformed to Ohio EPA protocols or with minor deviations.

オハイオ州EPAプロトコルに準拠した，もしくはやや逸脱した方法を用いて，1989年から1991年までConesville発電所付近で生物調査が行われた。

5. Water quality measurements were not made **as part of the survey**.
水質測定が調査の一部として行われなかった。
6. The following is a description of the problematic areas encountered **during this survey**.
以下に，この調査の間に生じる問題点の多いエリアが記述されている。
7. The following conclusions were established from the process **emissions survey** and bench-scale testing: 1)..., 2), ...
プロセス排気テストの調査，そしてベンチスケール実験から次の様な結論が確立された。1)…, 2)，…である。

★ Surveyの動詞としての使用例文を下に示します。

8. Ambient sites other than reference sites **should be surveyed** as part of the database.
参照地点以外の環境地点は，データベースの一部として調査されるべきである。

● **surveying**：調査すること

9. The **surveying of** ambient biota takes the form of either coordinated monitoring networks or a series of special studies.
環境生物相に対する調査は，調整された監視網もしくは一連の特別研究の形をとる。

Table：表

1. The data **in Table 1** suggest .....
   表1のデータは … を示唆している。
2. Results summarized **in Table 2** show that .....
   表2で要約した結果が，… のことを示している。
3. Examination **of Table 3** reveals that .....
   表3を検討すると … が明らかになる。
4. The third row **of Table 4** shows .....
   表4の3番目の行は … を示している。
5. In Columns (1) and (2) of **Table 5**, ...
   表5の列(1)と列(2)には，…
6. Some of the results **given in Table 6** have important implications with regard to the PCB distribution.
   表6に示された結果の一部は，PCBの分布に関する重要な含蓄を持っている。

   - **Table N shows** ..... ：表Nは … を示している
   - **Table N presents** ..... ：表Nは … を提示している
   - **Table N summarizes** ..... ：
     表Nは … を要約している；表Nは … をまとめたものである
   - **Table N indicates that** ..... ：表Nは … を示唆している
   - **Table N gives** ..... ：表Nは … を示している
   - **Table N highlights** ..... ：表Nは … を強調している
   - **Table N reports** ..... ：表Nに … が報告されている

7. **Table 1 shows** in part that .....
   表1は一部 … ことを示す。
8. **Table 2 presents** a summary of this comparison.
   表2が，この比較の要約を提示する。
9. **Table 3 summarizes** quantitative data on .....
   表3は … の量的なデータを要約する。
10. **Tables 1 through 6 summarize** the experimental and computed results of the five batch runs.

表1から表6までは，5回のバッチ実験と計算の結果をまとめたものである。

11. **Table 7 indicates that**, except for one sample, all samples contain between 1 and 20 weight percent of gypsum.
表7は，1つのサンプルを除いて，すべてのサンプルが重量で1から20パーセントの石膏を含んでいることを示している。
12. **Table 9 gives** some literature values for the vertical dispersion coefficient.
表9に，縦断拡散係数の文献値を若干示す。
13. **Table 8 reports** some values found in the literature.
表8には，文献にて見いだされた若干の値が報告されている。

    - **~ is given in Table N**：~は表Nに示されている
    - **~ is listed in Table N**：~を表Nにリストする
    - **~ is reported in Table N**：~が表Nに報告されている
    - **~ is shown in Table N**：~は表Nに示されている
    - **~ is displayed in Table N**：~は表Nに示されている

14. Results of the 1989 fishery surveys **are given in Table 1**.
1989年の漁業調査結果を表-1に示す。
15. The estimated rate constants **are listed in Table 2**.
推定速度定数を表2にリストする。
16. The results of the analysis of the eight samples from the monitoring wells **are listed in Table 3**.
モニタリング用の井戸からの8つのサンプルの分析の結果を表3にリストする。
17. A literature summary of longitudinal dispersion coefficients for streams and rivers **is reported in Table 4**.
小流と河川における縦方向拡散係数の文献値が要約され，表4に報告されている。
18. The derivation of the impairment for species richness **is shown** as an example **in Table 5**.
生物種豊富度についての損傷クライテリアの導出は，表5に例示している。

    - **As seen from Table N**：表Nから見られるように
    - **It is apparent from Table N that** .....：表Nから外見上明白なことは…

19. **As seen from Table 4**, the BIOMODEL estimates of biotic DDTR concentration correspond fairly closely to measured concentrations.
表4から見られるように，BIOMODELによって推定された生体内のDDTR濃度は測定された濃度にかなり近い。
20. **It is apparent from Table 3 that** a significant inhibitory effect was placed on the specific oxidation rate as a result of extreme winter temperatures.
極度な冬の温度が原因で相当な抑制効果が特定の酸化速度に及ぼしたことは，表

3から明白である。

Take(taking)：要する；取る；獲得する（取り入れること）

1. **How long will it take** a contaminated natural water system to recover?
   汚染された自然水系が回復するのにはどれほどの時間がかかる。
2. Sample processing and species proportional counts of a sample **should take** no more than 3 hr.
   サンプル処理とサンプルの生物種比率カウントに，3時間以上かけてはならない。
3. The purpose of this section is to present a description of some of these modifications and to illustrate the different directions many programs **have taken**.
   本項の目的は，これらの修正の一部を示すことと，多くのプログラムがとった異なる方向性を示すことである。

   ● **take for granted**：もちろんの事と思う；あたりまえだと思う

4. The widespread public concern about these problems draws attention to the fact that the quality of drinking water supplies **can** no longer **be taken for granted**.
   これらの問題に対する広い公共の関心が，供給飲料水の水質は，もはや安全とはいえないという事実に注意がむけられている。

   ● **care must be taken**(See **Care**)：
     配慮すべきである；注意を払うべきである(➡ **Care**)
   ● **care was taken**(See **Care**)：注意が払われた(➡ **Care**)
   ● **samples were taken**(See **Sample**)：
     サンプルが採取された（➡ **Sample**）
   ● **taken up**：取り込まれる

5. Naoki(1975) observed that the quantity of mercury **taken up** by the sludge did not alter over a wide range of concentrations. Neufeld(1975)
   Naoki(1975)の観察では，スラッジによって吸着された水銀の量は大きな濃度変化がなかった。

   ● **taking**：取り入れること

6. The advantage of **taking** this additional step is that it is now relatively easy to see the proportion of sites that do not meet their respective biocriterion scores.
   この追加措置による利点は，それぞれの生物クライテリアスコアを満たさない地

点の比率が比較的見やすくなることである。

- **~ take place**(See **Place**)：~が行われる；~が生じる(➡ **Place**)
- **~ is taking place**(See **Place**)：
  ~が行われている；~が起きている(➡ **Place**)
- **~ take account of** .....(See **Account**)：
  ~は … を考慮する(➡ **Account**)
- **taking into account** .....(See **Account**)：
  … を考慮しながら；… を考慮して(➡ **Account**)
- **~ take the form of** .....(See **Form**)：~は … の形をとる(➡ **Form**)

Target(Targeting)：目標；対象ターゲット；目標にする；ターゲットにする；的をしぼる(ターゲットにした)

1. Intensified research may be needed to enable proper selection of **target** organisms.
   目標生物の適切な選択を可能にするためには，強化研究が必要になるかもしれない。
2. The extent to which this subpopulation reflects the **target** population would elicit discussion.
   この部分母集団がどの程度ターゲット母集団を反映するかは，論議を呼ぶだろう。
3. Not all the information for refining the **target** population was available before field visits.
   現地視察の前には，ターゲット母集団を絞り込むための情報すべてが利用可能ではなかった。

★ **Target** の動詞としての使用例文を下に示します。

4. A few studies **have targeted** .....
   … に的を絞った研究は2, 3あるだけである。

   - **targeting**：ターゲットにした

5. Biosurveys **targeting** multiple species and assemblages and more likely to provide improved detection capability over a broader range as well as protection to a larger segment of the ecosystem than a single species or assemblage approach.
   複数の生物種および群がりをターゲットにした生物調査は，単一の生物種または群がり手法に比べて，より広範な探知能力向上をもたらす見込みが高いし，より

広範な生態系保護になるだろう。
6. The problem with using this type of framework for geographic assessment and **targeting** is that it does not depict areas that correspond to regions of similar ecosystems or even regions of similarity in the quality and quantity of water resources.
この種の枠組みを地理アセスメントおよびターゲット設定に用いるうえでの問題点は，類似生態系の地域や，水資源の質・量において類似した均一地域に符合する区域を描かないことである。

Task：作業；仕事

1. The process can be divided into three **major tasks** 1)...., 2)....., and 3).....
このプロセスは3つの主要な作業に分けられる。それらは1)…，2)…，3)…である。
2. Each of these three **tasks** is considered in more detail below.
これらの3つの仕事のそれぞれが下にもっと詳細に考慮されている。
3. Toxicity assessment of single chemical compounds and of complex industrial effluents has become an increasingly **difficult task**.
一つの化合物や複雑な産業廃水の毒性査定がますます難しい仕事になった。
4. The study of potential synergistic or antagonistic effects of a multitude of toxic substances present in many effluents becomes virtually **an impossible task**.
多くの排水中に存在している多数の毒性物質による潜在的相乗作用または拮抗作用を研究することは，事実上，不可能な仕事となる。

Technique：手法；テクニック

1. The previous volume dealt primarily with the use of **statistical techniques**.
前書では主に統計学的手法の使用について述べられた。
2. The **appropriate techniques** can easily be described and incorporated into a sampling protocol.
適切な技法を容易に記述し，サンプリングプロトコルへ盛り込むことができる。
3. The major assumption involved in this **estimation technique** is that .....
この概算手法に関与する主要な仮定は … である。
4. The important point is that the **same technique** should be used at each site that is to be compared.
重要な点は，比較対象の各地点で同一技法を用いることである。
5. Improved **detection techniques** have now shifted concern to the threat

posed by toxic chemicals.
改善された検出技術は，今や毒性化学物質による脅威に関心をうつらせた。
6. Clearly, much work needs to be done to evaluate and polish the **proposed technique**.
明らかにかなりの研究が，提案された手法を評価し，磨きをかけるためには必要である。
7. The approach chosen is based on dynamic mass balance and uses the **simulation techniques** described by Sasaki et al. (2001).
選択されたアプローチは，力学的物質収支に基づき，そしてSasakiら(2001)によって記述されたシミュレーション技巧を使う。
8. This concern has arisen lately as **analytical techniques** for detecting water contaminants have improved to the point where it is possible to detect contaminants at the part per-trillion level.
水汚染物質を検出するための分析技術が汚染物質をpptレベルで検出可能なまでに向上したため，最近この関心が高まった。

Tend：向かう；至る

● **tend to** .....：…の傾向にある

1. Reference streams **tend to** be similar to one another when compared to reference streams in adjacent regions.
参照水流は，隣接地域における参照水流と比べると，類似している傾向にある。
2. It is worth emphasizing that technology based performance standards **tend to** deemphasize receiving system condition and even human health.
技術ベースの実績基準が，システム状態と人間の健康状態さえ受けることに重点を置かない傾向がある，ということを強調する価値がある。
3. Leopold (1949) stated that "A thing is right when it **tends to** preserve the integrity, stability, and beauty of the biotic community. It is wrong when it **tends to** otherwise."
Leopold(1949)曰く，「生物群集の健全性，安定性および美を保つ傾向にあれば，その参照地点は正しい。そうでない傾向にあれば，それは誤っている」。

Tendency：傾向

1. There is **a tendency to** .....
…する傾向がある。
2. ～ have **a tendency to** .....

～は … する傾向がある。
3. Zinc showed **a tendency for** adsorption.
亜鉛が吸着の傾向を示した。
4. COD showed **an increasing tendency** with increasing TSS.
TSS が増えるとともに COD が増加する傾向を示した。
5. On the other hand, their low stiffness and **tendency** to creep are disadvantages that must be carefully weighed against their advantages.
他方，それらの剛度の低さおよびクリープの傾向は不利点であり，それらの利点と慎重に比較考察されなくてはならない。
6. We believe the latter approach may introduce some unintentional bias into the biological criteria calibration and derivation process because of the **inherent tendency** to select the best sites instead of a more representative, balanced cross section of sites that reflect both typical and exceptional communities.
筆者は，後者の手法が生物クライテリアの較正・導出過程に無意識な偏りを持ち込みかねないと考える。なぜなら，一般的生物群集と例外的生物群集の両方を反映した代表的かつ平衡的な地点見本の代わりに，最小地点を選ぶ傾向が内在するからである。

Term：用語；期間；呼ぶ

1. Setting **this term** equals to z yields .....
この項を z に等しいとおくと … となる。
2. The exponential **term** describes the settling while C'(x, y) describes the dispersion.
C'(x,y) が拡散を説明する一方，指数の項は沈降を示す。
3. **The first term** on the right-hand side of the equal sign is the general solution and the second term is the particular solution.
等号の右側の最初の項は一般的な解で，そして2番目の項は特定の解である。
4. These simplified models contain two types of **terms** describing the transport of suspended or dissolved material through the estuary.
これらの単純化されたモデルには，河口においての浮遊・溶解性物質の輸送を記述する2種の項が含まれている。
5. There is a wealth of technical literature, but it is written for use by scientists and engineers familiar with the **technical terms** and background.
専門的な文献は豊富にあるが，それは専門用語・背景に精通した科学者および技術者に使用されるように書かれている。

- **in terms of .....**：…の面で；…において

6. The central basin is the largest **in terms of** area $(16,317 \text{ km}^2)$ and has a mean depth of 25 m.
中央水域は面積では最大 $(16,317 \text{ km}^2)$ で，平均湖深が 25 m である。
7. Insufficient attention is directed toward assessing cumulative impact of these policies **in terms of** costs and outcomes.
費用と成果の面でこれらの政策の累積影響評価にあまり注意が払われていない。
8. Despite the large amount of effort that has been directed towards IBI development, much remains to be done, both **in terms of** generating new versions for different regions and habitat types, and **in terms of** validating existing versions.
IBI 案出へ向けて多大な労力が払われてきたにもかかわらず，様々な地域や生息場所種類について新バージョンを生じること，ならびに既存バージョンを確証することの両方においてまだ多くの課題が残されている。

- **term**：期間

9. **Short term** studies should be undertaken to evaluate existing source water monitoring programs.
短期の研究が，既存の源水モニタリングプログラムを評価するために着手されるべきである。
10. **Long-term** ecological effects due to shoreline oiling were highly unlikely.
海岸線での油田事業による長期の生態的影響はほとんどありそうもない。
11. The model is intended to compute **long term** trends.
このモデルでは，長期の傾向が計算されるようになっている。
12. The model represents a tool to estimate the **long-term** effort of human activities on the water environment.
このモデルは，水環境に対する人間活動の長期にわたる努力を推定するツールに相当する。
13. This work will aid in making predictions about **long-term** buildup of organic layers on the stone.
この研究は，石への長期にわたる有機層の蓄積を予測するのを助けるであろう。
14. Since the model is intended to compute **long-term** trends, average annual values are simulated and variability within the year is ignored.
このモデルは長期傾向を計算するようになっているので，年平均の値がシミュレートされ，年内の可変性が無視される。
15. The **long-term** fixed station is located on the mainstream approximately 1 mi (1.6 km) downstream from the confluence of the East and West Branches

and upstream of the Elyria wastewater treatment plant.
East 小流と West 小流の合流点から約 1mile (1.6km) 下流の本流と，Elyria 廃水処理場の上流に長期的な固定観測場所がある。

★**term** の動詞としての使用例文を下に示します。

16. These regions generally exhibit similarities in the mosaic of environmental resources, ecosystems, and effects of humans and **can** therefore **be termed** ecological regions or ecoregions.
これらの地域は，一般的に環境資源，生態系および人為的影響のモザイクにおいて類似性を示すので，生態学的地域，もしくは生態地域と呼ばれる。

Test (Testing)(See also Experiment)：
テスト；テストする（検定；テスト）(➡ Experiment)；実験

1. Figure 3 shows the result of the nitrification **test** for the media.
この培養液での硝化反応テストの結果を図3に示す。
2. Fig. 4 presents a typical reaction profile observed during **a test run**.
試運転の間に観察された典型的な反応プロフィールを図4に提示する。
3. **Repeated tests** over long periods are more significant.
長い期間にわたっての繰り返されたテストはいっそう重要である。
4. The mixture of chemicals had an EC10 of 3.7％ for the former and 5.2％ **for the latter test**.
混合化学物質では，EC10が，前のテストでは3.7％，後のテストでは5.2％であった。
5. Comparisons of toxicity data between **tests** conducted on-site to tests conducted off-site on samples have shown little variation.
サンプルでのオンサイトで行なわれたテストとオフサイトで行なわれたテストの間の毒性データの比較では，ほとんど変化がなかった。
6. A series of **batch tests** was carried out to **test** the hypothesis that the same enzymes are used to attack chlorophenols with the same position-specific dechlorination reactions, resulting in competition between these compounds.
位置特定の同じ脱塩反応において，同じ酵素が塩化フェノールを攻撃するために使われ，これらの化合物の間に競合反応をもたらす，という仮説をテストするために一連のバッチ試験が行われた。

● **testing**：検定；テスト

7. **Testing of** these questions yielded the following results: (1)......, (2)......

これらの問題への吟味により (Bode et al. 1990c), 以下の結果が生じた。(1)…, (2)…

8. During the course **of testing**, several modifications were made to the criteria.
検定中にクライテリアに対しいくつかの修正が行われた。
9. Kimoto et al (2004) have strengthened the concept through their interactions and **testing**.
Kimoto ら (2004) は,対話や検定を通じてこの概念を補強してくれた。
10. Case histories are presented below showing the outcome of impairment criteria **testing** in a variety of situations.
様々な状況での損傷クライテリア検定の結果を示した事例史が下に述べられている。
11. This report should serve to give the reader an idea of what is involved in **testing** a model.
この報告書は,モデルのテストには何が関与するのかといった考えを,読者に与えるのに役立つはずである。

★ **Test** の動詞としての使用例文を下に示します。

12. This hypothesis **is tested** in highly chelated seawater cultures.
この仮説は高度にキーレートした海水で培養した細菌でテストされる。
13. Over a 2-year period, the criteria **were tested**, focusing on the following questions: (1)....., (2).....
2年間にわたってクライテリアの検定が行われ,力点は,以下の諸問題に置かれていた。(1)…, (2)…
14. EPA **should have tested** the hypothesis of no differences in biological performance for reference sites on impounded and free-flowing sections.
EPAは,貯水部および流水部の参照地点について生物能力差に関する仮説をテストすべきだった。
15. The major aim of this investigation is **to test** the hypothesis that copper toxicity to algae and copper content of algal cells are functionally related to free cupric ion activity.
この調査の主要な目的は,藻類への銅毒性そして藻類細胞内の銅蓄積量は機能上遊離銅イオン活量と関係があるという仮説を検証する。

Than：よりも

1. On the basis of estimated rate constants, the oxidation rate of formic acid **is faster than** oxalic acid over a pH range of 4-8.
推定された速度定数によれば,ギ酸の酸化速度はpH4-8の範囲でシュウ酸より

速い。

2. Egawa (1997) demonstrated that fuel utilizing bacteria concentrated metal ions to **a lesser degree** when killed, **than** when metabolically active.
Egawa (1997) は，燃料 (ガソリン) を利用しているバクテリアが代謝活性にあるときより，死滅しているいるときのほうが，より少ない量の金属イオンを濃縮させていたことを明示した。

3. In many arid areas, spatial differences in subsurface watershed characteristics have a **stronger** influence on water quality **than** the size or characteristics of the surface watershed.
多くの乾燥区域において，地表下流域特性における空間差は，地表流域の規模または特性よりも水質に対し強い影響を及ぼす。

4. This challenge resulted in a search for numerical expressions in a form **simpler** to understand **than** long species lists and well-thought but lengthy technical expressions of the data.
この課題により，生物種の長大なリストや，綿密だが長々しいデータの技術的説明よりも理解しやすい形での数値表現が追求されるようになった。

5. Twice as many impaired waters have been discovered by using biological criteria and chemistry assessments together **than** were discovered using chemistry assessments alone.
化学アセスメントの単独利用に比べて，生物クライテリアと化学アセスメントの併用によって損傷水域が2倍も多く発見された。

- **than ever**：以前よりも

6. Two decades of Federal controls have sharply reduced the vast outflows of sewage and industrial chemicals into America's rivers and streams, yet the life they contain may be in deeper trouble **than ever**.
20年間にわたる連邦政府による防止策は，米国の河川・水流への下水および産業化学物質の大量流出を激減させてきたが，河川・水流に含有される生物は，以前よりも深いトラブルを抱え込んでいる。

- **better than** ..... (See **Better**)：より良い (➡ **Better**)
- **different than** ..... (See **Different**)：…と異なっている (➡ **Different**)
- **greater than** ..... (See **Greater**)：…より大きい (➡ **Greater**)
- **higher than** ..... (See **Higher**)：より高い (➡ **Higher**)
- **less than** ..... (See **Less**)：…以下 (➡ **Less**)
- **more than** ..... (See **More**)：…よりも；…以上 (➡ **More**)
- **rather than** ..... (See **Rather**)：…よりもむしろ (➡ **Rather**)

Thank：感謝する

1. **I thank** Wanda Murai and Toni Sano, who skillfully typed the manuscript.
   筆者は，Wanda Murai および Toni Sano が本稿を巧みにタイプしてくれたことに感謝する。
2. **I also would like to thank** Ichiro Sasaki and Jiro Nakajima for helping with formatting and editorial content.
   また，Ichiro Sasaki および Jiro Nakajima が構成や編集を手助けしてくれたことにも感謝したい。
3. Also, **I thank** the following individuals for contributing useful comments on a draft version: Kenji Watanabe, Jirou Aita, Chiyoko Suzuki, Tadashi Toda .
   また，わが草稿について有意義な意見を述べてくださった諸氏にも感謝したい。Kenji Watanabe, Makoto Aita, Chiyoko Suzuki, Tadashi Toda である。
4. First and foremost, **we thank** Dr. David Terata, USEPA for the lengthy discussions and his insights on the model development process and appropriate risk assessment approaches.
   まず第一に，我々が感謝を捧げる相手は，USEPAのDr. David Terataである。彼は，長時間にわたる論議や，モデル構築過程ならびに適切なリスクアセスメント手法に関する洞察を提供してくれた。
5. **Our thanks go out to** the many state agency biologists who have applied and refined the multimetric approach and have strengthened the concept through their interactions and testing.
   さらに，多数の州機関所属の生物学者にも感謝したい。彼らは，多メトリック手法を応用・改善したうえで，対話や試験を通じて，この概念を補強してくれた。
6. **The authors wish to thank** the following individuals for providing information on research in progress: Tomoyuki Kawano, Naomi Okada, Seiji Nakato.
   筆者らは，以下の各氏が現在進行中の研究に関する情報を提供してくれたことについて感謝したい。Tomoyuki Kawano, Naomi Okada, Seiji Nakato の各氏である。

Theoretical (Theoretically)：理論的（理論的に）

1. The results demonstrate the utility of using the **theoretical** range of the IBI to differentiate between and interpret different types of impacts.
   結果は，異なる影響種類を差別化するためにIBIの値域を利用することの有用性を実証している。
2. For the sake of clarity in presenting the test results, most of the details

concerning the **theoretical** background of the tests and the analytical procedures are provided in the Appendix B.
テスト結果の提出にあたってそれを明確にするために，試験と分析手法の理論的な背景の細部の多くを付録Bに示す。

3. Without a sufficient **theoretical** basis it would be very difficult, if not impossible, to develop meaningful measures and criteria for determining the condition of aquatic communities.
十分な理論的基礎がなければ，水生生物群集の状態を決定するための有意義な測度および基準（クライテリア）を設定することは，仮に不可能でなくても，非常に困難だろう。

4. Our first concern was to test where EXAMS was **theoretically sound**.
我々の最初の関心は，どのようなサイトでEXAMSが理論的に信頼できたか，テストすることだった。

Theory：理論

1. **In accordance with theory**, ...
理論のとおりに,…
2. There is considerable validity **in this theory**.
この理論にはかなりの妥当性がある。
3. There are many examples to **support this theory**.
この理論を支持する例が多くある。
4. There seems to be no established theory to explain this phenomenon.
この現象を説明する定説はないとおもわれる。
5. Ancillary to these assumptions are two others that affect the utility **of the theory**.
これらの仮定に対する補助として，他に，理論の有用に与える2つの影響がある。
6. This relationship is compliant with the "fixed demand" **theory** proposed by Wu and Yao (2001).
この関係は，Wu and Yao (2001)によって提案された「不変需要」理論に従う。

● **agreement with theory**：理論に合致

7. Quantitatively, **agreement with theory** is good at the lower flow-rates.
数量的に，流速の遅いときに，理論と良く合意する。
8. The results plotted in Figure 6 may be seen to be in general **agreement with the theory**.
図6にプロットされた結果は，その理論と一般的な合意にあるように見られるかもしれない。

Though：けれども；にもかかわらず

● **even though .....**：…であるが

1. A decision was made to record this information **even though** the eventual importance of its use was not immediately apparent.
   たとえその利用の最終的重要性が直ちに明らかでなくても，この情報を記録することが決定された．

Thought：考慮；考え；考慮する；考える；みなす

1. Our finding on .....**can be thought of** as .....
   …に関する我々の発見は，…とみなすことができる．
2. As discussed in the previous section, the status of biological criteria programs across the United States **can be thought of as** the sum of all the states in various stages of developing and implementing biological criteria.
   前節で述べたとおり，全米での生物クライテリアプログラムの現状は，生物クライテリア設定・実施の様々な段階における諸州すべての総和として考えうる．
3. This challenge resulted in a search for numerical expressions in a form simpler to understand than long species lists and **well-thought** but lengthy technical expressions of the data.
   この課題により，生物種の長大なリストや，綿密だが長々しいデータの技術的説明よりも理解しやすい形での数値表現が追求されるようになった．

● **without any thought**：考慮せずに

4. There is merit to Sudo's concern that an arbitrary mixing of variables, **without any thought** given to unintentional introductions of bias, **compounding**, and variance, is to be avoided.
   偏向，妥協および分散の意図せざる導入を考慮せずに変数の恣意的な混合が避けられるべきだ，というSudoの懸念にはメリットがある．

Thousand：千

1. **Many thousands of** miles of United States streams have been and continued to be degraded each year.
   米国の水流は，年に数千mileも劣化してきたし，今も劣化し続けている．
2. Humans have probably played a major role in shaping landscape pattern and molding ecosystem mosaics **for thousands of years**.

人間は，おそらく数千年にわたり，景観パターンの形成と生態系モザイクの生成において重要な役割を演じてきた。
3. Under these conditions, **several thousand years** would be required for removal of entire gypsum content of the tailings.
これらの条件の下で，選鉱くずに含まれる石膏の全部を撤去するには，数千年が必要だろう。

- **hundreds of thousands of** ..... ：数百ないし数千もの…

4. Sulfate occurs in the **hundreds to thousands of** milligram per liter range.
硫酸塩が数百 mg/L から数千 mg/L といった範囲で生じる。
5. Only a few rocks need to be sampled, because each has a periphyton assemblage with **hundreds of thousands of** individuals.
サンプリングに必要なのは，数個の岩石だけである。なぜなら，1個の岩石に個体数が数百ないし数千もの付着生物群がりがあるからである。

Throughout：すみからすみまで；すっかり；全体において

1. The problem is receiving increased attention **throughout** the country.
この問題は国全体にてさらに注目を集めている。
2. Core samples were collected at selected locations **throughout** the study area.
調査全地域の選択地点でコア (円筒型) サンプルが収集された。
3. Chromium was detected at higher than commonly expected values **throughout** the study area.
クロミウムが一般的予想値より高い値で調査地域全体で検出された。
4. The redesignation to WWH was upheld **throughout** the appeals process including the Illinois Supreme Court.
WWHへの指定変更は，イリノイ州最高裁など上訴過程全体で是認された。
5. California EPA has conducted biosurveys **throughout** the state for nearly 20 years.
カリフォルニア州EPAは，約20年間にわたり州全域で生物調査を行ってきた。
6. One notable exception is cadmium which was detected at higher than commonly expected values **throughout** the study area.
1つの顕著な例外はカドミウムで，一般的予想値より高い値が調査地域全体で検出された。
7. In Year 1 of MAHA, 266 sites were sampled **throughout** the mid-Atlantic highlands stretching from Virginia, through West Virginia and Maryland, into Pennsylvania.

MAHAの初年度に，バージニア州，ウェストバージニア州，メリーランド州からペンシルバニア州へ至る中部大西洋岸諸州高地全域の266地点でサンプリングが行われた。

Thus：したがって；このように

1. **Thus**, the model has a good foundation in the observation of actual environmental conditions and associated biological community responses.
   故に，このモデルは，実際の環境状態と関連の生物群種応答の観察において確かな基礎を持つ。
2. **Thus**, a weight-of-evidence approach can become increasingly obvious when hypothesis testing demonstrates the unnecessary usage of independent application
   それ故，証拠加重値手法は，仮説検定が独立適用の不要さを実証した場合には，ますます明らかになる。
3. **Thus**, loss of membrane integrity was not proven as a cause for the observed difference in removal between Tests 1 and 2.
   それゆえ，膜質の低下が，試験1と試験2の間で観察されたの除去率の相違の原因であるとは証明されなかった。
4. The results presented here reflect recent modifications of the global transport model, and, **thus**, the estimates are somewhat different than previously reported.
   ここで報告された結果は，地球上輸送モデルの最近の修正を反映している。それゆえ，予測値は前に報告された値より，よりいくぶん異なっている。
5. Predictive models can forecast future water quality, **thus** permitting trials of possible control strategies at very reasonable expense.
   予測モデルは将来の水質を予測することができるため，かなり妥当な費用で可能なコントロール戦略を試すことが出来る。
6. While we have dealt with most of these issues in Georgia, these and other issues will arise elsewhere, **thus** regionally consistency in achieving a resolution of these issues will be needed.
   筆者は，ジョージア州におけるこれらの問題の大半を扱ってきたが，これらおよび他の問題は，他地域でも生じることで，これらの問題の解決における地域的整合性が必要だろう。

- **thus far**：今までのところ

7. This project, **thus far**, has concentrated on two- and three-dimensional analyses of IBI, Miwb, and ICI metrics and selected subcomponents.

このプロジェクトでは，今までのところ，IBI, Miwb, ICI メトリックおよびいくつかの部分構成要素の二次元分析または三次元分析が重視されてきた。

Tier：段階；段

1. **The first tier** involved making comparisons between RBCA predictions and MEPAS predictions. **The second tier** of testing involved comparing MEPAS predictions with field data.
   第一段階では RBCA による予測と MEPAS による予測が比較され，第 2 段階のテストでは，MEPAS による予測が現地データと比較された。
2. This **three-tiered** scheme works well for groups like stoneflies in which all species are nearly equal in tolerance to organic enrichment pollution.
   この三段層方式は，カワゲラのようにすべての生物種が有機物濃縮汚濁への耐性度がほぼ等しい集団では，うまく機能する。
3. It is important to note that the **four-tiered** system has been widely used in biomonitoring because it recognizes the limitations of refinement and accuracy that can be ascribed to this science.
   この四段層方式が生物モニタリングで多用されている理由は，この科学にある改良および正確性の限界を認識している点にあることを念頭におくべきである。

Time：時間

1. Sample processing **time** decreases as the processor gains taxonomic expertise.
   サンプル処理時間は，処理担当者が分類群知識を得るにつれて短縮される。
2. Further investigations into the **time course** of metal uptake in activated sludge have revealed that this process occurs in two stages.
   活性汚泥中で時間とともに変わる金属摂取をさらに調査し，このプロセスが 2 つの段階で起こることを明らかにした。
3. The **time** spent by states in developing biological criteria includes both the bioassessments conducted and the activities related to the implementation of biological criteria within water quality standards.
   諸州が生物クライテリア設定に費やした時間は，生物アセスメントの実行と，水質基準における生物クライテリア実施に関わる活動の両方を含んでいる。

- **From time to time**：時折
- **for a long time period**：長期間
- **from one time to another**：時間ごとに

- **at one time or another**：すべてある時期に

4. **From time to time**, attempts to develop mathematical models of bacterial growth have appeared.
   時折，バクテリアの成長を表わす数学モデルを構築する試みがされてきた。
5. It is reasonable to expect that the process of pyrite oxidation and gypsum formation will continue to occur **for a long time period**.
   黄鉄鉱の酸化と石膏生成プロセスが長期間起こり続けると考えるのが自然だ。
6. Resource managers and scientists have come to realize that the nature of this resource varies in an infinite number of ways, from one place to another and **from one time to another**.
   資源管理者や科学者は，これらの資源の性格が，場所ごと，時間ごとに無限な様相で異なることを理解するようになっている。
7. Many are also heavily impacted by urbanization and industries, and all streams relative to watersheds of 30 mi2 or more have been channelized **at one time or another**.
   多くは，都市化や産業の影響を強く受けており，$30m^2$ 以上の流域に関わる水流は，すべてある時期に流路制御を受けてきた。

- **Since that time**：それ以降

8. **Since that time**, USEPA has sponsored over 40 workshops on approaches to biological assessments and frameworks for developing assessment approaches.
   それ以降，USEPAは，アセスメント手法を定めるための生物アセスメントおよび枠組みに関して40以上ものワークショップを後援してきた。
9. **Since that time**, a much greater understanding has developed of how each level of treatment manifests itself in the aquatic environment.
   それ以降，各水準の処理が水生環境でどれほど効果を発揮するかについて，理解がかなり深まった。

- **as a function of time**：時間の関数として

10. The model can be used to calculate in-lake concentration of total phosphorus **as a function of time**.
    モデルは，湖の全リン濃度を時間の関数で計算するのに使うことができる。

    - **at that time**：当時は
    - **at this time**：現時点では
    - **at the present time**：現在
    - **at the current time**：現在

- **at a particular point in time**：特定時点で
- **at the same time**：同時に
- **at some later time**：後の

11. No effort was made to partition background variability by using ecoregions since the technology was not ready for use **at that time**.
当時は，技術がまだ整っていなかったので，生態地域を用いて背景可変性を仕分けする試みはなされなかった。
12. Given tolerance and identification considerations, biotic indices may not be appropriate to use in developing countries **at this time**.
耐性および識別への考慮に鑑みて，生物指数は，現時点では開発途上諸国での利用に適さないだろう。
13. **At the present time**, there are several uncertainties.
現在，いくつかの不確実性がある。
14. Whereas many thematic maps are the products of interpretations that include consideration of seasonal and year-to-year differences, Landsat imagery and high-altitude aerial photography are snapshots of conditions **at a particular time**.
多くの主題地図は，季節差や年差への考慮といった解釈の産物である。他方，ランドサット画像や高高度航空写真は，特定時刻での状態の断片である。
15. **At the current time**, most model testing and validation studies must rely on a limited data base.
現在，たいていのモデルの試験・確証研究は，限定されたデータベースに頼らなくてはならない。
16. The models represent a snapshot of something **at a particular point in time**.
このモデルは，特定時点での何らかのスナップショットを表す。
17. **At the same time**, it can be determined if refinements and advancement in macroinvertebrate taxonomy have been sufficient enough to warrant further adjustments to ICI scoring categories or the metrics themselves.
同時に，大型無脊椎動物分類群における改善と前進がICIスコアリング区分またはメトリック自体のさらなる調整を保証するに足るほど十分か否か決定されうる。

- **at the time of .....**：…の時点では

18. **At the time of** writing, these New York State regulations have not yet been released.
この文献を書いている時点では，これらのニューヨーク州の規則はまだ発表されていなかった。

19. In all other respects, the autopsy revealed that theses animals were healthy **at the time of** collection.
他のすべての点で，検死によると，これらの動物が採取時には健康であったことを明らかにした。
20. **At the time of** this writing, a draft of ecoregions of Alaska, continent with the ecoregions of Canada, has been completed.
本稿執筆時点で，カナダの生態地域に従いアラスカ州の生態地域ドラフトが作られている。
21. Some of the state program documents used to develop information for this summary were in draft form **at the time of** our review and are representative of preliminary programs.
本論のための情報作成に用いられた州プログラム文書の一部は，我々による査読時点で草稿段階にあり，予備的プログラムを表している。

● **over time** ：経年的に；経時的に；時間的

22. The model reasonably predicted CP concentrations **over time** in batch tests fed 246–TCP.
このモデルは，246–TCP を投入したバッチテストにおいて CP の経時的濃度を適度に予測した。
23. For each index there has been a significant positive change **over time** at most sites.
各指数について，大半の地点で，経年的に有意なプラス変化が見られた。
24. These methods provide an indication of what specific environmental characteristics are changing **over time**.
これらの手法は，どの環境特性が経時的に変化しているか示す指標となる。
25. When used in a monitoring program, these methods provide an indication of what specific environmental characteristics are changing **over time**, and can be used to help diagnose the cause of changes in the aquatic system.
監視プログラムで用いた場合，これらは，どの環境特性が経時的に変化しているか示す指標となり，水生系における変化原因を判定するのに役立つ。

● **over a human time frame** ：人間対象の時間枠で

26. In contrast to many other human impacts, habitat loss can be essentially irretrievable **over a human time frame**.
他の多くの人為的影響と異なり，生息場所の消失は，基本的に人間対象の時間枠では取り返しがつかない。

● **over the time of .....** ：時間の …

27. **Over the time of** scale of interest, this assumption is reasonably valid and may be used as a basis for a preliminary analysis of the problem.
我々が関心ある時間のスケールにおいて，この仮定は適度に有効であり，この問題の予備解析の基盤として使用されるかもしれない。

- **..... of the time**：…の割合で

28. In the same Wisconsin results cited above, chemical and biosurvey results agreed 58％ **of the time**, and impairment was indicated by chemical but not biological criteria at only 6％ of the sites.
前記のウイスコンシン州でのアセスメント結果において，化学的調査結果と生物的調査結果は，58％で一致した。また，生物クライテリアでなく化学クライテリアで損傷が示された地点は，わずか6％にすぎなかった。

Timely：タイムリーな

1. This system is designed to promote more efficient use of ambient monitoring resources and to ensure **timely** results.
この手法の目的は，環境モニタリング資源の効率的利用を促すこと，タイムリーな結果を確保すること，である。
2. Remote, difficult to reach areas have also been monitored by citizen volunteers when they were too difficult for state personnel to reach on a **timely** basis.
カバーしづらい遠隔地も，州職員がタイムリーに到着しかねる場合には，市民ボランティアによって監視されてきた。

Total：合計；全部

- **a total of .....**：全部で…の

1. **A total of** 93 stream sites was sampled over a three-week period during the fall of 1991.
全部で93の水流地点が1991年秋の3週間にサンプリングされた。
2. Thus, **the total of** 60 study units will have been sampled after nine years.
故に，合計60の研究単位がすべてサンプリングされるのは，9年後になる。
3. **A total of** 27 individual samples were compiled for sites just upstream of Conesville Station；51 samples were taken within the entire WAP ecoregion.
Conesville発電所のすぐ上流の4地点について合計27のサンプルが作成され，WAP生態地域全体では51サンプルが採取された。

4. This program will focus its survey and assessment activities on **a total of** 60 study units, each study unit a targeted watershed.
同プログラムは，調査・評価活動の焦点を合計 60 の研究単位に絞ることになっており，各研究単位がターゲット流域に当たる。

★ **total of** .....の前には冠詞 a または the が付きます。

Transformation：変換

1. A specific aspect of transport and **transformation** is examined from a systems perspective.
輸送・変換の特定の側面が，システムの視点から検定される。
2. Virtually all of the petroleum industry is based on a series of catalytic **transformations**.
石油工業のほとんどすべては一連の触媒変換に基づいている。
3. This paper describes the construction and application of a numerical model representing mercury **transformations** in the aquatic environment.
この論文には，水生環境での水銀の変化を説明する数値モデルの構築と応用が記述されている。
4. Knowledge of the factors affecting the movement and **transformations of** these substances are of obvious importance in understanding and controlling the hazard.
これらの物質の移動と変化に影響を与える要因の知識は，その物質の危険性を理解しコントロールしていくために当然重要である。
5. The chlorophenol **transformation** kinetics are described in Eqn 3.
塩化フェノールの変化速度は式 3 で記述される。
6. In this paper we examine physical transport and chemical **transformation** processes.
この論文では，我々は物理的輸送と化学変化プロセスを検討している。

Translate：解釈する；変換する；翻訳する

- **translate into** .....：… に変換する
- **~ is translated into** .....：~ は … へ変換される

1. There is no evidence that any such effects **translate into** impacts on fish populations.
そのような効果が，魚類集団に対する影響につながるという証拠はない。
2. USEPA's User's Manual (USEPA, 1989) provides a good overview of how

reference site data **were translated into** regional water quality criteria.
USEPAの『Users Manual（利用者用マニュアル）』(USEPA, 1989)は，参照地点データが地域的水質クライテリアへ変換される様相に関する適切な概観を提供している。
3. Scores for each metric range from 1 to 5 and these scores **can** then **be translated into** descriptive site bioassessments such as excellent, good, fair, or poor.
各メトリックのスコアは，1点から5点までであり，これらのスコアが次に記述的な生物アセスメント（たとえば，優，良，可，劣）へ変換される。

Trend：傾向

1. The **trend for** whole body game fish is less obvious and not necessarily declining.
ゲーム魚一体による（魚の体内濃度）傾向は，明白さに欠け，そして必ずしも低下していない。
2. A need exists for prediction of **future trends** and control in contaminated areas.
汚染された地域に対し，これからの傾向の予測と管理の一つの必要性が存在する。
3. Other problems associated with chlorophyll a are high temporal variability and difficulty in interpreting **trends**.
クロロフィルaに関わる他の問題は，時間的可変性の強さと傾向解釈の難しさである。
4. There seems to be no similarity between the **trends** in arsenic profiles and the **trends** in solid-phase sulfate profiles.
ヒ素プロフィールの傾向と固体相の硫酸塩プロフィールの傾向の間に類似性がないようだ。
5. The model is intended to compute **long term trends**.
このモデルでは，長期の傾向が計算されるように意図されている。
6. Although all indices showed **a worsening trend**, none exceeded the criteria, and no significant impairment was indicated.
すべての指数が悪化傾向を示したものの，いずれもクライテリアを超えておらず，有意の損傷が認められなかった。

True：事実の；本当の

- **this is true**：これは本当である
- **this is probably true**：これはおそらく本当であろう
- **this is particularly true**：これは特に本当であろう

- **this is especially true**：これは特に事実である

1. In most instances, **this will be true**.
   大半の事例では、これは事実となるであろう。
2. In most instances, however, **this will not be true**.
   しかしながら、大半の事例では、これは事実にならないであろう。
3. **This is probably true** in the sense that .....
   これは … という意味でおそらく真実であろう。
4. **This is particularly true** if baseline data are available for the waterbody in question.
   これは、特に当該水域について基準データが利用可能な場合に当てはまる。
5. **This is especially true** in such cases where a waterbody may be perceived as being at risk due to new dischargers.
   ある水域が、新しく生じた排出のために危険な状態であるものと見なされるかもしれない場合、これは特に事実である。

- ～ **holds true**：～は正しい
- ～ **remains true**：～は依然として正しい
- **the reverse will be true**：逆も正当であろう
- **the same is true of** .....：… についても同様である
- ～ **is not too far from the truth**：～ は真実からそれほど遠くはない

6. If this assumption **holds true**, the following equation can be used.
   もしこの仮定が正しければ、次の式が使われる。
7. If this assumption **remains true**, we can predict the relationship between CO_2 and global warming.
   もしこの仮定が依然として正しいなら、我々は CO_2 と地球温暖化現象の間の関係を予測することができる。
8. **The reverse will be true** for compressive axial stresses.
   圧縮軸のストレスについては逆も正当であろう。
9. While it is possible for reference sites to double as upstream control sites, **the reverse is not always true**.
   参照地点を対照地点の2倍にすることが可能な場合でも、逆は必ずしも真でない。
10. **The same is true of** whole effluent toxicity criteria.
    これは、全流出毒性（WET）クライテリアについても同様である。

Turn（Turning）：になる；回る；返る；向きを変える（曲がること）

1. **By the turn of this century**, it is expected that .....

この新世紀の到来時までに，…が期待される。
2. The improvement is even more remarkable when conditions **since the turn of the century**, when the river lacked any fish life for a distance of nearly 40 mile, are considered.
この川で約 40 mile にわたり魚類が生息していなかった今世紀初頭移行の状態を考慮すれば，この改善は，一層明らかである。

- **～ turned out to .....**：～は結果的に…となった

3. The filed experiment based on the laboratory results **turned out to be** a failure.
実験室での実験結果を基にしての現地実験は，結果的に失敗に終わった。

- **turning**：曲がること

4. The Columbia River flows through the northern part of the Hanford Site, and **turning** south, it forms part of the site's eastern boundary.
コロンビア川は Hanford 地域の北の部分を通過し，そして南に曲がり，地域の東境界線の一部を形成する。

Type：タイプ

1. **The type of** data required depends to a larger extent on the chemical being tested.
必要とされるデータのタイプは，被試験化学物質に大きく左右される。
2. **The type of** error can be of significant concern in a regulatory environment.
誤差の種類が，規制環境においてかなりの懸念となりうる。
3. **These types of** streams and watersheds would comprise relatively undisturbed references for the region.
この種の水流および流域は，その地域について比較的未擾乱な参照地点からなるであろう。
4. **One such** waste **type** is radioactive waste.
そのような廃棄物タイプの1つが放射性廃棄物である。
5. Collection of **this type of** data will likely increase over the next few years.
この種のデータ収集は，今後数年間に増える見込みが強い。
6. Needs include **various types of** information concerning toxic substances in water supplies.
必要とするものの中には，供給水中の毒性物質に関する種々のタイプの情報が含まれる。
7. There are certain situations in which **each type** is appropriate for use.

8. Comparisons were made with **two types of** data sets.
   2種のデータセットで比較された。
9. These simplified models contain **two types of** terms describing the transport of suspended or dissolved material through the estuary.
   これらの単純化されたモデルには，河口においての浮遊・溶解性物質の輸送を記述する2種の項が含まれている。
10. Attempts are being made to integrate **all three types** (physical, chemical, and biological) of sampling data.
    3種類(物理的，化学的，生物的)のサンプリングデータを統合する試みがなされつつある。

Typically (Atypically)：通常 (不定型に)；一般的に

1. **Typically**, the area of interest is an entire state, or a particular river basin or region of a state or multiple regions.
   通常，関心区域は，1州全体，もしくは1州の特定の河川流域，または地域や複数州地域である。
2. Any noncombustible debris remaining **is typically** landfilled.
   残っているどんな不燃瓦礫も主として埋め立て地に処分される。
3. Laboratory data, although not **typically** viewed as useful in determining reference conditions, can be valuable.
   実験データは，通常の場合，参照状態設定に有用だとみなされないが，有益にもなりうる。
4. Although these indices have been regionally developed, they are **typically** appropriate over wide geographic areas with minor modification.
   これらの指数は，地域的に設定されてきたものだが，一般に，やや修正すれば広い地理的範囲に適している。
5. Ecological studies **typically** focus on a limited number of parameters that might include one or more of the following: (a), (b), and (c).
   生態学研究が主眼を置くのは，通常，限られた数のパラメータであり，これには次のパラメータのうち1つ以上が含まれる。
6. Upon further investigation it was learned that one site was downstream from an experimental agricultural conservation tillage demonstration plot where pesticide usage was **atypically** intensive.
   さらなる調査研究で，一地点が殺虫剤飼養の不定型に集中的な実験農業保全耕耘実証区画から下流にあることが判明した。

Uncertainty：不確実

1. Natural variability also contributes to **uncertainty**.
   自然的可変性も不確実性に寄与する。
2. The **uncertainty** of using the pore velocity of 12.6 cm/day should be discussed.
   12.6 cm/day の細孔速度の使用についての不確実性が論じられなければならない。
3. Because of methodological **uncertainties**, time–trend analysis has been limited to the period from 1940 to the present.
   方法論的不確実性のため，時間的傾向分析は1940から現在までの間に限定されてきた。
4. The likelihood of a false negative decision is higher, due to **uncertainty** regarding the vertical extent of soil contamination and the potential for groundwater contamination.
   縦方向の土壌汚染と地下水汚染の可能性に関して不確実であるために，誤った否定の決定になる可能性がより高い。
5. The **uncertainty** associated with the definition of probable conditions and the consequences of a false positive or false negative error has been considered as part of the decision selection process.
   ありそうな条件の定義と虚偽の肯定的であるか，あるいは虚偽の否定的なエラーの結果と結び付けられた不確実は決定セレクションプロセスの一部であると考えられた。

Unclear（See Clear）：不明確な（➡ **Clear**）

Under：もとに；下に

1. Control experiments show that the SQ dye does not degrade in TiO_2 suspensions **under** visible light in the absence of the TiO_2 particles.
   対照実験は，TiO_2 粒子が存在しない場合，可視光線下ではSQ染料が分解しないことを示している。

- **under study**(See **Study**)：研究対象の(➡ **Study**)
- **under the conditions of** .....(See **Condition**)：
  …の条件下で(➡ **Condition**)
- **under consideration**(See **Consideration**)：
  考慮中の(➡ **Consideration**)
- **under the definition of** .....(See **Definition**)：
  …の定義下(➡ **Definition**)
- **under development**(See **Development**)：
  目下案出中の(➡ **Development**)
- **under the direction of** .....(See **Direction**)：
  …の指導のもとで(➡ **Direction**)
- **under investigation**(See **Investigation**)：
  調査対象の(➡ **Investigation**)
- **under Section X**(See **Section**)：X条のもとで(➡ **Section**)

Underestimate (See Estimate)：
過小評価；過小に見積もる(➡ **Estimate**)

Undergo ： 経る；経験する

1. The manner in which biological data was analyzed also **underwent** changes.
   生物学的データのかつての分析方法も変化した。
2. As a follow-up note, this sewage treatment plant subsequently **underwent** a major upgrade, and later macroinvertebrate sampling revealed an improved community downstream.
   追記すれば，この下水処理場は，その後大幅な改善がなされ，後の大型無脊椎動物サンプリングで下流生物群集における向上が見られた。
3. Benthic metrics **have undergone** similar evolutionary developments and are documented in the Invertebrate Community Index(ICI), Rapid Bioassessment Protocols(RBPs), and the benthic IBI .
   底生生物メトリックは，似たような展開をたどってきており，無脊椎動物指数(ICI)，迅速生物アセスメントプロトコル(RBP)，ならびに底生生物IBIにおいて記述されている。

Underlying：根底にある；根本的

1. An **underlying** assumption of this approach is that the patch work of performance standards, when implemented as a whole, will be sufficient to restore and maintain the physical, chemical, and biological integrity of the water.
   この手法の根本的な前提は，実績標準の集成が全体として実施された場合，その水の物理的・化学的・生物的な完全度を十分に回復・維持するだろうというものである。
2. It is important to note that most models represent an idealization of actual field conditions and must be used with caution to ensure that the **underlying** model assumptions hold for the site-specific situation being modeled.
   ほとんどのモデルは，実際の現地状態を理想化していること，そしてモデルの対象となっている特定な状況に対してモデルの基礎仮定が有効であることを保証するため注意して使われなくてはならないということ，を指摘することが重要である。
3. The **underlying** concept of the biomarkers approach in biomonitoring is that contaminant effects occur at the lower levels of biological organization before more severe disturbances are manifested at the population or ecosystem level.
   生物モニタリングにおける生物標識手法の根底にある概念は，より重い擾乱が個体群レベルや生態系レベルで明らかになる前に，より低次の生物レベルで汚染影響が生じるというものである。

Underscore：強調する；重視する

1. The widespread contamination of water supply systems **underscores** the need to develop innovative technologies to reclaim groundwater systems.
   給水系の広範な汚染は，地下水系を改善する改新的技術の必要性を強調する。
2. The model results **underscore** the significance of chemical partitioning on chemical fate and highlight the importance.
   モデルの結果は，化学物質の衰退におけるその物質分配性の意義を強調し，そしてその重要性を強調する。
3. It also **underscores** the need to develop both technologies and management strategies to protect drinking water supplies.
   それはまた，飲料水供給を保護するための技術戦略と管理戦略の両方を開発する必要性を強調する。

Understand(Understanding)：理解する(理解すること)

- **to better understand** .....：…をもっと良く理解するために

1. We will use these results **to better understand** the cause of forest damage.
   我々はこれらの結果をもっと良く森林損害の原因を理解するために使う。
2. Clearly, additional work will need to be done **to better understand** how electricity generation affects BOD removal.
   明らかに追加研究が，発電作用がBOD除去に及ぼす影響をもっと良く理解するために必要だ。
3. They saw the need for research **to better understand** the process by which the regions are defined, and how quantitative procedures could be incorporated with the currently used, mostly qualitative methods to increase replicability.
   彼らは，各地域が定義される過程や，反復可能性を増すため目下用いられているおおむね質的な方法に量的手順がどれほど盛り込まれるか，を一層理解するための研究の必要性を認めた。

- **to help us understand** .....：…に関する我々の理解を助けるために
- **In order to fully understand** .....：…を十分に理解するために

4. **To help us understand** the earlier experiments, we are now running more controlled experiments.
   これまでの実験のよりよい理解のために，今我々はいっそう制御された実験を行なっている。
5. **In order to fully understand** the potential impact of local resource management decisions, data should be available on the status of all the waterbodies within a watershed.
   地方の資源管理決定による潜在的影響を十分に理解するため，データは，ある流域における水域すべての現状について利用可能とされるべきである。

- ～ **is poorly understood**：～はあまり理解されていない
- ～ **is only partially understood**：～はただ部分的に理解される
- ～ **is not completely understood**：～は完全には理解されていない
- ～ **is not entirely understood**：～は完全には理解されていない
- ～ **is yet to be understood**：～はまだ理解されていない
- ～ **is not yet thoroughly understood**：
  ～まだ完全には理解されていない
- **It must be understood that** .....：…は理解されねばならない

6. Many of these processes **are poorly understood** in marine organisms and especially bivalves.
   これらのプロセスの多くが，海洋生物と特に二枚貝ではあまり理解されていない。
7. The basic process and the cause and effect relations **are still poorly understood**.
   その基本的なプロセスとその因果関係は，まだ良く理解されていない。
8. The observed heavy metal content of waterways is a function of many variables, with the roles of some components **being only partially understood**.
   水路で検出された重金属含有量は，多くの変数により変動し，ただ部分的にだけ理解されているある成分としての役割がある。
9. The significance of these higher arsenic concentrations **is not completely understood** for most of the sites.
   これら比較的高濃度の砒素の重要性は，大部分の地域で完全には理解されていない。
10. Although the mechanisms of acoustic foam disintegration **are not entirely understood**, the following acoustic effects are believed to be most important in foam breakage.
    音波によって生じる泡の崩壊機構は完全に理解されていないが，次に示す音波効果は気泡崩壊過程上最も重要であると思われる。
11. Catalysis by fixed cell is quite complex and many basic aspects **are yet to be understood**.
    固定細胞による触媒作用は非常に複雑で，多くの基本的な側面がまだ理解されていない。
12. However, the kinetics of biological nitrification **is not yet thoroughly understood**.
    しかし，生物硝化反応の速度論はまだ完全には理解されていない。
13. **It must be understood that** reference sites do not represent "pristine" conditions.
    参照地点が「無傷」状態を意味しないことを理解しなければならない。

- ～ **well understood**：
  ～はよく理解されている；～は十分に理解されている
- **a well-understood** ..... ：十分に理解されている…

14. The role of trace metals in regulating the growth rates of marine phytoplankton **is not well understood**.
    海洋藻類の成長速度制御における微量金属の役割はよく理解されていない。
15. Seasonality is **a well-understood concept**, therefore, it is not necessary to sample in multiple seasons for the sake of data redundancy.

季節性は，十分に理解されている概念であるから，データ冗長性のために複数季節でサンプリングする必要がない。

- **Understanding .....**：…を理解することは
- **lack of understanding**：理解不足

16. **Understanding** the basis behind these differences is important.
    これらの差の背後に潜む基礎を理解することが重要である。
17. **Understanding** the reactions of PCBs with・OH is crucial to predicting the persistence of PCBs and the hazards.
    PCBsと・OHの反応を理解することは，PCBsの持続性と危険性を予測するのに極めて重要である。
18. Satisfying this requirement involves **understanding** the nature of variability that originates with sampling frequency and/or seasonal influences.
    この要件を満たすには，サンプリング頻度および／または季節的影響に由来した可変性を理解する必要がある。
19. These omissions were not due to **lack of understanding**.
    これらの手落ちは，理解不足によるものではなかった。

- **understanding of .....**：…の理解
- **detailed understanding of .....**：…の詳細な理解
- **a greater understanding of .....**：…をより良く理解すること
- **a full understanding of .....**：…の完全な理解
- **a clear understanding of .....**：…の明確な理解

20. **The understanding of** the dynamic behavior of chemical species in a given aquatic ecosystem is a prerequisite for the risk and hazard assessment of chemical pollution.
    所定の水生生態系での化学種の動的な行動を理解することは，化学汚染のリスクと危険アセスメントのための前提必要条件である。
21. **Such detailed understanding of** the interactions is impossible from steady-state models.
    このような相互作用の詳細な理解は定常モデルからは不可能である。
22. **A greater understanding of** kinetic behavior should lead to better predictions of the behavior of PCP during anaerobic bioremediation.
    運動力学的挙動をより良く理解することは，嫌気性生物による修復中のPCPの挙動をもっと良く予測することにつながる。
23. Watershed studies are a necessity, but equally important is the **development of an understanding of** the spatial nature of ecosystems, their components, and the stress we humans put upon them.

流域研究は必要であるが，同じく重要なのは，生態系の空間的性質，その構成要素，そして我々人間が生態系に与えるストレスへの理解を育むことである。

24. By calibrating over large areas, decisions about level of protection can be made with **a clear understanding of** the differences among areas over which differing criteria might be established.
広い地域での較正による保護レベルに関する決定は，異なるクライテリアが設定されうる各地域の差異を明確に理解したうえで下されるようになる。

> ★ understanding を名詞として使う場合には，一般に冠詞 a または the が付きますが little を伴う場合は冠詞は付きません。
>
> ● **there is little understanding of** …… ：
>   … はほとんど理解されていない。

25. Presently, **there is little understanding of** the interaction of surfactants and these compounds.
現在，界面活性剤とこれらの化合物の相互作用はほとんど理解されていない。

> ● **develop an understanding of** …… ：… への理解を生じさせる
> ● **develop a clearer understanding of** …… ：
>   … への明確な理解を進展させる
> ● **a much greater understanding has developed** ：
>   理解がかなり深まった

26. We must **develop an understanding of** ecosystem regionalities at all scales, in order to make meaningful extrapolations from site-specific data collected from studies or watershed studies, or whatever they are called.
我々は，ケーススタディまたは分水界研究，もしくは名称を問わずあらゆる研究で収集された地点特定データからの推定を有意義にするために，すべての縮尺での生態系地域性を理解しなければならない。

27. The field investigations were focused on the local study areas in order to obtain sufficient data **to develop an understanding of** the hydrogeochemical conditions.
水理地質学化学的状態を理解するのに十分なデータを得るために，この野外調査は近郊の研究地域に絞られた。

28. As we increase our awareness that a holistic ecosystem approach to environmental resource assessment and management is necessary, we must also **develop a clearer understanding of** ecosystems and their regional patterns.
我々は，環境資源の評価・管理への全体論的生態系手法が必要だと深く認識する

につれて，生態系とその地域パターンも明確に理解しなければならない。

29. In order to maximize the meaningfulness of extrapolations from these studies, we must **develop a clear understanding of** ecosystem regionalities.
これら研究からの推定の意義を最大化するため，我々は生態系地域性へのより明確な理解を進展させるべきである。

30. Since that time, **a much greater understanding has developed** of how each level of treatment manifests itself in the aquatic environment.
それ以降，各水準の処理が水生環境でどれほど効果を発揮するかについて，理解がかなり深まった。

- **to gain an understanding of .....** ：… を理解するために
- **to gain a further understanding of .....** ：
  … をいっそう理解するために

31. Levin (1992) stressed that **to gain an understanding of** the patterns of ecosystems in time and space and the causes and consequences of patterns, we must develop the appropriate measures and quantify these patterns.
Levin (1992) は，時間・空間における生態系パターンや，パターンの原因・結果を理解するために，我々が適切な測度を定め，これらのパターンを数量化しなければならないことを強調した。

32. More extensive and detailed investigations are necessary **to gain a further understanding of** these processes.
いっそう大規模かつ詳細な調査が，これらのプロセスをいっそう理解するために必要である。

- **～ lead to a better understanding of .....** ：
  ～は… をより深く理解するのに役立つ

33. This approach **leads** itself **to a better understanding of** the nature of the impairment, including which elements or processes of the community are most affected.
この手法は，生物集団のどの要素や過程が最も影響されるかなど，損傷の性質をより深く理解するのに役立つ。

34. The increasing number of investigations concerning the fate of aromatic xenobiotics in the marine environment **has led to a better understanding of** the effects of these compounds.
海洋環境における芳香族生体異物の衰退に関する調査数が増加しており，これらの化合物の影響をより深く理解するのに役立ってきた。

- **in understanding of .....** : … を理解するうえで

35. In most areas, the use of watersheds is an obvious necessity **in** defining and **understanding** special patterns of aquatic ecosystem quality and addressing ecological risk.
大半の区域において，流域の利用は，水生生態系の質での空間パターンの画定・理解，ならびに生態学的リスクへの対処において明らかに必要不可欠である。

36. Knowledge of the factors affecting the movement and transformations of these substances are of obvious importance **in understanding** and controlling the hazard.
これらの物質の移動・変化に影響を与えている要因に対する知識は，危険を理解し，またコントロールするうえで明らかに重要である。

Undertaken：着手される

1. Research **has been undertaken** to increase the level of understanding of the effects of heavy metals.
重金属の影響に関する理解力レベルを向上させるために研究が着手された。

2. In 1970–1971, a series of studies by the U.S. EPA (1973) **were undertaken** to document and characterize the discharge of tailings to Whitewood Creek.
1970 から 1971 までに，Whitewood Creek への選鉱廃物の排出を文書化し，特徴づけするための一連の研究が U.S. EPA (1973) によって着手された。

3. In developing biological criteria for water quality programs, states **have undertaken** a wide range of efforts to improve bioassessment methods, transform biological monitoring activities into programs using biological criteria, and incorporate biological criteria into water quality standards or water resource management programs.
水質プログラムのための生物クライテリア設定において，諸州は，生物アセスメント方法を改良し，生物モニタリング活動を生物クライテリア利用プログラムへ変化させ，生物クライテリアを水質基準または水資源管理プログラムに盛り込むため，広範な努力を払ってきた。

4. In contrast, many other states **did not undertake** substantial bioassessment programs because they felt the cost of bioassessment outweighed the benefits.
対照的に，他の多くの州は，生物アセスメントの費用が便益をはるかに上回ると考えたせいで，実質的な生物アセスメントプログラムを行わなかった。

5. The first systematic studies **undertaken** by EPA in 1964 quantified cyanide loading to Rock Creek.

1964年に EPA によって着手された最初の組織的な研究によって，Rock Creek へのシアン化物の負荷量が数量化された。

Underway：進行中

1. **Work is underway** to develop methods for Lake Erie river mouth and near shore areas.
   Erie 湖の河口・沿岸区域において，方法を策定する作業が進行中である。
2. **Work is underway** to address water quality criteria for these areas within the next 3 to 5 years.
   今後3年間ないし5年間，これらの区域についての水質クライテリアに取り組む活動が進行中である。
3. Evaluation and optimization of specific benthic invertebrate indicators **is** currently **underway**.
   具体的な底生無脊椎動物指標の評価と最適化が目下進行中である。
4. Substantial national **efforts are underway** to provide consistency in the implementation of biological criteria.
   生物クライテリアの実施における整合性をもたらすために相当の全米的な努力が払われている。
5. To our knowledge, **efforts are** currently **underway** to generate IBI version for streams in the coastal plain of Maryland and Delaware.
   筆者の知るところでは，メリーランド州およびデラウェア州の海岸平野における水流について，IBI バージョンを生じるための努力が目下，払われている。

Uniqu(Uniquely)：ユニークな(珍しく；独特な)

1. Each element of material flows downstream in a **unique** discrete fashion.
   物質の要素それぞれが，独特の断続的な形で下流に流れる。
2. This report is **unique** in that it is based on Hanford-specific agreements discussed in the facility agreement.
   それが施設合意書に論じられている Hanford 特定協定に基づいているという点で，この報告はユニークである。
3. Therefore, permitting experiences for AEP facilities are **unique** and cannot be considered representative of Alabama's regulated industry as a whole.
   したがって，AEP 施設の許可経験は，独自のものであり，アラバマ州の被規制業界全体を代表しているとみなすことができない。
4. Simply referring to a waterbody as a high-quality or significant resource is inadequate given the penchant for these characterizations to have **unique**

attributes according to the individual making the pronouncements.
意見表明者によれば，これらの特性が独自な属性を帯びがちな傾向に鑑みて，ある水域を単に高質または有意資源と評するだけでは不十分である。

- **uniquely**：珍しく；独特に

5. In extensively disturbed regions and **uniquely** undisturbed regions, the method of reference site selection will likely be less of an issue because of relatively homogeneous conditions.
はなはだしく攪乱された地域や，珍しく攪乱されなかった地点では，比較的同質な状態の故に，参照地点選定方法がさほど問題にならないだろう。

Unless：…でない限り；もし…でなければ

1. Underground tanks used only for storage of products have not been included **unless** a release is known or suspected to have occurred.
流出が知られている場合，あるいは流出が起こったという疑いがある場合でない限り，生産物の貯蔵のためにだけ使われた地下タンクは含まれなかった。

- **unless otherwise noted**：注意書きが無い場合は
- **unless otherwise stated**：述べられていない場合は

2. **Unless otherwise noted**, all information and procedures for working with worksheet apply to macro sheets.
注意書きが無い場合は，ワークシートで作業するためのすべての情報・手順がマクロシートに適用される。

Unlike (See Like)：と異なり (→ Like)

Unlikely (See Likely)：と異なり (→ Likely)

Until：…まで

1. Samples were collected and stored at 4 ℃ **until** use.
サンプルは集められ，使うまで4℃で保管された。
2. Groundwater modeling is a relatively new field that was not extensively pursued **until** about 1965.
地下水モデリングは比較的新しい分野で，およそ1965までは幅広く追究追され

なかった。
3. For a period of approximately 100 years, from the original discovery of gold at Deadwood Gulch in 1875 **until** the late 1970's, huge volumes of mining and milling wastes were discharged into Whitewood Creek.
1875年にDeadwood Gulchでの最初の金の発見から1970年代後期までおよそ100年の間，大量の鉱山採掘廃棄物と粉砕廃棄物がWhitewood Creekに放出された。
4. However, **until** more reference sites are established, the Division of Water Pollution Control (DPC) is using upstream reference sites to assess stream impacts on a case-by-case basis.
しかしながら，より多くの参照地点が定められる日まで，水質汚濁防止局(DPT)は，上流の参照地点を用いて水流影響をケースバイケースで評価している。
5. For example, the proportion of sites with IBI scores <30 is illustrated by following the vertical line originating at 30 on the x axis to its intersection with the CDF, the following the horizontal line **until** it intersects the y axis.
例えば，IBIスコアが30点未満の地点の比率は，x軸の30点で始まる垂直線をCDFとの交差点まで辿った後，それがy軸を横切るまで水平線を辿れば示される。

Upheld：是認される

1. So long as the science is sound and well supported, the findings likely **will be upheld** in the courts.
科学が堅実で十分に裏づけられる限り，その所見は，法廷で是認されるだろう。
2. The redesignation to WWH **was upheld** throughout the appeals process including the North Dakota Supreme Court.
WWHへの指定変更は，ノースダコタ州最高裁など上訴過程全体で是認された。

Upon：に基づいて

1. **Upon completion of** the initial revision of ecoregions and delineation of subregions, sets of reference sites are identified for each subregion.
生態地域の初回改良，ならびに小地域の画定の完了時に，各小地域について参照地点セットが確認される。
2. **Upon further investigation** it was learned that one site was downstream from an experimental agricultural conservation tillage demonstration plot where pesticide usage was atypically intensive.
さらなる調査研究で，一地点が殺虫剤飼養の不定型に集中的な実験農業保全耕耘実証区画から下流にあることが判明した。

- ~ **is based on**(**upon**)…..(See **Base**)：
  ~は … に基づいている(➡ **Base**)
- ~ **is dependent on**(**upon**)…..(See **Dependent**)：
  ~は … に依存している(➡ **Dependent**)
- **depending on**(**upon**)…..(See **Dependent**)：
  … に応じ(➡ **Dependent**)

Use(Using)： 使用；利用；使う；用いる(用いて；使って)

1. Arsenic has a long history of **use for** its toxic and medicinal properties.
   ヒ素は毒として，また薬として長い歴史を持つ。

   - **the use of** ….. ： … の利用は
   - **the widespread use of** ….. ： … の広範囲にわたる使用
   - ~ **is of use in** ….. ： ~は … の役に立つ

2. **The use of** probabilities offers certain advantages for implementation.
   確率の利用は，実施上の利点をいくつかもたらす。

3. **The use of** larger-scale (covering smaller areas) materials is also expensive and they must be carefully evaluated for representativeness.
   大縮尺(狭い面積をカバー)資料の利用は，高価でもあり，代表性に入念に配慮しなければならない。

4. A major problem preventing **the widespread use of** ozone is the high cost of treatment.
   オゾンの広範囲にわたる使用を妨げている主要な問題は，処理コストが高いことにある。

5. This research provides background information which **would be of use in** interpreting the effects of heavy metals in wastewater treatment systems.
   この研究は，廃水処理システムにおいて重金属の影響を解釈するのに役立つであろうバックグラウンド情報を提供する。

   - ~ **make use of**….. ： ~は … を活用する
   - ~ **make full use of**….. ： ~は … を十分に活用する

6. Design of such a repository would **make use of** multiple engineered and natural barriers.
   このような貯蔵所の設計は，技術によるバリア(遮断物)と自然なバリアを複式で活用している。

7. Multivariate analyses are useful because they **make use of** the information in assemblages of quantitatively infer ecological characteristics.

多変量分析は，生態特性を量的に推論するために群がり情報を活用しているので，有用である。

8. A modeling strategy which **makes full use of** available data must be devised.
   利用可能なデータを十分に活用するモデリング戦略が考案されなくてはならない。

   ★ Use の動詞としての使用例文を下に示します。

   - **We will use** …：我々は…を使うであろう
   - ～ **has been widely used in**…..：～は…で多用されている

9. **We will use** these results to better understand the cause of forest damage.
   我々は，森林の損傷の原因をいっそう理解するために，これらの結果を使うであろう。

10. It is important to note that the four-tiered system **has been widely used in** biomonitoring because it recognizes the limitations of refinement and accuracy that can be ascribed to this science.
    この四段層方式が生物モニタリングで多用されている理由は，この科学に帰されうる改良および正確性の限界を認識している点にあることを念頭におくべきである。

11. The plant discharges mercury **used** in the processing from early years.
    早年から，プラントは使用済みの水銀を排出している。

12. The method **used** here may be a new approach not only for gaining better results, but also for acquiring new knowledge and insights about toxic behavior.
    ここで使われた方法は，もっと良い結果を得ることだけではなく，毒物の挙動について新しい知識と洞察を得するのに，新しいアプローチであるかもしれない。

    - **using** …..：…を用いて；…を使って

13. **Using** all the estimated rate constants, the calculated accumulation of CO_2 is plotted in Figure 2.
    すべての推定速度定数を使って計算された CO_2 の蓄積量を図2に図示する。

14. One advantage of **using** a database such as RF3 is that it enumerates the resource.
    RF3といったデータベース使用の利点は，資源を列挙することである。

15. The approach in **using** ultrasound is obviously applicable to organic compounds which contain halides.
    超音波を使うアプローチは，ハロゲンを含む有機化合物に対し明らかに適用できる。

16. There have been some attempts to estimate copper toxicity in seawater **using** bioassay.

バイオアッセイによって，海水での銅の毒性を推測する試みが若干されてきた。
17. Data were gathered from a number of experiments **using** a pilot-scale reactor.
試験規模の反応器を使った多数の実験から，データが収集された。
18. The solution of these equations is accomplished **using** finite difference approximations in identical fashion to the Water Quality Analysis Simulation Program (WASP).
これらの方程式は水質解析シミュレーションプログラム (WASP) と同一方法，有限差分近似を使って解かれている。

Useful (Usefulness)：有用な（有用性）

1. This application **can be useful in** a number of ways.
この適用は，多くの点で有用となりうる。
2. Laboratory data, although not typically viewed as **useful in** determining reference conditions, can be valuable.
実験データは，通常の場合，参照状態の設定に有用だとみなされないが，有益にもなりうる。
3. Physical characteristics of individuals that **may be useful for** assessing chemical contaminants would result from microbial or viral infection, some sort of tetragenic or carcinogenic effects during development of that individual.
化学的汚染アセスメントに有用だと思われる個体の身体特性は，細菌・ウイルス感染や，その個体の発生中における何らかの催奇形影響または発癌影響に起因する。
4. While model applications may differ greatly in scope and purpose, it is hoped that the representative data derived from pilot studies **will be useful** in this process.
モデルのアプリケーションが応用範囲と目的で大いに異なるかもしれない一方，予備試験的研究にむける代表的データがこのプロセスに有用であろうことが期待される。

- **~ have proven useful for .....**：
〜は … に有用であることが判明している

5. Periphyton **have proven useful for** environmental assessment.
付着生物は，環境アセスメントに有用であることが判明している。
6. Such information **has proven very useful for** assessing changes in climate, salinity, pH, and nutrients.
こうした情報は，気候，塩度，pH，養分における変化を評価するため非常に有

7. As a tool for identifying problems, advocating for mitigation and enforcement of regulations, volunteer data **have proven very useful**.
   問題点を特定するためのツールとして，規制の緩和・施行を求める擁護活動や，ボランティアデータは，非常に有用だと判明している。

    ● **It is especially useful to .....**：…することが特に重要である

8. **It is especially useful to** select panelists and reviewers with opposing biases and different professional backgrounds to ensure different points of view, and to increase the credibility of the product in the eyes of the public.
   様々な視点を確保し，判断結果を一般市民が眺めた際の信頼度を高めるには，反対意見や異なる専門的背景を持つ委員および査読者を選ぶことが特に重要であろう。

    ● **useful .....**：有用な…

9. **Useful** information on farming practices had been gathered for the Iowa River study.
   農業に関する有用な情報がIowa川の研究のために集められていた。

10. Once field data are collected, processes, and finalized, the next step is to reduce the data to scientifically and managerially **useful** information.
    現地データが収集・処理・最終確定されたならば，次のステップは，データをまとめて科学的・管理的に有用な情報にすることである。

11. Also, I thank the following individuals for contributing **useful** comments on a draft version: .....
    また，わが草稿について有意義な意見を述べてくださった次の諸氏にも感謝したい。…

    ● **usefulness**：有用性

12. **The usefulness of** this to watershed planning is obvious since quantifiable estimates of land use and riparian corridor compatible with IBI values that stain a particular use or goal can be made with a known degree of certainty.
    流域立案におけるこの有用性は，明瞭である。なぜなら，特定の使用または目的を達成するIBI値と両立する土地利用および水辺地帯の計量可能な推定は，既知の確実度でもって得られるからである。

13. The data lead to some direct conclusion about **the usefulness of** the three models in safety assessment of new chemicals.
    このデータは，新化学物質の安全性アセスメントでの3つのモデルの有用性について，ある直接的結論に導く。

14. Additional work is needed to more precisely define and set expectations for

these two metrics, and to test **their usefulness** in field applications.
これら2つのメトリックについて期待値をより正確に設定し，現地適用におけるその有用性を試すため，さらなる研究が必要とされる。

15. To date, qualitative methods, although used for many applications where **the usefulness of** the results is more important than the scientific rigor of the technique used, have not been widely accepted.
今日までのところ，質的方法は，技法の科学的厳密性よりも結果の有用性が重要な多くの適用例で用いられているが，まだ広く容認されてはいない。

Utilize：活用する；利用する；役立たせる

1. The method **utilizes** the qualitative, natural substrate collection procedure, which necessitates one site visit and minimal laboratory analysis.
この方法は，1回の地点訪問と最小限の実験室分析を必要とする質的な自然底質採集手順を活用してくる。

2. Future work is needed **to better utilize** the ADV as part of resource value assessment.
資源価値アセスメントの一部としてADVをより活用するには，今後の作業が必要である。

3. Precipitation gages and evaporation pans **can be utilize**d with sufficient accuracy.
降水量計器と蒸発パンが十分な正確さで使用できる。

Valid：有効な

1. This is a **valid** concept.
   これは有効な概念である。

   - **~ is valid to .....**：~することは妥当である
   - **~ has been found valid for .....**：
     ~が … で有効であることが見いだされた

2. This assumption is reasonably **valid** and may be used as a basis for a preliminary analysis of the problem.
   この仮定はかなり有効であって，この問題の事前分析の基礎として使用されるかもしれない。

3. Just because a model **has been found valid for** one use does not mean it is appropriate for some other use.
   ただモデルが1つの使用で有効であることが見いだされたからといって，それが何か他の使用において適切であることを意味しない。

   - **It is valid to .....**：… することは妥当である

4. **It is valid to** directly apply a model.
   モデルを直接適用することは妥当である。

5. **It may be valid to** argue that the solar spectrum is sufficiently constant.
   太陽光線スペクトルが十分に（条件を満たし）一定であるという点について議論することは妥当であろう。

Validate：実証する

1. Few attempts have been made **to validate** these models.
   これらのモデルを実証する試みがほとんどされなかった。

2. **To validate** the above hypothesis, the experiments were repeated with a catalyst concentration of 300 ppm.
   上記の仮説を実証するため 300 ppm の触媒濃度で実験がくりかえし行われた。

3. We recommend that the highest priority for future experimental work is **to**

validate the existing models.
これからの実験研究の最優先事項として我々は，既存のモデルを実証することを推奨する。

- ～ **have not been validated**：～は実証されていない
- ～ **has as yet not been properly validated**：
 ～は，まだ適切に確証されていない

4. This hypothesis **has not been validated**, and this phenomenon remains an open question.
 この仮説は実証されていないし，それにこの現象は未解決の問題のままである。
5. Finally, and perhaps most importantly, many existing versions **have as yet not been properly validated** with independent data.
 最後に，そしておそらく最も重要なことに，多くの既存バージョンは，独立データで適切に実証されていない。

Validity：妥当性

1. There is considerable **validity** in this theory.
 この理論にはかなりの妥当性がある。
2. **The validity of** these predictions could be determined by an examination of the biotic condition of the stream.
 これらの予測の正当性は水流の生物状態の検定によって決定されえる。
3. **The valldlty of** an integrated assessment using multiple metrics is supported by the use of measurements of biological attributes firmly rooted in sound ecological principles.
 多メトリックを用いた統合的アセスメントの妥当性は，確固たる生態学的原理に根ざした生物学的属性を測定することで確保される。

Velocity：速度

1. **A settling velocity of** 1.2 m/d is assigned as characteristics of the James River solids.
 1.2 m/d といった沈降速度がジェームズ川の固形物の特徴として使われている。
2. Aggregation effects are taken into account as an increase of **sedimentation velocities of** the particles.
 凝集効果が粒子の堆積速度の増加の原因として考慮されている。
3. Cavitations and the shock waves can accelerate solid particles to **high velocities**.

キャビテーションと衝撃波は，固体の微片を高速度に加速することがでる。
4. To estimate **the average velocity**, use the plot and identify the velocity that corresponds to the average annual discharge.
平均速度を推定するためにプロットを使い，そして平均年排水量に対応する速度を特定してください。
5. Riffles and runs, with **current velocities of** 10 to 20 cm/sec, are best because biomass is least variable in these habitats.
流速10〜20 cm/sの瀬および急流が最適である。なぜなら，これらの生息場所では，生物量が最も弱いからである。

Variability：ばらつき；変動

1. Sample **variability** can result as a function of substrate angle and positioning.
サンプルのバラツキは，底質の角度や配置に連関して生じうる。
2. Due to **variability** in microbial densities between samples, these changes were not statistically significant.
微生物の密度はサンプル間でばらつきであるため，これらの変動は統計的に有意ではない。
3. Differences among sites would be much greater than year−to−year **variability** at the same site.
地点間の差の方が同一地点での年の変動よりも大きいのである。
4. Other problems associated with chlorophyll a are high temporal **variability** and difficulty in interpreting trends.
クロロフィルaに関わる他の問題は，時間的変動の強さと傾向解釈の難しさである。
5. The most accurate way to decrease sample **variability** is to collect from only one type of habitat within a reach and to composite many samples within that habitat.
サンプルのばらつきを減じるための最も正確な方法は，ある河区において1種類だけの生息場所から採集し，その生息場所における多数のサンプルを複合することである。
6. Satisfying this requirement involves understanding the nature of **variability** that originates with sampling frequency and/or seasonal influences.
この要件を満たすには，サンプリング頻度および／または季節的影響に由来した変動特性を理解する必要がある。

Variable：可変性の；変数

1. The observed heavy metal content of waterways is a function of many **variable**s.
   水路にて観察された重金属含有量は，多くの変数の間数である。
2. Discrimination analysis provides a weight for each **variable** in the form of a coefficient.
   判別分析は，係数の形で各変数に加重値を与える。
3. Since the **variable**s have different information content, the weight of each **variable** used in classification is expected to be different.
   変数は，それぞれ異なる情報内容を持つので，分類で用いられる各変数の加重値がそれぞれ異なると予想される。
4. Comparisons were made between process operational **variable**s and effluent quality to assess the influence of the heavy metals.
   重金属の影響を評価するため，プロセス操作上の変改と流出の水質の比較を行った。
5. Although this is a substantial change in species, the assessment of stream condition based on indices is much less **variable**.
   これは，相当大幅な生物種変化だか，指数に基づく水流状態アセスメントは，可変性がはるかに弱まる。
6. Work has shown that highly **variable** and unpredictable flow regimes can have strong influences on fish assemblages.
   研究によれば，可変性が強くて予測不能な流量状況は，魚群に強い影響を及ぼしうる。
7. This is probably attributable to the **variable** nature of activated sludge.
   これはおそらく活性汚泥の可変的性質に帰することができる。

Variety：多種多様

● **a variety of .....**：さまざまな…；いろいろな…

1. **A variety of** stochastic and deterministic soil leaching models have been developed.
   さまざまな推計的・決定論的な土壌汚染浸出モデルが構築されてきた。
2. **A variety of** decisions must be made, not all of which are technical or scientific in nature.
   このためには様々な決定が必要なのだが，そのすべてが技術的決定や科学的決定というわけではない。

3. Toxic metals enter waste waters from **a variety of** sources.
   毒性金属がいろいろな源から下水に流入する。
4. Rapid assessment approaches are intended to identify water quality problems and to classify aquatic habitats according to **a variety of** water resource criteria.
   迅速アセスメント手法の意図は，水質上の問題点を確認すること，様々な水資源クライテリアに従って水生生息地を分類することである。
5. It is anticipated that contaminated plant leaves may be exposed to **a variety of** acidic or basic conditions from rain water, prior to analysis.
   分析の以前に，汚染された植物の葉が，雨水から様々な酸性，あるいはアルカリ性の状態にさらされるかもしれないことが予想される。
6. Case histories are presented below showing the outcome of impairment criteria testing in **a variety of** situations.
   次に，様々な状況での損傷クライテリア検定の結果を示した事例史を述べる。

Verify（Verification）：立証する（立証）

1. Because of the lack of data, the methodology **was not fully verified**.
   データの欠如のために，この方法論は十分に立証されなかった。
2. Thus, metrics reflecting biological characteristics may be considered as appropriate in the programs if their relevance can be demonstrated, response range **is verified** and documented, and the potential for application in water resource programs exists.
   それ故，生物学的特性を反映するメトリックは，その関連性が実証され，応答値域が検証・文書化され，水資源プログラムにおける応用可能性が存在する限り，このプログラムに適しているとみなされよう。

　　● **verification of** ……：…の立証

3. This will provide further **verification of** the results obtained.
   これは得られた結果を更に立証することになろう。

Vary（Variation）：異なる（変化；変動）

1. In this figure, the dashed line represents how A **varies with** B.
   この図において，ダッシュラインはBの変化に対してAがどのように変化するかを示している。

　　● **vary from A to B**：AからBまで異なる

2. The factors that are more or less important **vary from** one place **to** another at all scales.
多かれ少なかれ重要な因子は，あらゆる縮尺で場所ごとに異なる。
3. The removal efficiency of metals **varied** significantly **from** metal **to** metal and plant **to** plant for the same metal.
金属の除去率は金属ごとで，および同金属においてもプラントごとで大幅に異なる。

- **vary among** …… ：… の間で異なる

4. The extent and form of biological survey data **vary** widely **among** states.
生物調査データの程度と形式は，州ごとで大幅に異なる。
5. Although every program provides some form of educational benefits, the extent of educational opportunities **varies among** the programs, some of which exist almost entirely to provide education, some for which education is secondary to action or data collection, and others where the objectives of education and data collection are equally important.
どのプログラムも何らかの啓発効果を有するものの，啓発機会の度合は，プログラムごとに異なっている。例えば，現存プログラムの一部は，もっぱら啓発を提供していたり，啓発が行動やデータ収集の従属的なものだったり，啓発目標とデータ収集目標が同等に重要だったりする。

- **Variation** ：変化；変動

6. The available field data were not sufficiently detailed to allow an assessment of spatial **variations**.
入手可能な野外データは，空間的変動の評価を可能にするまでには詳細さに欠けていた。
7. Figure 3 shows **the variation of** oxalic acid concentration with irradiation time.
図3は照射時間でのシュウ酸の濃度変化を示す。
8. Different classes of algae have different proportions of internal structure occupied by vacuoles, which are important sources of **variation** in measurements of biovolume.
異なる綱の藻類は，液胞に占められた内部構造の比率が異なっており，それは，生物体積測定における変動の主因となる。

via ：によって

1. This could be accomplished **via** a redirection of existing state agency resources.

これは，州機関の既存資源の方向転換によって達成されうる。
2. Efforts to assess, research, and manage the ecosystems are normally carried out **via** extrapolation from data gathered from single-medium/single-purpose research.
生態系を評価・研究・管理する努力は，通常，単一媒体／単一目的の研究で収集されたデータからの推定によって行われる。

View：視点；見方；観点

1. One **view** is that .....
   一つの見方は … ということである。
2. If this **view** is adopted, ...
   もしこの見方が採用されるなら，…
3. It should be noted that our soil data comparisons are an extremely conservative **view of** the data.
   我々の土壌データの比較がデータの極めて保守的な見方であることは指摘されるべきである。

   - **In view of** ..... ：… を考慮したうえで；… の見地から
   - **with a view of** ..... ：… という視点では

4. **In view of** the impracticability of .....
   … が実行不可能であることから判断して
5. 36Cl was the radionuclide of primary interest **in view of** its importance for post-closure radiological safety assessment.
   施設閉鎖後の放射線学的安全性アセスメントの重要性の見地からして，36Clは主に興味深い放射性核種であった。
6. This is somewhat surprising, **in view of** the order of magnitude higher water column decay rate of DDE relative to Lindane.
   Lindaneと比較して，水柱でのDDEの桁違い高い分解速度の見地から，これはいくぶん驚きである。
7. It is likely that interventions of man **with a view of** improving groundwater quality in the alluvium would not cause much improvement.
   沖積層での地下水の水質改善という視点からは，人間が介入しても多くの改良は期待できない。

   - **point of view** ：視点;考え方
   - **from a point of view**：
     … の視点から；… の見地から；… の観点から見ると

8. The **point of view** we take in this paper is that .....
   我々がこの論文でとっている考え方は，… ということである。
9. It is especially useful to select panelists and reviewers with opposing biases and different professional backgrounds to ensure different **points of view**, and to increase the credibility of the product in the eyes of the public.
   様々な視点を確保し，判断結果を一般市民が眺めた際の信頼度を高めるには，反対意見や異なる専門的背景を持つ委員および査読者を選ぶことが特に重要であろう。
10. Lakes are looked at more **from a** dynamic and mechanistic **point of view**, rather than a descriptive one.
    湖が，記述的な見地よりどちらかと言うと，いっそう動勢・機械論的見地から見られる。
11. This study determines the overall quality both **from an** ecological **point of view** and also **from a** user's **point of view**.
    この研究では，生態的な見地とユーザーの見地両方から総体的な質が決められる。

    ★ View の動詞としての使用例文を下に示します。

    ● ～ is viewed as ..... ：～は … とみなされる

12. The procedure **can be viewed as** part of the regionalization process.
    この手順は，地域分け過程の一部とみなしうる。
13. Coralville Reservoir does not thermally stratify to any great extent, so the failure to include a hypolimnion compartment **is not viewed as** a serious problem.
    Coralville 貯水池は大きくは温度層化しない，それゆえ（このモデルに）深層部が含まれていないということが重大な問題だとは見なされない。
14. Each area **can be viewed as** a discrete system which has resulted from the mesh and interplay of the geologic, landform, soil, vegetative, climatic, wildlife, water and human factors which may be present.
    各領域は，個別的なシステムとみなされ，地質，地形，土壌，植生，気候，野生生物，水，そして潜在的な人間因子の調和および相互作用から生じてきた。
15. Laboratory data, although not typically **viewed as** useful in determining reference conditions, can be valuable.
    実験データは，通常の場合，参照状態設定に有用だとみなされないが，有益にもなりうる。

Viewpoint ：立場；観点；見地

1. From the **viewpoint of** .....

…の立場からすると
2. From the ….' s **viewpoint**,
   …の視点から，
3. The following is a list of his criticisms and a response to **a potentially limited viewpoint of** the IBI.
   以下に，彼の批判と，IBI の潜在的に限られた観点への反応を列挙していく。

vice versa ： 逆もまた同様に

1. For example, what are the antidegradation ramifications where water quality exceeds levels necessary to protect existing uses, but uses do not exist, or **vice versa**?
   例えば，現行使用を保護するのに必要なレベルを水質が上回るけれど使用が存在しない場合，もしくは逆の場合，劣化防止の結果は，何なのか？

viz. ： すなわち；換言すると

1. This can be attributed to the fact that the optimum catalyst concentration value will be a strong function of the geometry of the system and operating conditions **viz.**, incident intensity of UV light and concentration and type of the pollutant.
   これは，触媒の最適濃度が実験装置の形状ならびに操作条件，すなわち，紫外線強度，汚濁物質の濃度と種類，に強く関連するであろうという事実に帰因するかもしれない。
2. The merits of combining two advanced oxidation processes, **viz.**, sonolysis and photocatalysis, have been evaluated by investigating the degradation of PCBs using a high-frequency ultrasonic generator and a UV-lamp.
   二つの促進酸化法，すなわち超音波反応と光触媒反応，を結合させるメリットは，高周波数超音波発生器と UV ランプによる PCB の分解を研究することによって評価されてきた。

viz はラテン語の **videlicet** の略語で，**namely** また **i.e.** が通常使われています。

Volume ： 容積；書

1. It is assumed that 10 % of **the total volume of** constructive material is disposed of as non-hazardous waste.
   建設材料総量の 10 % が非危険廃棄物として処分されると想定している。

2. For a period of approximately 100 years, from the original discovery of gold at Deadwood Gulch in 1875 until the late 1970's, **huge volumes of** mining and milling wastes were discharged into Whitewood Creek.
1875 年に Deadwood Gulch での最初の金の発見から 1970 年代後期までおよそ 100 年の間，大量の鉱山採掘廃棄物と粉砕廃棄物が Whitewood Creek に放出された．

3. A sample solution of 65 mL was sonicated in a cylindrical glass vessel of 50 mm i.d. with **a total volume of** 150 mL.
内径 50 mm，全容量 150 mL の円筒状のガラス容器内で，65 mL のサンプル液を超音波処理した．

4. The lake has a total area of 25,300 km^2, **total volume of** 470 km^3, length of 386 km, mean width of 17 km, and is divided into three major, distinct subbasins.
この湖は，全面積が 25,300 km^2，全容積が 470 km^3，湖長が 386 km，平均湖幅が 17 km あり，三つの主要な水域に明確に分けられる．

★ Volume（液体容積）に関連する表現例文を下に示します．

5. A sample solution of 65 mL was sonicated in a cylindrical glass vessel of 50 mm i.d. with a total volume of 150 mL.
内径 50 mm，全容量 150 mL の円筒状のガラス容器内で，65 mL のサンプル液を超音波処理した．

6. To study the kinetics of destruction, 10−mL samples were periodically withdrawn from the reactor for analysis during the course of experiments.
分解速度を研究するために，実験中にサンプル 10−mL を反応器から定期的に持取した．

7. A 10−mL sample of calibration standards was contacted with 1 mol of hexane in conical bottom test tubes followed by shaking for 2−5 minutes by hand.
コニカルボトム試験管で，10 mol の較正標準溶液サンプルに 1 mol のヘキセンを加えてから，2−5 分間手で振った．
(See also **Aliquot**) (➡ **Aliquot**)

★ volume（書）として使用する場合の例文を下に示します．

8. **The previous volume** dealt primarily with the use of statistical techniques.
前書では主に統計学のテクニックの使用について述べた．

9. More detailed discussions of water quality criteria development for the States of Maine, New York, and Texas can be found elsewhere **in this volume**.
メイン州，ニューヨーク州，テキサス州での水質クライテリア設定に関する詳細な論考は，本書のほかのところにも記されている．

Way：方法

- **Another way to .....**：…するための別方法
- **One way to.....**：…するための一つの方法
- **The most accurate way to .....**：…するための最も正確な方法

1. **Another way to** visualize these trends is to examine changes in the cumulative frequency distribution (CFD) of biological index scores between each time period.
   これらの傾向を視覚化するための別方法は，各時期間の生物学的指数値の累積度数分布 (CFD) における変化を検討することである。
2. **Another way to** describe the attributes of ambient biological data for characterizing different types of environmental impacts is with a conceptual model.
   様々な種類の環境影響を特性把握するため環境生物学的データの属性を述べるもう一つの方法は，概念モデルによるものである。
3. **One way to** describe the attributes of ambient biological data for characterizing different types of environmental impacts is with a conceptual model.
   様々な種類の環境影響を特性把握するため環境生物学的データの属性を述べる一つの方法は，概念モデルによるものである。
4. **The most accurate way to** decrease sample variability is to collect from only one type of habitat within a reach and to composite many samples within that habitat.
   サンプル可変性を減じるための最も正確な方法は，ある河区において1種類だけの生息場所から採集し，その生息場所における多数のサンプルを複合することである。
5. It is also **a way to** learn if some unknown stressor is acting on a system.
   また，未知のストレッサーがある系に影響しているか否か知るための方法でもある。

- **in the same way**：同様に
- **in a similar way**：同様に
- **in three ways**：3つの点で

- **in a number of ways**：多くの点で
- **in various ways**：様々な形で
- **in a defensible way**：弁護可能な形で
- **in one way or another**：何らかの方法で
- **as a way of .....**：…の方法として

6. Water quality criteria could be viewed **in the same way**.
   水質クライテリアも同様にみなすことができる。
7. Mathematical models can contribute to lake water quality control **in two ways**.
   数学的なモデルが，2つの点で湖の水質管理に貢献することができる。
8. This application can be useful **in a number of ways**.
   この適用は，多くの点で有用となりうる。
9. However, well founded the legal basis for biocriteria and their implementation, it is almost certain that they will be challenged **in various ways**.
   生物クライテリアとその実施についての法的根拠がいかに十分に確立していても，様々な形で異議申立てされることは，ほぼ確実である。
10. The protocol provided the means for obtaining such an answer **in a defensible way**, and the results have been used to make regulation decisions.
    このプロトコルは，そうした回答を弁護可能な形で導くための手段となった。そして，結果を用いて規制決定が下された。

Weigh (Outweigh)：計量する；比較検討する（にまさる；を上回る）

1. Fish **are weighed** and measured and comparisons are made between fish in contaminated and "clean" areas.
   魚類を計量・測定し，汚染区域と「清浄」区域における魚類の比較を行う。
2. Their low stiffness and tendency to creep are disadvantages that must be carefully **weighed** against their advantages.
   それらの剛度の低さおよびクリープの傾向は不利点であり，それらの利点と慎重に比較考察されなくてはならない。

- **outweigh .....**：…にまさる；…を上回る（→ **outweigh**）

Weight：重量；加重値；加重する

1. Discrimination analysis provides **a weight** for each variable in the form of a

coefficient.
判別分析は，係数の形で各変数に加重値を与える。

2. Since the variables have different information content, **the weight of** each variable used in classification is expected to be different.
変数は，それぞれ異なる情報内容を持つので，分類で用いられる各変数の加重値がそれぞれ異なると予想される。

3. To determine the tolerance value, ICI scores at all locations where the taxon was collected **were weighted by** the abundance data of that taxon.
耐性値を決定するため，分類群が採集されたすべての場所でのICIスコアにこの分類群の数度データが加重された。

- **more weight should be given to ..... :**
  … に，もっとウエートが置かれるべきである

4. In evaluating the accuracy of the estimated parameter, **more weight should be given to** the ability of the model to fit the data points near segment D than to its ability to fit the points near segment D.
推定パラメータの正確さを評価することにおいて，モデルが川の切片Gの近くのデータポイントで合う能力により，切片Dの近くでデータポイントに合うことができる能力に，もっとウエートが置かれるべきである。

Well：ずっと；より更に；よく

1. All other values are at concentrations **well** below this risk level.
すべての他の値は，このリスクレベルをずっと下まわる濃度にある。

2. The concentration of arsenic is **well** above the upper range commonly found in soils.
ヒ素の濃度は，一般に土壌に見られる（濃度）範囲の上限より更に高い。

3. The equation for a loading to a **well-mixed** lake is ....
よく混合している湖への汚濁負荷は … の式で示される。

4. The model value corresponds reasonably **well** with the experimental data.
モデル値は適度によく実験的なデータと一致している。

- **as well**：同様に

5. Under the conditions of the present experiments, chloride was the major ionic product, and small amounts of $HCOO^-$ were detected **as well**.
現在の実験条件下では，塩化物が主要なイオン生成物であった，そして同様に小量の $HCOO^-$ も検出された。

6. Patterns in human activities must be considered **as well**, and these often

vary as a function of ownership or political unit, as well as ecoregion.
人間の活動におけるパターンも考慮すべきであり，それは，しばしば所有単位や政治単位，そして生態地域と関連して異なる。

7. An advantage of examining the trophic status of component populations is that it also provides information on functional aspects of the community **as well**.
構成要素たる個体群の栄養状況を検討することの利点は，その生物群集の機能面に関する情報も提供することである。

- **as well as ...**：… と同様に；… も

8. The present model structure may apply to other compounds **as well as** to other species.
現在のモデル構造は，他種と同様に他の化合物に当てはまるかもしれない。

9. Scales are collected so that growth rates **as well as** general size can be calculated and compared.
鱗を採集して生長度や一般的体寸が計算・比較されえる。

10. The development and adoption of biocriteria into state water quality standards poses challenges **as well as** opportunities.
州水質基準における生物クライテリアの設定・採択は，課題ばかりでなく，機会をもたらす。

11. Recent advances in AOP treatment technology **as well as** their use in industrial applications are summarized in the literature.
AOP処理技術における最近の進歩は，産業的な応用同様にこの文献に要約されている。

12. There has also been concern for a mechanism to structure the assessment and management of nonpoint source pollution **as well as** a variety of environmental resource regulatory programs.
面源汚濁の評価・管理を構成するメカニズムへの関心や，多様な環境資源規制プログラムへの関心も存在してきた。

13. Volunteer monitoring programs have been developed to provide education and advocacy opportunities **as well as** basic data collection.
ボランティアモニタリングプログラムが設けられた目的は，啓発機会や擁護機会を提供して，基礎データ収集を可能にすることであった。

- **~ is well documented**(See Document)：
  ~は文書化されている(➡ **Document**)
- **~ is well established**(See Establish)：
  ~は十分に確立されている(➡ **Establish**)

- ~ is well known (See **Known**)：~は，よく知られている（➡ **Known**）
- well into the future (See **Future**)：
  未来に向けて；ずっと将来まで（➡ **Future**）
- well understood (See **Understand**)：
  よく理解されている；十分に理解されている（➡ **Understand**）
- how well (See **How**)：どれほどよく（➡ **How**）

What：何が；のことは

1. The question now arises: **What is** the ecological reference site?
   ここに疑問が生じる。すなわち生態参照地域とはなんであろうか。
2. **What are** the antidegradation ramifications where water quality exceeds levels necessary to protect existing uses, but uses do not exist, or vice versa?
   現行使用を保護するのに必要なレベルを水質が上回るけれど使用が存在しない場合，もしくは逆の場合，劣化防止の結果は，何なのか？
3. One of two regulatory questions is generally posed: 1) **what are** the maximum allowable discharges of toxic chemicals to a water system, and 2) how long will it take a contaminated natural water system to recover?
   規定に関する質問の2つのうち1つが一般に提出される。それら2つの質問とは1)毒性化学物質が水系に放出されうる最大許容量はいくらか，ならびに2)汚染された自然水系が回復するのにはどれほどの時間を要するのか，である。
4. More specifically, questions such as, "**What** proportion of the stream miles (and how many stream miles) are affected by point source discharges?" can be addressed.
   より具体的には，「水流 mile 数のどれほどの比率（どれほどの水流 mile 数）が点源排出に影響されるか？」といった疑問に対処できる。

- **what is so surprising is** .....：驚くべきことは … である
- **what we are aiming toward is** .....：
  我々が狙いを定めているものは … である
- **what seems to be lacking is** .....：
  欠けていると思われることは，… である
- **what is more likely is that** .....：
  … をもたらす見込みが高いのは … である

5. **What is so surprising** on Easter Island is the high abundance of corals.
   イースター島についてとても驚くべきことは，サンゴチュウの豊富度が高いことである。

6. **What we are aiming toward is** establishment of the validity of the concept of controlled catalytic biomass.
我々が狙いを定めているものは制御触媒性のある生物量概念の正当性を立証することである。
7. **What seems to be lacking is** a consideration of the ecological background rather than water chemistry background.
欠けていると思われることは，水質化学の背景より生態学の背景を考慮することである。
8. **What is more likely is that** biosurveys targeting multiple species and assemblages provide improved detection capability over a broader range as well as protection to a larger segment of the ecosystem than a single species or assemblage approach.
複数の生物種および群がりをターゲットにした生物調査は，単一の生物種または群がり手法に比べて，より広範な探知能力向上をもたらす見込みが高いし，より広範な生態系保護になるだろう。

- **what is expected for .....**：…について何が期待しうるか
- **what is known of .....**：…に関する知見
- **what need to be done**：何がされなければならないか
- **what has been discussed**：論じられてきたこと
- **From what has been discussed above**：論じられてきたこと

9. Data are evaluated within the ecological context (waterbody type and size, season, geographic location, and other elements) that defines **what is expected for** similar waterbodies.
データは，生態学的文脈（水域の種類・規模，季節，地理的位置，および他の要素）において評価され，これによって類似の水域について何が期待しうるかが決まる。
10. The condition of the stream can be inferred from the taxa present and **what is known of** their requirements and tolerance.
水流状態は，常在分類群や，その要件・耐性に関する知見から推断されうる。
11. At each step in the application process, we will first explain **what need to be done**.
その適用過程での段階それぞれにおいて，何がされなければならないか，を我々は最初に説明するでしょう。
12. **From what has been discussed** above, we can conclude that .....
上に論じられてきたことにより…と結論つけることができる。

- **To what extent**：どの程度まで

- **on what basis**：何に基づいて

13. **To what extent** is mercury in the form of mercuric sulfide available for biological methylation?
    硫化水銀として水銀は，生物によってどの程度までメチル化されるのか？
14. However, **on what basis** were the decisions to select a baseline numerical criterion value for each index made?
    しかし，各指標について，ベースライン数量的クライテリア値を選ぶ際，何に基づいて決定が下されたのか？

- **～ of what is.....**：…かといった～

15. This report should serve to give the reader an idea **of what is** involved in testing a model.
    この報告書は，モデルのテストには何が関与するのかといった考えを，読者に与えるのに役立つはずである。
16. These methods provide an indication **of what** specific environmental characteristics **are** changing over time.
    これらの手法は，どの環境特性が経時的に変化しているか示す指標となる。

Whatever：が何であるとしても

- **Whatever the reasons for .....**：…の理由が何であるとしても
- **Whatever the explanation**：説明が何であるとしても

1. **Whatever the reasons for** the differences, it is important to keep in mind that .....
   この相違の理由が何であるとしても，…を念頭におくことは重要である。
2. **Whatever the explanation**, the weak and inconsistent detection of naphthalene did not allow estimation of a threshold concentration by the method used here and elsewhere.
   説明が何であるとしても，微弱，かつ不規則なナフタリンの検出は，ここで使われた方法，そしてほかのところでも使われた方法による濃度閾の推測を難しくした。

When：の時；の場合；の際

1. **When** the SQM analysis was used, agreement improved.
   SQM 分析が用いられた場合では，合致度は上昇した。
2. The problem stems from the period 1947 to 1970, **when** a private DDT plant discharged wastes containing DDT residues into a drainage ditch.
   この問題は 1947 から 1970 の間に生じ，そしてその期間，私的な DDT 工場が

DDT 残物を含んでいる廃棄物を排水溝に放出した。
3. The PCB issue dates back to 1973, **when** state fish and game officials detected unusually high levels of PCBs in the river's striped bass population.
   PCB 問題は，1973 年にさかのぼる。その当時，州の fish and game 当局者は川にいるストライプバスの個体群に異常に高いレベルの PCB を検出していた。
4. EPA (2004) found good agreement between results in about 20 % of cases **when** effluent toxicity was assessed, and in about 30 % of cases **when** mixing zone toxicity was measured.
   EPA は，流出毒性を評価した際には約 20 ％の事例結果で，また，混合区域毒性を測定した際には約 30 ％の事例結果で，かなりの一致を認めた。

★ When__ed の形で使用される場合の例文を下に掲げます。

5. **When considered** on the basis of agencywide water programs, this percentage is approximately 6%.
   機関全体の水関連プログラムベースで考えた場合，この割合は，約 6 ％である。
6. **When used** in a monitoring program, these methods can be used to help diagnose the cause of changes in the aquatic system.
   監視プログラムで用いた場合，これらは，水生系における変化原因を判定するのに役立つ。
7. Approaches for assessing the response of stress proteins in organisms **when exposed** to chemicals are currently under development.
   化学物質に曝された際の生物体内ストレス蛋白質の応答を評価するための手法を目下開発中である。

★ When__ing の形で使用される場合の例文を下に示します。

8. **When flying** over the Southern Rocky Mountains along the Wyoming/Colorado state line in mid-1980, this author noticed differences in timber management practices between states.
   1980 年代中期，ワイオミング州／コロラド州の境界に沿った Southern Rocky 山脈の上空を飛んだ時に，著者は，両州での材木管理慣行の相違に気づいた。
9. **When selecting** these sites one must account for the fact that minimally disturbed conditions often vary considerably from one region to another.
   これらの地点を選ぶ際には，最小擾乱状態が地域ごとに大幅に異なりうるという事実を考慮しなければならない。

● **only when .....**：… のときにだけ

10. The hydraulic conductivity comes into play **only when** quantitative discharge

calculations are made.
流体の伝達速度は，排出量を計算するときにだけ，有用になる。
11. A consistently high percentage of agreement occurred **only when** both the volunteer and professional analyses rated the site in the fair/poor ranges.
高率の合致が一貫して生じる事例は，ボランティアと専門家の両方が可／不可の値域にある地点を対象とした場合に限られていた。

- **when compared to** .....(See **Compare**)：
  … に比べる場合（➡ **Compare**）
- **when comparing A with B**(See **Compare**)：
  AとBとの比較では（➡ **Compare**）

Whenever：するときはいつでも

1. **Whenever** possible, base the area on natural vs. political, boundaries to maximize applicability and the probability of selecting the least disturbed sites possible.
可能なら，常に最小擾乱地点選定の可能性および確率を最大限に高めるため，自然的境界および政治的境界に基づくべきである。

Whereas：一方；に対し

1. North of the line in Wyoming, patterns of logging activity were apparent, **whereas** south of the line they were not.
ワイオミング州境の北では伐採活動パターンが顕著だったのに対し，州境の南ではパターンが認められなかった。
2. Some of the original models have been changed in nearly every subsequent version, **whereas** others have been largely retained.
オリジナルモデルのいくつかは，その後のバージョンほぼすべてで変更されたが，そのまま保たれたモデルも多い。
3. **Whereas** many thematic maps are the products of interpretations that include consideration of seasonal and year-to-year differences, Landsat imagery and high-altitude aerial photography are snapshots of conditions at a particular time.
多くの主題地図は，季節差や年差を考慮に入れた解釈の産物である。他方，ランドサット画像や高高度航空写真は，特定時刻での状態の断片である。

Wherein：そこで；この場合

1. Some states are instituting comprehensive biological monitoring networks based on a rotational basin approach, **wherein** water body assessments rotate among watersheds on regular intervals.
いくつかの州は，交代流域手法に基づき総合的な生物モニタリングネットワークの構築に取り組んでいる。この場合，水域アセスメントは，定期間隔で対象流域が交代するのである。

Whether：かどうか；有るか否か

1. Unknown is **whether** there is free product on the water table.
地下水面上にフリープロダクト（浮上油）があるかどうかはわかっていない。
2. There is insufficient information to determine **whether** contamination present at this source area may present an unacceptable risk.
この汚染源での汚染が，容認出来ない危険性を引き起こすかもしれないかどうか決定するには情報が不十分である。
3. The question arose **whether** these individual deviations could be the result of measurable differences between the fishes.
これらの個別の逸脱が魚の間の測定可能な相違の結果で有るのか否かという質問が起こった。
4. To predict **whether** a behavioral response to a chemical pollutant will occur, one must ask **whether** the organism can detect the pollutant at concentrations likely to be encountered in field situations.
化学汚染物質に対する挙動反応が起こるかどうか予測するには，現場の状況で起こりえる濃度で生物が汚染物を検知できるかどうか尋ねなくてはならない。
5. Vermont uses biological criteria from ambient stream data to determine **whether** two different types of biological standards are being met.
バーモント州は，環境水流データからの生物クライテリアを用いて，2種類の生物基準が満たされているか否か決定している。

● whether or … : … かどうか

6. **Whether** the models we create are good models **or** poor models depends on the extent to which they aid us in developing the understanding which we seek.
我々が作るモデルが良いモデルか貧弱なモデルであるかどうかは，我々が求める理解力をつけるのにモデルがどれほど助けになるかのその程度による。

7. An initial decision faced in identifying stream populations is **whether** to describe the condition of streams in terms of stream segments **or** the total length of streams.
   水流母集団の確認で直面する初期決定は，水流の状態を，水流区間において述べるか，水流全長において述べるかである。

   ● **whether or not** ..... ：…しているか否か

8. **Whether or not** petroleum entering the marine environment will have substantial or minimal impact depends on interactions among complex variables that are only now beginning to be understood.
   海の環境に流入している石油が本質的な影響，または最小影響を持つであろうかどうかは，複雑な変数の間での相互作用に依存する。

9. Definite patterns in biological community data exist and can be used in the determination of **whether or not** a waterbody is attaining its designated use and in identifying the predominant associated causes of impairment.
   生物群集データにおける明確なパターンは，現に存在し，また，ある水域が指定使用を達成しているか否か決定するため，ならびに優勢な関連の損傷原因を確認するため利用できる。

10. A robust statistical treatment of these data often can help determine **whether or not** the biological response is attributable to the measured pollutant.
    これらのデータへの確かな統計処理は，生物応答が測定された汚濁物質に帰せられるか否か決定するのに役立つことが多い。

    ● **as to whether** ..... ：…するかどうか；…についての判断は
    ● **as to whether A or B** ：Aによるのか，またはBによるのか

11. Determination **as to whether** a release has occurred is generally made at the time a waste line is decommissioned.
    放出が起こったかどうかについての判断は，一般的に廃物ラインが使用済みになる時に下される。

12. Opinions differ among states **as to whether** formal incorporation of biological criteria into state water quality standards should be the ultimate goal of all biological criteria programs.
    生物クライテリアを州の水質基準へ正式に盛り込むことが生物クライテリアプログラムの究極目標か否かに関して，諸州で意見の相違がある。

13. The data do not permit a determination **as to whether** the toxicity was due to a slug of something toxic in the water **or** was continuously present.
    この毒性が，水に短時的に存在する毒性物質によるのか，または継続して存在す

る毒物によるのか，このデータでは決められない。

- **no matter whether** ..... ：… かどうかにかかわらず

14. Naphthalene produced this inhibition **no matter whether** the melanin was fully aggregated or more less dispersed at the time of initial exposure.
最初にさらされた時にメラニンが完全に凝結していたか，あるいはある程度分散していたかの状態にはかかわらず，ナフタリンはこの抑制効果を呈した。

- **question of whether** .....（See **Question**）：
… かどうかという質問（➡ **Question**）
- **question of whether or not** .....（See **Question**）：
… しているか否かという質問（➡ **Question**）

Which ：どちら；どれ；どちらの；どれの

1. An equation **which** is of considerable importance in such application is .....
このような適用においてかなりの重要性を持っている式は … である。
2. It has not yet been determined **which** level would be most appropriate for the water quality standard.
どのレベルが水質基準に最も適しているかは，まだ未決定である。
3. In this study, attempts were made to ascertain **which** parameters were causing toxicity.
この研究で，どのパラメータが毒性を起こしていたか確認する試みがされた。
4. The approach in using ultrasound is obviously applicable to organic compounds **which** contain halides.
超音波を使うアプローチは，ハロゲンを含む有機化合物に対し明らかに適用できる。
5. There is also a need to develop models **which** can be used to predict water quality changes within distribution systems.
また，配水システム内での水質変化を予測するために使うことができるモデル構築の必要性もある。
6. All values were obtained for a time period of 1900 to 1960 except for the Detroit River flow **which** was averaged over the period from 1939 to 1960.
すべての値は，1939から1960までの期間にわたって平均されたデトロイト川の流量以外は，1900から1960までの期間に収集された。
7. A 40-mile segment has been sampled repeatedly over multiple years, **which** is a prerequisite for trend analysis.
40mileの区間で，複数年度にわたり繰り返しサンプリングされた。これは，傾向分析の必須条件である。

8. We appreciated critical review comments from three anonymous reviewers, **which** greatly improved an earlier draft of this manuscript.
   筆者は，3人の匿名査読者からの批評的論評に感謝する。それは，本章の草稿を大幅に改善してくれた。

   - **by which**：…による
   - **for which**：…のために
   - **from which**：…から
   - **all of which**：…のすべて

9. The procedure **by which** this is done can be viewed as part of the regionalization process.
   このための手順は，地域分け過程の一部とみなしうる。

10. Stream classification provides relatively homogeneous classes of streams **for which** biocriteria may differ among the classes.
    水流分類は，比較的同質な水流等級を生じさせるが，水流階級ごとに生物クライテリアが異なることもある。

11. The best and most representative sites for each stream class are selected and represent the best set of reference sites **from which** the reference condition is established.
    各水流等級について最善かつ最も代表的な地点が選ばれて，参照地点セットにされ，それに基づき参照状態が定められる。

12. A variety of decisions must be made, not **all of which** are technical or scientific in nature.
    このためには様々な決定が必要なのだが，そのすべてが技術的決定や科学的決定というわけではない。

    - **over which**：…の範囲で

13. It is important to define the regional extent **over which** a particular biocriterion is applicable.
    ある特定の生物クライテリアが適用可能な地域的度合を定めることが重要である。

14. Examining the variance structure can give insight into the extent **over which** particular biocriteria might be applicable.
    変動構造を検討すれば，ある生物クライテリアが適用可能な度合に関する洞察が得られよう。

15. A critical issue is to determine the regional extent **over which** a particular biological attribute is applicable.
    重大問題は，ある特定の生物学的属性が適用可能な地域範囲を決定することである。

16. It is better to identify large areas **over which** calibrations will be performed

rather than small areas.
較正を行うに当たり，狭い地域よりも広い地域を確認した方が良い。

17. By calibrating over large areas, decisions about level of protection can be made with a clear understanding of the differences among areas **over which** differing criteria might be established.
広い地域での較正による保護レベルに関する決定は，異なるクライテリアが設定されうる各地域の差異を明確に理解したうえで下されるようになる。

- **to which** ：…に対する

18. A second decision is to define the set of sites **to which** a biocriterion applies.
第二の決定は，生物クライテリアが適用される地点セットを定めることである。

19. This distinction is made necessary by the widespread degree **to which** macrohabitats have been altered among the headwater and wadable streams in the HELP ecoregion.
この弁別が広範に必要になった理由は，HELP 生態地域における源流・徒渉可能水流で，マクロ生息地が大幅に改変されたことである。

- **extent to which** …..(See **Extent**)：…する度合(➡ **Extent**)

While ：の間；する一方；の場合でも

1. **While** it is possible for reference sites to double as upstream control sites, the reverse is not always true.
参照地点を上限の対照地点の2倍にすることが可能な場合でも，逆は必ずしも真でない。

2. **While** a number of HSPF applications are in progress, few studies are complete.
多数の HSPF アプリケーションが進行中である一方，ほとんど研究が完成していない。

3. **While** these classifications were based on ecological attributes, the criteria associated with each were entirely chemical/physical.
これらの分類は，生態的属性に基づいていたが，それぞれに関連するクライテリアは全く化学的/物理的なものであった。

4. **While** it is recognized that individual waterbodies differ to varying degree, the basis for having regional reference sites is the similarity of watersheds within defined geographical regions.
個々の水域が様々に異なることは，認識されるものの，地域参照地点を設けるこ

との基礎は，所定の地理的地域内での流域の類似性にある。

5. **While** we have dealt with most of these issues in California, these and other issues will arise elsewhere, thus regionally consistency in achieving a resolution of these issues will be needed.
   筆者は，カリフォルニア州におけるこれらの問題の大半を扱ってきたが，これらおよび他の問題は，他地域でも生じることで，これらの問題の解決における地域的整合性が必要だろう。

6. The western basin is the shallowest, with a mean depth of 11 m. The central basin is the largest in terms of area (16, 317 km^2) and has a mean depth of 25 m, **while** the eastern basin is the deepest, with a maximum depth of 64 m.
   西水域は最も浅く，平均湖深が 11 m である。中央水域は面積では最大 (16,317 km^2) で，平均湖深が 25 m である。一方，東水域は最も深く，平均湖深が 64 m である。

7. The difficulty of dealing with a large mass of information is that of the many interactions occurring within the community, some may be related to water quality **while** others may not.
   大量の情報を取り扱う場合の困難は，生物群集において多くの相互作用が生じることであり，それらは，水質に関係したり，関係しなかったりする。

Whole：全体の；すべての

1. The trend for **whole** body game fish is less obvious and not necessarily declining.
   釣魚全体よる (魚の体内濃度) 傾向は，明白さに欠け，そして必ずしも低下していない。

2. Relative abundance of taxa refers to the number of individuals of one taxon as compared to that of the **whole** community.
   分類群の相対豊富度とは，生物群集全体の個体数に比した一分類群の個体数を意味する。

3. Thickness of tailings deposits varies considerably and cannot be accurately predicted over the **whole of** the study area.
   選鉱滓堆積物の厚さはかなり変わり，調査全地域では正確に予測することができない。

   ● **as a whole**：全体として

4. Therefore, permitting experiences for AEP facilities are unique and cannot be considered representative of Texas's regulated industry **as a whole**.
   したがって，AEP 施設の許可経験は，独自のものであり，テキサス州の統制さ

れた業界全体を代表しているとみなすことができない。
5. An underlying assumption of this approach is that the patch work of performance standards, when implemented **as a whole**, will be sufficient to restore and maintain the physical, chemical, and biological integrity of the water.
この手法の根本的な前提は，実績標準の集成が全体として実施された場合，その水の物理的・化学的・生物的な完全度を十分に回復・維持するだろうというものである。

Why：なぜ

1. However, the charge that the IBI (and similar multimetric indices) are too ambiguous to determine **why** the index is high or low is refuted by the discussion of the Biological Response Signatures and the conceptual model of community response.
しかし，IBI（および，類似の多メトリック指数）が曖昧すぎ，なぜ指数が高いか低いか理由を決定できないという非難は，生物応答サインおよび生物群集応答の概念モデルに関する論考によって反駁される。

- **There is no reason why .....**：…の理由は存在しない

2. **There is no reason why** these same principles would not apply here.
これらの原則がここで適用されないという理由は，見あたらない。
3. In concept, **there is no legal reason why** NPDES permit cannot be used to implement biocriteria.
概念上，NPDES 許可を用いて生物クライテリアを実施してはならないという法的理由は，存在しない。

Word：言葉

1. **The word** is relatively new in mathematics and would seem to have been borrowed from earlier usage because of the analogy between the mathematical model and the scale model.
この言葉は数学の領域では比較的新しく，そして数学的モデルとスケールモデルの間の類似性のためにより以前の使用法から借りられたように思われる。

- **In other words**：すなわち；換言すれば

2. A statistical analysis of the initial and four week data indicates that the reduction was significant at the 0.001 probability level. **In other words,**

there is a 99.9 % probability that the difference between the initial and four-week TPH concentrations can be attributed to treatment.
初期データと4週間データの統計分析は，減少が0.001の確率レベルで有意であったことを示す。換言すれば，初期のTPH濃度と4週間のTPH濃度の間の相違が処理に帰因しうる確率が99.9％である。

Work：研究；作業；材能する

1. **Work** over the last decade by investigators, such as Noguchi (1993), have emphasized the importance of riparian areas for maintaining the quality and function of the hyporheic zones of streams.
Noguchi (1993) などの調査者による過去10年間の研究は，水流のhyporheic地帯の質・機能の維持にとっての水辺区域の重要性を強調した。

　　● **work on** ..... ：… に関する研究

2. In his early **work on** .....
… に関する彼の早期研究において

3. Despite the large amount of current and past **work on** IBI versions for wadable warmwater streams, much remains to be done.
徒渉可能温水水流についてのIBIバージョンに関する現在・過去の研究が大量であるにもかかわらず，残された課題は多い。

　　● **work is underway to** ..... ：… する作業が進行中である

4. **Work is underway to** develop methods for Lake Erie river mouth and nearshore areas.
Erie湖の河口・沿岸区域において，方法を策定する作業が進行中である。

5. **Work is underway to** address water quality criteria for these areas within the next 3 to 5 years.
今後3年間ないし5年間，これらの区域についての水質クライテリアに取り組む活動が進行中である。

　　● **much work needs to be done to** ...... ：
　　　かなりの研究が … には必要である
　　● **additional work will need to be done to** ..... ：
　　　明らかに追加研究が … には必要であろう
　　● **much less work has been done on** ..... ：
　　　… に関する研究はあまりされてこなかった

6. Clearly, **much work needs to be done** to evaluate and polish the proposed

technique.
明らかに，提案された技巧を評価して，そしてそれに磨きをかけるためにはかなりの研究が必要である。

7. Clearly, **additional work will need to be done to** better understand how electricity generation affects BOD removal.
明らかに，発電作用がBOD除去に及ぼす影響をもっと良く理解するために追加研究が必要であろう。

8. **Much less work has been done on** the activation of less reactive metals.
反応性がより低い金属の励起に関する研究はあまりされてこなかった。

- **very little experimental work has been done on** ..... ：
  … については，ほとんど実験研究がされてこなかった
- **relatively little work has been directed towards** ..... ：
  … に関する研究は比較的少なかった
- **very little work has been completed in** ..... ：
  … の研究はほとんど完成していない
- **little work has been published for** ..... ：
  … に関する研究はあまり発表されていない

9. Relatively **little work has been directed towards** modifying the IBI for use on rivers too large to sample by wading.
徒渉サンプリングには大きすぎる河川でIBIを用いるための修正に関する研究は，比較的少なかった。

10. **Little work has been published for** lakes and reservoirs, but significant efforts are ongoing as part of the USEPA effort to develop consistent bioassessment protocols.
湖沼および堰止め湖に関する研究は，あまり発表されていないが，整合的なバイオアセスメントプロトコルを設定するUSEPA活動の一環としてかなりの努力が払われつつある。

11. **Very little work has been completed in** the development of estuarine version of the IBI.
IBIの河口域バージョンの案出は，ほとんど完成していない。

12. **Very little experimental work has been done on** the vertical dispersion coefficient.
垂直拡散係数については，ほとんど実験研究がされてこなかった。

- **work has shown that** ..... ：研究によれば … である
- **recent work by** ..... ：… による最近の研究
- **work has emphasized** ..... ：研究は … を強調してきた

13. **Work has shown that** highly variable and unpredictable flow regimes can have strong influences on fish assemblages.
    研究によれば，可変性が強くて予測不能な流量状況は，魚類群がりに強い影響を及ぼしうる。
14. **Recent work by** Kouno (1990), and even articles in the New York Times (Morikawa, 1993) have reported on the widespread degradation to the nation riverine resources.
    Kino (1990) による最近の研究論文や，『New York Times』掲載記事 (Morikawa, 1993) でさえ，全米の水辺資源の広範な悪化を報じている。

    - **additional work is needed to .....**：
      … にはさらなる研究が必要とされる
    - **future work is needed .....**：… には今後の研究が必要である

15. **Additional work is needed to** more precisely define and set expectations for these two metrics, and to test their usefulness in field applications.
    これら2つのメトリックについて期待値をより正確に設定し，現地適用におけるその有用性を試すため，さらなる研究が必要とされる。
16. The very early time periods were suggestive of rapid loss of a small amount of PCP ; however, **additional work is needed to** confirm this.
    非常に早い時間の期間は小量のPCPが速い速度で分解することを意図する。しかしながら，更に追加の研究がこれを確証するために必要である。
17. **Future work is needed**, however, to better utilize the ADV as part of resource value assessment such as NRDAs and in assessing other types of environmental damage claims.
    しかし，NRDAsといった資源価値アセスメントの一部として，また，他の種類の環境損害請求の評価においてADVをより活用するには，今後の作業が必要である。

    ★workの動詞としての使用例文を下に示します。

    - **～ work well**：～はうまく機能する

18. This three-tiered scheme **works well** for groups like stoneflies in which all species are nearly equal in tolerance to organic enrichment pollution.
    この三段層方式は，カワゲラのようにすべての生物種が有機物濃縮汚濁への耐性度がほぼ等しい集団では，うまく機能する。

Worth (Worthwhile) : 価値（価値がある）

- **It is worth .....** : …する価値がある
- **It is worthwhile .....** : …する価値がある

1. **It is worth** emphasizing that technology based performance standards tend to deemphasize receiving system condition and even human health.
技術ベースのパフォーマンス基準が，受系状態ならびに人間の健康状態さえにも重点を置かない傾向があるということを強調する価値がある。

Y

Year：年

- **in 2004**： 2004年に
- **in the late 1992**： 1992年後半
- **in the early 1990's**： 1990年代初期に
- **in the late 1800s and early 1900s**： 1800年代後期と1900年代初期に
- **in the period from about 1970 to 1975**： およそ1970年から1975年までの期間に.
- **in the 1980−1990 period**： 1980年から1990年の期間に
- **in the mid−1990s**： 1990年代中期に
- **in mid−1980**： 1980年中期に

1. **In 1968**, the total world production of mercury was estimated to be 8000 tons.
   1968年に，世界の水銀の全生産量は8000トンであると推定された。
2. These methods were first used in the United States **in the early 1940's**.
   これらの手法が1940年代初期に合衆国で最初に使われた。
3. This class of models was born **in the late 60's and early 70's**.
   この種のモデルは60代後期と70代初期につくられた。
4. The facilities were constructed **in the late 1800s and early 1900s**.
   この施設は1800年代後期と1900年代初期に建設された。
5. The work was performed **in the 1980−1990 period**.
   この研究は1980年から1990年の期間に行なわれた。
6. A qualitative/narrative system of evaluating biological data **in the 1970s and early 1980s** shifted to more quantitative/numerical framework **in the mid−1980s**.
   1970年代から1980年代前期までの質的/記述的な生物学的データ評価方式は，1980年代中期により量的/数量的な枠組みへ移った。
7. When flying over the Southern Rocky Mountains along the Wyoming/Colorado state line **in mid−1980**, this author noticed differences in timber management practices between states.

1980年代中期，ワイオミング州／コロラド州の境界に沿ったSouthern Rocky山脈の上空を飛んだ時に，著者は，両州での材木管理慣行の相違に気づいた。

- **in recent years**：近年
- **in previous years**：それまで数年間
- **in the years preceding**：よりも数年前
- **in the last few years**：過去2，3年に
- **in the first of five years of .....**：5箇年の…の初年度

8. **In recent years** there has been an increasing awareness that effective research, inventory, and management of environmental resources must be undertaken with an ecosystem perspective.
近年，環境資源の効果的な研究・調査一覧作成・管理は，生態系を視野に入れて行うべきだ，との認識が強まってきた。

9. Improvements in the treatment process were made in the year before this sampling, and macroinvertebrate changes were not as great as **in previous years**.
サンプリング前年に処理過程の改善が行われ，大型無脊椎動物の変化は，それまで数年間ほど大幅でなかった。

10. **In the years preceding** these recommendations, Karr and Dudley (1981) and Karr et al. (1986) developed an operational definition of biological integrity.
これらの勧告よりも数年前，KarrとDudley(1981)ならびにKarrら(1986)が生物健全性の持性的定義を定めた。

11. The National Water Quality Assessment (NAWQA) program is **in the first of three years of** intensive data collection that include measures of fish, invertebrate, and algal communities.
全米水質アセスメント（NAWQA）プログラムは，魚類，無脊椎動物，藻類という測度を含む3箇年の集中的データ収集の初年度である。

- **for the first three years,**：最初の3年間
- **for the last several years**：ここ数年の間
- **for nearly 10 years**：約10年間にわたり
- **for thousands of years**：数千年にわたり
- **for a period of approximately 50 years**：およそ50年の間

12. **For the first ten years**, the primary emphasis was on groundwater flow modeling.
最初の10年間，地下水流のモデリングに重点がおかれた。

13. The environmental concern with endocrine disruptors has been growing **for**

the last several years.
ここ数年の間，内分泌撹乱物質について環境的関心が高まってきている。

14. Michigan EPA has conducted biosurveys throughout the state **for nearly 20 years**.
ミシガン州EPAは，約20年間にわたり州全域で生物調査を行ってきた。

15. Humans have probably played a major role in shaping landscape pattern and molding ecosystem mosaics **for thousands of years**.
人間は，おそらく数千年にわたり，景観パターンの形成と生態系モザイクの生成において重要な役割を演じてきた。

16. **For a period of approximately 100 years**, from the original discovery of gold at Deadwood Gulch in 1875 until the late 1970's, huge volumes of mining and milling wastes were discharged into Whitewood Creek.
1875年にDeadwood Gulchでの最初の金の発見から1970年代後期までおよそ100年の間，大量の鉱山採掘廃棄物と粉砕廃棄物がWhitewood Creekに放出された。

- **during the last fifty years**：これまでの50年間
- **during the past twenty years**：過去20年間に
- **during recent years**：近年の間に

17. This has resulted in a steady increase in the production of heavy metals **during the last fifty years**.
これまでの50年間，これは重金属の生産の着実な増加をもたらした。

18. Significant improvement in the biological indices has been observed **during the past twenty years**.
生物指標における大幅な改善が過去20年間に観察されている。

19. While the quality of some regions has improved **during recent years**, there is still great concern regarding the marine environment.
近年の間に若干の地域環境が良くなってきた一方，海洋環境に関してまだ大きい懸念がある。

- **during 2002 to 2004**：2002年から2004年にかけて
- **during the 30 years of .....**：30年の…期間中
- **during the past 50 to 100 years**：過去50年間ないし100年間
- **during the latter part of 1982**：1982年後半の間に
- **during the early to mid-1970's**：1970年代前期から中期の間
- **during the 7 years of .....**：7年の…期間中

20. **During 1988 to 1990**, TPWD joined TNRCC to sample an additional 66 streams.
1988年から1990年にかけて，TPWDは，TNRCCと共同で，さらに66水流でサ

ンプリングを行った。

21. It is unlikely that extreme care was taken to prevent deposition of DDT into the ABC River **during the 27 years of** plant operation.
27年の工場創業期間中，ABC川へのDDTの堆積を妨ぐために出来るだけの注意が払われたという事は，ありそうもない。

22. The landscape and aquatic ecosystems of New York have been significantly altered **during the past 150 to 200 years**.
ニューヨーク州の景観と水生生態系は，過去150年間ないし200年間で大幅に変化してきた。

   - **during the period 1990–1995** ： 1990–1995期間中に
   - **during the period between June 10 and June 25, 2004** ： 2004年の6月10日から6月25日の期間中に
   - **during the period from 1950 to 2000** ： 1950年から2000年の間に

23. **During the period 1930–1970**, the total mercury mined in the United States was 31,800 tons.
1930–1970期間中に，合衆国で採鉱された水銀は全部で31,800トンであった。

24. **During the** first "high flow" sampling **period between** June 15 **and** June 24, 2000, ...
6月15日から2000年6月24日の間の，最初の「流量がおおい」時のサンプリングの期間中に，…

   - **from 1965 to 1970** ： 1965年から1970年まで
   - **from the period 1987 to 1990** ： 1987年から1990年までの期間
   - **from the late 1800s to the present** ： 1800年代後期から現在まで

25. The problem stems **from the period 1947 to 1970**, when a private DDT plant discharged wastes containing DDT residues into a drainage ditch.
この問題は，そしてその期間，私的なDDT工場がDDT残物を含んでいる廃棄物を排水溝に放出した。1947から1970の間に生じた。

26. Data are available **from the late 1800s to the present**.
1800年代後期から現在まで，データが得られる。

   - **from early years**：早年から
   - **from year to year** ：年ごとに

27. The plant discharges mercury used in the processing **from early years**.
早年から，この工場は工程で使われた水銀を放出している。

28. The periphyton assemblage may change up to 30% in taxa present **from year to year** within the spring–summer period.

付着生物群集は，春夏の時期で年ごとに30％もの部類群変化が生じうる。

- **over the last 1000 years**：これまでの1000年にわたって
- **over a 5-year period**：5年間にわたって
- **over the period from 1950 to 1970**：
  1950年から1970年までの期間にわたって
- **over the 100-year-period ending in 1998**：
  1998年を最後に，100年間にわたって

29. In any case, there are no continuous records of ecological change **over the last 2000 years**.
いずれにしても，これまでの2000年にわたって生態的変化の継続した記録がない。
30. **Over a 2-year period**, the criteria were tested, focusing on the following questions: (1)....., (2).....
2年間にわたってクライテリアがテストされ，焦点は，以下の諸問題に置かれていた。(1)…, (2)…
31. All values were obtained for a time period of 1900 to 1960 except for the Detroit River flow which was averaged **over the period from 1939 to 1960**.
1939から1960までの期間にわたって平均されたデトロイト川の流量以外のすべての値，1900から1960までの期間に収集された。
32. **Over the 100-year-period ending in 1977**, many millions of tons of tailings were discharged into Whitewood Creek.
1977年を最後に，100年間にわたって，何百万トンという選鉱廃物がWhitewood Creekに放出された。

- **over multiple years**：複数年度にわたり
- **over the next few years**：今後数年間に
- **over the last few years**：過去2, 3年にわたって
- **over the past few years**：過去2, 3年にわたって

33. A 40-mile segment has been sampled repeatedly **over multiple years**, which is a prerequisite for trend analysis.
40mileの区間で，複数年度にわたり繰り返しサンプリングされた。これは，傾向分析の必須条件である。
34. Collection of this type of data will likely increase **over the next few years** as biological criteria are developed in each state and as USEPA refines the guidance on how to incorporate biological community parameters into state standards.
各州で生物クライテリアが設定され，USEPAが生物群集パラメータを州基準に

盛り込む方法に関する指針を調整するにつれて，この種のデータ収集は，今後数年間に増える見込みが強い。

- **beginning about 1975 to 1980**：
  およそ 1975 年から 1980 年に始まって
- **since the late 1970s**：1970 年代後半以降
- **until about 1965**：およそ 1965 年までは
- **within the next 3 to 5 years**：今後 3 年ないし 5 年いないに

35. **Beginning about 1973 to 1975**, groundwater quality models have been emerging in the literature.
およそ 1973 年から 1975 年に始まって，地下水水質モデルが文献に出現していた。
36. The State of Texas Environmental Protection Agency has been intensively monitoring the condition of Texas's surface waters **since the late 1970s**.
テキサス州環境保護局は，1970 年代後半以降，テキサス州の地表水域状態の集中的モニタリングを行ってきた。
37. Groundwater modeling is a relatively new field that was not extensively pursued **until about 1965**.
地下水モデリングは比較的新しい分野で，およそ 1965 年までは幅広く追究されなかった。
38. Work is underway to address water quality criteria for these areas **within the next 3 to 5 years**.
今後 3 年ないし 5 年いないに，これらの区域についての水質クライテリアに取り組む活動が進行中である。

- **as of 2004**：2004 年現在
- **later in 1990 and into 1995**：後の 1990 年から 1995 年にかけて
- **by 1980**：1980 年までに
- **〜 dates back to 1977**：〜は 1977 年にさかのぼる

39. This chapter has attempted to describe the status of biological criteria efforts in the United States by indicating the diversity of programs and activities in the states **as of 1994**.
本章では，1994 年現在の諸州での計画・活動の多様性を示すことにより，米国における生物クライテリアについての取りくみの現状を述べようと試みてきた。
40. The PCB issue **dates back to 1973**, when state fish and game officials detected unusually high levels of PCBs in the river's striped bass population.
PCB 問題は 1973 年にさかのぼる。その当時，州の fish and game 当局者は川にいるストライプトバスの個体群に異常に高いレベルの PCB を検出していた。

41. **Later in 1992 and into 1993** discussions were initiated in order to determine the feasibility of and possible way to restore WWH use attainment.
後の 1992 年から 1993 年にかけて，WWH 使用達成復活の実行可能性とその方法をめぐる討論が開始された。

    - **several thousand years**：数千年
    - **year-to-year differences**：年差
    - **year-round**：一年中
    - **each year**：年ごとに
    - **5 or 10 years ago**：5 または 10 年前

42. Under these conditions, **several thousand years** would be required for removal of entire gypsum content of the tailings.
これらの条件の下で，選鉱くずに含まれる石膏の全部を撤去するには，数千年が必要とされるであろう。

43. Whereas many thematic maps are the products of interpretations that include consideration of seasonal and **year-to-year differences**, Landsat imagery and high-altitude aerial photography are snapshots of conditions at a particular time.
多くの主題地図は，季節差や年差を考慮した解釈の産物である。他方，ランドサット画像や高高度航空写真は，特定時刻での状態の断片である。

44. Seasonal variability was tested by monthly sampling **year-round** at two streams, and showed that between-month comparisons should not be done, but upstream-downstream sampling on the same date is valid **year-round**.
季節変動は，2つの水流での通年サンプリング（毎月）により検討されて，月ごとの比較は，すべきでないが，同一日の上流地点-下流地点サンプリングが一年中有効であることを示した。

45. As a result many thousands of miles of United States streams have been and continued to be degraded **each year**.
その結果，米国の水流は，年にのべ数千マイルも悪化してきたし，今も悪化しつづけている。

46. Several new approaches, including multivariate analysis and weighted average metrics, make periphyton analysis a better indicator than it was **5 or 10 years ago**.
多変量分析や加重平均メトリックといった新手法のおかげで，付着生物分析は，5～10 年前よりも優れた指標となっている。

Yet：まだ

1. Catalysis by fixed cell is quite complex and many basic aspects are **yet** to be understood.
   固定細胞による触媒作用は非常に複雑で，多くの基本的な側面がまだ理解されていない。
2. Three states have not **yet** begun to formulate their bioassessment approach, but have initiated discussion within their own agencies and with the USEPA.
   3州は，まだ生物アセスメント手法の作成を開始していないが，州機関内での討論やUSEPAとの論議に着手した。
3. All things are somewhat different, **yet** some things are more similar than others, offering a possible solution to the apparent chaos.
   あらゆる事物は，それぞれ若干異なるが，あるものは他のものよりも類似性が強く，見かけの混沌に対する潜在的解決策を提示する。
4. Two decades of Federal controls have sharply reduced the vast outflows of sewage and industrial chemicals into America's rivers and streams, **yet** the life they contain may be in deeper trouble than ever.
   20年間にわたる連邦政府による防止策は，米国の河川・水流への下水および産業化学物質の大量流出を激減させてきたが，河川・水流に含有される生物は，以前よりも深いトラブルを抱え込んでいる。

- **as yet**：まだ

5. ....., but there is **as yet** no clear evidence of .....
   …，しかしまだ…の明確な証拠がない。
6. Little **as yet** known about the mechanism of carbon dioxide fixation by growing cells.
   成長している細胞によって二酸化炭素が固定されるメカニズムについてほとんどまだ知られていない。

- **～ is not yet .....**：～がまだ…ない
- **It has not yet been .....**：まだ…でない

7. Data **are not yet** available to examine this possibility quantitatively.
   この可能性を定量的に検討するためのデータがまだ得られていない。
8. The kinetics of biological nitrification **is not yet** thoroughly understood.
   生物硝化反応の速度論は，まだ完全には理解されていない。
9. **It has not yet been** determined which level would be most appropriate for the water quality standard.

どのレベルが水質基準に最も適しているかは，まだ未決定である。

Yield：生じる；もたらす；得られる

1. Integrating this equation for the interval 0 to t **yields** ......
   この式を0からtまで間で積分すると，… となる。
2. Testing of these questions **yielded** the following results: (1) ....., (2) .....
   これらの質問への試験が，以下の結果をもたらした。それらは，(1)…(2)…
3. Data gathering efforts for the Tone River **yielded** adequate stream flow.
   利根川でのデータ収集の努力で十分な流量データが得られた。
4. From these reactions, it can be seen that trisodium phosphate **yields** more sodium hydroxide than disodium phosphate.
   これらの化学反応から，リン酸三ナトリウムはリン酸二ナトリウムよりも多くの水酸化ナトリウムを生じさせることがわかる。
5. A linear regression of measurement and predicted values **yielded** an R^2 of 0.75, indicating a reasonable agreement.
   測定値と予測値の線形回帰が0.75のR^2をもたらし，妥当な合意を示した。

■著者紹介■

CHIKASHI SATO

Environmental Engineering Program
College of Engineering
Idaho State University
Pocatello, Idaho, U.S.A.

佐藤元志（Chikashi Sato）

1951年，福島県いわき市(旧平市)に生まれる。1971年，福島国立工業高等専門学校工業化学科卒業。1976年，カンサス大学環境衛生工学専攻修士課修了。1981年，アイオア大学環境工学専攻博士課修了。ポリテクニック大学工学部助教授，パシフィックノースウエスト国立研究所研究員を経歴。現在，アイダホ州立大学工学部順教授として環境工学大学院生の指導にあたっている。米国特許を二つ持ち，研究論文多数。2002〜2003年には日本学術振興会フェローとして，独立行政法人土木研究所にて河川・湖沼環境の調査にあたる。

英語論文表現例集 with CD-ROM
—すぐに使える5,800の例文—

定価はカバーに表示してあります。

2009年4月10日　1版1刷発行	ISBN978-4-7655-3014-9 C3040
2015年6月20日　1版2刷発行	

著　者　　佐　藤　元　志
発行者　　長　　　滋　彦
発行所　　技報堂出版株式会社

〒101-0051　東京都千代田区神田神保町1-2-5
電　話　　営　業　（03）（5217）0885
　　　　　編　集　（03）（5217）0881
F A X　　　　　　（03）（5217）0886
振替口座　00140-4-10
http://www.gihodoshuppan.co.jp/

日本書籍出版協会会員
自然科学書協会会員
土木・建築書協会会員

Printed in Japan

© CHIKASHI SATO, 2009　　装幀　ジンキッズ　印刷・製本　昭和情報プロセス

落丁・乱丁はお取り替えいたします。
本書の無断複写は，著作権法上での例外を除き，禁じられています。